S0-AIB-105

SOLUTIONS AND PROBLEM-SOLVING MEGAMANUAL
FOR

Genetics:
A Conceptual Approach
Second Edition

JUNG H. CHOI
MARK E. MCCALLUM

Interactive Genetics:
A Step-by-Step Approach to Problem-Solving CD-Rom

LIANNA JOHNSON
JOHN MERRIAM

The New Genetics:
Mining Genomes

MARK S. WILSON

W. H. FREEMAN AND COMPANY
NEW YORK

ISBN-13: 978-0-7167-6665-0
ISBN-10: 0-7167-6665-5

© 2005 by W. H. Freeman and Company

No part of this book may be reprinted by any mechanical, photographic, or electronic process, or in the form of a phonographic recording, nor may it be stored in a retrieval system, transmitted, or otherwise copied for public or private use, without written permission from the publisher.

Printed in the United States of America

Second Printing

W. H. Freeman and Company
41 Madison Avenue
New York, NY 10010
www.whfreeman.com

Contents

Chapter One: Introduction to Genetics

COMPREHENSION QUESTIONS

1. Outline some of the ways in which genetics is important to each of us.
Genetics directly influences our lives and is fundamental to what and who we are. For example, genes affect our appearance (e.g., eye color, height, weight, skin pigmentation, and hair color). Our susceptibility to diseases and disorders is affected by our genetic makeup. Genes also influence our intelligence and personality.

*2. Give at least three examples of the role of genetics in society today.
Genetics plays important roles in the diagnosis and treatment of hereditary diseases, in breeding plants and animals for improved production and disease resistance, and in producing pharmaceuticals through genetic engineering.

3. Briefly explain why genetics is crucial to modern biology.
Genetics is crucial to modern biology in that it provides unifying principles: All organisms use nucleic acid as their genetic material, and all organisms encode genetic information in the same manner. The study of many other biological disciplines such as developmental biology, ecology, and evolutionary biology is supported by genetics.

4. All organisms have the same universal genetic system. What are the implications of this universal genetic system?
One implication of the universal genetic system is that all organisms share a common genetic ancestor. In other words, all living organisms have a common origin. Another implication is that the study of one organism's genes and genetic processes often provides information and principles that apply to other organisms. A final implication is that genes will often function in foreign cells, which makes genetic engineering possible and also makes it possible for viruses to reproduce in their host cells.

*5. List the three traditional subdivisions of genetics and summarize what each covers.
Transmission genetics: inheritance of genes from one generation to the next, gene mapping.
Molecular genetics: structure, organization, and function of genes at a molecular level.
Population genetics: genes and changes in genes in populations.

6. What are some characteristics of model genetic organisms that make them useful for genetic studies?
Model genetic organisms have relatively short generation times, produce numerous progeny, are amenable to laboratory manipulations, and can be maintained and propagated inexpensively.

7. When and where did agriculture first arise? What role did genetics play in the development of the first domesticated plants and animals?
Agriculture first arose 10,000 to 12,000 years ago in the area now referred to as the Middle East (i.e., Turkey, Iran, Iraq, Syria, Jordan, and Israel). Early farmers selectively bred individual wild plants or animals that had useful characteristics with others that had similar useful traits. The farmers then selected for offspring that contained those useful features. Early farmers did not completely understand genetics, but they clearly understood that breeding individual plants or animals with desirable traits would lead to offspring that contained these same traits. This selective breeding led to the development of domesticated plants and animals.

*8. Outline the notion of pangenesis and explain how it differs from the germ-plasm theory.
Pangenesis theorizes that information originating from all parts of the body is carried through the reproductive organs to the embryo at conception. Pangenesis allows changes in parts of the body to then be conveyed to the reproductive organs and to the next generation. The germ-plasm theory, in contrast, states that the reproductive cells possess all of the information required to make the complete body; the rest of the body contributes no information to the next generation.

*9. What does the concept of the inheritance of acquired characteristics propose and how is it related to the theory of pangenesis?
The theory of inheritance of acquired characteristics postulates that traits acquired during one's lifetime can be transmitted to offspring. It developed from pangenesis, which postulates that information from all parts of one's body is transmitted to the next generation. Thus, for example, learning acquired in the brain or larger arm muscles developed through exercise could be transmitted to offspring.

*10. What is preformationism? What does it have to say about how traits are inherited?
Preformationism is the theory that the adult form is already preformed in the sperm or the egg. All traits thus would be inherited from only one parent, either the father or the mother, depending on whether the homunculus (the preformed miniature adult) resided in the sperm or the egg.

11. Define blending inheritance and contrast it with preformationism.
The theory of blending inheritance proposes that the egg and sperm from two parents contains material that blends upon conception, influencing the development of the offspring. This theory indicates that the offspring is an equal blend of the two parents. In preformationism, the offspring inherits all of its traits from one parent.

12. How did developments in botany during the seventeenth and eighteenth centuries contribute to the rise of modern genetics?
Botanists of the seventeenth and eighteenth centuries developed new techniques for crossing plants and creating plant hybrids. These early experiments provided essential background work for Mendel's plant crosses. Mendel's work laid the foundation for the study of modern genetics.

13. How did developments in cytology in the nineteenth century contribute to the rise of modern genetics?
In the nineteenth century, Robert Brown described the nucleus. Others, including Theodor Schwann and Matthis Jacob Schleiden, developed cell theory. The background work provided by these and other cytologists stimulated biologists to examine how traits are inherited from cell to cell during cell division.

*14. Who first discovered the basic principles that laid the foundation for our modern understanding of heredity?
Gregor Mendel

15. List some of advances in genetics that have occurred in the twentieth century.
1902 Proposal that genes are located on chromosomes by Walter Sutton
1910 Discovery of the first genetic mutation in a fruit fly by Thomas Hunt Morgan
1930 The foundation of population genetics by Ronald A. Fisher, John B. S. Haldane, and Sewall Wright
1940s The use of viral and bacterial genetic systems
1953 Three-dimensional structure of DNA described by Watson and Crick
1966 Deciphering of the genetic code
1973 Recombinant DNA experiments
1977 Chemical and enzymatic methods for DNA sequencing developed by Walter Gilbert and Frederick Sanger
1986 PCR developed by Kary Mullis
1990 Gene therapy

*16. Briefly define the following terms: (a) gene; (b) allele; (c) chromosome; (d) DNA; (e) RNA; (f) genetics; (g) genotype; (h) phenotype; (i) mutation; (j) evolution.
(a) Gene: the fundamental unit of heredity, a unit of information that determines an inherited characteristic
(b) Allele: a form of the gene
(c) Chromosome: a structure consisting of DNA and associated proteins that carries a linear array of genes
(d) DNA: deoxyribonucleic acid, the molecule that encodes genetic information through the sequence of bases A, C, G, and T
(e) RNA: ribonucleic acid; encodes genetic information through the sequence of bases A, C, G, and U
(f) Genetics: the science of heredity
(g) Genotype: the genetic information that an individual possesses that determines a trait
(h) Phenotype: a trait expressed by an individual
(i) Mutation: heritable alteration in the genotype of the individual, brought about by permanent alteration in the DNA
(j) Evolution: genetic change in a species or population

17. What are the two basic cell types (from a structural perspective) and how do they differ?
The two basic cell types are prokaryotic and eukaryotic. Prokaryotic cells do not have a nucleus, and their chromosomes are found within the cytoplasm. They do not possess membrane-bound cell organelles. Eukaryotic cells possess a nucleus and membrane-bound cell organelles.

18. Outline the relations between genes, DNA, and chromosomes.
Genes are composed of DNA nucleotide sequences and are located on the chromosomes.

APPLICATION QUESTIONS AND PROBLEMS

*19. Genetics is said to be both a very old science and a very young science. Explain what is meant by this statement.
Genetics is old in the sense that hereditary principles have been applied at least since the beginning of agriculture and the domestication of plants and animals. It is very young in the sense that the fundamental principles were not uncovered until Mendel's time, and the advent of molecular biology and recombinant DNA has revolutionized genetics.

20. Match the theory or concept on the left with the correct description on the right.

Preformationism *b*	a. each reproductive cell contains a complete set of genetic information
Pangenesis *d*	b. all traits inherited from one parent
Germplasm theory *a*	c. genetic information may be altered by use of a feature
Inheritance of acquired characteristics *c*	d. different genetic information occurs in cells of different tissues

*21. For each of the following genetic topics, indicate whether it focuses on transmission genetics, molecular genetics, or population genetics.

a. Analysis of pedigrees to determine the probability of someone inheriting a trait
Transmission genetics

b. Study of the genetic history of people on a small island to determine why a genetic form of asthma is so prevalent on the island.
Population genetics

c. The influence of nonrandom mating on the distribution of genotypes among a group of animals.
Population genetics

d. Examination of the nucleotide sequences found at the ends of chromosomes.
Molecular genetics

e. Mechanisms that ensure a high degree of accuracy during DNA replication.
Molecular genetics

f. Study of how the inheritance of traits encoded by genes on sex chromosomes (sex-linked traits) differs from the inheritance of traits encoded by genes on nonsex chromosomes (autosomal traits).

Transmission genetics

22. The following concepts were widely believed at one time, but are no longer accepted as valid genetic theories. What experimental evidence suggests that these theories are incorrect and what theories have taken their place: (a) pangenesis; (b) the inheritance of acquired characteristics; (c) preformationism; (d) blending inheritance?
A series of experiments and observations put to rest these theories. Key among these experiments and observations was the work of Mendel and other early geneticists, as well as the work of several cytologists.
    (a) *Pangenesis*
    The theory of pangenesis has been replaced by Mendel's rules. By statistically analyzing the results of his pea-breeding experiments, Mendel demonstrated that the hereditary material was carried in the germ cells of the plants. The rediscovery of Mendel, and further work by Carl Correns and Hugo de Vries, indicated that the theory of pangenesis was incorrect.
    (b) *The inheritance of acquired characteristics*
    August Weismann's experiments with mice led to his proposal of the germ-plasm theory and signaled the end of the theory of inheritance of acquired characteristics. By cutting the tails off of mice for 22 consecutive generations and observing the presence and lengths of tails in offspring from matings of the tailless mice, he demonstrated that the offspring did not acquire the characteristic.
    (c) *Preformationism*
    The observation of nuclei in plant cells by the botanist Robert Brown followed by similar observations in animal cells by Theodor Schwann led to the proposal of the cell theory by Schwann and Scleiden. This new theory replaced the idea that the sperm or egg contained the homunculus (or perfectly formed adult).
    (d) *Blending inheritance*
    Observations of plant hybrid crosses by Joseph Gottleib Kolreuter and Mendel's observations of the inheritance of dominant and recessive traits in pea plants eliminated the theory of blending inheritance. It was replaced by Mendel's laws.

CHALLENGE QUESTIONS

23. Describe some of the ways in which your own genetic makeup affects you as a person. Be as specific as you can.
Answers will vary, but should include observations similar to those in the following example: Genes affect my physical appearance; for example, they probably have largely determined the fact that I have brown hair and brown eyes. Undoubtedly, genes have affected my height of five feet, seven inches, which is quite close to the height of my father and mother, and my slim build. My dark complexion mirrors the skin color of my mother. I have inherited susceptibilities to certain diseases and disorders, which tend to run in my family; these include asthma, a slight tremor of the hand, and vertigo.

*24. Suppose that life exists elsewhere in the universe. All life must contain some type of genetic information, but alien genomes might not consist of nucleic acids and have the same features as those found in the genomes of life on Earth. What do you think might be the common features of all genomes, no matter where they exist?
All genomes must have the ability to store complex information and must have the capacity to vary. The blueprint for the entire organism is contained within the genome of each reproductive cell. The information has to be in the form of a code that can be used as a set of instructions for assembling the components of the cells. The genetic material of any organism must be stable, be replicated precisely, and be transmitted faithfully to the progeny.

25. Pick one of the following ethical or social issues and give your opinion on this issue. For background information, you might read one of the articles on ethics marked with an asterisk and listed in the Suggested Readings at the end of the chapter.
(a) Should a person's genetic makeup be used in determining his or her eligibility for life insurance?
Arguments pro: Genetic susceptibility to certain types of diseases or conditions is relevant information regarding consequences of exposure to certain occupational hazards. Genes that will result in neurodegenerative diseases, such as Huntington disease, Alzheimer's, or breast cancer, could logically be considered preexisting conditions. Insurance companies have a right, and arguably a duty to their customers, to exclude people with genetic preconditions so that insurance rates can be lowered for the general population.
Arguments con: The whole idea of insurance is to spread the risk and pool assets. Excluding people based on their genetic makeup would deny insurance to people who need it most. Indeed, as information about various genetic risks accumulates, more people would become excluded until only a small fraction of the population is insurable.
(b) Should biotechnology companies be able to patent newly sequenced genes?
Pro: Patenting genes provides companies with protection for their investment in research and development of new drugs and therapies. Without such patent protection, companies would have less incentive to expend large amounts of money in genetic research, and thus would slow the pace of advancement of medical research. Such a result would be detrimental to everyone.
Con: Patents on human genes would be like allowing companies to patent a human arm. Genes are integral parts of our selves, and how can a company patent something that every human has?
(c) Should gene therapy be used on people?
Pro: Gene therapy can be used to cure previously incurable or intractable genetic disorders and to relieve the suffering of millions of people.
Con: Gene therapy may lead to genetic engineering of people for unsavory ends. Who determines what is a genetic defect? Is short stature a genetic defect?
(d) Should genetic testing be made available for inherited conditions for which there is no treatment or cure?

Pro: Information will provide relief from unnecessary anxiety (if the test is negative). Even if the test result is positive for a genetic disorder, it provides the individual, the family, and friends with information and time to prepare. Information about one's own genetic makeup is a right; every person should be able to make his or her own choice as to whether he or she wants this information.

Con: If there is no treatment or cure, a positive test result can have no good consequences. It's like receiving a death sentence or sentence of extended punishment. It will only engender feelings of hopelessness and depression and may cause some people to terminate their own lives prematurely. Applied to the unborn as a prenatal test, it may lead to abortion.

(e) Should governments outlaw the cloning of people?

Pro (for outlawing human cloning): There is no medical necessity for human cloning. There is no good research objective that can be attained solely via human cloning. Human cloning creates human beings for a purpose and may be a form of slavery. As such, it violates our sense that all humans should have free will. The risk of birth defects and complicated pregnancies is too great to justify these experiments with humans.

Con (against outlawing human cloning): Cloning people is just another method of assisted reproduction for infertile couples. People should be free to do whatever they want with their own bodies. Research on human cloning may lead to stem cell lines and methods that can be used medically to grow organ replacements.

26. We now know as much or more about the genetics of humans as any other organism, and humans are the foscus of many genetic studies. Do you think humans should be considered a model genetic organism? Why or why not?

Although human genetics has been intensively studied, humans should not be considered a model genetic organism because they lack the characteristics of model genetic organisms. Humans have a long time between generations, usually bear only one offspring per mating, and are expensive to maintain! Most importantly, humans are not amenable to laboratory manipulation; the ethical barriers to controlling human matings and subjecting humans to experiments are insurmountable. Ultimately, the purpose of model organisms is to enable experimental investigations that are not possible with humans.

Chapter Two: Chromosomes and Cellular Reproduction

COMPREHENSION QUESTIONS

1. Give some genetic differences between prokaryotic and eukaryotic cells.

Prokaryotic Cell	*Eukaryotic Cells*
No nucleus *No paired chromosomes (haploid)* *Typically single circular chromosome consisting of a single origin of replication* *Single chromosome is replicated with each copy moving to opposite sides of the cell* *No histone proteins complexed to DNA*	*Nucleus present* *Paired chromosomes common (diploid)* *Typically multiple linear chromosomes consisting of centromeres, telomeres, and multiple origins of replication* *Chromosomes are replicated but require mitosis or meiosis to ensure that chromosome migrates to the proper location* *Histone proteins are complexed to DNA*

2. Why are the viruses that infect mammalian cells useful for studying the genetics of mammals?

   It is thought that viruses must have evolved after their host cells, because a host is required for viral reproduction. Viral genomes are closely related to their host genomes. The close relationship between a mammalian virus and its mammalian cell host, along with the simpler structure of the viral particle, makes it useful in studying the genetics of mammals. The viral genome will have a similar structure to the mammalian cell host, but because it has fewer genes it will be easier to decipher the interactions and regulation of the viral genes.

*3. List three fundamental events that must take place in cell reproduction.

   (1) A cell's genetic information must be copied.
   (2) The copies of the genetic information must be separated from one another.
   (3) The cell must divide.

4. Outline the process by which prokaryotic cells reproduce.

   (1) Replication of the circular chromosome takes place.
   (2) The two replicated chromosomal copies attach to the plasma membrane.
   (3) The plasma membrane grows, which results in the separation of the two chromosomes.
   (4) A new cell wall is formed between the two chromosomes, producing two cells, each with its own chromosome.

5. Name three essential structural elements of a functional eukaryotic chromosome and describe their functions.

   (1) Centromere: serves as the point of attachment for the spindle fibers (microtubules).
   (2) Telomeres or the natural ends of the linear eukaryotic chromosome: serve to stabilize the ends of the chromosome; may have a role in limiting cell division.
   (3) Origins of replication: serve as the starting place for DNA synthesis.

*6. Sketch and label four different types of chromosomes based on the position of the centromere.

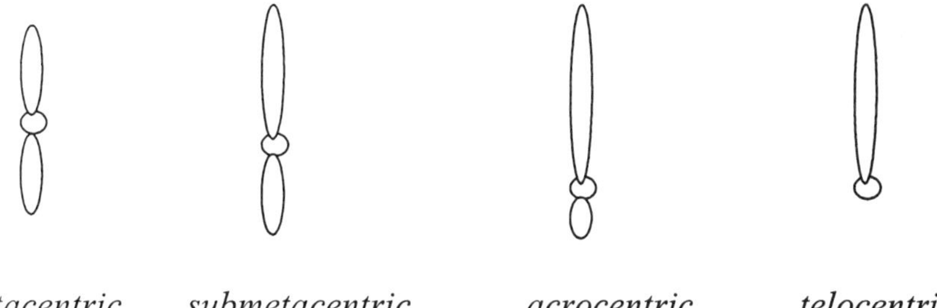

metacentric *submetacentric* *acrocentric* *telocentric*

7. List the stages of interphase and the major events that take place in each stage.

   Three predominant stages are found in interphase of cells active in the cell cycle.
   (1) G_1 (Gap 1). In this phase, the cell grows and synthesizes proteins necessary for cell division. During G_1, the G_1/S checkpoint takes place. Once the cell has passed this checkpoint, it is committed to divide.
   (2) S phase. During S phase, DNA replication takes place.
   (3) G_2 (Gap 2). In G_2, additional biochemical reactions take place that prepare the cell for mitosis. A major checkpoint in G_2 is the G_2/M checkpoint. Once the cell has passed this checkpoint, it enters into mitosis.

   A fourth stage is frequently found in cells prior to the G_1/S checkpoint. Cells may exit the active cell cycle and enter into a nondividing stage called G_0.

*8. List the stages of mitosis and the major events that take place in each stage.

   (1) Prophase: The chromosomes condense and become visible, and the centrosomes move apart along with the formation of microtubule fibers from the centrosomes.
   (2) Prometaphase: The nucleoli disappear and the nuclear envelope begins to disintegrate, allowing for the cytoplasm and nucleoplasm to join. The sister chromatids of each chromosome are attached to microtubles from the opposite centrosomes.
   (3) Metaphase: The spindle microtubules are clearly visible and the chromosomes arrange themselves on the equatorial plane of the cell.

(4) *Anaphase: The sister chromatids separate at the centromeres after the breakdown of cohesin protein, and the newly formed daughter chromosomes move to the opposite poles of the cell.*
(5) *Telophase: The nuclear envelope reforms around each set of daughter chromosomes. Nucleoli reappear. Spindle microtubules disintegrate.*

9. Briefly describe how the chromosomes move toward the spindle poles during anaphase.

Due to the actions of the microtubule subunits attached to the kinetochores of the chromosome and motor proteins such as kinesin, the chromosomes are pulled toward the spindle poles during anaphase. The spindle fibers are composed of tubulin protein subunits. As the tubulin subunits are removed from the "-" end of the microtubule, the chromosome is pulled (or "reeled in") toward the spindle pole as the microtubule is shortened. While at the "+" end, the kinetochore is removing tubulin subunits of the microtubule attached to the kinetochore with the net effect being the movement of the chromosome closer to the spindle pole. Molecular motor proteins, such as kinesin, are responsible for removing the subunits at the "+" and "-" ends of the microtubules and thus generate the force needed to move the chromosomes.

*10. What are the genetically important results of the cell cycle?

In this process, one cell produces two cells that contain the same genetic information. In other words, the cells are identical to each other and to the mother cell.

11. Why are the two cells produced by the cell cycle genetically identical?

The two cells are genetically identical because during S phase an exact copy of each DNA molecule was created. These exact copies give rise to the two identical sister chromatids. Mitosis ensures that each new cell receives one copy of the two identical sister chromatids. Thus, the newly formed cells will contain identical daughter chromosomes.

12. What are checkpoints? List some of the important checkpoints in the cell cycle. What two general classes of compounds regulate progression through the cell cycle?

Checkpoints function to ensure that all the cellular components, such as important proteins and chromosomes, are present and functioning before the cell moves to the next stage of the cell cycle. If components are missing or not functioning, the checkpoint will prevent the cell from moving to the next stage. The checkpoints prevent defective cells from replicating and malfunctioning.

These checkpoints occur throughout the various stages of the cell cycle. Important checkpoints include the G_1/S checkpoint, which occurs during G_1 prior to the S phase; the G_2/M checkpoint, which occurs in G_2 prior to mitosis; and the spindle-assembly checkpoint, which occurs during mitosis.

Two types of proteins are responsible for movement through the cell cycle: cyclin proteins and cyclin-dependent kinases.

13. What are the stages of meiosis and what major events take place in each stage?

Meiosis I: Separation of homologous chromosomes

Prophase I: The chromosomes condense and homologous pairs of chromosomes undergo synapsis. While the chromosomes are synapsed, crossing over occurs. The nuclear membrane disintegrates and the meiotic spindle begins to form.

Metaphase I: The homologous pairs of chromosomes line up on the equatorial plane of the metaphase plate.

Anaphase I: Homologous chromosomes separate and move to opposite poles of the cell. Each chromosome possesses two sister chromatids.

Telophase I: The separated homologous chromosomes reach the spindle poles and are at opposite ends of the cell.

Meiosis I is followed by cytokinesis, resulting in the division of the cytoplasm and the production of two haploid cells. These cells may skip directly into meiosis II or enter interkinesis, where the nuclear envelope reforms and the spindle fibers break down.

Meiosis II: Separation of sister chromatids

Prophase II: Chromosomes condense, the nuclear envelope breaks down, and the spindle fibers form.

Metaphase II: Chromosomes line up at the equatorial plane of the metaphase plate.

Anaphase II: The centromeres split, which results in the separation of sister chromatids.

Telophase II: The daughter chromosomes arrive at the poles of the spindle. The nuclear envelope reforms, and the spindle fibers break down. Following meiosis II, cytokinesis takes place.

*14. What are the major results of meiosis?

Meiosis involves two cell divisions, thus resulting in the production of four new cells (in many species). The chromosome number of a haploid cell produced by meiosis is half the chromosome number of the original diploid cell. Finally, the cells produced by meiosis are genetically different from the original cell and genetically different from each other.

15. What two processes unique to meiosis are responsible for genetic variation? At what point in meiosis do these processes take place?

(1) Crossing over, which begins during the zygotene stage of prophase I and is completed near the end of prophase I.

(2) The random distribution of chromosomes to the daughter cells takes place in anaphase I of meiosis.

*16. List similarities and differences between mitosis and meiosis. Which differences do you think are most important and why?

Mitosis	*Meiosis*
A single cell division produces two genetically identical progeny cells.	*Two cell divisions usually result in four progeny cells that are not genetically identical.*
Chromosome number of progeny cells and original cell remain the same.	*Daughter cells are haploid and have half the chromosomal complement of the original diploid cell as a result of the separation of homologous pairs during anaphase I.*
Daughter cells and original cell are genetically identical. No separation of homologous chromosomes or crossing over takes place.	*Crossing over in prophase I and separation of homologous pairs during anaphase I produce daughter cells that are genetically different from each other and from the original cell.*
Homologous chromosomes do not synapse.	*Synapsis of homologous chromosomes takes place during prophase I.*
In metaphase, individual chromosomes line up on the metaphase plate.	*In metaphase I, homologous pairs of chromosomes line up on the metaphase plate. Individual chromosomes line up in metaphase II.*
In anaphase, sister chromatids separate.	*In anaphase I, homologous chromosomes separate. Separation of sister chromatids takes place in anaphase II.*

A key difference is that mitosis produces cells genetically identical to each other and to the original cell, resulting in the orderly passage of information from one cell to its progeny. In contrast, by producing progeny that do not contain pairs of homologous chromosomes, meiosis results in the reduction of chromosome number from the original cell. Meiosis also allows for genetic variation through crossing over and the random assortment of homologs.

17. Briefly explain why sister chromatids remain together in anaphase I but separate in anaphase of mitosis and anaphase II of meiosis.

During the S phase, the cohesin protein complex forms and holds together the sister chromatids throughout the early stages of mitosis or meiosis I. In mitosis at the end of metaphase, the cohesin molecules that connect the sister chromatids at the centromeres are cleaved by the protein separase, which allows the sister chromatids to separate. Prior to this point, separase is kept inactive by a protein called securin. Securin is broken down at the end of metaphase allowing separase to become active.

In meiosis a similar process occurs. The cohesin complexes form at the centromeres of the sister chromatids during the S phase. At the beginning of meiosis, cohesin molecules are also found along the entire length of the chromosome arms assisting in the formation of the synaptonemal complex and holding together the two homologs. During anaphase I of meiosis, the cohesin molecules along the arms are cleaved by activated separase allowing the homologs to separate. However, the cohesin complexes at the centromeres of the sister chromatids are protected from the action of separase and are unaffected. The result is that sister chromatids remained attached during anaphase I. At the end of metaphase II, the protection of the cohesin molecules at the centromeres is lost, and the separase proteins can now cleave the cohesin complex, which allows the sister chromatids to separate.

Also during metaphase I, proteins called monopolins function to allow only the two kinetochores of sister chromatids to orient toward the same spindle pole and attach to microtubules of the same pole. The end result is that the attached sister chromatids move toward the same pole during meiosis.

18. Outline the process by which male gametes are produced in plants. Outline the process of female gamete formation in plants.

Plants alternate between a multicellular haploid stage called the gametophyte and a multicellular diploid stage called the sporophyte. Meiosis in the diploid sporophyte stage of plants produces haploid spores that develop into the gametophyte. The gametophyte produces gametes by mitosis.

In flowering plants, the microsporocytes found in the stamen of the flower undergo meiosis to produce four haploid microspores. Each microspore divides by mitosis to produce the pollen grain, or the microgametophyte. Within the pollen grain are two haploid nuclei. One of the haploid nuclei divides by mitosis to produce two sperm cells. The other haploid nucleus directs the formation of the pollen tube.

Female gamete production in flowering plants takes place within the megagametophyte. Megasporocytes found within the ovary of a flower divide by meiosis to produce four megaspores. Three of the megaspores disintegrate, while the remaining megaspore divides mitotically to produce eight nuclei that form the embryo sac (or female gametophyte). Of the eight nuclei, one will become the egg.

19. Outline the process of spermatogenesis in animals. Outline the process of oogenesis in animals.

In animals spermatogenesis occurs in the testes. Primordial diploid germ cells divide mitotically to produce diploid spermatogonia that can either divide repeatedly by mitosis or enter meiosis. A spermatogonium that has entered prophase I of meiosis is called a primary spermatocyte and is diploid. Upon completion of meiosis I, two haploid cells, called secondary spermatocytes, are produced. Upon completing meiosis II, the secondary spermatocytes produce a total of four haploid spermatids.

Female animals produce eggs through the process of oogenesis. Similar to what takes place in spermatogenesis, primordial diploid cells divide mitotically to produce diploid oogonia that can divide repeatedly by mitosis, or enter meiosis. An oogonium that has entered prophase I is called a primary oocyte and is diploid. Upon completion of meiosis I, the cell divides, but unequally. One of the newly produced haploid cells receives most of the cytoplasm and is called the secondary oocyte. The other haploid cell receives only a small portion of the cytoplasm and is called the first polar body. Ultimately, the secondary oocyte will complete meiosis II and produce two haploid cells. One cell, the ovum, will receive most of the cytoplasm from the secondary oocyte. The smaller haploid cell is called the second polar body. Typically, the polar bodies disintegrate, and only the ovum is capable of being fertilized.

APPLICATION QUESTIONS AND PROBLEMS

20. A certain species has three pairs of chromosomes: an acrocentric pair, a metacentric pair, and a submetacentric pair. Draw a cell of this species as it would appear in metaphase of mitosis.

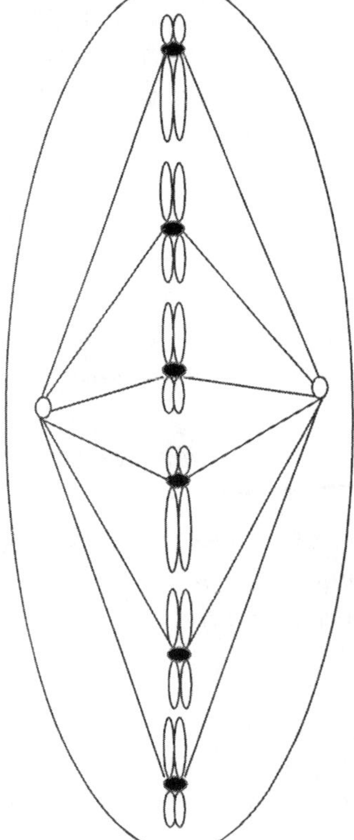

21. A biologist examines a series of cells and counts 160 cells in interphase, 20 cells in prophase, 6 cells in prometaphase, 2 cells in metaphase, 7 cells in anaphase, and 5 cells in telophase. If the complete cell cycle requires 24 hours, what is the average duration of M phase in these cells? Of metaphase?

To determine the average duration of M phase in these cells, the proportion of cells in interphase or in each stage of M phase should be calculated by dividing the number of cells in each stage by the total number of cells counted. To calculate the time required for a given phase, multiply 24 hours by the proportion of cells at that stage. This will give the average duration of each stage in hours.

Stage	*Number of cells counted*	*Proportion of cells at each stage*	*Average duration (hours)*
Interphase	*160*	*0.80*	*19.2*
Prophase	*20*	*0.10*	*2.4*
Prometaphase	*6*	*0.03*	*0.72*
Metaphase	*2*	*0.01*	*0.24*
Anaphase	*7*	*0.035*	*0.84*
Telophase	*5*	*0.025*	*0.6*
Totals	*200*	*1.0*	*24*

The average duration of M phase can be determined by adding up the hours spent in each stage of mitosis. In these cells, M phase lasts 4.8 hours. The table shows that metaphase requires 0.24 hours, or 14.4 minutes.

*22. A cell in G_1 of interphase has 12 chromosomes. How many chromosomes and DNA molecules will be found per cell when this original cell progresses to the following stages?

The number of chromosomes and DNA molecules depends on the stage of the cell cycle. Each chromosome consists of only one centromere, but during certain times in the cell cycle could consist of two DNA molecules. After the completion of S phase, but prior to anaphase of mitosis or cytokinesis of meiosis I, each chromosome will consist of two DNA molecules.

(a) *G_2 of interphase*

G_2 of interphase occurs after S phase, when the DNA molecules are replicated. Each chromosome now consists of two DNA molecules. So a cell in G_2 will contain 12 chromosomes and 24 DNA molecules.

(b) *Metaphase I of meiosis*

Neither homologous chromosomes nor sister chromatids have separated by metaphase I of meiosis. Therefore, the chromosome number is 12, and the number of DNA molecules is 24.

(c) *Prophase of mitosis*

This cell will contain 12 chromosomes and 24 DNA molecules.

(d) *Anaphase I of meiosis*
During anaphase I of meiosis, homologous chromosomes separate and begin moving to opposite ends of the cell. However, sister chromatids will not separate until anaphase II of meiosis. The number of chromosomes is still 12, and the number of DNA molecules is 24.

(e) *Anaphase II of meiosis*
Homologous chromosomes were separated and migrated to different daughter cells at the completion of meiosis I. However, in anaphase II of meiosis, sister chromatids separate, resulting in a temporary doubling of the chromosome number in the now haploid daughter cell. The number of chromosomes and the number of DNA molecules present will both be 12.

(f) *Prophase II of meiosis*
The daughter cells in prophase II of meiosis are haploid. The haploid cells will contain six chromosomes and 12 DNA molecules.

(g) *After cytokinesis following mitosis*
After cytokinesis following mitosis the daughter cells will enter G_1. Each cell will contain 12 chromosomes and 12 DNA molecules.

(h) *After cytokinesis following meiosis II*
After cytokinesis following meiosis II, the haploid daughter cells will contain six chromosomes and six DNA molecules.

*23. All of the following cells, shown in various stages of mitosis and meiosis, come from the same rare species of plant. What is the diploid number of chromosomes in this plant? Give the names of each stage of mitosis or meiosis shown.

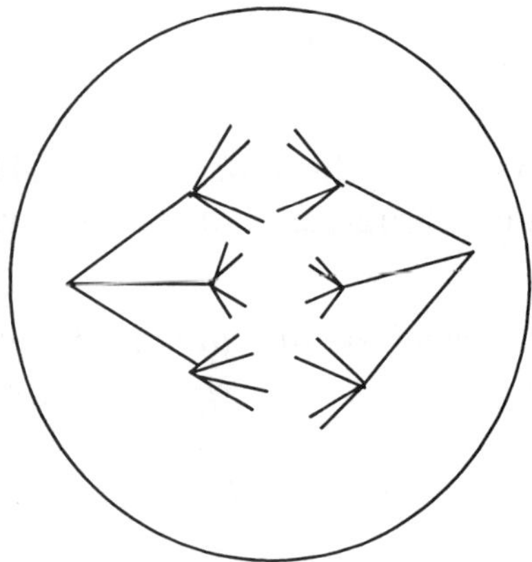

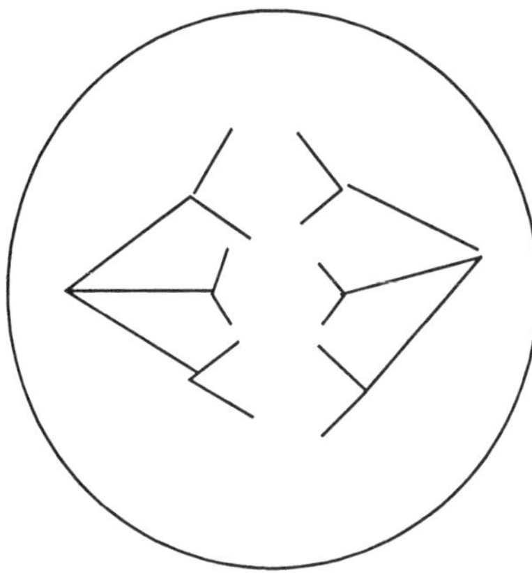

To determine the diploid chromosome number in this plant, the number of centromeres present within a cell that contains homologous pairs of chromosomes must be determined. Remember, each chromosome possesses a single centromere. The location and presence of a centromere are determined by the attachment of the spindle fibers to the chromosome, which occurs at the centromere in the above diagram. Only the cell in stage (a) clearly has homologous pairs of chromosomes. So, the diploid chromosome number for cells of this species of plant is six.

(a) This cell is undergoing anaphase of meiosis I, as indicated by the separation of the homologous pairs of chromosomes.

(b) In the diagram, the cell contains six chromosomes, the diploid chromosome number for this species. Also in this cell, sister chromatids have separated, resulting in a doubling of the chromosome number within the cell from six to 12. Based on the number of chromosomes, the separation of sister chromatids in this cell must be occurring during anaphase of mitosis.

(c) Again sister chromatids are being separated, but the number of chromosomes present in the cell is only six. This indicates that no homologs are present within the cell, so in this cell the separation of sister chromatids is occurring in anaphase II of meiosis.

24. A cell has x amount of DNA in G_1 of interphase. How much DNA (in multiples or fractions of x) will be present per cell at the following stages?

The amount of DNA in the cell will be doubled after the completion of S-phase in the cell cycle and prior to cytokinesis in either mitosis or meiosis I. At the completion of cytokinesis following meiosis II, the amount of DNA will be halved.

(a) G_2 takes place directly after the completion of S-phase, so the amount of DNA is 2x.

(b) During anaphase of mitosis the amount of DNA in the cell is 2x.

(c) Prophase II takes place after the cytokinesis associated with meiosis I and results in the daughter cells receiving only half the DNA found in their mother cell. In prophase II of meiosis, the amount of DNA in each cell is 1x, because each chromosome still consists of two DNA molecules.

(d) Following cytokinesis associated with meiosis II, each daughter cell will contain only ½x the amount of DNA of a mother cell found in G_1 of interphase. By the completion of cytokinesis associated with meiosis II, both homologous pairs of chromosomes and sister chromatids have been separated into different daughter cells. Therefore each daughter cell will contain only ½x the amount of DNA of the original cell in G_1.

25. Indicate where in mitosis and/or meiosis the following events occur. Give all possible stages.

Event	Stage(s)
Independent assortment	*Anaphase I of meiosis*
Separation of chromatids	*Anaphase of mitosis and anaphase I of meiosis I*
Crossing over	*Prophase I of meiosis*
Bivalent pairs line up on metaphase plate	*Metaphase I of meiosis*

26. A cell in prophase II of meiosis contains 12 chromosomes. How many chromosomes would be present in a cell from the same organism if it were in prophase of mitosis? Prophase I of meiosis?

A cell in prophase II of meiosis will contain the haploid number of chromosomes. For this organism, 12 chromosomes represent the haploid chromosome number of a cell, or one complete set of chromosomes.

A cell from the same organism that is undergoing prophase of mitosis would contain a diploid number of chromosomes, or two complete sets of chromosomes, which means that homologous pairs of chromosomes are present. So, a cell in this stage should contain 24 chromosomes.

Homologous pairs of chromosomes have not been separated by prophase I of meiosis. During this stage, a cell of this organism will contain 24 chromosomes.

*27. The fruit fly *Drosophila melanogaster* has four pairs of chromosomes, whereas the house fly *Musca domestica* has six pairs of chromosomes. Other things being equal, in which species would you expect to see more genetic variation among the progeny of a cross? Explain your answer.

The progeny of an organism whose cells contain more homologous pairs of chromosomes should be expected to exhibit more variation. The number of different combinations of chromosomes that are possible in the gametes is 2^n, where n is equal to the number of homologous pairs of chromosomes. For the fruit fly with four pairs of chromosomes, the number of possible combinations is $2^4 = 16$. For Musca domestica with six pairs of chromosomes, the number of possible combinations is $2^6 = 64$.

*28. A cell has two pairs of submetacentric chromosomes, which we will call chromosomes I_a, I_b, II_a, and II_b (chromosomes I_a and I_b are homologs, and chromosomes II_a and II_b are homologs). Allele *M* is located on the long arm of chromosome I_a, and allele *m* is located at the same position on chromosome I_b. Allele *P* is located on the short arm of chromosome I_a, and allele *p* is located at the same position on chromosome I_b. Allele *R* is located on chromosome II_a and allele *r* is located at the same position on chromosome II_b.

(a) Draw these chromosomes, labeling genes *M, m, P, p, R, and r,* as they might appear in metaphase I of meiosis. Assume that there is no crossing over.

Metaphase I

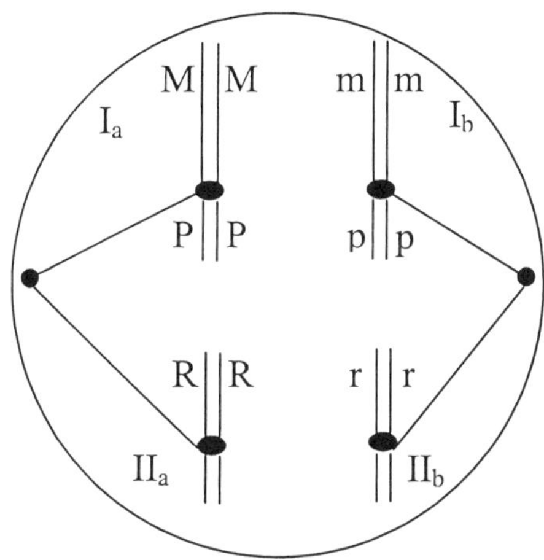

(b) Considering the random separation of chromosomes in anaphase I, draw the chromosomes (with labeled genes) present in all possible types of gametes that might result from this cell going through meiosis. Assume that there is no crossing over.

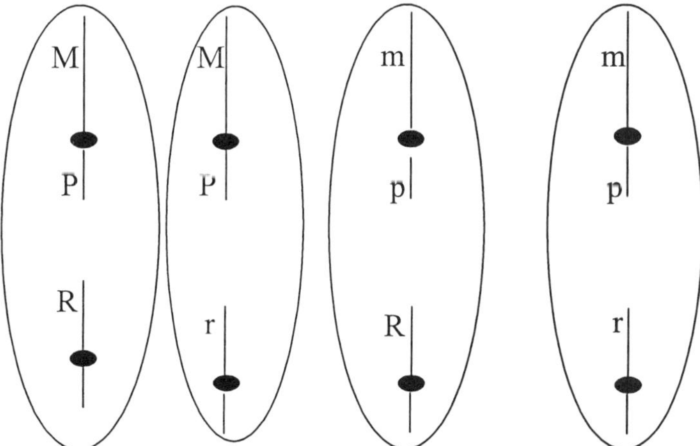

29. A horse has 64 chromosomes and a donkey has 62 chromosomes. A cross between a female horse and a male donkey produces a mule, which is usually sterile. How many chromosomes does a mule have? Can you think of any reasons for the fact that most mules are sterile?

The haploid egg produced by the female horse contains 32 chromosomes. The haploid sperm produced by the male donkey contains 31 chromosomes. The union of the horse and donkey gametes will produce a zygote containing 63 chromosomes. From the zygote, the adult mule will develop and will contain cells with a chromosome number of 63. Because an odd number of chromosomes in the mule's cells are present, at least one chromosome will not have a homolog. During the production of gametes by meiosis when pairing and separation of homologous chromosomes occurs, the odd chromosome will be unable to pair up. Furthermore, the mule's chromosomes, which are contributed by the horse and donkey, are from two different species. Not all of the mule's chromosomes may be able to find a suitable homolog during meiosis I, and thus may not synapse properly during prophase I of meiosis. If improper synapsis or no synapsis occurs during prophase I, this will result in faulty segregation of chromosomes to the daughter cells produced at the conclusion of meiosis I. This leads to gametes that have abnormal numbers of chromosomes. When these abnormal gametes unite, the resulting zygote has an abnormal number of chromosomes and will be nonviable.

30. Normal somatic cells of horses have 64 chromosomes (2n = 64). How many chromosomes and DNA molecules will be present in the following types of horse cells?

Cell Type	Number of Chromosomes	Number of DNA Molecules
a. Spermatogonium	*64*	*64*

Assuming the spermatogonium is in G_1 prior to the production of sister chromatids in S phase, the chromsome number will be the diploid number of chromosomes.

Cell Type	Number of Chromosomes	Number of DNA Molecules
b. First polar body	*32*	*64*

The first polar body is the product of meiosis I so it will be haploid, but the sister chromatids have not separated so each chromosome will consist of two sister chromatids.

Cell Type	Number of Chromosomes	Number of DNA Molecules
c. Primary oocyte	*64*	*128*

The primary oocyte has stopped in prophase I of meiosis. So, the homologs have not yet separated and each chromosome consists of two sister chromatids.

Cell Type	Number of Chromosomes	Number of DNA Molecules
d. Secondary spermatocyte	*32*	*64*

The secondary spermatocyte is a product of meiosis I and has yet to enter meiosis II. So, the secondary spermatocyte will be haploid because the homologous pairs were separated in meiosis I, but each chromosome is still composed of two sister chromatids.

CHALLENGE QUESTIONS

31. Eighty to ninety percent of the most common chromosome abnormalities in humans arise because the chromosomes fail to divide properly in *female* oogenesis. Can you think of a reason why failure of chromosome division might be more common in female gametogenesis than male gametogenesis?

Male gametogenesis or spermatogenesis in human males occurs regularly. Once the spermatogonium begins meiosis, the process quickly goes to completion resulting in the formation of four spermatids, which can mature into sperm cells. Female gametogenesis or oogenesis in human females is more complicated. Each oogonium enters meiosis I, but stops at prophase I generating a primary oocyte. This primary oocyte remains frozen in prophase I until ovulation begins and continues through meiosis I. Only if the egg is fertilized will meiosis II be completed. Because the primary oocyte is present at birth, the completion of meiosis I, by a primary oocyte may not occur for many years (35 to 40 years or more). The length of time could lead to degradation or damaging of the meiotic machinery (such as the meiotic spindle fibers or cohesin complex). The damaged meiotic machinery could result in an improper separation of homologous pairs or of sister chromatids during the meiotic process. The spermatogenesis process does not have this time delay, which may protect the process from age-induced damage to the meiotic machinery.

32. On average, what proportion of the genome in the following pairs of humans would be exactly the same if no crossing over occurred? (For the purposes of this question only, we will ignore the special case of the X and Y sex chromosomes and assume that all genes are located on nonsex chromosomes.)

(a) Father and child

The father will donate ½ of his chromosomes to his child. Therefore, the father and child will have ½ of their genomes that are similar.

(b) Mother and child

The mother will donate ½ of her chromosomes to her child. Therefore, the mother and child will have ½ of their genomes that are similar.

(c) Two full siblings (offspring who have the same two biological parents)

The parents can contribute only ½ of their genome to each offspring. So, it is likely that the siblings share ¼ of their genes from one parent. Because each sibling would share ¼ of their genes from each parent, their total relatedness is ½ (or ¼ + ¼).

(d) Half siblings (offspring that have only one biological parent in common)

Half siblings share only ¼ of their genomes with each other because they have only one parent in common.

(e) Uncle and niece

An uncle would share ½ of his genomes with his sibling, who would share ½ of his or her genome with his or her child. So, an uncle and niece would share ¼ of their genomes (½ × ½).

(f) Grandparent and grandchild

The grandparent and grandchild would share ¼ of their genomes because the grandchild would share ½ of her genome with her parent and the parent would share ½ of her genome with the child's grandparent.

33. Females bees are diploid and male bees are haploid. The haploid males produce sperm and can successfully mate with diploid females. Fertilized eggs develop into females and unfertilized eggs develop into males. How do you think the process of sperm production in male bees differs from sperm production in other animals?

Most male animals produce sperm by meiosis. In haploid male bees, meiosis will not occur, since meiosis can only occur in diploid cells. Male bees can still produce sperm but only through mitosis. Haploid cells that divide mitotically produce more haploid cells.

34. Rec8 is a protein that is found in yeast chromosome arms and centromeres. Rec8 persists throughout meiosis I but breaks down at anaphase II. When the gene that encodes Rec8 is deleted, sister chromatids separate in anaphase I.

(a) From these observations, propose a mechanism for the role of Rec8 in meiosis that helps to explain why sister chromatids normally separate in anaphase II but not anaphase I.

Rec8 belongs to a family of proteins called cohesins that holds the sister chromatids together. Rec8 persists throughout meiosis I, which accounts for why sister chromatids fail to separate in anaphase I. Rec8 breaks down in anaphase II, which allows separation of sister chromatids in anaphase II.

(b) Make a prediction about the presence or absence of Rec8 during the various stages of mitosis.

Although Rec8 protein has not been shown to be associated with mitosis, similar cohesin proteins are. The function of these proteins is similar to Rec8 protein in that they hold the sister chromatids together. During prophase, prometaphase, and metaphase the cohesins are present. The cohesins disappear during anaphase, allowing the sister chromatids to separate.

Chapter Three: Basic Principles of Heredity

COMPREHENSION QUESTIONS

*1. Why was Mendel's approach to the study of heredity so successful?

Mendel was successful for several reasons. He chose a plant, Pisum sativum, *that was easy to cultivate, grew relatively rapidly, and produced many offspring, which allowed Mendel to detect mathematical ratios. The seven characteristics he chose to study were also important because they exhibited only a few distinct phenotypes and did not show a range of variation. Finally, by looking at each trait separately and counting the numbers of the different phenotypes, Mendel adopted an experimental approach and applied the scientific method. From his observations, he proposed hypotheses that he was then able to test empirically.*

2. What is the relation between the terms *allele, locus, gene,* and *genotype*?

We have defined a gene as a genetic factor that determines a characteristic. An allele is one of the alternative forms of the gene. The locus refers to the specific place on a chromosome where a gene or an allele is located. Finally, the genotype refers to the set of genes (or alleles) that an individual possesses.

*3. What is the principle of segregation? Why is it important?

The principle of segregation, or Mendel's first law, states that an organism possesses two alleles for any one particular trait and that these alleles separate during the formation of gametes. In other words, one allele goes into each gamete. The principle of segregation essentially explains that homologous chromosomes segregate during anaphase I of meiosis.

4. What is the concept of dominance? How does dominance differ from incomplete dominance?

The concept of dominance states that when two different alleles are present in a genotype, only the dominant allele is expressed in the phenotype. Incomplete dominance occurs when different alleles are expressed in a heterozygous individual, and the resulting phenotype is intermediate to the phenotypes of the two homozygotes.

5. Give the phenotypic ratios that may appear among the progeny of simple crosses and the genotypes of the parents that may give rise to each ratio.

Phenotypic ratio	*Parental genotype*	*Type of dominance*
3:1	Aa × Aa	*Dominance*
1:2:1	Aa × Aa	*Incomplete dominance*
1:1	Aa × aa Aa × aa	*Dominance* *Incomplete dominance*

Uniform progeny	AA × AA aa × aa AA × aa Aa × AA	*Any type of dominance* *Any type of dominance* *Any type of dominance* *Dominance*

6. Give the genotypic ratios that may appear among the progeny of simple crosses and the genotypes of the parents that may give rise to each ratio.

Genotypic ratio	*Parental genotype*
1:2:1	Aa × Aa
1:1	Aa × aa
Uniform progeny	AA × AA aa × aa AA × aa

*7. What is the chromosome theory of inheritance? Why was it important?

Walter Sutton developed the chromosome theory of inheritance. The theory states that genes are located on the chromosomes. The independent segregation of homologous chromosomes in meiosis provides the biological basis for Mendel's principles of heredity.

8. What is the principle of independent assortment? How is it related to the principle of segregation?

According to the principle of independent assortment, genes for different characteristics and at different loci segregate independently of one another. Essentially, the principle of independent assortment is an extension of the principle of segregation. The principle of segregation indicates that the two alleles at a locus separate; the principle of independent assortment indicates that the separation of alleles at one locus is independent of the separation of other pairs at other loci.

9. How is the principle of independent assortment related to meiosis?

In anaphase I of meiosis, each pair of homologous chromosomes separates independently of all other pairs of homologous chromosomes. This assortment of homologs explains how genes located on different pairs of chromosomes will separate independently of one another.

10. How is the goodness-of-fit chi-square test used to analyze genetic crosses? What does the probability associated with a chi-square value indicate about the results of a cross?

The goodness-of-fit chi-square test is a statistical method used to evaluate the role of chance in causing deviations between the observed and the expected numbers of offspring produced in a genetic cross. The probability value obtained from the chi-

square table refers to the probability that random chance produced the deviations of the observed numbers from the expected numbers.

APPLICATION QUESTIONS AND PROBLEMS

11. What characteristics of an organism would make it well suited for studies of the principles of inheritance? Can you name several organisms that have these characteristics?

    <u>Useful characteristics</u>
    - *Are easy to grow and maintain,*
    - *grow rapidly producing many generations in a short period,*
    - *produce large numbers of offspring, and*
    - *have distinctive phenotypes that are easy to recognize.*

    <u>Examples of organisms that meet these criteria are</u>
    - *Neurospora, a fungus,*
    - *Saccharomyces cerevisiae, a yeast,*
    - *Arabidopsis, a plant,*
    - *Caenorhabditis elegans, a nematode,*
    - *Drosophilia melanogaster, and a fruit fly.*

*12. In cucumbers, orange fruit color (*R*) is dominant over cream fruit color (*r*). A cucumber plant homozygous for orange fruits is crossed with a plant homozygous for cream fruits. The F_1 are intercrossed to produce the F_2.

(a) Give the genotypes and phenotypes of the parents, the F_1, and the F_2.

The cross of a homozygous cucumber plant that produces orange fruit (RR) *with a homozygous cucumber plant that produces cream fruit* (rr) *will result in an F_1 generation heterozygous for the orange fruit phenotype.*

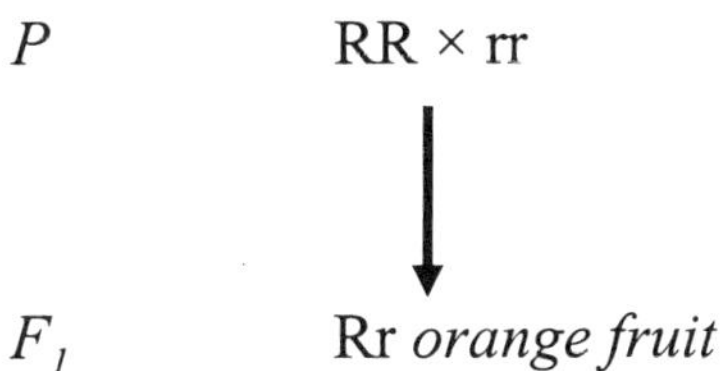

Intercrossing the F_1, will produce F_2 that are expected to show a 3:1 orange to cream fruit phenotypic ratio.

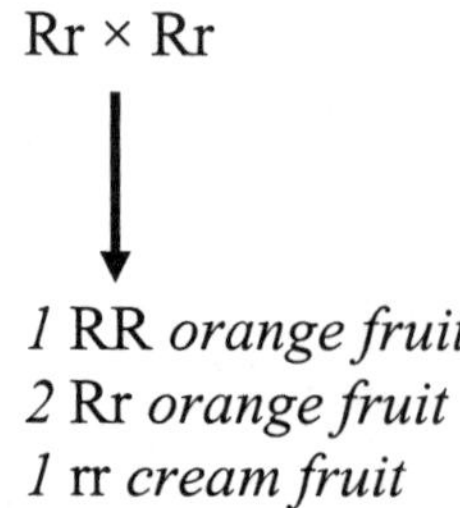

Rr × Rr

1 RR *orange fruit*
2 Rr *orange fruit*
1 rr *cream fruit*

(b) Give the genotypes and phenotypes of the offspring of a backcross between the F_1 and the orange parent.
The backcross of the F_1 orange offspring (Rr) *with homozygous orange parent* (RR) *will produce progeny that all have the orange fruit phenotype. However, ½ of the progeny will be expected to be homozygous for orange fruit and ½ of the progeny will be expected to be heterozygous for orange fruit.*

Rr *(F_1)* × RR *(orange parent)*

↓

½ RR *orange fruit*
½ Rr *orange fruit*

(c) Give the genotypes and phenotypes of a backcross between the F_1 and the cream parent.
The backcross of the F_1 offpspring (Rr) *with the cream parent* (rr) *is also a testcross. The product of this testcross should produce progeny, ½ of which are heterozygous for orange fruit and ½ of which are homozygous for cream fruit.*

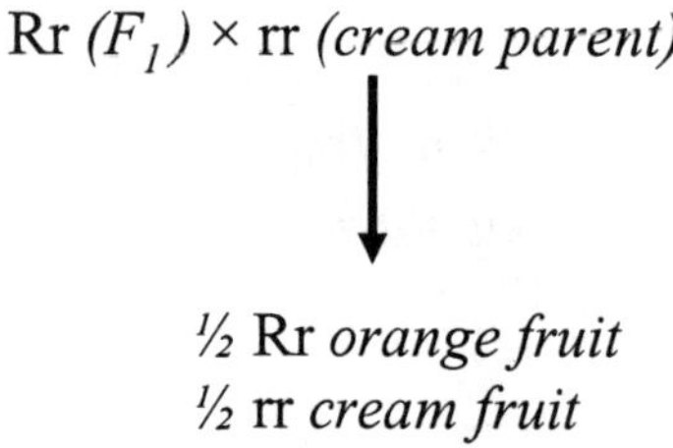

Rr *(F_1)* × rr *(cream parent)*

½ Rr *orange fruit*
½ rr *cream fruit*

*13. In rabbits, coat color is a genetically determined characteristic. Some black females always produce black progeny, whereas other black females produce black progeny and white progeny. Explain how these outcomes occur.
Because some black female rabbits produce black and white progeny, they must be heterozygous for the black coat color. These heterozygous black rabbits possess a white coat color allele that is recessive to the black coat color allele. Most likely, the black rabbits that produce only black progeny are homozygous for the black coat color allele.

*14. In cats, blood type A results from an allele (I^A) that is dominant over an allele (i^B) that produces blood type B. There is no O blood type. The blood types of male and female cats that were mated and the blood types of their kittens follow. Give the most likely genotypes for the parents of each litter.

	Male parent	Female parent	Kittens
(a)	blood type A	blood type B	4 kittens with blood type A, 3 with blood type B
(b)	blood type B	blood type B	6 kittens with blood type B
(c)	blood type B	blood type A	8 kittens with blood type A
(d)	blood type A	blood type A	7 kittens with blood type A, 2 with blood type B
(e)	blood type A	blood type A	10 kittens with blood type A
(f)	blood type A	blood type B	4 kittens with blood type A, 1 with blood type B

(a) Male with blood type A × Female with blood type B
Because the female parent has blood type B, she must have the genotype i^Bi^B. *The male parent could be either* I^AI^A *or* I^Ai^B. *However, some of the offspring are kittens with blood type B, the male parent must have contributed an* i^B *allele to these kittens. Therefore, the male must have the genotype of* I^Ai^B.

(b) Male with blood type B × Female with blood type B
Because blood type B is caused by the recessive allele i^B, *both parents must be homozygous for the recessive allele or* i^Bi^B. *Each contributes only the* i^B *allele to the offspring.*

(c) Male with blood type B × Female with blood type A
Again, the male with type B blood must be i^Bi^B. *A female with type A blood could have either the* I^AI^A *or* I^Ai^B *genotypes. Because all of her kittens have type A blood, this suggests that she is homozygous for the for* I^A *allele* (I^AI^A) *and contributes only the* I^A *allele to her offspring. It is possible that she is heterozygous for type A blood, but unlikely that chance alone would have produced eight kittens with blood type A.*

(d) Male with blood type A × Female with blood type A
Because kittens with blood type A and blood type B are found in the offspring, both parents must be heterozygous for blood type A, or I^Ai^B. *With both parents being heterozygous, the offspring would be expected to occur in a 3:1 ratio of blood type A to blood type B, which is close to the observed ratio.*

(e) Male with blood type A × Female with blood type A
Only kittens with blood type A are produced, which suggests that each parent is homozygous for blood type A (I^AI^A), *or that one parent is homozygous for blood type A* (I^AI^A), *and the other parent is heterozygous for blood type A* (I^Ai^B). *The data from the offspring will not allow us to determine the precise genotype of either parent.*

(f) Male with blood type A × Female with blood type B
On the basis of her phenotype, the female will be i^Bi^B. *In the offspring, one kitten with blood type B is produced. This kitten would require that both parents contribute an* i^B *to produce its genotype. Therefore, the male parent's genotype is* I^Ai^B. *From this cross the number of kittens with blood type B would be expected to be similar to the number of kittens with blood type A. However, due to the small number of offspring produced, random chance could have resulted in more kittens with blood type A than kittens with blood type B.*

15. Joe has a white cat named Sam. When Joe crosses Sam with a black cat, he obtains ½ white kittens and ½ black kittens. When the black kittens are interbred, they produce all black kittens. On the basis of these results, would you conclude that white or black coat color in cats is a recessive trait. Explain your reasoning.

The black coat color is likely recessive. When Sam was crossed with a black cat, ½ the offspring were white and ½ were black. This ratio potentially indicates that one of the parental cats is heterozygous dominant while the other parental cat is homozygous recessive—a testcross. The interbreeding of the black kittens produced only black kittens, indicating that the black kittens are likely to be homozygous, and thus the black coat color is the recessive trait.

If the black allele was dominant, we would have expected the black kittens to be heterozygous, containing a black coat color allele and a white coat color allele. Under this condition, we would expect ¼ of the progeny from the interbred black kittens to have white coats. Because this did not happen, we can conclude that the black coat color is recessive.

16. In sheep, lustrous fleece (*L*) results from an allele that is dominant over an allele for normal fleece (*l*). A ewe (adult female) with lustrous fleece is mated with a ram (adult male) with normal fleece. The ewe then gives birth to a single lamb with normal fleece. From this single offspring, is it possible to determine the genotypes of the two parents? If so, what are their genotypes? If not, why not?

Yes, it is possible to determine the genotype of each parent assuming that the dominant lustrous allele (L) *exhibits complete penetrance. The ram and the single lamb must be homozygous for the normal allele* (l) *because both have the normal fleece phenotype. Because the lamb receives only a single allele* (l) *from the ram, the ewe must have contributed the other recessive* l *allele. Therefore, the ewe must be heterozygous for lustrous fleece.*

In summary:

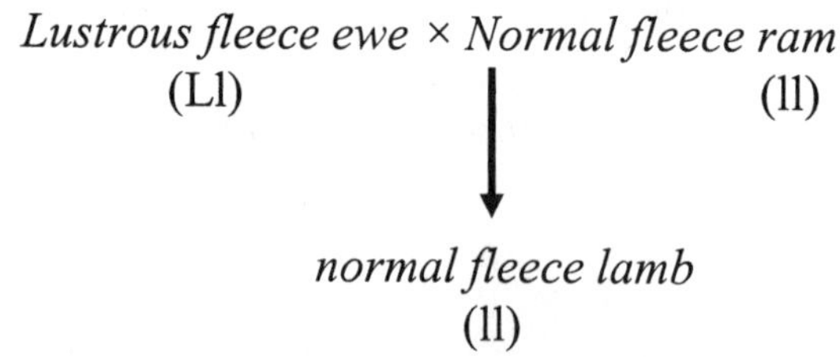

*17. In humans, alkaptonuria is a metabolic disorder in which affected persons produce black urine (see the introduction to this chapter). Alkaptonuria results from an allele (*a*) that is recessive to the allele for normal metabolism (*A*). Sally has normal metabolism, but her brother has alkaptonuria. Sally's father has alkaptonuria, and her mother has normal metabolism.

(a) Give the genotypes of Sally, her mother, her father, and her brother.

Sally's father, who has alkaptonuria, must be aa. *Her brother, who also has alkaptonuria, must be* aa *as well. Because both parents contributed one* a *allele to her brother, Sally's mother, who is phenotypically normal, must be heterozygous* (Aa). *Sally, who is normal, received the* A *allele from her mother but could have received an* a *allele only from her father.*

The genotypes of the individuals are: Sally (Aa), *Sally's mother* (Aa), *Sally's father* (aa), *and Sally's brother* (aa).

(b) If Sally's parents have another child, what is the probability that this child will have alkaptonuria?

Sally's father (aa) × *Sally's mother* (Aa)

Sally's mother has a ½ chance of contributing the a allele to her offspring. Sally's father can contribute only the a allele. The probability of an offspring with genotype aa and alkaptonuria is therefore ½ × 1 = ½.

(c) If Sally marries a man with alkaptonuria, what is the probability that their first child will have alkaptonuria?

Since Sally is heterozygous (Aa), *she has a ½ chance of contributing the* a *allele. Her husband with alkaptonuria* (aa) *can only contribute the* a *allele. The likelihood of their first child having alkaptonuria* (aa) *is ½ × 1 = ½.*

18. Suppose that you are raising Mongolian gerbils. You notice that some of your gerbils have white spots, whereas others have solid coats. What type of crosses could you carry out to determine whether white spots are due to a recessive or a dominant allele?

If white spots are recessive, then any gerbil with white spots must be homozygous for white spots (ww), *and a cross between two white-spotted gerbils* (ww × ww) *should produce offspring with only white spots. If white spots are dominant to solid, then a cross between a gerbil with white spots and a gerbil with a solid coat should produce either progeny all having solid coats* (WW × ww → Ww) *or progeny half having solid coats and half having white spots* (Ww × ww → ½ Ww ½ ww).

*19. Hairlessness in American rat terriers is recessive to the presence of hair. Suppose that you have a rat terrier with hair. How can you determine whether this dog is homozygous or heterozygous for the hairy trait?

We will define the hairless allele as h *and the dominant allele for the presence of hair as* H. *Because the allele for hair is dominant over the allele for hairless, a rat terrier with hair could be either homozygous* (HH) *or heterozygous* (Hh). *To determine which condition is present in the rat terrier with hair, a testcross with a hairless rat terrier* (hh) *should be performed. If the terrier is homozygous* (HH) *for the presence of hair, then no hairless offspring will be produced by the testcross. However, if the terrier is heterozygous for the presence of hair* (Hh), *then ½ of the offspring will be expected to be hairless.*

20. In snapdragons, red flower color (*R*) is incompletely dominant over white flower color (*r*); the heterozygotes produce pink flowers. A red snapdragon is crossed with a white snapdragon, and the F_1 are intercrossed to produce the F_2.

(a) Give the genotypes and phenotypes of the F_1 and F_2, along with their expected proportions.

Because the red flower color (R) *is not completely dominant to the white flower color* (r) *the F_1 heterozygotes* (Rr) *will all be pink.*

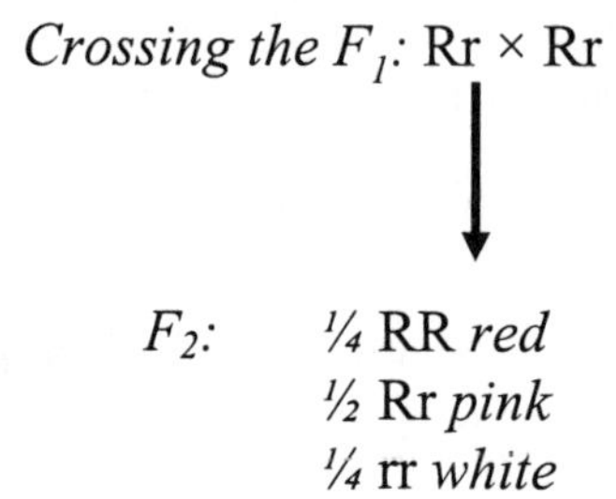

(b) If the F_1 are backcrossed to the white parent, what will the genotypes and phenotypes of the offspring be?

Crossing the F_1 to the white parent: Rr × rr
Offspring: ½ Rr *pink and* ½ rr *white*

(c) If the F_1 are backcrossed to the red parent, what are the genotypes and phenotypes of the offspring?

Crossing the F_1 to the red parent: Rr × RR
Offspring: ½ Rr *pink and* ½ RR *red*

21. What is the probability of rolling one six-sided die and obtaining the following numbers?

(a) 2

Because 2 is only found on one side of a six-sided die, then there is a 1/6 chance of rolling a two.

(b) 1 or 2

The probability of rolling a 1 on a six-sided die is 1/6. Similarly, the probability of rolling a 2 on a six-sided die is 1/6. Because the question asks what is the probability of rolling a 1 or *a 2, and these are mutually exclusive events, we should use the additive rule of probability to determine the probability of rolling a 1 or a 2:*
(p *of rolling a 1)* + (p *of rolling a 2)* = p *of rolling either a 1 or a 2.*

1/6 + 1/6 = 2/6 = 1/3 probability of rolling either a 1 or a 2.

(c) An even number

The probability of rolling an even number depends on the number of even numbers found on the die. A single die contains three even numbers (2, 4, 6). The probability of rolling any one of these three numbers on a six-sided die is 1/6. To determine the probability of rolling either a 2, a 4, or a 6, we apply the additive rule: 1/6 + 1/6 + 1/6 = 3/6 = ½.

(d) Any number but a 6

The number 6 is found only on one side of a six-sided die. The probability of rolling a 6 is therefore 1/6. The probability of rolling any number but 6 is (1 – 1/6) = 5/6.

*22. What is the probability of rolling two six-sided dice and obtaining the following numbers?

(a) 2 and 3

To calculate the probability of rolling two six-sided dice and obtaining a 2 and a 3, we will need to use the product and additive rules. There are two possible ways in which to obtain the 2 and the 3 on the dice.

There is 1/6 chance of rolling a 2 on the first die and a 1/6 chance of rolling a 3 on the second die. The probability of this taking place is therefore 1/6 × 1/6 = 1/36.

There is also a 1/6 chance of rolling a 3 on the first die and a 1/6 chance of rolling a 2 of the second die. Again, the probability of this taking place is 1/6 × 1/6 = 1/36. So, the probability of rolling a 2 and a 3 would be 1/36 + 1/36 = 2/36 or 1/18.

(b) 6 and 6

There is only one way to roll two 6's on a pair of dice: the first die must be a 6 and the second die must be a 6. The probability is 1/6 × 1/6 = 1/36.

(c) At least one 6

To determine the probability of rolling at least one 6, we need to know the probability of rolling a 6 on a given die and the probability of not rolling a 6 on that same die. The probability of rolling a 6 is 1/6, while the probability of not rolling a 6 is 1 – 1/6 = 5/6.
(probability of a 6 on first die) + [(probability of no 6 on first die) × (probability of 6 on second die)] = the probability of at least one 6.
(1/6) + (5/6)(1/6) = 6/36 + 5/36 = 11/36 chance of rolling at least one 6.

(d) Two of the same number (two 1's, or two 2's, or two 3's, etc.)

There are several ways to roll two of the same number. You could roll two 1's, two 2's, two 3's, two 4's, two 5's, or two 6's. Using the multiplication rule, the probability of rolling two 1s is 1/6 × 1/6 = 1/36. The same is true of two 2's, two 3's, two 4's, two 5's, and two 6's. Using the addition rule, the probability of rolling either two 1's, two 2's, two 3's, two 4's, two 5's, and two 6's is 1/36 + 1/36 +1/36 +1/36 +1/36 + 1/36 = 6/36 = 1/6.

(e) An even number on both dice

Three out of the six sides of a die are even numbers, so there is a 3/6 probability of rolling an even number on each of the dice. The chance of having an even number on both dice is (3/6)(3/6) = 9/36 or ¼.

(f) An even number on at least one die

Three out of the six sides of a die are even numbers, so the probability of rolling an even number on the one die is 3/6. The probability of not rolling an even number is 3/6. An even number on at least one die could be obtained by rolling (a) an even on the first but not on the second die (3/6 × 3/6 = 9/36), (b) an even on the second die but not on the first (3/6 × 3/6 = 9/36), or (c) an even on both dice (3/6 × 3/6 = 9/36). Using the addition rule to obtain the probability of either a or b or c, we obtain 9/36 + 9/36 + 9/36 = 27/36 = ¾.

*23. In a family of seven children, what is the probability of obtaining the following numbers of boys and girls?

(a) All boys

$(½)^7$ *= 1/128*

(b) All children of the same sex

The children could be all boys or all girls:
$(½)^7$ *chance of being all boys and* $(½)^7$ *chance of being all girls.*

1/128 + 1/128 = 2/128 or 1/64 chance of being either all boys or all girls.

Parts c–e require the use of the binomial expansion. Let a *equal the probability of being a girl and* b *equal the probability of being a boy. The probabilities of* a *and* b *are ½.*

$$(a + b)^7 = a^7 + 7a^6b + 21a^5b^2 + 35a^4b^3 + 35a^3b^4 + 21a^2b^5 + 7ab^6 + b^7$$

(c) Six girls and one boy

The probability for part (c) is provided for by the term $7a^6b$*. Because the probabilities of* a *and* b *are ½, then the overall probability is* $7(\frac{1}{2})^6(\frac{1}{2}) = 7/128$*.*

(d) Four boys and three girls

This probability is provided for by the term $35a^3b^4$*. The overall probability is* $35(\frac{1}{2})^3(\frac{1}{2})^4 = 35/128$*.*

(e) Four girls and three boys

Using the term $35a^4b^3$*, we see that the overall probability is* $35(\frac{1}{2})^4(\frac{1}{2})^3 = 35/128$*.*

24. Phenylketonuria (PKU) is a disease that results from a recessive gene. Two normal parents produce a child with PKU.

Because *the two normal parents have a child with PKU, each parent must be heterozygous. We will define the recessive PKU allele as p and the dominant normal allele as P. Therefore, both parents have the genotype Pp.*

(a) What is the probability that a sperm from the father will contain the PKU allele?

The father has a ½ chance of donating a sperm with the PKU allele.

(b) What is the probability that an egg from the mother will contain the PKU allele?

The mother's egg has a ½ chance of containing the PKU allele.

(c) What is the probability that their next child will have PKU?

Each parent has a ½ chance of donating the p *allele to the child. So, the child has a ½ × ½ = ¼ chance of having PKU.*

(d) What is the probability that their next child will be heterozygous for the PKU gene?

Each parent has a ½ chance of donating the P *allele or a ½ chance of donating the* p *allele to the child. Therefore, the child has a (½ × ½) + (½ × ½) = ½ chance of being heterozygous.*

*25. In German cockroaches, curved wing (*cv*) is recessive to normal wing (*cv*+). A homozygous cockroach having normal wings is crossed with a homozygous cockroach having curved wings. The F_1 are intercrossed to produce the F_2. Assume that the pair of chromosomes containing the locus for wing shape is metacentric. Draw this pair of chromosomes as it would appear in the parents, the F_1, and each class of F_2 progeny at metaphase I of meiosis. Assume that no crossing over takes place. At each stage, label a location for the alleles for wing shape (*cv* and *cv*+) on the chromosomes.

Parents:

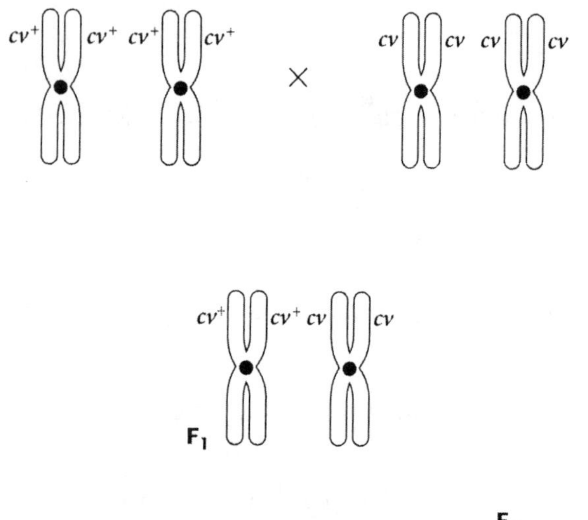

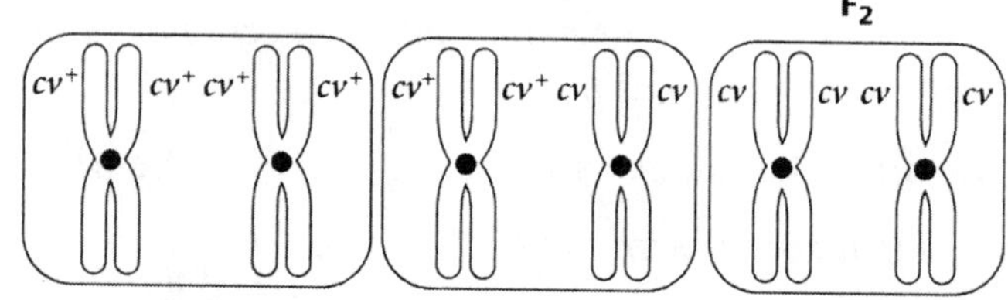

*26. In guinea pigs, the allele for black fur (*B*) is dominant over the allele for brown (*b*) fur. A black guinea pig is crossed with a brown guinea pig, producing five F_1 black guinea pigs and six F_1 brown guinea pigs.

(a) How many copies of the black allele (*B*) will be present in *each* cell from an F_1 black guinea pig at the following stages: G_1, G_2, metaphase of mitosis, metaphase I of meiosis, metaphase II of meiosis, and after the second cytokinesis following meiosis? Assume that no crossing over takes place.

The cross of a black guinea pig with a brown guinea pig produced black and brown guinea pigs in the offspring. Because the brown guinea pig is homozygous (bb), *the black guinea pig must be heterozygous* (Bb).

Black guinea pigs (Bb) × *Brown guinea pigs* (bb) → *F_1 five black guinea pigs* (Bb) *and six brown guinea pigs* (bb).

To determine the number of copies of the B *allele or the* b *allele at the different stages of the cell cycle, we need to remember that following the completion of S phase and prior to anaphase of mitosis and anaphase II of meiosis, each chromosome will consist of two sister chromatids.*

In the F_1 black guinea pigs (Bb), *only one chromosome possesses the black allele, so we would expect in G_1 one black allele; G_2, two black alleles; metaphase of mitosis, two black alleles; metaphase I of meiosis, two black alleles; after cytokinesis of meiosis, one black allele but only in ½ of the cells produced by meiosis. (The remaining ½ will not contain the black allele.)*

(b) How may copies of the brown allele (*b*) will be present in each cell from an F_1 brown guinea pig at the same stages? Assume that no crossing over takes place.

In the F_1 brown guinea pigs (bb), *both homologs possess the brown allele, so we would expect in G_1 two brown alleles; G_2, four brown alleles; metaphase of mitosis, four brown alleles; metaphase I of meiosis, four brown alleles; metaphase II, two brown alleles; and after cytokinesis of meiosis, one brown allele.*

27. In watermelons, bitter fruit (*B*) is dominant over sweet fruit (*b*), and yellow spots (*S*) are dominant over no spots (*s*). The genes for these two characteristics assort independently. A homozygous plant that has bitter fruit and yellow spots is crossed with a homozygous plant that has sweet fruit and no spots. The F_1 are intercrossed to produce the F_2.

(a) What will be the phenotypic ratios in the F_2?

P: homozygous bitter fruit, yellow spots (BBSS) × *homozygous sweet fruit and no spots* (bbss)

F_1: All progeny have bitter fruit and yellow spots (BbSs)
The F_1 are intercrossed to produce the F_2: BbSs × BbSs
The F_2 phenotypic ratios are as follows:

9/16 Bitter fruit and yellow spots
3/16 Bitter fruit and no spots
3/16 Sweet fruit and yellow spots
1/16 Sweet fruit and no spots

(b) If an F_1 plant is backcrossed with the bitter, yellow spotted parent, what phenotypes and proportions are expected in the offspring?

The backcross of a F_1 plant (BbSs) *with the bitter, yellow-spotted parent (BBSS) will produce all bitter, yellow-spotted offspring.*

(c) If an F_1 plant is backcrossed with the sweet, nonspotted parent, what phenotypes and proportions are expected in the offspring?

The backcross of a F_1 plant (BbSs) *with the sweet, nonspotted parent* (bbss) *will produce the following phenotypic proportions in the offspring:*

¼ Bitter fruit and yellow spots
¼ Bitter fruit and no spots
¼ Sweet fruit and yellow spots
¼ Sweet fruit and no spots

28. In cats, curled ears (*Cu*) result from an allele that is dominant over an allele for normal ears (*cu*). Black color results from an independently assorting allele (*G*) that is dominant over an allele for gray (*g*). A gray cat homozygous for curled ears is mated with a homozygous black cat with normal ears. All the F_1 cats are black and have curled ears.

(a) If two of the F_1 cats mate, what phenotypes and proportions are expected in the F_2?

If F_1 cats mated, GgCucu × GgCucu, *then following proportions and phenotypes are expected in the F_2:*

9/16 Black with curly ears
3/16 Black with normal ears
3/16 Gray with curly ears
1/16 Gray with normal ears

(b) An F_1 cat mates with a stray cat that is gray and possesses normal ears. What phenotypes and proportions of progeny are expected from this cross?

The mating of an F_1 cat (GgCucu) *with a gray cat with normal ears* (ggcucu) *is a testcross in which we would expect to produce equal numbers of all the different progeny classes:*

¼ Black with curly ears
¼ Black with normal ears
¼ Gray with curly ears
¼ Gray with normal ears

*29. The following two genotypes are crossed: *AaBbCcddEe* × *AabbCcDdEe*. What will the proportion of the following genotypes be among the progeny of this cross?

The simplest procedure for determining the proportion of a particular genotype in the offspring is to break the cross down into simple crosses and consider the proportion of the offspring for each cross.

AaBbCcddEe × AabbCcDdEe
Locus 1: Aa × Aa = ¼ AA, ½ Aa, ¼ aa
Locus 2: Bb × bb = ½ Bb, ½ bb
Locus 3: Cc × Cc = ¼ CC, ½ Cc, ¼ cc

Locus 4: dd × Dd = ½ Dd, ½ dd
Locus 5: Ee × Ee = ¼ EE, ½ Ee, ¼ ee

(a) *AaBbCcDdEe:* ½ (Aa) × ½ (Bb) × ½ (Cc) × ½ (Dd) × ½ (Ee) = *1/32*

(b) *AabbCcddee: ½ (Aa) × ½ (bb) × ½ (Cc) × ½ (dd) × ¼ (ee) = 1/64*

(c) *aabbccddee: ¼ (aa) × ½ (bb) × ¼ (cc) × ½ (dd) × ¼ (ee) = 1/256*

(d) *AABBCCDDEE: Will not occur. The* AaBbCcddEe *parent cannot contribute a* D *allele, the* AabbCcDdEe *parent cannot contribute a* B *allele. Therefore, their offspring cannot be homozygous for the* BB *and* DD *gene loci.*

30. In mice, an allele for apricot eyes (a) is recessive to an allele for brown eyes (a^+). At an independently assorting locus, an allele for tan (t) coat color is recessive to an allele for black (t^+) coat color. A mouse that is homozygous for brown eyes and black coat color is crossed with a mouse having apricot eyes and a tan coat. The resulting F_1 are intercrossed to produce the F_2. In a litter of eight F_2 mice, what is the probability that two will have apricot eyes and tan coats?

The F_1 will have brown eyes and tan coats, and the genotype $a^+a\ t^+t$.
The F_2 will be produced by intercrossing the F_1: $a^+a\ t^+t \times a^+a\ t^+t$. *By considering each locus individually with a simple cross we can easily calculate the proportion of any offspring class in the F_2.*

Locus 1: $a^+a \times a^+a$ = ¼ a^+a^+, ½ a^+a, ¼ aa

Producing the phenotypic ratio: ¾ *brown eyes* (a^+a^+ *or* a^+a)
¼ *apricot eyes* (aa)

Locus 2: $t^+t \times t^+t$ = ¼ t^+t^+, ½ t^+t, ¼ tt

Producing the phenotypic ratio: ¾ *black coat* (t^+t^+, t^+t)
¼ *tan coat* (tt)

To determine the probability that, out of a litter of eight mice, two will have apricot eyes and a tan coat, we first need to determine the likelihood of an apricot mouse with a tan coat being produced from the mating of the F_1:
aatt: ¼ (aa) × ¼ (tt) = *1/16.*

The probability of two mice with this phenotype then can be determined using the binomial expansion defining "a" as the probability that 1/16 of the mice will have apricot eyes and tan coats while defining "b" as the probability that 15/16 will have another phenotype.

$$(a + b)^8 = a^8 + 8a^7b + 28a^6b^2 + 56a^5b^3 + 70a^4b^4 + 56a^3b^5 + 28a^2b^6 + 8ab^7 + b^8$$

The probability of having two apricot mice with tan coats is provided by the term $28a^2b^6$ in the binomial. The probability of "a" is 1/16, while the probability of "b" is 15/16. So the overall probability is $28(1/16)^2(15/16)^6 = 0.074$.

31. In cucumbers, dull fruit (*D*) is dominant over glossy fruit (*d*), orange fruit (*R*) is dominant over cream fruit (*r*), and bitter cotyledons (*B*) are dominant over nonbitter cotyledons (*b*). The three characters are encoded by genes located on different pairs of chromosomes. A plant homozygous for dull, orange fruit and bitter cotyledons is crossed with a plant that has glossy, cream fruit and nonbitter cotyledons. The F_1 are intercrossed to produce the F_2.

All of the F_1 plants have dull, orange fruit and bitter cotyledons (DdRrBb). By intercrossing the F_1, the F_2 are produced. The expected phenotypic ratios in the F_2 can be calculated more easily by examining the phenotypic ratios produced by the individual crosses of each gene locus.

F_1 are intercrossed: DdRrBb × DdRrBb
Locus 1: Dd × Dd = ¾ dull (DD and Dd); ¼ glossy (dd)
Locus 2: Rr × Rr = ¾ orange (RR and Rr); ¼ cream (rr)

Locus 3: Bb × Bb = ¾ bitter (BB and Bb); ¼ nonbitter (bb)

(a) Give the phenotypes and their expected proportions in the F_2.

dull, orange, bitter: *¾ dull × ¾ orange × ¾ bitter = 27/64*
dull, orange, nonbitter: *¾ dull × ¾ orange × ¼ nonbitter = 9/64*
dull, cream, bitter: *¾ dull × ¼ cream × ¾ bitter = 9/64*
dull, cream, nonbitter: *¾ dull × ¼ cream × ¼ nonbitter = 3/64*
glossy, orange, bitter: *¼ glossy × ¾ orange × ¾ bitter = 9/64*
glossy, orange, nonbitter: *¼ glossy × ¾ orange × ¼ nonbitter = 3/64*
glossy, cream, bitter: *¼ glossy × ¼ cream × ¾ bitter = 3/64*
glossy, cream, nonbitter: *¼ glossy × ¼ cream × ¼ nonbitter = 1/64*

(b) An F_1 plant is crossed with a plant that has glossy, cream fruit and nonbitter cotyledons. Give the phenotypes and expected proportions among the progeny of this cross.

Intercrossing the F_1 with a plant that has glossy, cream fruit, and nonbitter cotyledons is an example of a testcross. All progeny classes will be expected in equal proportions, because the phenotype of the offspring will be determined by the alleles contributed by the F_1 parent.

DdRrCc (F_1) × ddrrcc (tester)
F_1 Locus 1 (Dd): ½ D and ½ d
F_1 Locus 2 (Rr): ½ R and ½ r
F_1 Locus 3 (Cc): ½ C and ½ c
dull, orange, bitter: ½ dull × ½ orange × ½ bitter = 1/8
dull, orange, nonbitter: ½ dull × ½ orange × ½ nonbitter = 1/8
dull, cream, bitter: ½ dull × ½ cream × ½ bitter = 1/8
dull, cream, nonbitter: ½ dull × ½ cream × ½ nonbitter = 1/8
glossy, orange, bitter: ½ glossy × ½ orange × ½ bitter = 1/8
glossy, orange, nonbitter: ½ glossy × ½ orange × ½ nonbitter = 1/8
glossy, cream, bitter: ½ glossy × ½ cream × ½ bitter = 1/8
glossy, cream, nonbitter: ½ glossy × ½ cream × ½ nonbitter = 1/8

*32. *A* and *a* are alleles located on a pair of metacentric chromosomes. *B* and *b* are alleles located on a pair of acrocentric chromosomes. A cross is made between individuals having the following genotypes: *AaBb* × *aabb*.

(a) Draw the chromosomes as they would appear in each type of gamete produced by the individuals of this cross.

Gametes from *AaBb* individual:

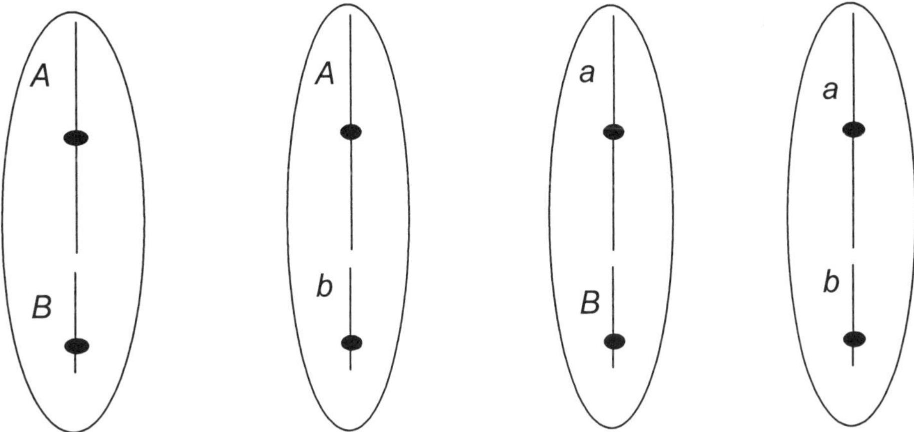

Gametes from *aabb* individual:

(b) For each type of progeny resulting from this cross, draw the chromosomes as they would appear in a cell at G_1, G_2, and metaphase of mitosis.

Progeny at G_1

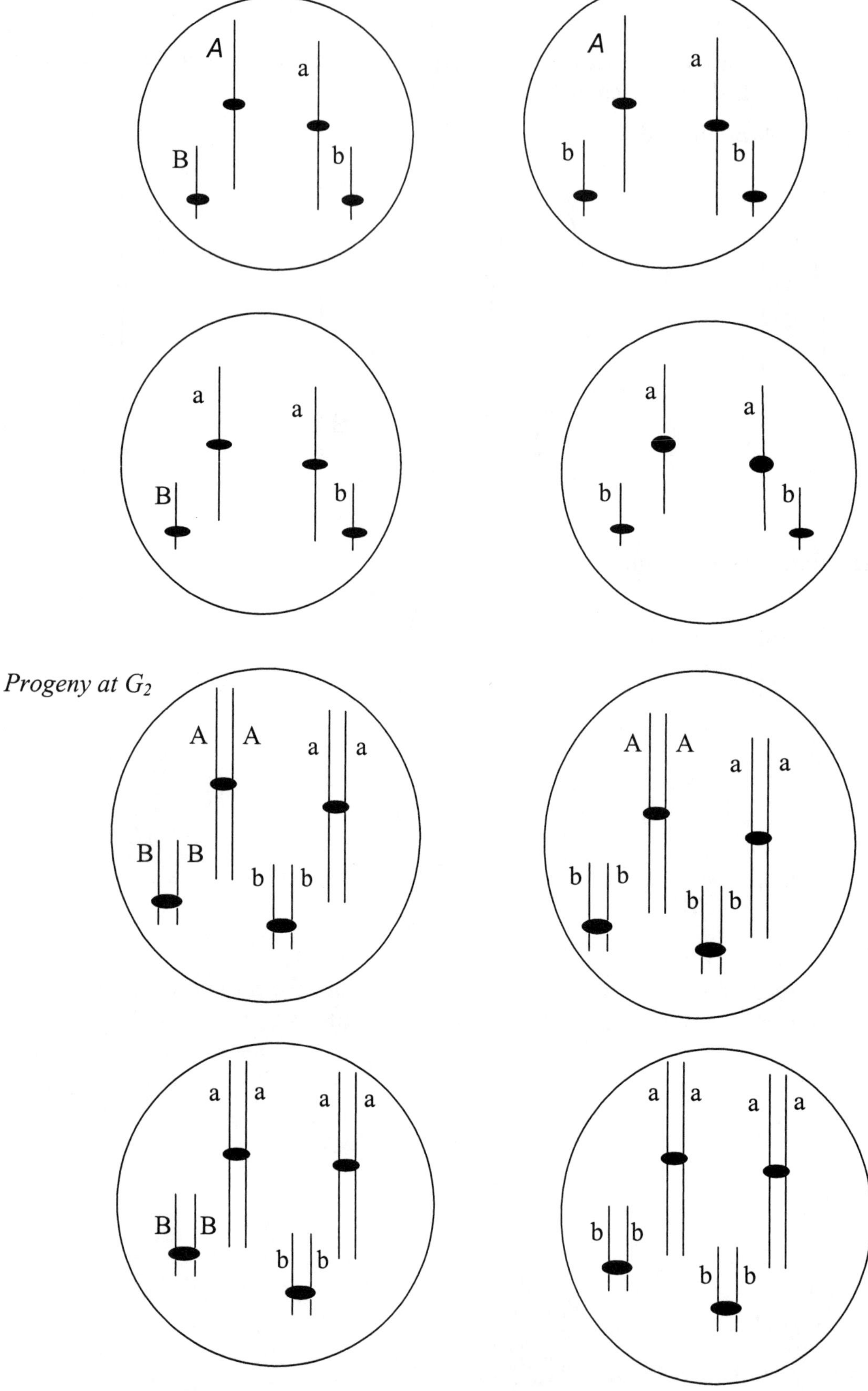

Progeny at metaphase of mitosis:

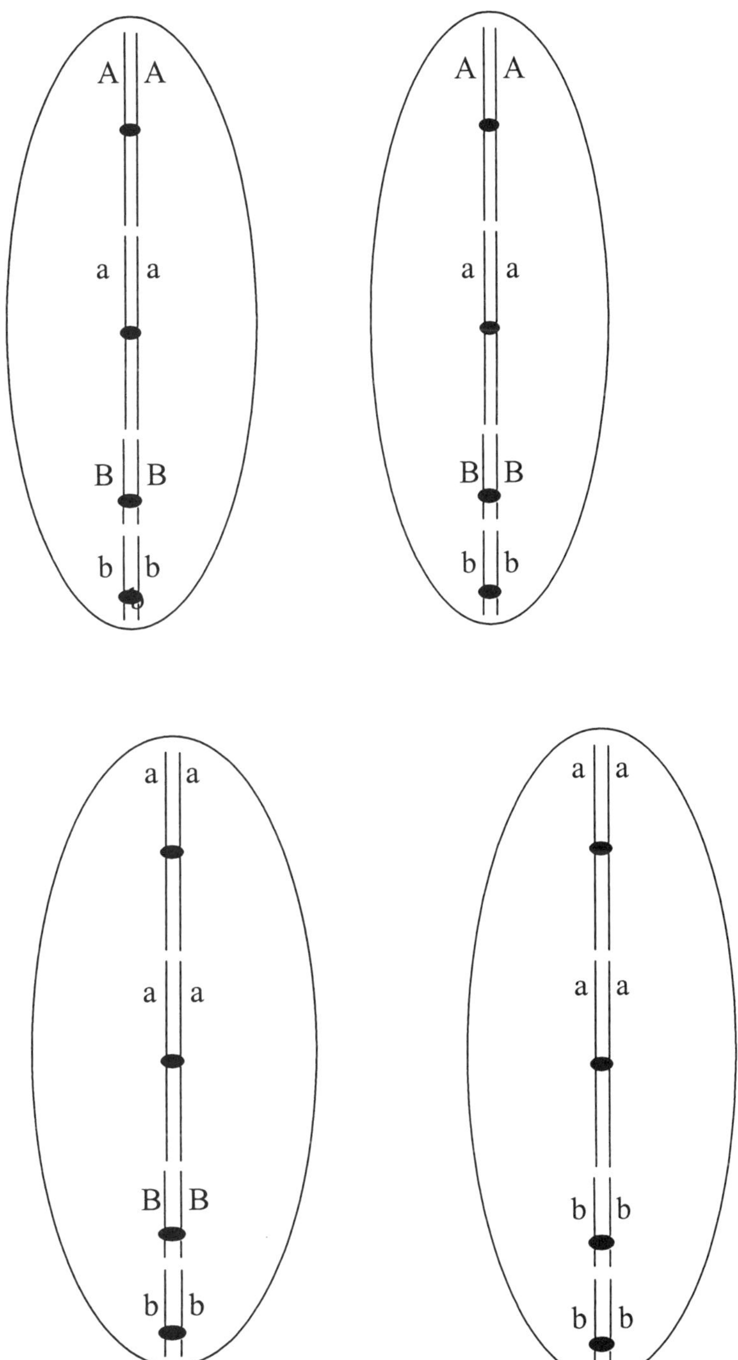

The order of chromosomes on metaphase plate can vary.

33. In sailfin mollies (fish), gold color is due to an allele (*g*) that is recessive to the allele for normal color (*G*). A gold fish is crossed with a normal fish. Among the offspring, 88 are normal and 82 are gold.

(a) What are the most likely genotypes of the parents in this cross?

Because we know that gold (g) *is recessive to normal* (G), *we can figure out that the gold parent fish has the* aa *genotype. We also know that the normal parent fish must have at least one* G *allele, but because there are gold offspring* (gg), *we can predict that the normal parent must be heterozygous* Gg.

In summary: gold fish (gg) × *normal fish* (Gg) → *88 normal fish* (Gg)
82 gold fish (gg)

(b) Assess the plausibility of your hypothesis by performing a chi-square test.

If the parents are gold (gg) × *normal* (Gg), *we would expect the offspring to have a phenotypic ratio of ½ gold to ½ normal. By using the goodness-of-fit chi-square test, we can determine if the deviations are likely to have been produced by chance.*

Phenotype	*Observed (O)*	*Expected (E)*	*$(O–E)^2/E$ or (X^2)*
Gold	*82*	*85*	*0.11*
Normal	*88*	*85*	*0.11*
Total	*170*	*170*	*0.21*

Degrees of freedom (df) = (number of phenotypic classes) – 1.
Because there are two phenotypic classes, the degrees of freedom (df) is 1. From the chi-square table, we can see that the calculated chi-square value falls between 0.211 (P *of .9) and 0.51 (* P *of .975). The probability is relatively high that differences between what we expected and what we observed was generated by chance and that our parents are as predicted,* (gg × Gg).

34. In guinea pigs, the allele for black coat color (*B*) is dominant over the allele for white coat color (*b*). At an independently assorting locus, an allele for rough coat (*R*) is dominant over an allele for smooth coat (*r*). A guinea pig that is homozygous for black color and rough coat is crossed with a guinea pig that has a white and smooth coat. In a series of matings, the F_1 are crossed with guinea pigs having white, smooth coats. From these matings, the following phenotypes appear in the offspring: 24 black, rough guinea pigs; 26 black, smooth guinea pigs; 23 white, rough guinea pigs; and 5 white, smooth guinea pigs.

(a) Using a chi-square test, compare the observed numbers of progeny with those expected from the cross.

Phenotype	*Observed (O)*	*Expected (E)*	*$(O - E)^2 / E$ or (X^2)*
Black, smooth	*24*	*19.5*	*1.04*
Black, rough	*26*	*19.5*	*2.17*
White, Rough	*23*	*19.5*	*0.63*
White, Smooth	*5*	*19.5*	*10.9*
Total	*78*	*19.5*	*14.74*

Degrees of freedom = 4 – 1 = 3. The chi-square value falls below 12.838 for a probability value less than .005 or 0.5 % that random chance produced the observed ratio of guinea pigs.

(b) What conclusions can you draw from the results of the chi-square test?

From the chi-square value, we can see that it is unlikely that random variations produced the observed ratio. Some other phenomena must be acting.

(c) Suggest an explanation for these results.

Two possible explanations are:

(1) The white parents were actually BbRr, Bbrr, *and* bbRr. *In this case, both dominant alleles would be exhibiting incomplete penetrance.*

(2) The combination of bbrr *is sublethal. In other words, guinea pigs with the homozygous recessive phenotype are less viable.*

CHALLENGE QUESTIONS

35. Dwarfism is a recessive trait in Hereford cattle. A rancher in western Texas discovers that several of the calves in his herd are dwarfs, and he wants to eliminate this undesirable trait from the herd as rapidly as possible. Suppose that the rancher hires you as a genetic consultant to advise him on how to breed the dwarfism trait out of the herd. What crosses would you advise the rancher to conduct to ensure that the allele causing dwarfism is eliminated from the herd?

To eliminate the recessive dwarfism allele from the herd, you will need to rid the herd of any heterozygous individuals—assuming that heterozygous individuals have a phenotype similar to the homozygous normal individuals. In essence, the farmer wants to create a homozygous normal cattle population. The first step is to advise the farmer to cull from the herd any bulls and cows that when mated produced a dwarf calf. Because the dwarf calf must be homozygous for the dwarf allele, each parent had to contribute the dwarf allele to the calf; thus each parent is heterozygous.

Next, a possible way to determine if any of the remaining cows are heterozygous is to perform a series of testcrosses using a dwarf bull. In the progeny produced by such a cross, ½ are expected to be normal and ½ are expected to be dwarfs if the cow was heterozygous. Any cows that produce dwarf calves should be eliminated from the herd. If the cow is homozygous, no dwarf calves will be produced. Unfortunately, due to the limited number of offspring (typically one) produced by each cow for each mating, several matings for each cow would be necessary to determine if she is heterozygous.

The farmer may not think it practical to purchase a dwarf bull, or even that the dwarf bull would be able to mate with the cows. A second method would be to cross the cows with a bull that is known to be heterozygous normal. For a cross between this bull and a heterozygous cow, we would expect ¾ normal offspring and ¼ dwarfs. A homozygous cow mated with a heterozygous bull would not produce any dwarf offspring. Again several matings would be necessary to determine if the cow is heterozygous. The farmer will not want to keep the progeny of these crosses because if the cow is heterozygous, the chance that her normal offspring are carriers of the dwarf allele is 50%.

In either scenario, the process for building a pure breeding herd will take several years and careful monitoring of the offspring.

36. A geneticist discovers an obese mouse in his laboratory colony. He breeds this obese mouse with a normal mouse. All the F_1 mice from this cross are normal in size. When he interbreeds two F_1 mice, eight of the F_2 mice are normal in size and two are obese. The geneticist then intercrosses two of his obese mice, and he finds that all of the progeny from this cross are obese. These results lead the geneticist to conclude that obesity in mice results from a recessive allele.

A second geneticist at a different university also discovers an obese mouse in her laboratory colony. She carries out the same crosses as the first geneticist did and obtains the same results. She also concludes that obesity in mice results from a recessive allele. One day the two geneticists meet at a genetics conference, learn of each other's experiments, and decide to exchange mice. They both find that, when they cross two obese mice from the different laboratories, all the offspring are normal; however, when they cross two obese mice from the same laboratory, all the offspring are obese. Explain their results.

The first geneticist has identified an obese allele that he believes to be recessive. We will define his allele as o_1 *and the normal allele as* O_1. *The obese allele appears to be recessive based on the series of crosses he performed.*

Cross 1 with possible genotype:
Obese (o_1o_1) × *Normal* (O_1O_1) → *F_1 All normal* (O_1o_1)
Cross 2 with possible genotypes:
F_1 normal (Oo_1) × *F_1 normal* (O_1o_1) → *F_2* *8 normal* (O_1O_1 and O_1o_1)
2 obese (o_1o_1)
Cross 3 with possible genotypes:
Obese (o_1o_1) × *Obese* (o_1o_1) → *All Obese* (o_1o_1)

A second geneticist also finds an obese mouse in her colony and performs the same types of crosses, which indicate to her that the obese allele is recessive. We will define her obese allele as o_2 *and the normal allele as* O_2.

The cross of obese mice between the two different laboratories produced only normal mice. These different alleles are both recessive. However, they are located at different gene loci. Essentially, the obese mice from the different labs have separate obesity genes that are independent of one another.

The likely genotypes of the obese mice are as follows:

Obese mouse 1 ($o_1o_1O_2O_2$) × *Obese mouse 2* ($O_1O_1o_2o_2$)
→ F_1 *All normal* ($O_1o_1O_2o_2$)

37. Albinism is a recessive trait in humans. A geneticist studies a series of families in which both parents are normal and at least one child has albinism. The geneticist reasons that both parents in these families must be heterozygotes and that albinism should appear in ¼ of the children of these families. To his surprise, the geneticist finds that the frequency of albinism among the children of these families is considerably greater than ¼. Can you think of an explanation for the higher-than-expected frequency of albinism among these families?

    The geneticist has indeed identified parents who are heterozygous for the albinism allele. However, by looking only at parents who have albino children, he is missing parents who are heterozygous and have no albino children. Since most parents are likely to have only a few children, the result is that the frequency of albino children produced by parents with an albino child will be higher than what would be predicted. If he were to consider the offspring of normal heterozygous parents with no albino children, along with the parents who have albino children, the expected frequency of albino offspring would most likely approach ¼.

38. Two distinct phenotypes are found in the salamander *Plethodon cinereus*: a red form and a black form. Some biologists have speculated that the red phenotype is due to an autosomal allele that is dominant over an allele for black. Unfortunately, these salamanders will not mate in captivity; so the hypothesis that red is dominant over black has never been tested.

    One day a genetics student is hiking through the forest and finds 30 female salamanders, some red and some black, laying eggs. The student places each female and her eggs (about 20–30 eggs per female) in separate plastic bags and takes them back to the lab. There, the student successfully raises the eggs until they hatch. After the eggs have hatched, the student records the phenotypes of the juvenile salamanders, along with the phenotypes of their mothers. Thus, the student has the phenotypes for 30 females and their progeny, but no information is available about the phenotypes of the fathers.

Explain how the student can determine whether red is dominant over black with this information on the phenotypes of the females and their offspring.

To determine whether red is dominant over black the student will need to examine the colors and phenotypic ratios of the colors found in the offspring of each salamander. Certain trial assumptions will have to be made.

If black is recessive, the following assumptions about the phenotypic ratios in the offspring of the black females can be made:

Black female × black male → All black offspring
Black female × red male (heterozygous) → ½ black offspring; ½ red offspring
Black female × red male (homozygous) → All red offspring

If red is dominant, then the following assumptions about the phenotypic ratios in the offspring of the red females can be made:

Red female (homozygous) × black male → All red offspring
Red female (heterozygous) × black male → ½ red offspring; ½ black offspring
Red female (homozygous or heterozygous) × red male (homozygous) → All red
Red female (homozygous) × red male (homozygous or heterozygous) → All red
Red female (heterozygous) × red male (heterozygous) → ¾ red offspring; ¼ black offspring

The key ratio will be any female salamander that produces offspring with a 3:1 phenotypic ratio. If in the offspring of a red salamander, a ratio of 3:1 red is produced, then the red allele is dominant over the black. If, however, in the offspring of a black salamander, a 3:1 phenotypic ratio of black to red is observed, then black is the dominant allele.

Chapter Four: Sex Determination and Sex-Linked Characteristics

COMPREHENSION QUESTIONS

*1. What is the most defining difference between males and females?
Males produce relatively large gametes; females produce larger gametes.

2. How do monoecious organisms differ from dioecious organisms?
Monoecious organisms exist as only one form, which has male and female reproductive structures; the same organism produces male and female gametes. Dioecious organisms exist as organisms of two distinct genders, one producing male gametes, the other producing female gametes.

3. Describe the XX-XO system of sex determination. In this system, which is the heterogametic sex and which is the homogametic sex?
In the XX-XO sex determination system, females have two copies of the sex-determining chromosome, whereas males have only one copy. Males must be considered heterogametic because they produce two different types of gametes with respect to the sex chromosome: either containing an X or not containing an X.

4. How does sex determination in the XX-XY system differ from sex determination in the ZZ-ZW system?
In the XX-XY system, males are heterogametic and produce gametes with either an X chromosome or a Y chromosome. In the ZZ-ZW system, females are heterogametic and produce gametes with either a Z or a W chromosome.

*5. What is the pseudoautosomal region? How does the inheritance of genes in this region differ from the inheritance of other Y-linked characteristics?
The pseudoautosomal region is a region of similarity between the X and Y chromosomes that is responsible for pairing the X and Y chromosomes during meiotic prophase I. Genes in this region are present in two copies in males and females, and thus are inherited like autosomal genes, whereas other Y-linked genes are passed on only from father to son.

*6. How is sex determined in insects with haplodiploid sex determination?
Diploid individuals are female, whereas haploid individuals are male. Eggs that are fertilized by a sperm develop into females, and eggs that are not fertilized develop as males.

7. What is meant by genic sex determination?
In organisms that follow this system, there is no recognizable difference in the chromosome contents of males and females. Instead of a sex chromosome that differs between males and females, alleles at one or more loci determine the sex of the individual.

8. How does sex determination in *Drosophila* differ from sex determination in humans?

In humans, the presence of a functional Y chromosome determines maleness. People with XXY and XXXY are phenotypically male. In Drosophila, *the ratio of X chromosome material to autosomes determines the sex of the individual, regardless of the Y chromosome. Flies with XXY are female.*

9. Give the typical sex chromosomes found in the cells of people with Turner syndrome, Klinefelter syndrome, and androgen insensitivity syndrome, as well as in poly-X females.
Turner syndrome: XO
Klinefelter syndrome: XXY (rarely XXXY or XXYY)
Androgen insensitivity: XY
Poly-X females: XXX (rarely XXXX or even XXXXX)

*10. What characteristics are exhibited by an X-linked trait?
Males show the phenotypes of all X-linked traits, regardless of whether the X-linked allele is normally recessive or dominant. Males inherit X-linked traits from their mothers, pass X-linked traits to their daughters, and subsequently on to their grandsons, but not to their sons.

11. Explain how Bridges's study of nondisjunction in *Drosophila* helped prove the chromosome theory of inheritance.
Bridges showed that in crosses with white-eyed flies, a sex-linked trait, exceptional progeny had abnormal inheritance of sex chromosomes. In matings of white-eyed females with red-eyed males, most of the progeny followed the expected pattern of white-eyed males and red-eyed females. Exceptional red-eyed male progeny were XO and exceptional white-eyed females were XXY. These karyotypes were exactly as Bridges predicted with his hypothesis that the exceptional red-eyed males inherited their X chromosome with the red-eye allele from their red-eyed fathers, and were male because they did not inherit an X chromosome from their mothers, resulting in an XO condition that is phenotypically male but sterile. Moreover, the exceptional white-eyed females inherited two X chromosomes from their white-eyed mothers, as a result of nondisjunction in meiosis I of the female, and none from their red-eyed fathers, receiving instead a Y chromosome to make them XXY females. Calvin Bridges linked exceptional inheritance of a sex-linked trait to exceptional inheritance of the X chromosome: the linked exceptions proved the rule that genes reside on chromosomes.

12. What are some of its characteristics that make *Drosophila melanogaster* a good model genetic organism?
These fruit flies have a relatively short generation time of about 10 days, and produce large numbers of progeny, with females producing 400 to 500 eggs in a 10-day period. They are easily and inexpensively cultured in the laboratory, and being small, require little laboratory space. Nevertheless they have a complex life cycle and morphology and are large enough that males and females are easily distinguished, and many morphological mutations may be observed with just a hand lens or dissecting scope. Drosophila melanogaster also has very large polytene chromosomes in the salivary glands, which have facilitated cytological studies of chromosomes. At a molecular level, the relatively small genome of Drosophila, amounting to 180 million base pairs of DNA

or only about 5% of the size of the human genome, has been completely sequenced, and techniques have been developed for facile genetic engineering of the Drosophila genome.

13. Explain why tortoiseshell cats are almost always female and why they have a patchy distribution of orange and black fur.
Tortoiseshell cats have two different alleles of an X-linked gene: X^+ *(non-orange, or black) and* X^o *(orange). The patchy distribution results from X-inactivation during early embryo development. Each cell of the early embryo randomly inactivates one of the two X chromosomes, and the inactivation is maintained in all of the daughter cells. So each patch of black fur arises from a single embryonic cell that inactivated the* X^o*, and each patch of orange fur arises from an embryonic cell that inactivated the* X^+*.*

14. What is a Barr body? How is it related to the Lyon hypothesis?
Barr bodies are darkly staining bodies in the nuclei of female mammalian cells. Mary Lyon hypothesized that Barr bodies are inactivated (condensed) X chromosomes. By inactivating all X chromosomes beyond one, female cells achieve dosage compensation for X-linked genes.

*15. What characteristics are exhibited by a Y-linked trait?
Y-linked traits appear only in males and are always transmitted from fathers to sons, thus following a strict paternal lineage. Autosomal male-limited traits also appear only in males, but they can be transmitted to boys through their mothers.

APPLICATION QUESTIONS AND PROBLEMS

*16. What is the sexual phenotype of fruit flies with the following chromosomes?

	Sex chromosomes	Autosomal chromosomes	Sexual phenotype
(a)	XX	All normal	*Female (X:A = 1.0)*
(b)	XY	All normal	*Male (X:A = 0.5)*
(c)	XO	All normal	*Male, sterile (X:A = 0.5)*
(d)	XXY	All normal	*Female (X:A = 1.0)*
(e)	XYY	All normal	*Male (X:A = 0.5)*
(f)	XXYY	All normal	*Female (X:A = 1.0)*
(g)	XXX	All normal	*Metafemale (X:A > 1.0)*
(h)	XX	Four haploid sets	*Male (X:A = 0.5)*
(i)	XXX	Four haploid sets	*Intersex (X:A between 0.5 and 1.0)*
(j)	XXX	Three haploid sets	*Female (X:A = 1.0)*
(k)	X	Three haploid sets	*Metamale, sterile (X:A < 0.5)*
(l)	XY	Three haploid sets	*Metamale (X:A < 1.0)*
(m)	XX	Three haploid sets	*Intersex (X:A between 0.5 and 1.0)*

In fruit flies, the X to autosome ratio determines sex: males have X:A ratios of 0.5 (metamales have X:A ratios less than 0.5), and females have X:A ratios of 1.0 (metafemales have ratios greater than 1.0).

17. For parts a through g in problem 15, what would be the human sexual phenotype (male or female)?
In humans, the Y chromosome determines maleness. Although extra sex chromosomes may be tolerated, extra sets of autosomes are lethal.
(a) female
(b) male
(c) female (Turner syndrome)
(d) male (Klinefelter syndrome)
(e) male
(f) male (Klinefelter syndrome)
(g) female (triple-X syndrome)

18. A normal female *Drosophila* produces abnormal eggs that contain all (a complete diploid set) of her chromosomes. She mates with a normal male *Drosophila* that produces normal sperm. What will be the sex of the progeny from this cross?

In this cross, all the eggs will have two X chromosomes and two sets of autosomes. These eggs will be fertilized by two kinds of sperm produced in equal proportions: half the sperm will have one X chromosome and one set of autosomes, and the other half will have one Y chromosome and one set of autosomes. Thus, half the progeny will have 3 X chromosomes and 3 sets of autosomes, and will be female because the X:autosome ratio will be 1. The other half will have 2 X chromosomes and 3 sets of autosomes, for an X:autosome ratio of 2/3, and will be intersex (see answer to problem 16m, above).

19. Hemophilia results from a recessive X-linked gene. Jill is a female who has hemophilia. She marries Bill, a male who has normal blood clotting. What proportion of their children is expected to have hemophilia?
Because hemophilia is recessive, Jill must be homozygous for the hemophilia mutation; both of her X chromosomes have the mutation (X^hX^h). Bill must have a wild- type X chromosome (X^+Y).

$$X^hX^h \text{ mated with } X^+Y \rightarrow \tfrac{1}{2}\, X^h X^+ \text{ and } \tfrac{1}{2}\, X^h Y$$

Thus ½ of the children, or all the sons, will have hemophilia.

*20. Joe has classic hemophilia, which is an X-linked recessive disease. Could Joe have inherited the gene for this disease from the following persons?

	Yes	No
(a) His mother's mother	*X*	
(b) His mother's father	*X*	
(c) His father's mother		*X*
(d) His father's father		*X*

X-linked traits are passed on from mother to son. Therefore, Joe must have inherited the hemophilia trait from his mother. His mother could have inherited the trait from either her mother (a) or her father (b). Because Joe could not have inherited the trait from his father (Joe inherited the Y chromosome from his father), he could not have inherited hemophilia from either (c) or (d).

*21. In *Drosophila,* yellow body is due to an X-linked gene that is recessive to the gene for gray body.

(a) A homozygous gray female is crossed with a yellow male. The F_1 are intercrossed to produce F_2. Give the genotypes and phenotypes, along with the expected proportions, of the F_1 and F_2 progeny.

We will use X^+ *as the symbol for the dominant gray body color, and* X^y *for the recessive yellow body color. Male progeny always inherit the Y chromosome from the male parent and either of the two X chromosomes from the female parent. Female progeny always inherit the X chromosome from the male parent and either of the two X chromosomes from the female parent.*

F_1 males inherit the Y chromosome from their father, and X^+ *from their mother; hence, their genotype is* X^+Y *and they have gray bodies.*

F_1 females inherit X^y *from their father and* X^+ *from their mother; hence, they are* X^+X^y *and also have gray bodies.*

When the F_1 progeny are intercrossed, the F_2 males again inherit the Y from the F_1 male, and they inherit either X^+ *or* X^y *from their mother. Therefore, we should get ½* X^+Y *(gray body) and ½* X^yY *(yellow body). The F_2 females will all inherit the* X^+ *from their father and either* X^+ *or* X^y *from their mother. Therefore, we should get ½* X^+X^+ *and ½* X^+X^y *(all gray body).*

In summary:

P X^+X^+ *(gray female)* × X^yY *(yellow male)*

F_1 ½ X^+Y *(gray males)*

½ X^+X^y *(gray females)*

F_2 ¼ X^+Y *(gray males)*

¼ X^yY *(yellow males)*

¼ X^+X^y *(gray females)*

¼ X^+X^+ *(gray females)*

The net F_2 phenotypic ratios are ½ gray females, ¼ gray males, and ¼ yellow males.

This problem can also be analyzed using a Punnett square.

	X^+X^+ *(gray females)*	X^+Y *(gray males)*
	X^+X^y *(gray females)*	X^yY *(yellow males)*

(b) A yellow female is crossed with a gray male. The F_1 are intercrossed to produce the F_2. Give the genotypes and phenotypes, along with the expected proportions, of the F_1 and F_2 progeny.

The yellow female must be homozygous X^yX^y *because yellow is recessive, and the gray male, having only one X chromosome, must be* X^+Y*. The* F_1 *male progeny are all* X^yY *(yellow) and the* F_1 *females are all* X^+X^y *(heterozygous gray).*

P X^yX^y *(yellow female)* × X^+Y *(gray male)*
F_1 ½ X^yY *(yellow males)*
½ X^+X^y *(gray females)*
F_2 ¼ X^+Y *(gray males)*
¼ X^yY *(yellow males)*
¼ X^+X^y *(gray females)*
¼ X^yX^y *(yellow females)*

(c) A yellow female is crossed with a gray male. The F_1 females are backcrossed with gray males. Give the genotypes and phenotypes, along with the expected proportions, of the F_2 progeny.

If the F_1 X^+X^y *females are backcrossed to* X^+Y *gray males, then*

F_2 ¼ X^+Y *(gray males)*
¼ X^yY *(yellow males)*
¼ X^+X^+ *(gray females)*
¼ X^+X^y *(gray females)*

(d) If the F_2 flies in part (b) mate randomly, what are the expected phenotypic proportions of flies in the F_3?

The outcome of F_2 *flies from b mating randomly should be equivalent to random union of the male and female gametes. We need to predict the overall male and female gamete types and their frequencies.*

As a result of meiosis, half of the male gametes will have the Y chromosome. Because there are equal numbers of males with either the X^+ *or* X^y*, the X-bearing male gametes will be split equally: (½ with* X^+*)(½ with* X^+*) = ¼* X^+*; (½ with* X^+*)(½ with* X^y*) = ¼* X^y

Male gametes: ½Y, ¼ X^+*, ¼* X^y

Half the F_2 *females in part (b) are homozygous* X^yX^y*, so all their gametes will be* X^y*: ½* X^y*. The other half are heterozygous and will produce equal proportions of* X^+ *and* X^y *gametes: ½(½*X^+*) = ¼* X^+*; ½(½* X^y*) = ¼* X^y*.*

Female gametes: ¼ X^+*, ¾* X^y

Now using a Punnett square:

	½ *Y*	¼ X^+	¼ X^y
¼ X^+	*1/8* X^+Y	*1/16* X^+X^+	*1/16* X^+X^y
¾ X^y	*3/8* X^yY	*3/16* X^+X^y	*3/16* X^yX^y

Overall genotypic ratios are 1/8 X^+Y*, 3/8* X^yY*, 1/16* X^+X^+*, 4/16* X^+X^y*, 3/16* X^yX^y*. Overall phenotypic ratios are 1/8 gray males, 3/8 yellow males, 5/16 gray females, and 3/16 yellow females.*

*22. John and Cathy have normal color vision. After 10 years of marriage to John, Cathy gives birth to a color-blind daughter. John files for divorce, claiming he is not the father of the child. Is John justified in his claim of nonpaternity? Explain why. If Cathy had given birth to a color-blind son, would John be justified in claiming nonpaternity?

Assuming the color blindness is a recessive trait, the color-blind daughter must be homozygous recessive. If the color blindness trait here is not *X-linked, then John has no justification because John and Cathy could be carriers. If the color blindness is the X-linked red-green colo-blindness, then John has grounds for suspicion. Normally, their daughter would have inherited John's X chromosome. Because John is not colo rblind, he could not have transmitted a color-blind X chromosome to the daughter.*

A remote alternative possibility is that the daughter is XO, having inherited a recessive color-blind allele from her mother and no sex chromosome from her father. In that case, the daughter would have Turner syndrome.

If Cathy had a color-blind son, then John would have no grounds for suspicion. The son would have inherited John's Y chromosome and the color-blind X chromosome from Cathy.

23. Red-green color blindness in humans is due to an X-linked recessive gene. A woman whose father was color-blind possesses one eye with normal color vision and one eye with color blindness.

(a) Propose an explanation for this woman's vision pattern.

One of the two X chromosomes in the woman's zygotes must have been inactivated during early embryogenesis. If one eye derived exclusively from a progenitor cell that inactivated the normal X, then that eye would be color blind, whereas the other eye may be derived from a progenitor cell that inactivated the color-blind X, or is a mosaic with sufficient normal retinal cells to permit color vision.

(b) Would it be possible for a man to have one eye with normal color vision and one eye with color blindness?

The only way would be for the man to be XXY.

*24. Bob has XXY chromosomes (Klinefelter syndrome) and is colorblind. His mother and father have normal color vision, but his maternal grandfather is colorblind. Assume that Bob's chromosome abnormality arose from nondisjunction in meiosis. In which parent and in which meiotic division did nondisjunction occur? Explain your answer.

Because Bob must have inherited the Y chromosome from his father, and his father has normal color vision, there is no way a nondisjunction event from the paternal lineage could account for Bob's genotype. Bob's mother must be heterozygous X^+X^c *because she has normal color vision, and she must have inherited a color-blind X chromosome from her color-blind father. For Bob to inherit two color-blind X chromosomes from his mother, the egg must have arisen from a nondisjunction in meiosis II. In meiosis I, the homologous X chromosomes separate, so one cell has the* X^+ *and the other has* X^c*. Failure of sister chromatids to separate in meiosis II would then result in an egg with two copies of* X^c*.*

25. In certain salamanders, it is possible to alter the sex of a genetic female, making her into a functional male; these salamanders are called sex-reversed males. When a sex-reversed

male is mated with a normal female, approximately ⅔ of the offspring are female and ⅓ are male. How is sex determined in these salamanders? Explain the results of this cross.
The 2:1 ratio of females to males could be explained if a quarter of the progeny are embryonic lethals. The sex-reversed male has the same chromosome complement as normal females. If females were homogametic, then matings between sex-reversed males and normal females must result in all females: XX (sex-reversed male) crossed to XX (normal female) results in all XX (female) progeny. However, if females are heterogametic (ZW), then ZW (sex-reversed male) crossed to ZW (normal female) results in ¼ ZZ (male), ½ ZW (female), and ¼ WW (embryonic lethal). The net result is a 2:1 ratio of females to males.

26. In some mites, males pass genes to their grandsons, but they never pass genes to male offspring. Explain.
A system in which males are haploid and females are diploid would explain these results. Haploid males pass genes only to female progeny. The female progeny can then generate haploid male grandsons that contain the grandfather's genes.

27. The Talmud, an ancient book of Jewish civil and religious laws, states that if a woman bears two sons who die of bleeding after circumcision (removal of the foreskin from the penis), any additional sons that she has should not be circumcised. (The bleeding is most likely due to the X-linked disorder hemophilia.) Furthermore, the Talmud states that the sons of her sisters must not be circumcised, while the sons of her brothers should. Is this religious law consistent with sound genetic principles? Explain your answer.
Yes. If a woman has a son with hemophilia, then she is a carrier. Any of her sons have a 50% chance of having hemophilia. Her sisters may also be carriers. Her brothers, if they do not themselves have hemophilia (because they survived circumcision, they most likely do not have hemophilia), cannot be carriers, and therefore there is no risk of passing hemophilia on to their children.

*28. Miniature wings (X^m) in *Drosophila* result from an X-linked allele that is recessive to the allele for long wings (X^+). Give the genotypes of the parents in the following crosses:

	Male parent	Female parent	Male offspring	Female offspring
(a)	Long	Long	231 long, 250 miniature	560 long
(b)	Miniature	Long	610 long	632 long
(c)	Miniature	Long	410 long, 417 miniature	412 long, 415 miniature
(d)	Long	Miniature	753 miniature	761 long
(e)	Long	Long	625 long	630 long

The genotype of the male parent is the same as his phenotype for an X-linked trait. Because the male progeny get their X chromosomes from their mother, the phenotypes of the male progeny give us the genotypes of the female parents.

(a) Male parent is $X^{+}Y$. Because the male offspring are 1:1 long:miniature, the female parent must be $X^{+}X^{m}$. You can use a Punnett square to verify that all the female progeny from such a cross will have long wings (they get the dominant X^{+} from the father).
(b) Male parent is $X^{m}Y$. Because the male offspring are all long, the female parent must be $X^{+}X^{+}$.
(c) Male parent is $X^{m}Y$; female parent is $X^{+}X^{m}$.
(d) Male parent is $X^{+}Y$; female parent is $X^{m}X^{m}$.
(e) Male parent is $X^{+}Y$; female parent is $X^{+}X^{+}$.

*29. In chickens, congenital baldness results from a Z-linked recessive gene. A bald rooster is mated with a normal hen. The F_1 from this cross are interbred to produce the F_2. Give the genotypes and phenotypes, along with their expected proportions, among the F_1 and F_2 progeny.

For species with the ZZ-ZW sex-determination system, the females are heterogametic ZW. So a bald rooster must be $Z^{b}Z^{b}$ (where Z^{b} denotes the recessive allele for baldness), and a normal hen must be $Z^{+}W$.

P $Z^{b}Z^{b} \times Z^{+}W$
F_1 ½ $Z^{b}Z^{+}$ *(normal males)*
½ $Z^{b}W$ *(bald females)*
F_2 *Using a Punnett square:*

	$Z^{+}Z^{b}$ *(normal roosters)*	$Z^{+}W$ *(normal hens)*
	$Z^{b}Z^{b}$ *(bald roosters)*	$Z^{b}W$ *(bald hens)*

*30. How many Barr bodies would you expect to see in human cells containing the following chromosomes?
(a) XX—*1 Barr body*
(b) XY—*0*
(c) XO—*0*
(d) XXY—*1*
(e) XXYY—*1*
(f) XXXY—*2*
(g) XYY—*0*
(h) XXX—2
(i) XXXX—*3*

Human cells inactivate all X chromosomes beyond one. The Y chromosome has no effect on X-inactivation.

31. A woman with normal chromosomes mates with a man who also has normal chromosomes.
(a) Suppose that during oogenesis, the woman's sex chromosomes undergo nondisjunction in meiosis I; the man's chromosomes separate normally. Give all possible combinations of sex chromosomes that might occur in this couple's children and the number of Barr bodies you would expect to see in each.

Eggs produced by nondisjunction in meiosis I: XX and O (nullo)
Sperm produced by normal meiosis: X and Y
Children: XXX (two Barr bodies), XO (no Barr body), XXY (1 Barr body), YO (no Barr body)

(b) What chromosome combinations and numbers of Barr bodies would you expect to see if the chromosomes separate normally during oogenesis, but nondisjunction of the sex chromosomes occurs in meiosis I of spermatogenesis.

Eggs produced by normal meiosis: X
Sperm produced by nondisjunction in meiosis I: XY and O (nullo)
Children: XXY (1 Barr body) and XO (no Barr body)

32. Red-green color blindness is an X-linked recessive trait in humans. Polydactyly (extra fingers and toes) is an autosomal dominant trait. Martha has normal fingers and toes and normal color vision. Her mother is normal in all respects, but her father is colorblind and polydactylous. Bill is colorblind and polydactylous. His mother has normal color vision and normal fingers and toes. If Bill and Martha marry, what types and proportions of children can they produce?

The first step is to deduce the genotypes of Martha and Bill. Because the two traits are independent, we can deal with just one trait at a time.

Starting with the X-linked color-blind trait, Bill must be X^cY *because he is colorblind. Bill's mother must be a carrier* (X^+X^c). *Martha must be* X^+X^c, *a carrier for color blindness because her father is color blind* (X^cY).

For polydactyly, Bill must be Dd *(*D *denotes the dominant polydactyly allele). Because his mother has normal fingers* (dd), *he cannot be homozygous* DD. *Martha, with normal fingers, must be* dd.

If Martha (dd,X^+X^c) *marries Bill* (Dd,X^cY), *then we can predict the types and probability ratios of children they could produce.*

For polydactyly, ½ of children will be polydactylous, and ½ will have normal fingers.

For color blindness, ¼ of children will be color-blind girls, ¼ will be girls with normal vision but carrying the color blindness allele, ¼ will be color-blind boys, and ¼ will be boys with normal vision.

Combining both traits, then:
1/8 color-blind girls with normal fingers
1/8 color-blind girls with polydactyly
1/8 girls with normal vision and normal fingers
1/8 girls with normal vision and polydactyly
1/8 color-blind boys with normal fingers
1/8 color-blind boys with polydactyly
1/8 boys with normal vision and normal fingers
1/8 boys with normal vision and polydactyly
This analysis can also be carried out with a Punnett square.

33. The following two genotypes are crossed: $AaBbCcX^+X^r \times AaBBccX^+Y$, where a, b, and c represent autosomal genes and X^+ and X^r represent X-linked alleles in an organism with

XY sex determination. What is the probability of obtaining genotype $aaBbCcX^+X^+$ in the progeny?

We have to assume that the autosomal genes a, b, and c assort independently of each other as well as of the sex chromosomes. Given independent assortment, we can calculate the probability of the genotype for each gene separately, and then multiply the probabilities to calculate the probability of the combined genotype for all four genes.

For gene a: Aa × Aa → ¼ aa

For gene b: Bb × BB → ½ Bb

For gene c: Cc × cc → ½ Cc

For the sex-linked gene r: $X^+X^r \times X^+Y \rightarrow$ ¼ X^+X^+

Combined probability of genotype $aaBbCcX^+X^+$ = ¼ × ½ × ½ × ¼ = 1/64

*34. Miniature wings in *Drosophila melanogaster* result from an X-linked gene (X^m) that is recessive to an allele for long wings (X^{m+}). Sepia eyes are produced by an autosomal gene (*s*) that is recessive to an allele for red eyes (s^+).

(a) A female fly that has miniature wings and sepia eyes is crossed with a male that has normal wings and is homozygous for red eyes. The F_1 are intercrossed to produce the F_2. Give the phenotypes and their proportions expected in the F_1 and F_2 flies from this cross.

The female parent (miniature wings, sepia eyes) must be X^mX^m,ss.

The male parent (normal wings, homozygous red eyes) is $X^{m+}Y,s^+s^+$.

F_1 *males are* X^mY,s^+s *(miniature wings, red eyes)*

females are $X^{m+}X^m,s^+s$ *(long wings, red eyes)*

F_2: *We can analyze the expected outcome of this cross with either a branch diagram or with a Punnett square.*

First, the branch diagram:

½ male (Y) →	½ X^{m+} normal wings →	¾ red (¾ s^+)	= ½ × ½ × ¾ = $^3/_{16}$ male, normal, red
		¼ sepia (¼ ss)	= ½ × ½ × ¼ = $^1/_{16}$ male, normal, sepia
	½ X^m miniature →	¾ red	= ½ × ½ × ¾ = $^3/_{16}$ male, miniature, red
		¼ sepia	= ½ × ½ × ¼ = $^1/_{16}$ male, miniature, sepia
½ female (X^m) →	½ X^{m+} normal →	¾ red	= ½ × ½ × ¾ = $^3/_{16}$ female, normal, red
		¼ sepia	= ½ × ½ × ¼ = $^1/_{16}$ female, normal, sepia
	½ X^m miniature →	¾ red	= ½ × ½ × ¾ = $^3/_{16}$ female, miniature, red
		¼ sepia	= ½ × ½ × ¼ = $^1/_{16}$ female, mini, sepia

Explanation: ½ of the F_2 *progeny are males because they inherit Y from the* F_1 *male. The other ½ are females that inherit the* X^m *from the* F_1 *male. In each case, ½ of the* F_2 *males and females inherit* X^{m+} *from the* F_1 *female and have normal wings, whereas the other ½ inherit* X^m *and have miniature wings. Finally, ¾ of the progeny will have the dominant red eyes, and ¼ will have sepia eyes.*

Now the Punnett square:

	¼ $X^m s^+$	¼ X^m s	¼ Y s^+	¼ Y s
¼ $X^{m+} s^+$	$X^{m+}X^m s^+s^+$ Long wings, red eyes	$X^{m+}X^m s^+s$ Long wings, red eyes	$X^{m+}Y s^+s^+$ Long wings, red eyes	$X^{m+}Y s^+s$ Long wings, red eyes
¼ X^{m+} s	$X^{m+}X^m s^+s$	$X^{m+}X^m$ ss	$X^{m+}Y s^+s$	$X^{m+}Y$ ss

	Long wings, red eyes	*Long wings, sepia eyes*	*Long wings, red eyes*	*Long wings, sepia eyes*
¼ $X^m s^+$	$X^mX^m s^+s^+$ *Mini wings, red eyes*	$X^mX^m s^+s$ *Mini wings, red eyes*	$X^mY s^+s^+$ *Mini wings, red eyes*	$X^mY s^+s$ *Mini wings, red eyes*
¼ $X^m s$	$X^mX^m s^+s$ *Mini wings, red eyes*	$X^mX^m ss$ *Mini wings, sepia eyes*	$X^mY s^+s$ *Mini wings, red eyes*	$X^mY ss$ *Mini wings, sepia eyes*

(b) A female fly that is homozygous for normal wings and has sepia eyes is crossed with a male that has miniature wings and is homozygous for red eyes. The F_1 are intercrossed to produce the F_2. Give the phenotypes and proportions expected in the F_1 and F_2 flies from this cross.

Parents $X^{m+}X^{m+}$,ss *and* X^mY,s^+s^+

F_1 $X^{m+}X^m$,s^+s *and* $X^{m+}Y$,s^+s

F_2

Sex chromosome inherited From male	*From female*	*Autosomal phenotype*	*Combined phenotype*
½ Y male	½ X^{m+}	*¾ red*	*3/16 long wings, red eyes*
		¼ sepia	*1/16 long wings, sepia eyes*
	½ X^m	*¾ red*	*3/16 mini wings, red eyes*
		¼ sepia	*1/16 mini wings, sepia eyes*
½ X^{m+} female	X^{m+} *or* X^m	*¾ red*	*3/8 long wings, red eyes*
		¼ sepia	*1/8 long wings, sepia eyes*

Note that in this case, the X chromosome the F_2 females inherit from the mother does not affect their phenotype because they all have a dominant X^{m+} *for long wings from their father.*

35. In organisms with the ZZ, ZW sex determining system, a female can inherit her Z chromosome from which of the following:

	Yes	No
Her mother's mother	___	x
Her mother's father	___	x
Her father's mother	x	___
Her father's father	x	___

A female inherits her W chromosome from her mother, and her Z chromosome from her father. A male inherits one Z from his mother and one Z from his father.

36. Suppose that a recessive gene that produces a short tail in mice is located in the pseudoautosomal region. A short-tailed male is mated with a female mouse that is homozygous for a normal tail. The F_1 from this cross are intercrossed to produce the F_2. What will the phenotypes and proportions of the F_1 and F_2 mice be from this cross?

If the gene is in the pseudoautosomal region, then it must be present on the X and Y chromosome. We will use X^+ *and* Y^+ *for the normal alleles, and* X^s *and* Y^s *for the short-tail alleles. The short-tailed male must be* X^sY^s*, and the normal female is* X^+X^+*. The expected* F_1 *would then be:*

X^+Y^s *males with long tails and* X^+X^s *females with long tails.*

F_2 *from the intercross will be:*
¼ X^+Y^s *males with long tails*
¼ X^sY^s *males with short tails*
¼ X^+X^+ *females with long tails*
¼ X^+X^s *females with long tails*

So all the females will have long tails, and equal proportions of the males will have short and long tails.

*37. A color-blind female and a male with normal vision have three sons and six daughters. All the sons are color blind. Five of the daughters have normal vision, but one of them is color blind. The color-blind daughter is 16 years old, is short for her age, and has never undergone puberty. Propose an explanation for how this girl inherited her color blindness.

The trivial explanation for these observations is that this form of color blindness is an autosomal recessive trait. In that case, the father would be a heterozygote, and we would expect equal proportions of color-blind and normal children, of either sex.

If, on the other hand, we assume that this is an X-linked trait, then the mother is X^cX^c *and the father must be* X^+Y*. Normally, all the sons would be color blind, and all the daughters should have normal vision. The only way to have a daughter that is color blind would be for her not to have inherited an* X^+ *from her father. The observation that the color-blind daughter is short in stature and has failed to undergo puberty is consistent with Turner syndrome (XO). The color-blind daughter would then be* X^cO.

CHALLENGE QUESTIONS

38. Antibiotics kill the *Wolbachia* bacteria that sometimes infect isopods and cause ZZ males to become females (see Sex Wars in Isopods at beginning of the chapter). A biologist collects some isopods from a natural population that exhibits a female-biased sex ratio. She adds antibiotics to the isopod's food to kill any bacteria. She crosses several male and female isopods and rears their offspring in the laboratory. To her surprise, the offspring of many of the crosses are all male. Can you explain her result?

If the female bias in the population of isopods was indeed caused by Wolbachia infection, then many of the phenotypic females in the matings were actually genotypically ZZ. Only ZZ progeny would result from matings of ZZ males with ZZ Wolbachia-infected phenotypic females. With antibiotics in the larval food, the Wolbachia would be eliminated, and all the ZZ progeny would develop as males.

39. On average, what proportion of the X-linked genes in the first individual is the same as that in the second individual?

(a) A male and his mother

A male inherits his single X chromosome from his mother, so 100% of his X-linked genes are the same as those in his mother.

(b) A female and her mother

A female inherits one of her two X chromosomes from her mother, so 50% of her X-linked genes are the same as those in her mother.

(c) A male and his father

A male inherits only the Y chromosome from his father, so none of his X-linked genes are the same as those in his father.

(d) A female and her father

A female inherits one of her two X chromosomes from her father, so 50%.

(e) A male and his brother

The chance that a male's brother inherited the same X chromosome of the two possible maternal X chromosomes is ½, or 50%.

(f) A female and her sister

A female and her sister share at least 50% because they inherited the same X chromosome from their father. In addition, there is another ½ chance that they inherited the same X chromosome from their mother, so overall it is 50% + ½(50%) = 75%.

(g) A male and his sister

The chance that a sister inherited the same X chromosome from her mother as her brother did is ½, or 50%.

(h) A female and her brother

Of the two X chromosomes in the female, the one inherited from the father cannot be present in her brother. Then there is a ½, or 50%, chance that the second X chromosome is shared.

40. A geneticist discovers a male mouse in his laboratory colony with greatly enlarged testes. He suspects that this trait results from a new mutation that is either Y-linked or autosomal dominant. How could he determine if the trait is autosomal dominant or Y-linked?

Because testes are present only in males, enlarged testes could either be a sex-limited autosomal dominant trait or a Y-linked trait. Assuming the male mouse with enlarged testes is fertile, mate it with a normal female. If the trait is autosomal dominant, and the parental male is heterozygous, only half the male progeny will have enlarged testes. If the trait is Y-linked, all the male progeny will have enlarged testes.

With either outcome, however, the results from this first cross will not be conclusive. If all the male progeny do express the trait, the trait may still be autosomal dominant if the parental male was homozygous. If only some of the male progeny express the trait, the possibility still remains that the trait is Y-linked but incompletely penetrant. In either case, more conclusive evidence is needed.

Mate the female progeny (F_1 females) with normal males. If the trait is autosomal dominant, some of the male F_2 progeny will have enlarged testes, proving that the trait can be passed through a female. If the trait is Y-linked, all the male F_2 progeny will have normal testes, like their normal male father.

41. Amanda is a genetics student at a small college in Connecticut. While counting her fruit flies in the laboratory one afternoon, she observed a strange species of fly in the room. Amanda captured several of the flies and began to raise them. After having raised the flies for several generations, she discovered a mutation in her colony that produces yellow eyes, in contrast with normal red eyes, and Amanda determined that this trait is definitely X-linked recessive. Because yellow eyes are X-linked, she assumed that either this species has the XX-XY system of sex determination with genic balance similar to *Drosophila* or it has the XX-XO system of sex determination.
How can Amanda determine whether sex determination in this species is XX-XY or XX-XO? The chromosomes of this species are very small and hard for Amanda to see with her student microscope so she can conduct crosses only with flies having the yellow-eye mutation. Outline the crosses Amanda should conduct and explain how they will prove XX-XY or XX-XO sex determination in this species.
To distinguish between these two possibilities, Amanda has to look for situations where the presence of the Y chromosome makes a difference in the outcome of a cross. The XX-XY and XX-XO systems behave similarly in normal matings. However, the outcomes can be very different in exceptional circumstances, such as nondisjunction of the X chromosomes in female meiosis.

Amanda should cross yellow-eyed females with red-eyed males. With either sex determination system, the progeny of such a cross should be all yellow-eyed males and all red-eyed females. However, she should be able to observe rare (~1/1000) progeny that are yellowed-eyed females or red-eyed males, as a result of meiotic nondisjunction, as Calvin Bridges observed in a similar cross with Drosophila melanogaster *white-eyed mutants. Nondisjunction at either meiosis I or II will produce female gametes with two* X^m *(bearing the yellow-eye mutant allele) or nullo-X. Upon fertilization of the nullo-X egg by an* X^+*-bearing sperm, exceptional red-eyed* X^+O *males will be produced with either sex determination system. However, fertilization of the* X^mX^m *egg with a Y-bearing sperm will result in exceptional yellow-eyed females with* X^mX^mY *in the case of XX-XY sex determination. For the XX-XO sex determination system, fertilization of the* X^mX^m *egg by a nullo-X bearing sperm will produce* X^mX^m *females.*

If these exceptional yellow-eyed females are now mated again to normal red-eyed males, then the two systems give different predicted outcomes. Meiosis in X^mX^mY *females will be complicated by the presence of three sex chromosomes and should yield relatively high proportions of* X^mY *and* X^m *gametes along with* X^mX^m *and Y. In contrast, meiosis in* X^mX^m *females should proceed normally, and produce just* X^m *gametes, excepting the rare nondisjunction events. Therefore, the progeny produced from mating these yellow-eyed exceptional females to normal red-eyed males will be very different, as shown in the following table:*

P: $X^mX^m \times X^+Y$
F_1: X^+X^m *females,* X^mY *males*
Rare: X^mX^mY *females*
X^+O *males*
Cross X^mX^mY *female* × X^+Y *male*

P: $X^mX^m \times X^+O$
F_1: X^+X^m *females,* X^mO *males*
Rare: X^mX^m *females*
X^+O *males*
Cross X^mX^m *females* × X^+O *male*

If sex determination is XX-XY		
Gametes from $X^m X^m Y$	X^+Y *male gametes* X^+	*Y*
X^m	X^+X^m *red female*	X^mY *yellow male*
X^mY	X^+X^mY *red female*	X^mYY *yellow male*
X^mX^m	$X^mX^mX^+$ *lethal*	X^mX^mY *yellow female*
Y	X^+Y *red male*	*YY* *lethal*

If sex detemination is XX-XO		
Gametes from X^mX^m	X^+O *male gametes* X^+	*O*
X^m	X^+X^m *red female*	X^mO *yellow male*

If the cross of the exceptional yellow-eyed females results in a relatively high frequency (at least several percent, rather than 1/1000) of red-eyed males and yellow-eyed females, then these flies use an XX-XY system. If the matings produce the usual phenotypic ratios, then the flies use an XX-XO system.

42. Occasionally, a mouse X chromosome is broken into two pieces and each piece becomes attached to a different autosomal chromosome. In this event, only the genes on one of the two pieces undergo X-inactivation. What does this observation indicate about the mechanism of X-chromosome inactivation?
The X-inactivation mechanism must require or recognize a specific region or locus on the X-chromosome and must inactivate chromatin attached to this center of inactivation. When the X-chromosome breaks, the fragment containing the X-inactivation locus or center becomes inactivated. The other fragment escapes inactivation because it is no longer attached.

Chapter Five: Extensions and Modifications of Basic Principles

COMPREHENSION QUESTIONS

*1. How do incomplete dominance and codominance differ?
Incomplete dominance means the phenotype of the heterozygote is intermediate to the phenotypes of the homozygotes. Codominance refers to situations in which both alleles are expressed and both phenotypes are manifested simultaneously.

*2. What is incomplete penetrance and what causes it?
Incomplete penetrance occurs when an individual with a particular genotype does not express the expected phenotype. Environmental factors, as well as the effects of other genes, may alter the phenotypic expression of a particular genotype.

3. Explain how dominance and epistasis differ.
Dominance is an allelic *interaction of two alleles of a single gene or locus, in which the phenotype corresponds only to the dominant allele. Epistasis results from the interaction of two or more genes or loci, in which the phenotype of the organism is governed by the genotype at the epistatic gene or locus, masking the genotype of the other.*

4. What is a recessive epistatic gene?
Recessive epistasis occurs when homozygous recessiveness of the epistatic gene masks the interacting gene or genes. In the example from the text, being homozygous recessive at the locus for deposition of color in hair shafts (ee) *completely masked the effect of the color locus regardless of whether it had the dominant black* (B-) *or recessive brown* (bb) *allele.*

*5. What is a complementation test and what is it used for?
Complementation tests are used to determine whether different recessive mutations affect the same gene or locus (are allelic) or whether they affect different genes. The two mutations are introduced into the same individual by crossing homozygotes for each of the mutants. If the progeny show a mutant phenotype, then the mutations are allelic (in the same gene). If the progeny show a wild-type (dominant) phenotype, then the mutations are in different genes and are said to complement each other because each of the mutant parents can supply a functional copy (or dominant allele) of the gene mutated in the other parent.

6. What is genomic imprinting?
Genomic imprinting refers to different expression of a gene depending on whether it was inherited from the male parent or the female parent.

7. What characteristics do you expect to see in a trait that exhibits anticipation?
Traits that exhibit anticipation become stronger or more pronounced, or are expressed earlier in development, as they are transmitted to each succeeding generation.

*8. What characteristics are exhibited by a cytoplasmically inherited trait?
Cytoplasmically inherited traits are encoded by genes in the cytoplasm. Because the cytoplasm usually is inherited from a single (most often the female) parent, reciprocal crosses do not show the same results. Cytoplasmically inherited traits often show great variability because different egg cells (female gametes) may have differing proportions of cytoplasmic alleles from random sorting of mitochondria (or plastids in plants).

9. What is the difference between genetic maternal effect and genomic imprinting?
In genetic maternal effects, the phenotypes of the progeny are determined by the genotype of the mother only. The genotype of the father and the genotype of the affected individual have no effect. In genomic imprinting, the phenotype of the progeny differs based on whether a particular allele is inherited from the mother or the father. The phenotype is therefore based on both the individual's genotype and the paternal or maternal origins of the genotype.

10. What is the difference between a sex-influenced gene and a gene that exhibits genomic imprinting?
For a sex-influenced gene, the phenotype is influenced by the sex of the individual bearing the genotype. For an imprinted gene, the phenotype is influenced by the sex of the parent *from which each allele was inherited.*

*11. What are continuous characteristics and how do they arise?
Continuous characteristics, also called quantitative characteristics, exhibit many phenotypes with a continuous distribution. They result from the interaction of multiple genes (polygenic traits), the influence of environmental factors on the phenotype, or both.

APPLICATION QUESTIONS AND PROBLEMS

*12. Palomino horses have a golden yellow coat, chestnut horses have a brown coat, and cremello horses have a coat that is almost white. A series of crosses between the three different types of horses produced the following offspring:

Cross	**Offspring**
palomino × palomino	13 palomino, 6 chestnut, 5 cremello
chestnut × chestnut	16 chestnut
cremello × cremello	13 cremello
palomino × chestnut	8 palomino, 9 chestnut
palomino × cremello	11 palomino, 11 cremello
chestnut × cremello	23 palomino

(a) Explain the inheritance of the palomino, chestnut, and cremello phenotypes in horses.
The results of the crosses indicate that cremello and chestnut are pure-breeding traits (homozygous). Palomino is a hybrid trait (heterozygous) that produces a 2:1:1 ratio when palominos are crossed with each other. The simplest hypothesis

consistent with these results is incomplete dominance, with palomino as the phenotype of the heterozygotes resulting from chestnuts crossed with cremellos.

(b) Assign symbols for the alleles that determine these phenotypes and list the genotypes of all parents and offspring given in the preceding table.
Let C^B = chestnut, C^W = cremello, C^BC^W = palomino.

Cross	**Offspring**
palomino × palomino	*13 palomino, 6 chestnut, 5 cremello*
$C^BC^W \times C^BC^W$	C^BC^W C^BC^B C^WC^W
chestnut × chestnut	*16 chestnut*
$C^BC^B \times C^BC^B$	C^BC^B
cremello × cremello	*13 cremello*
$C^WC^W \times C^WC^W$	C^WC^W
palomino × chestnut	*8 palomino, 9 chestnut*
$C^BC^W \times C^BC^B$	C^BC^W C^BC^B
palomino × cremello	*11 palomino, 11 cremello*
$C^BC^W \times C^WC^W$	C^BC^W C^WC^W
chestnut × cremello	*23 palomino*
$C^BC^B \times C^WC^W$	C^BC^W

*13. The L^M and L^N alleles at the MN blood group locus exhibit codominance. Give the expected genotypes and phenotypes and their ratios in progeny resulting from the following crosses:

(a) $L^ML^M \times L^ML^N$
½ L^ML^M (type M), ½ L^ML^N (type MN)
(b) $L^NL^N \times L^NL^N$
All L^NL^N (type N)
(c) $L^ML^N \times L^ML^N$
½ L^ML^N (type MN), ¼ L^ML^M (type M), ¼ L^NL^N (type N)
(d) $L^ML^N \times L^NL^N$
½ L^ML^N (type MN), ½ L^NL^N (type N)
(e) $L^ML^M \times L^NL^N$
All L^ML^N (type MN)

14. Ptosis (droopy eyelid) may be inherited as a dominant human trait. Among 40 people who are heterozygous for the ptosis allele, 13 have ptosis and 27 have normal eyelids.
(a) What is the penetrance for ptosis?
The penetrance is 13/40 = 0.325 = 32.5%
(b) If ptosis exhibited variable expressivity, what would that mean?
Variable expressivity of ptosis would mean that individuals that are heterozygous for this allele would be affected to varying degrees; some would have very droopy eyelids, others would have less droopy eyelids.

15. In the eastern mosquito fish (*Gambusia affinis holbrooki*), which has the XX-XY sex determination, spotting is inherited as a Y-linked trait. The trait exhibits 100%

penetrance when the fish are raised at 22°C, but the penetrance drops to 42% when the fish are raised at 26°C. A male with spots is crossed with a female without spots, and the F1 are intercrossed to produce the F2. If all the offspring are raised at 22°C, what proportion of the F1 and F2 will have spots? If all the offspring are raised at 26°C, what proportion of the F1 and F2 will have spots.

Because spotting is Y-linked, the parental genotypes are: XY^s and XX, where Y^s denotes the spotted allele on the Y chromosome. The F1 genotypes will be: ½ XY^s and ½ XX, like the parents. The F2 genotypes will also be ½ XY^s and ½ XX. Note that incomplete penetrance and expressivity do not affect genotypic ratios. At 22°C, where penetrance is 100%, the phenotypic ratios will be all spotted males and all unspotted females in both the F1 and F2 progeny. At 26°C, where penetrance is only 42%, then 42% of the XY^s males will be spotted, in the F1 and F2.

16. In the pearl millet plant, color is determined by three alleles at a single locus: Rp^1 (red), Rp^2 (purple), and *rp* (green). Red is dominant over purple and green, and purple is dominant over green (Rp^1>Rp^2>*rp*). Give the expected phenotypes and ratios of offspring produced by the following crosses:

(a) $Rp^1/Rp^2 \times Rp^1/rp$

We expect ¼ Rp^1/Rp^1 *(red),* ¼ Rp^1/rp *(red),* ¼ Rp^2/Rp^1 *(red),* ¼ Rp^2/rp *(purple), for overall phenotypic ratio of* ¾ *red,* ¼ *purple.*

(b) $Rp^1/rp \times Rp^2/rp$

¼ Rp^1/Rp^2 *(red),* ¼ Rp^1/rp *(red),* ¼ Rp^2/rp *(purple),* ¼ rp/rp *(green), for overall phenotypic ratio of* ½ *red,* ¼ *purple,* ¼ *green.*

(c) $Rp^1/Rp^2 \times Rp^1/Rp^2$

This cross is equivalent to a two-allele cross of heterozygotes, so the expected phenotypic ratio is ¾ *red,* ¼ *purple.*

(d) $Rp^2/rp \times rp/rp$

Another two-allele cross of a heterozygote with a homozygous recessive. Phenotypic ratio is ½ *purple,* ½ *green.*

(e) $rp/rp \times Rp^1/Rp^2$

½ Rp^1/rp *(red),* ½ Rp^2/rp *(purple)*

*17. Give the expected genotypic and phenotypic ratios for the following crosses for ABO blood types:

(a) $I^Ai \times I^Bi$

¼ I^AI^B *(AB),* ¼ I^Ai *(A),* ¼ I^Bi *(B),* ¼ ii *(O)*

(b) $I^AI^B \times I^Ai$

¼ I^AI^A *(A),* ¼ I^Ai *(A),* ¼ I^AI^B *(AB),* ¼ I^Bi *(B)*

(c) $I^AI^B \times I^AI^B$

¼ I^AI^A *(A),* ½ I^AI^B *(AB),* ¼ I^BI^B *(B)*

(d) $ii \times I^Ai$

½ I^Ai *(A),* ½ ii *(O)*

(e) $I^AI^B \times ii$

½ I^Ai *(A),* ½ I^Bi *(B)*

18. If there are five alleles at a locus, how many genotypes may there be at this locus? How many different kinds of homozygotes will there be? How many genotypes and homozygotes would there be with eight alleles?

Mathematically, this question is the same as asking how many different groups of two (diploid genotypes have two alleles for each locus) are possible from n *objects (alleles). Assign numbers 1, 2, 3, 4, and 5 to each of the five alleles, and group the possible genotypes according to the following table:*

1,1					
1,2	*2,2*				
1,3	*2,3*	*3,3*			
1,4	*2,4*	*3,4*	*4,4*		
1,5	*2,5*	*3,5*	*4,5*	*5,5*	

Such an arrangement allows us to easily see that the number of genotypes for any n *number of alleles is simply* $\Sigma(1, 2, 3 \ldots n) = n(n+1)/2$. *Looking at the table, we see that the number of filled boxes (genotypes) is equal to half the number of boxes in a rectangle of dimensions* $n \times (n+1)$. *So, the number of genotypes* $= n(n+1)/2$. *For five alleles* ($n = 5$), *we get 15 possible genotypes and five homozygotes. For eight alleles, there are* $8(8+1)/2 = 36$ *possible genotypes.*

19. Turkeys have black, bronze, or black-bronze plumage. Examine the results of the following crosses:

Parents	**Offspring**
Cross 1: black and bronze	All black
Cross 2: black and black	¾ black, ¼ bronze
Cross 3: black-bronze and black-bronze	All black-bronze
Cross 4: black and bronze	½ black, ¼ bronze, ¼ black-bronze
Cross 5: bronze and black-bronze	½ bronze, ½ black-bronze
Cross 6: bronze and bronze	¾ bronze, ¼ black-bronze

Do you think these differences in plumage arise from incomplete dominance between two alleles at a single locus? If yes, support your conclusion by assigning symbols to each allele and providing genotypes for all turkeys in the crosses. If your answer is no, provide an alternative explanation and assign genotypes to all turkeys in the crosses.

The results of Cross 2 tell us that black is dominant to bronze. Similarly, the results of Cross 6 tell us that bronze is dominant to black-bronze. We can use B^L *for black,* B^R *for bronze, and* b *for black-bronze.*

Parents	***Offspring***
Cross 1: black (B^LB^L) × *bronze* (B^RB^R)	*All black* (B^LB^R)
Cross 2: black (B^LB^R) × *black* (B^LB^R)	¾ *black* (B^L--), ¼ *bronze* (B^RB^R)
Cross 3: black-bronze (bb) × *black-bronze* (bb)	*All black-bronze* (bb)
Cross 4: black (B^Lb) × *bronze* (B^Rb)	½ *black* (B^L--), ¼ *bronze* (B^Rb), ¼ *black-bronze* (bb)
Cross 5: bronze (B^Rb) × *black-bronze* (bb)	½ *bronze* (B^Rb), ½ *black-bronze* (bb)
Cross 6: bronze (B^Rb) × *bronze* (B^Rb)	¾ *bronze* (B^R--), ¼ *black-bronze* (bb)

20. In rabbits, an allelic series helps to determine coat color: *C* (full color), c^{ch} (chinchilla, gray color), c^h (Himalayan, white with black extremities), and *c* (albino, all white). The *C* allele is dominant over all others, c^{ch} is dominant over c^h and *c*, c^h is dominant over *c*, and *c* is recessive to all the other alleles. This dominance hierarchy can be summarized as $C>c^{ch}>c^h>c$. The rabbits in the following list are crossed and produce the progeny shown. Give the genotypes of the parents for each cross.

(a) full color × albino → ½ full color, ½ albino

Cc × cc. *1:1 phenotypic ratios in the progeny result from a cross of a heterozygote with a homozygous recessive. Because albino is recessive to all other alleles, the full-color parent must have an albino allele, and the albino parent must be homozygous for the albino allele.*

(b) himalayan × albino → ½ himalayan, ½ albino

c^hc × cc. *Again, the 1:1 ratio of the progeny indicate the parents must be a heterozygote and a homozygous recessive.*

(c) full color × albino → ½ full color, ½ chinchilla

Cc^{ch} × cc. *This time, we get a 1:1 ratio, but we have chinchilla progeny instead of albino. Therefore, the heterozygous full-color parent must have a chinchilla allele as well as a dominant full-color allele. The albino parent has to be homozygous albino because albino is recessive to all other alleles.*

(d) full color × himalayan → ½ full color, ¼ himalayan, ¼ albino

Cc × c^hc. *The 1:2:1 ratio in the progeny indicates that both parents are heterozygotes. Both must have an albino allele because the albino progeny must have inherited an albino allele from each parent.*

(e) full color × full color → ¾ full color, ¼ albino

Cc × Cc. *The 3:1 ratio indicates that both parents are heterozygous. Both parents must have an albino allele for albino progeny to result.*

21. In this chapter, we discussed Joan Barry's paternity suit against Charlie Chaplin and how, on the basis of blood types, Chaplin could not have been the father of her child.

(a) What blood types are possible for the father of Barry's child?

Because Barry's child inherited an I^B *allele from the father, the father could have been B or AB.*

(b) If Chaplin had possessed one of these blood types, would that prove that he fathered Barry's child?

No. Many other men have these blood types. The results would have meant only that Chaplin cannot be eliminated as a possible father of the child.

*22. A woman has blood type A MM. She has a child with blood type AB MN. Which of the following blood types could *not* be that of the child's father? Explain your reasoning.

George	O	NN
Tom	AB	MN
Bill	B	MN
Claude	A	NN
Henry	AB	MM

The child's blood type has a B allele and an N allele that could not have come from the mother and must have come from the father. Therefore, the child's father must have a B and an N. George, Claude, and Henry are eliminated as possible fathers because they lack either a B or an N.

23. Allele *A* is epistatic to allele *B*. Indicate whether each of the following statements is true or false. Explain why.

(a) Alleles *A* and *B* are at the same locus.
False. If they were at the same locus, they would be allelic, not epistatic.

(b) Alleles *A* and *B* are at different loci.
True. Epistasis applies only to genes at different loci.

(c) Alleles *A* and *B* are always located on the same chromosome.
False. Epistatic genes can show independent assortment.

(d) Alleles *A* and *B* may be located on different, homologous chromosomes.
True. If the two genes are linked on the same chromosome, then allele A for the epistatic gene can be either on the same chromosome as allele B (in a coupling relationship) or on different, homologous chromosomes (in a repulsion relationship), with no difference in epistasis.

(e) Alleles *A* and *B* may be located on different, nonhomologous chromosomes.
True. Epistatic genes can and often do show independent assortment.

*24. In chickens, comb shape is determined by alleles at two loci (*R*, *r* and *P*, *p*). A walnut comb is produced when at least one dominant allele *R* is present at one locus and at least one dominant allele *P* is present at a second locus (genotype *R_P_*). A rose comb is produced when at least one dominant allele is present at the first locus and two recessive alleles are present at the second locus (genotype *R_pp*). A pea comb is produced when two recessive alleles are present at the first locus and at least one dominant allele is present at the second (genotype *rrP_*). If two recessive alleles are present at the first and at the second locus (*rrpp*), a single comb is produced. Progeny with what types of combs and in what proportions will result from the following crosses?

(a) *RRPP* × rrpp
All walnut (RrPp)

(b) *RrPp × rrpp*
¼ walnut (RrPp), *¼ rose* (Rrpp), *¼ pea* (rrPp), *¼ single (rrpp)*
(c) *RrPp × RrPp*
9/16 walnut (R_P_), *3/16 rose* (R_pp), *3/16 pea* (rrP_), *1/16 single* (rrpp)
(d) *Rrpp × Rrpp*
¾ rose (R_pp), *¼ single* (rrpp)
(e) *Rrpp × rrPp*
¼ walnut (RrPp), *¼ rose* (Rrpp), *¼ pea* (rrPp), *¼ single* (rrpp)
(f) *Rrpp × rrpp*
½ rose (Rrpp), *½ single* (rrpp)

*25. Eye color of the Oriental fruit fly (*Bactrocera dorsalis*) is determined by a number of genes. A fly having wild-type eyes is crossed with a fly having yellow eyes. All the F_1 flies from this cross have wild type eyes. When the F_1 are interbred, $^9/_{16}$ of the F_2 progeny have wild-type eyes, $^3/_{16}$ have amethyst eyes (a bright, sparkling blue color), and $^4/_{16}$ have yellow eyes.

(a) Give genotypes for all the flies in the P, F_1, and F_2 generations.

The F_1 progeny tell us that wild type is dominant over yellow. The F_2 progeny show a modified dihybrid ratio: Instead of 9:3:3:1, we see a 9:3:4. These results suggest that two genes are interacting, and the F_1 is heterozygous for both genes.

Let's start by working through the F_1 heterozygous cross, assigning A *and* a *as the dominant and recessive alleles for one gene and* B *and* b *as dominant and recessive alleles for the second gene.*

F_1 AaBb × AaBb *all wild type*
F_2 *9/16* A-B- *red (like F_1)*
3/16 A-bb *amethyst*
3/16 aaB- *yellow*
1/16 aabb *yellow*

The assignment of phenotypes to genotypes is arbitrary (after all, the genotype symbols were picked arbitrarily).

The parents then must have been AABB *wild type ×* aabb *yellow. These genotypes will result in all* AaBb *wild-type F_1 and are consistent with the phenotype:genotype assignments in the F_2.*

(b) Does epistasis account for eye color in Oriental fruit flies? If so, which gene is epistatic and which gene is hypostatic?

Yes, gene a *exhibits recessive epistasis because the* aa *genotype masks the expression of genes at the B locus. Genes* B *and* b *are hypostatic to* a *because their expression is masked by the presence of* aa.

26. A variety of opium poppy (*Papaver somniferum L.*) having lacerate leaves was crossed with a variety that has normal leaves. All the F_1 had lacerate leaves. Two F_1 plants were interbred to produce the F_2. Of the F_2, 249 had lacerate leaves and 16 had normal

leaves. Give genotypes for all the plants in the P, F_1, and F_2 generations. Explain how lacerate leaves are determined in the opium poppy.
The F_1 progeny tell us that lacerate is dominant over normal leaves. In the F_2, 249:16 does not come close to a 3:1 ratio. Let's see if these numbers fit a dihybrid ratio. Dividing 265 total progeny by 16 (because dihybrid ratios are based on 16ths), we see that 1/16 of 265 is 16.56. Therefore, the F_2 progeny are very close to 15/16 lacerate, 1/16 normal, a modified dihybrid ratio. If we symbolize the two genes as A *and* B, *then:*

F_1 AaBb × AaBb *all lacerate*
F_2 *9/16* A-B- *lacerate (like F_1)*
3/16 A-bb *lacerate*
3/16 aaB- *lacerate*
1/16 aabb *normal*

A dominant allele at either gene A *or gene* B, *or both, results in lacerate leaves. Finally, the parents must have been* AABB *lacerate* × aabb *normal. Note that only* AABB *for the lacerate parent would result in F_1 that are* AaBb.

*27. A dog breeder liked yellow and brown Labrador retrievers. In an attempt to produce yellow and brown puppies, he bought a yellow Labrador male and a brown Labrador female and mated them. Unfortunately, all the puppies produced in this cross were black. (See page 113 for a discussion of the genetic basis of coat color in Labrador retrievers.)

(a) Explain this result.
Labrador retrievers vary in two loci, B *and* E. *Black dogs have dominant alleles at both loci* (B-E-), *brown dogs have* bbE-, *and yellow dogs have* B-ee *or* bbee. *Because all the puppies were black, they must all have inherited a dominant* B *allele from the yellow parent, and a dominant* E *allele from the brown parent. The brown female parent must have been* bbEE, *and the yellow male must have been* BBee. *The black puppies were all* BbEe.

(b) How might the breeder go about producing yellow and brown Labradors?
Simply mating yellow with yellow will produce all yellow Labrador puppies. Mating two brown Labradors will produce either all brown puppies, if at least one of the parents is homozygous EE, *or ¾ brown and ¼ yellow if both parents are heterozygous* Ee.

28. When a yellow female Labrador retriever was mated with a brown male, half of the puppies were brown and half were yellow. The same female, when mated to a different brown male, produced all brown males. Explain these results.
The first brown male was heterozygous for the E *locus, hence he was* bbEe. *The yellow female has to be* bbee. *The puppies from this first mating were therefore ½* bbEe *(brown) and ½* bbee *(yellow). The second brown male was homozygous* bbEE. *Thus, all the puppies from the second mating were* bbEe *(brown).*

*29. In summer squash, a plant that produces disc-shaped fruit is crossed with a plant that produces long fruit. All the F_1 have disc-shaped fruit. When the F_1 are intercrossed, F_2

progeny are produced in the following ratio: $^{9}/_{16}$ disc-shaped fruit: $^{6}/_{16}$ spherical fruit: $^{1}/_{16}$ long fruit. Give the genotypes of the F_2 progeny.

The modified dihybrid ratio in the F_2 indicates that two genes interact to determine fruit shape. Using generic gene symbols A *and* B *for the two loci, the F_1 heterozygotes are* AaBb.

The F_2 are:
9/16 A-B- *disc-shaped (like F_1)*
3/16 A-bb *spherical*
3/16 aaB- *spherical*
1/16 aabb *long*

30. In sweet peas, some plants have purple flowers and other plants have white flowers. A homozygous variety of pea that has purple flowers is crossed with a homozygous variety that has white flowers. All the F_1 have purple flowers. When these F_1 are self-fertilized, the F_2 appear in a ratio of $^{9}/_{16}$ purple to $^{7}/_{16}$ white.

(a) Give genotypes for the purple and white flowers in these crosses.

The F_2 ratio of 9:7 is a modified dihybrid ratio, indicating two genes interacting. Using A *and* B *as generic gene symbols, we can start with the F_1 heterozygotes:*

F_1 AaBb *purple self-fertilized*
F_2 *9/16* A-B- *purple (like F_1)*
3/16 A-bb *white*
3/16 aaB- *white*
1/16 aabb *white*

Now we see that purple requires dominant alleles for both genes, so the purple parent must have been AABB, *and the white parent must have been* aabb *to give all purple F_1.*

(b) Draw a hypothetical biochemical pathway to explain the production of purple and white flowers in sweet peas.

White precursor 1 → *white intermediate 2* → *purple pigment*
Enzyme A *Enzyme B*

31. For the following questions, refer to page 114 for a discussion of how coat color and pattern are determined in dogs.

(a) Explain why Irish setters are reddish in color.

According to the information in Table 5.3, Irish setters are BBCCDDeeSStt, *and* A *or* a^t. *The* ee *genotype prevents black color on the body coat, except on the nose and in the eyes. The other genes* B, C, *and* D *allow expression of underlying pigment, which happens to be reddish in setters.*

(b) Will a cross between a beagle and a Dalmatian produce puppies with ticking? Why or why not?

Beagles are a^sa^sBBCCDDs^ps^ptt, *and Dalmatians are* A^sA^sCCDDEEs^ws^wTT. *A cross will produce puppies that are all* A^sa^sB-CCDDE-s^ps^wTt. *These puppies will have piebald spots because of the* s^p *allele that is dominant to* s^w*, and because of the* T *allele the white areas should have ticking.*

(c) Can a poodle crossed with any other breed produce spotted puppies? Why or why not?
Poodles are SStt. *Because the dominant* S *allele prevents spotting, no puppies from matings with poodles will have spotting.*

(d) If a St. Bernard is crossed with a Doberman, will the offspring have solid, yellow, saddle, or bicolor coats?
St. Bernards are a^ya^yBBCCDDtt, *and Dobermans are* a^ta^tCCEESStt. *The offspring will be of genotype* a^ya^tB-CCD-E-S-tt. *Because* a^y *specifying yellow is dominant over* a^t*, and the* E *allele allows expression of the* A *genotype throughout, the offspring will have yellow coats.*

(e) If a Rottweiler is crossed with a Labrador Retriever, will the offspring have solid, yellow, saddle, or bicolor coats?
Rottweilers are a^ta^tBBCCDDEESStt, *and Labrador Retrievers are* A^SA^SCCDDSStt. *The offspring will be* A^Sa^tB-CCDDE-SStt. *The combination of the dominant* A^S *and* E *alleles should create solid coats.*

*32. When a Chinese hamster with white spots is crossed with another hamster that has no spots, approximately ½ of the offspring have white spots and ½ have no spots. When two hamsters with white spots are crossed, 2/3 of the offspring possess white spots and 1/3 have no spots.

(a) What is the genetic basis of white spotting in Chinese hamsters?
The 2:1 ratio when two spotted hamsters are mated suggests lethality, and the 1:1 ratio when spotted hamsters are mated to hamsters without spots indicates that spotted is a heterozygous phenotype. Using S *and* s *to symbolize the locus responsible for white spotting, spotted hamsters are* Ss, *solid-colored hamsters are* ss. *One quarter of the progeny expected from a mating of two spotted hamsters is* SS, *embryonic lethal, and missing from those progeny, resulting in the 2:1 ratio of spotted to solid progeny.*

(b) How might you go about producing Chinese hamsters that breed true for white spotting?
Because spotting is a heterozygous phenotype, it should not be possible to obtain Chinese hamsters that breed true for spotting, unless the locus that produces spotting can somehow be separated from the lethality.

33. Male-limited precocious puberty results from a rare, sex-limited autosomal gene (P) that is dominant over the allele for normal puberty (p) and is expressed only in males. Bill undergoes precocious puberty, but his brother Jack and his sister Beth underwent puberty at the usual time, between the ages of 10 and 14. Although Bill's mother and father underwent normal puberty, two of his maternal uncles (his mother's brothers) underwent precocious puberty. All of Bill's grandparents underwent normal puberty. Give the most likely genotypes for all the relatives mentioned in this family.
Since precocious puberty is dominant, all the males who experienced normal puberty, such as Jack, Bill's father, and Bill's grandfathers, must be pp. *Bill and his two*

maternal uncles, who all experienced precocious puberty, are Pp. *We know they are heterozygotes not only because* P *is a rare allele but also because these individuals all had fathers that are* pp. *This means Bill inherited* P *from his mother, who must have been* Pp. *Bill's sister Beth could be either* Pp *or* pp.

*34. Pattern baldness in humans is a sex-influenced trait that is autosomal dominant in males and recessive in females. Jack has a full head of hair. JoAnn also has a full head of hair, but her mother is bald. (In women, pattern baldness is usually expressed as a thinning of the hair.) If Jack and JoAnn marry, what proportion of their children is expected to be bald?

We will use H^b *to denote the baldness allele that is dominant in males and* H^+ *to denote the full hair allele. Jack must be* H^+H^+ *because he has the recessive phenotype for males. JoAnn's mother must be* H^bH^b *because she has the recessive phenotype for females. JoAnn, therefore, although she has a full head of hair, must be* H^bH^+ *because she must have inherited an* H^b *from her mother. Therefore, a marriage between Jack and JoAnn would produce:*

$H^+H^+ \times H^bH^+ \rightarrow$ ¼ H^+H^+ *males with full hair*
¼ H^+H^+ *females with full hair*
¼ H^bH^+ *males with pattern baldness*
¼ H^bH^+ *females with full hair*

Therefore, ¼ of their children will be bald.

35. In goats, a beard is produced by an autosomal allele that is dominant in males and recessive in females. We'll use the symbol B^b for the beard allele and B^+ for the beardless allele. Another independently assorting autosomal allele that produces a black coat (*W*) is dominant over the allele for white coat (*w*). Give the phenotypes and their proportions expected for the following crosses:

(a) B^+B^b *Ww* male × B^+B^b *Ww* female

Because beardedness and coat color independently assort, we can treat them independently. The difference between this cross and a dihybrid cross is that the bearded allele B^b *is dominant in males and recessive in females. So we deal with male and female progeny separately. For each sex, then, we should get a typical dihybrid ratio. In males, the dominant phenotype is bearded, so we should get ¾ bearded, ¼ beardless. In females the dominant phenotype is beardless, so we should get ¾ beardless and ¼ bearded. Each sex will have ¾ black and ¼ white coats.*

Males:		*Females:*	
	9/16 bearded, black		*9/16 beardless, black*
	3/16 bearded, white		*3/16 beardless, white*
	3/16 beardless, black		*3/16 bearded, black*
	1/16 beardless, white		*1/16 bearded, white*

(b) B^+B^b *Ww* male × B^+B^b *ww* female

Here the males will again be ¾ bearded and ¼ beardless, and the females will be ¾ beardless and ¼ bearded. This time half the progeny of either sex will be black, and half will be white.

Males:	*3/8 bearded, black*	*Females:*	*3/8 beardless, black*
	3/8 bearded, white		*3/8 beardless, white*
	1/8 beardless, black		*1/8 bearded, black*
	1/8 beardless, white		*1/8 bearded, white*

(c) B^+B^+ Ww male × B^bB^b Ww female

In this cross, all of the male progeny will be bearded, and all of the female progeny will be beardless. All will be ¾ black, ¼ white.

Males:	*¾ bearded, black*	*Females:*	*¾ beardless, black*
	¼ bearded, white		*¼ beardless, white*

(d) B^+B^b Ww male × B^+B^bww female

Males and females will be ½ bearded, ½ beardless and ½ black, ½ white.

Males:	*¼ bearded, black*	*Females:*	*¼ beardless, black*
	¼ bearded, white		*¼ beardless, white*
	¼ beardless, black		*¼ bearded, black*
	¼ beardless, white		*¼ bearded, white*

36. In the snail *Limnaea peregra,* shell coiling results from a genetic maternal effect. An autosomal allele for a right-handed shell (s^+), called dextral, is dominant over the allele for a left-handed shell (s), called sinistral. A pet snail called Martha is sinistral and reproduces only as a female (the snails are hermaphroditic). Indicate which of the following statements are true and which are false. Explain your reasoning in each case.

(a) Martha's genotype *must* be *ss*.

False. For maternal effect genes, the phenotype of the individual is determined solely by the genotype of the individual's mother. So we know Martha's mother must have been ss *because Martha is sinistral. If Martha was produced as a result of self-fertilization, then Martha must indeed be* ss. *But if Martha was produced by cross-fertilization, then we cannot know Martha's genotype without more information.*

(b) Martha's genotype cannot be s^+s^+.

True. As explained in the answer to part (a), Martha's mother is ss, *so Martha must be either* s^+s *or* ss.

(c) All the offspring produced by Martha *must* be sinistral.

False. Because we do not know Martha's genotype, we cannot yet predict the phenotype of her offspring.

(d) At least some of the offspring produced by Martha *must* be sinistral.

False. If Martha is s^+s, *then all her children will be dextral. If Martha is* ss, *then all her children will be sinistral.*

(e) Martha's mother *must* have been sinistral.

False. Martha's mother's phenotype is determined by the genotype of her *mother (Martha's maternal grandmother). We know Martha's mother's genotype must have been* ss, *so her mother's mother had at least one* s *allele. But we cannot know if she was a heterozygote or homozygous* ss.

(f) All Martha's brothers *must* be sinistral.

True. Because Martha's mother must have been ss, *all her progeny must be sinistral.*

37. Hypospadias, a birth defect in male humans in which the urethra opens on the shaft instead of the tip of the penis, results from an autosomal dominant gene in some families. Females who carry the gene show no effects. This is an example of: (a) an X-linked trait, (b) a Y-linked trait, (c) a sex-limited trait, (d) a sex influenced trait, or (e) genetic maternal effect? Explain your answer.
Knowing that the condition arises from an autosomal gene, we can eliminate either (a) an X-linked trait or (b) a Y-linked trait. Dominant inheritance also eliminates (e) maternal effect. If it were (d) a sex influenced trait, females would be affected to a lesser degree or differently than males. Because females who carry the gene show no effects, this condition is (c) a sex-limited trait.

38. In unicorns, two autosomal loci interact to determine the type of tail. One locus controls whether a tail is present at all; the allele for a tail (*T*) is dominant over the allele for tailless (*t*). If a unicorn has a tail, then alleles at a second locus determine whether the tail is curly or straight. Farmer Baldridge has two unicorns with curly tails. When he crosses these two unicorns, ½ of the progeny have curly tails, ¼ have straight tails, and ¼ do not have a tail. Give the genotypes of the parents and progeny in Farmer Baldridge's cross. Explain how he obtained the 2:1:1 phenotypic ratio in his cross.
We are given the symbols T *for dominant tailed and* t *for recessive tailless. We are not given any information about dominance or recessiveness for the second locus. We will use* S *and* s *for the second locus that determines whether the tail is curly or straight. Although two genes are interacting, we can analyze one locus at a time. Farmer Baldridge crossed two unicorns with tails and got a 3:1 ratio of tailed to tailless. Therefore, the two unicorns were heterozygous for the tail locus:* Tt. *The parents were both curly, and the progeny were both curly and straight, in a 2:1 ratio of curly:straight. The fact that he got straight-tailed progeny indicates that the curly tailed parents were heterozygous* Ss. *The fact that he got a 2:1 ratio instead of a 3:1 ratio indicates that this locus may not have a dominant:recessive relationship. A 2:1 ratio may be obtained if ¼ of the progeny, one of the homozygote classes, are missing (because of embryonic lethality).*

Having deduced the genotypes of the parents, we can determine the expected genotypes of the progeny:
P: curly tailed TtSs × *curly tailed* TtSs
Using a branch diagram:

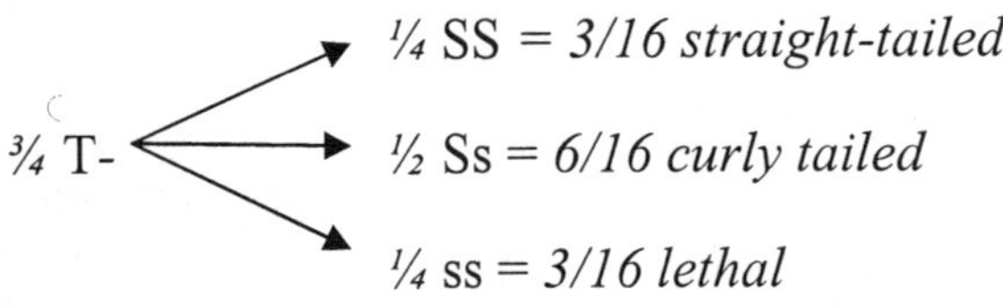

and

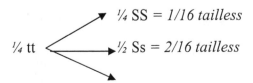

¼ tt → ¼ SS = *1/16 tailless*
→ ½ Ss = *2/16 tailless*
→

¼ ss = *1/16 lethal*

Overall, the surviving progeny would be 6/16 curly tailed, 3/16 straight-tailed, and 3/16 tailless, which reduces to a 2:1:1 ratio of curly tailed to straight-tailed to tailless.

If S *were dominant over* s *without lethality, we would expect the following* F_1:

9/16 T-S- *curly tailed*
3/16 T-ss *straight-tailed*
3/16 ttS- *tailless*
1/16 ttss *tailless*

However, these predictions do not fit the observed 2:1:1 ratio.

39. In 1983, a sheep farmer in Oklahoma noticed a ram in his flock that possessed increased muscle mass in his hindquarters. Many of the offspring of this ram possessed the same trait, which became known as the callipyge mutant (*callipyge* is Greek for "beautiful buttocks"). The mutation that caused the callipyge phenotype was eventually mapped to a position on the sheep chromosome 18.

When the male callipyge offspring of the original mutant ram were crossed with normal females, they produced the following progeny: ¼ male callipyge, ¼ female callipyge, ¼ male normal, and ¼ female normal. When female callipyge offspring of the original mutant ram were crossed with normal males, all of the offspring were normal. Analysis of the chromosomes of these offspring of callipyge females showed that half of them received a chromosome 18 with the callipyge gene from their mother. Propose an explanation for the inheritance of the callipyge gene. How might you test your explanation?

Here we get different results depending on the sex of the parent with the callipyge mutation. Because we know the gene is on chromosome 18, we can eliminate sex linkage as a possible cause. The phenotype is also not sex-limited because male callipyge sire equal proportions of male and female callipyge offspring. We can further eliminate maternal inheritance for the same reason. That leaves imprinting as a possible explanation. We note that half the progeny are callipyge if the father has the mutation, but none express the callipyge mutation if the mother has the mutation. We can therefore hypothesize that maternal alleles of this gene undergo imprinting and are silenced, so that the embryo expresses only the paternal allele.

We can test this hypothesis by mating the phenotypically normal male and female progeny that inherited the chromosome 18 with the callipyge gene from their mother. The hypothesis predicts that males will have normal and callipyge progeny if mated to either a normal female or a callipyge female. Conversely, the females will have all normal progeny if mated to a normal male, and both normal and callipyge progeny if mated to a callipyge male. In short, the progeny will reflect the genotype of the father, and the genotype of the mother will not be expressed.

CHALLENGE QUESTION

40. Suppose that you are tending a mouse colony at a genetics research institute and one day you discover a mouse with twisted ears. You breed this mouse with twisted ears and find that the trait is inherited. Male and female mice have twisted ears, but when

you cross a twisted-eared male with normal-eared female, you obtain different results from those you obtained when you cross a twisted-eared female with normal-eared male—the reciprocal crosses give different results. Describe how you would go about determining whether this trait results from a sex-linked gene, a sex-influenced gene, a genetic maternal effect, a cytoplasmically inherited gene, or genomic imprinting. What crosses would you conduct and what results would be expected with these different types of inheritance?

Each of these is a distinct pattern of inheritance. Because male and females have twisted ears, Y-linkage is eliminated. X-linked genes are passed from mother to son and from father to daughter. A sex-influenced gene shows a different phenotype depending on the sex but is inherited autosomally. A genetic maternal effect depends only on the genotype of the mother; the genotype of the zygote is immaterial. A cytoplasmically inherited trait is serially perpetuated from mother to all her progeny. Genomic imprinting results in the gene of only one of the parents being expressed.

To distinguish among these possibilities, you will need pure-breeding lines of mice with twisted ears and normal ears. Perform reciprocal crosses of males with twisted ears to females with normal ears (Cross A) and males with normal ears to females with twisted ears (Cross B).

	A: te male × normal female		*B: normal male × te female*	
	F_1 males	*F_1 females*	*F_1 males*	*F_1 females*
Sex-linked	*All normal*	*All dominant*	*All te*	*All dominant*
Sex-influenced	*Het male*	*Het female*	*Het male*	*Het female*
Maternal	*Normal*	*Normal*	*te*	*Te*
Cytoplasmic	*Normal*	*Normal*	*te*	*Te*
Imprinting pat	*Normal*	*Normal*	*te*	*Te*
Imprinting mat	*te*	*Te*	*Normal*	*Normal*

What we see from the table above is that if the trait is sex-linked, Cross A and Cross B give different phenotypes for the F_1 males, which match the phenotypes of their mothers. The F_1 females have the same dominant phenotype in either cross. If the trait is sex-influenced (heterozygous males have a different phenotype than heterozygous females), these reciprocal crosses with pure-breeding parents give the same results.

Both of these results are distinct from the results with maternal inheritance, cytoplasmic inheritance, or paternal imprinting, which all give the same results: no difference between male and female F_1 progeny, but the two crosses result in opposite phenotypes.

A further cross is needed to distinguish among maternal effect, cytoplasmic inheritance, and paternal imprinting. For these modes of inheritance, the phenotypes of the progeny depend solely on the maternal contribution, and no phenotypic differences are expected among male and female progeny. The F_1 female progeny from Cross A and Cross B should have the same genotype (heterozygous), but they have different phenotypes. The three remaining modes of inheritance predict different phenotypes of F_2 progeny from these females, as shown in the table below.

	Phenotypes of progeny of normal male × F_1 female from:	
Mode of inheritance	*Cross A (normal ears)*	*Cross B (twisted ears)*
Maternal	*Dominant phenotype*	*Dominant phenotype*
Cytoplasmic	*Normal ears*	*Twisted ears*
Paternal imprinting	*1:1 normal:twisted*	*1:1 normal:twisted*

In the case of maternal inheritance, the progeny depend on the genotype of the mother, and because the F_1 females from both crosses have the same heterozygous genotype, their progeny will have the same phenotype: normal ears or twisted ears, whichever is dominant.

For cytoplasmic inheritance, the phenotype of the progeny will be the same as the phenotype of the mother. Because the F_1 females have different phenotypes, their progeny will have different phenotypes.

For paternal imprinting, only the maternal genes are expressed in the progeny. Because the mother is heterozygous, the progeny should have 1:1 ratio of normal ears and twisted ears.

Other solutions are possible; this is just one.

Chapter Six: Pedigree Analysis and Applications

COMPREHENSION QUESTIONS

*1. What three factors complicate the task of studying the inheritance of human characteristics?

(1) Mating cannot be controlled, so it is not possible to set up controlled mating experiments.

(2) Humans have a long generation time, so it takes a long time to track inheritance of traits over more than one generation.

(3) The number of progeny per mating is limited, so phenotypic ratios are uncertain.

*2. Describe the features that will be exhibited in a pedigree in which a trait is segregating with each of the following modes of inheritance: autosomal recessive, autosomal dominant, X-linked recessive, X-linked dominant, and Y-linked inheritance.

Pedigrees with autosomal recessive traits will show affected males and females arising with equal frequency from unaffected parents. The trait often appears to skip generations. Unaffected people with an affected parent will be carriers.

Pedigrees with autosomal dominant traits will show affected males and females arising with equal frequency from a single affected parent. The trait does not usually skip generations.

X-linked recessive traits will affect males predominantly and will be passed from an affected male through his unaffected daughter to his grandson. X-linked recessive traits are not passed from father to son.

X-linked dominant traits will affect males and females and will be passed from an affected male to all his daughters, but not to his sons. An affected woman (usually heterozygous for a rare dominant trait) will pass on the trait equally to half her daughters and half her sons.

Y-linked traits will show up exclusively in males, passed from father to son.

*3. What are the two types of twins and how do they arise?

The two types of twins are monozygotic and dizygotic. Monozygotic twins arise when a single fertilized egg splits into two embryos in early embryonic cleavage divisions. They are genetically identical. Dizygotic twins arise from two different eggs fertilized at the same time by two different sperm. They share, on the average, 50% of the same genes.

4. Explain how a comparison of concordance in monozygotic and dizygotic twins can be used to determine the extent to which the expression of a trait is influenced by genes or environmental factors.

Monozygotic twins have 100% genetic identity, whereas dizygotic twins have 50% genetic identity. Any trait that is completely genetically determined will therefore be 100% concordant in monozygotic twins and 50% concordant in dizygotic twins. Conversely, any trait that is completely environmentally determined will have the same degree of concordance in monozygotic and dizygotic twins. To the extent that a trait has greater concordance in monozygotic twins than in dizygotic twins, the trait is genetically influenced. Environmental influences will reduce the concordance in monozygotic twins below 100%.

5. How are adoption studies used to separate the effects of genes and environment in the study of human characteristics?
Studies of adoptees, their biological parents, and their adoptive parents separate environmental and genetic influences on traits. Adoptees share similar environments with their adoptive parents (because they live in the same house and eat similar foods), but they share 50% of their genes with each of their biological parents. If adoptees have greater similarity for a trait with their adoptive parents, then the trait is environmentally influenced. If the adoptees have greater similarity for the trait with their biological parents, then the trait is genetically influenced.

*6. What is genetic counseling?
Genetic counseling provides assistance to clients by interpreting results of genetic testing and diagnosis; providing information about relevant disease symptoms, treatment, and progression; assessing and calculating the various genetic risks that the person or couple faces; and helping clients and family members cope with the stress of decision-making and facing up to the drastic changes in their lives that may be precipitated by a genetic condition.

7. Briefly define newborn screening, heterozygote screening, presymptomatic testing, and prenatal diagnosis.
Newborn screening: Newborn infants are tested for various treatable genetic disorders by sampling a few drops of their blood soon after birth.

Heterozygote screening: Normal or asymptomatic individuals in a population or community are tested for recessive disease alleles to determine the frequency of the disease allele in the population and to identify carriers, particularly if there is a relatively high incidence of the disease in the population or community.

Presymptomatic testing: People known to be at higher risk for a disease that occurs later in life are tested while they are still asymptomatic.

Prenatal diagnosis: Results from prenatal testing for any of a number of genetic conditions. Techniques, such as amniocentesis or chorionic villus sampling, are used to obtain tissue samples of the still developing fetus, or fetal protein or cells in the maternal circulation are characterized.

*8. What are the differences between amniocentesis and chorionic villus sampling? What is the purpose of these two techniques?
Amniocentesis samples the amniotic fluid by inserting a needle into the amniotic sac, usually performed at about 16 weeks of pregnancy. Chorionic villus sampling can be performed several weeks earlier (10th or 11th week of pregnancy) and samples a small piece of the chorion by inserting a catheter through the vagina. The purpose of these techniques is to obtain fetal cells for prenatal genetic testing.

9. What is preimplantation genetic diagnosis?
Preimplantation genetic diagnosis may be performed on embryos created through in vitro fertilization. The embryos are cultured until they reach the 8–16 cell stage, and one cell is removed from each embryo for genetic testing.

APPLICATION QUESTIONS AND PROBLEMS

*10. Joe is color blind. His mother and father have normal vision, but his mother's father (Joe's maternal grandfather) is color blind. All Joe's other grandparents have normal color vision. Joe has three sisters—Patty, Betsy, and Lora, all with normal color vision. Joe's oldest sister, Patty, is married to a man with normal color vision; they have two children, a 9-year-old color-blind boy and a 4-year-old girl with normal color vision.

(a) Using correct symbols and labels, draw a pedigree of Joe's family.

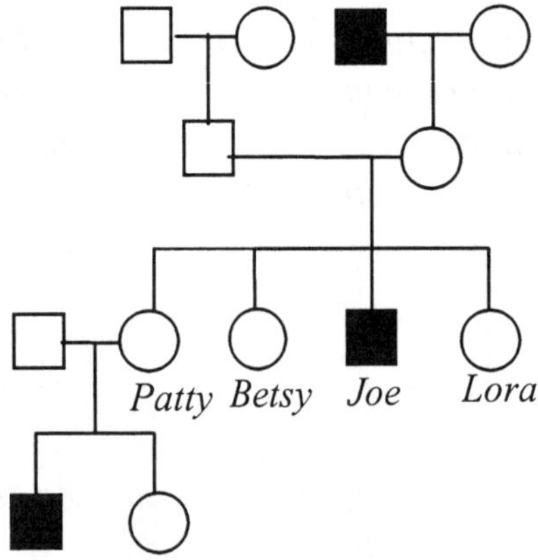

(b) What is the most likely mode of inheritance for color blindness in Joe's family?
X-linked recessive. Only males have the trait, and they inherit the trait from their mothers, who are carriers. The trait is never passed from father to son.

(c) If Joe marries a woman who has no family history of color blindness, what is the probability that their first child will be a color blind boy?
Barring a new mutation or nondisjunction, zero. Joe cannot pass his color-blind X chromosome to his son.

(d) If Joe marries a woman who is a carrier of the color blind allele, what is the probability that their first child will be a color blind boy?
The probability is ¼. There is ½ probability that their first child will be a boy, and there is an independent ½ probability that the first child will inherit the color-blind X chromosome from the carrier mother. ½(½) = ¼.

(e) If Patty and her husband have another child, what is the probability that it will be a color blind boy?
Again, ¼. Patty is a carrier because she had a color-blind son. The same reasoning applies as in part (d). Each child is an independent event.

11. A man with a specific unusual genetic trait marries an unaffected woman and they have four children. Pedigrees of this family are shown in parts (a) through (e), but the presence or absence of the trait in the children is not indicated. For each type of inheritance, indicate how many children of each sex are expected to express the trait by filling in the appropriate circles and squares. Assume that the trait is rare and fully penetrant.

(a) Autosomal recessive trait—*none*

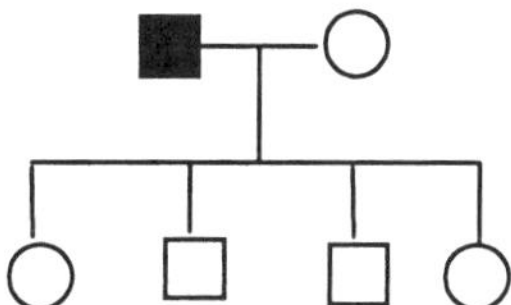

(b) Autosomal dominant trait—*½ of each sex*

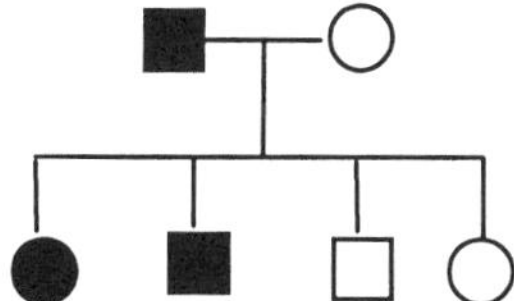

(c) X-linked recessive trait—*none*

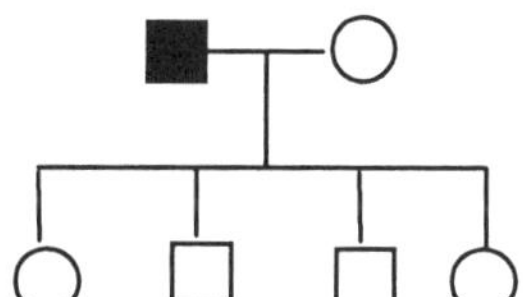

(d) X-linked dominant trait—*all the female children*

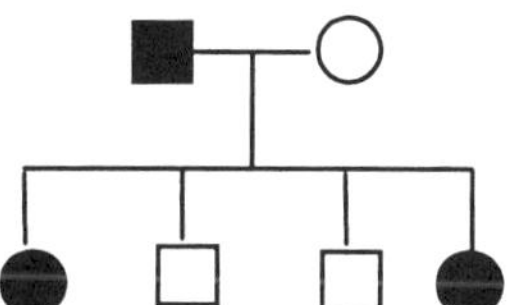

(e) Y-linked trait—*all the male children*

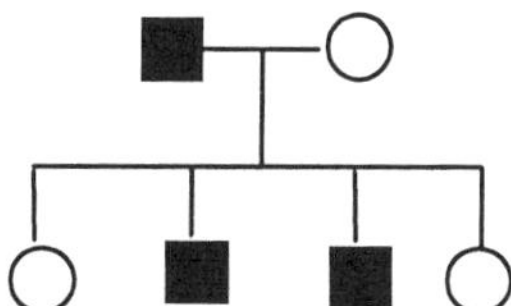

*12. For each of the following pedigrees, give the most likely mode of inheritance, assuming that the trait is rare. Carefully explain your reasoning.

(a)

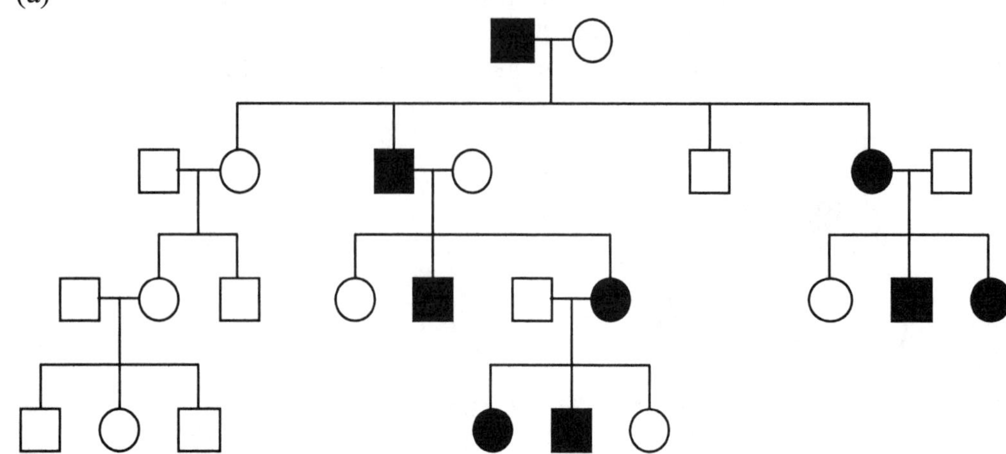

Autosomal dominant. Males and females are affected and can pass on the trait to sons and daughters. So the trait must be autosomal and dominant because affected children are produced in matings between affected and unaffected individuals. For rare traits, we can assume that unaffected individuals are not carriers. Therefore, individuals affected with recessive traits marrying unrelated unaffected individuals would be expected to have all unaffected children.

(b)

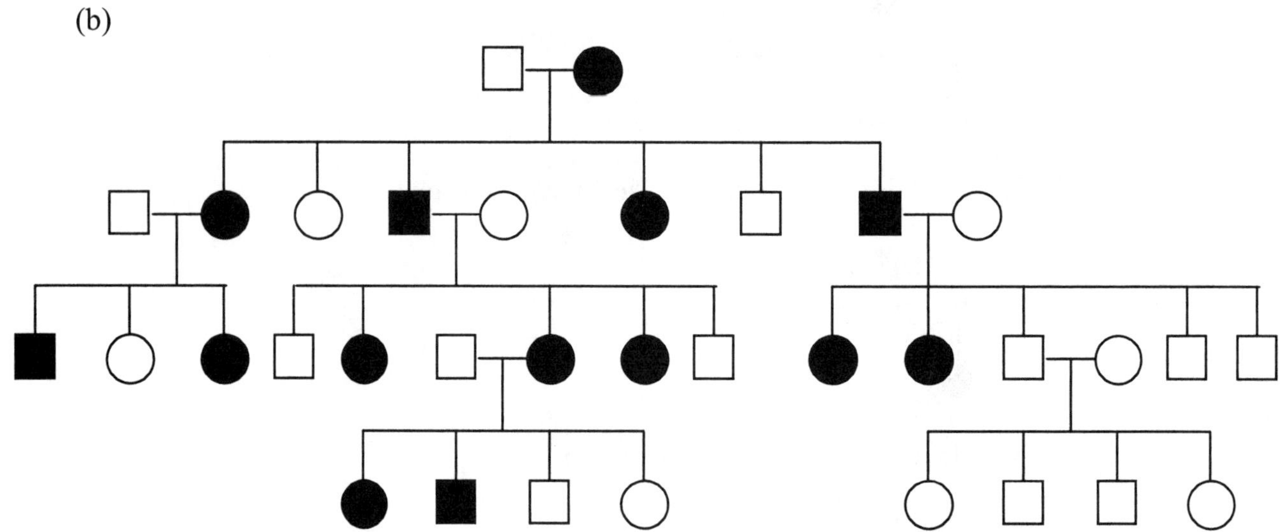

X-linked dominant. Superficially this pedigree appears similar to the pedigree in part (a) in that both males and females are affected, and it appears to be a dominant trait. However, closer inspection reveals that whereas affected females can pass on the trait to either sons or daughters, affected males pass on the trait only to all daughters.

(c)

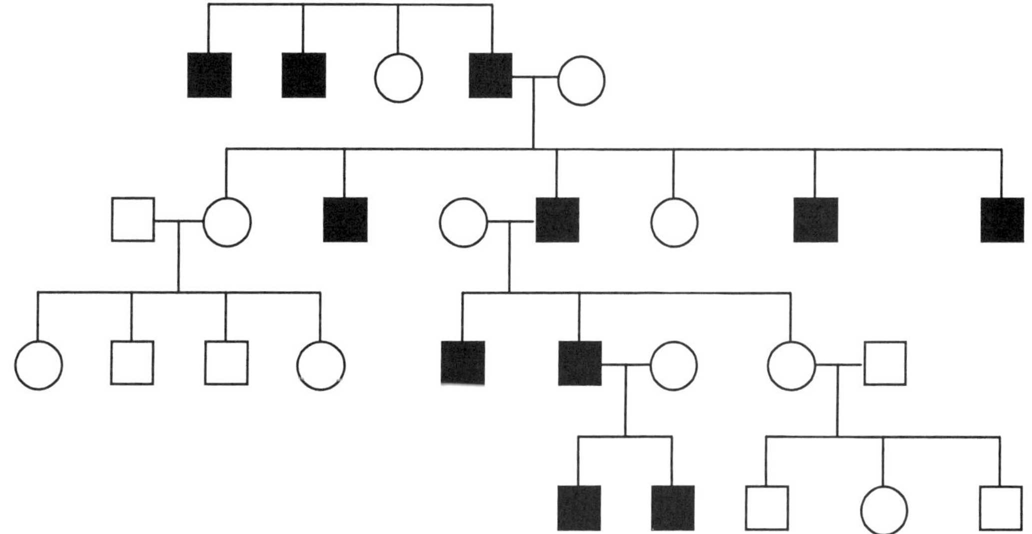

Y-linked. The trait affects only males and is passed from father to son. All sons of an affected male are affected.

(d)

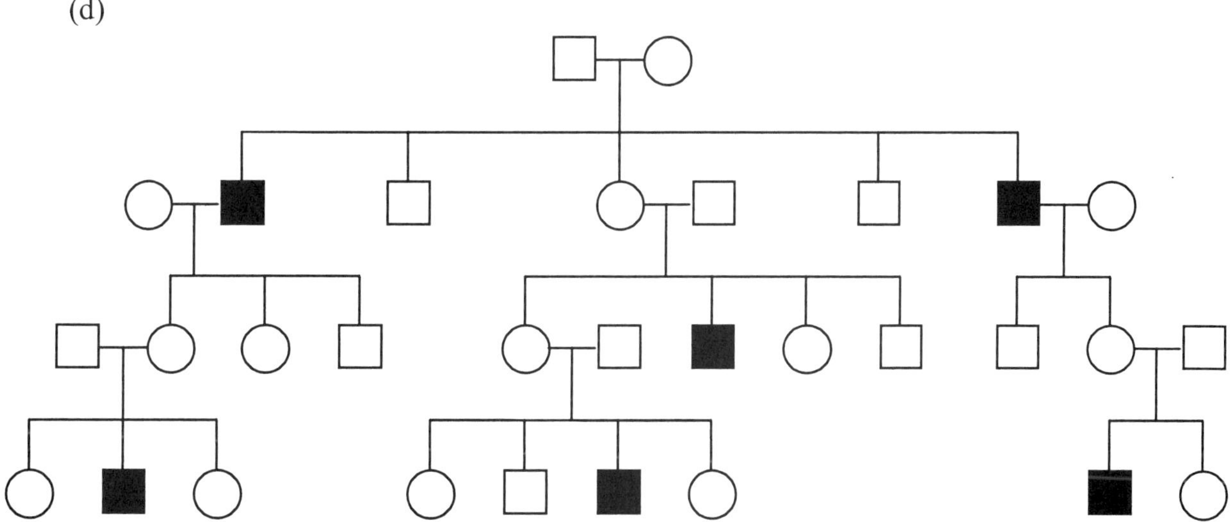

X-linked recessive or sex-limited autosomal dominant. Because only males show the trait, the trait could be X-linked recessive, Y-linked, or sex-limited. We can eliminate Y-linkage because affected males do not pass on the trait to their sons. X-linked recessive inheritance is consistent with the pattern of unaffected female carriers producing both affected and unaffected sons and affected males producing unaffected female carriers, but no affected sons. Sex-linked autosomal dominant inheritance is also consistent with unaffected heterozygous females producing affected heterozygous sons, unaffected homozygous recessive sons, and unaffected heterozygous or homozygous recessive daughters. The two remaining possibilities of X-linked recessive versus sex-limited autosomal dominant could be distinguished

if we had enough data to determine whether affected males could have both affected and unaffected sons, as expected from autosomal dominant inheritance, or whether affected males can have only unaffected sons, as expected from X-linked recessive inheritance. Unfortunately, this pedigree shows only two sons from affected males. In both cases, the sons are unaffected, consistent with X-linked recessive inheritance, but two instances are not enough to conclude that affected males cannot produce affected sons.

(e)

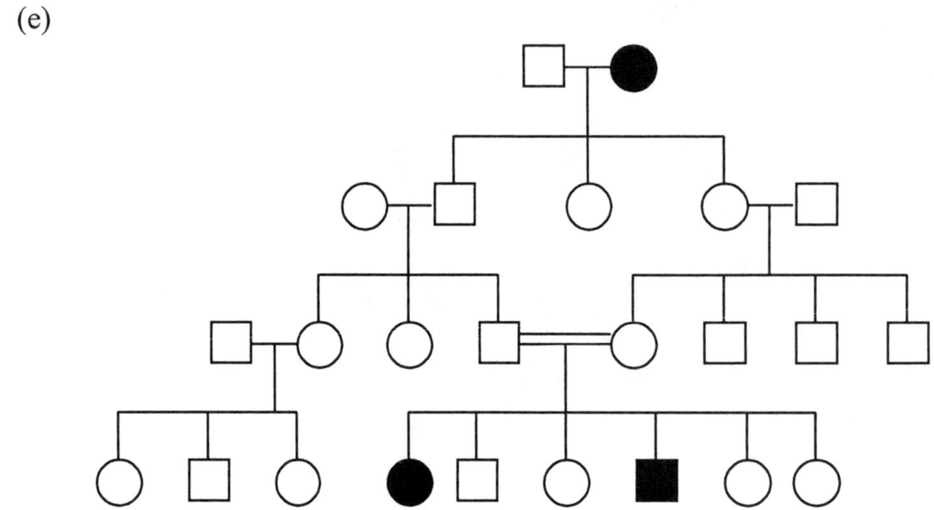

Autosomal recessive. All the children of the original affected female were carriers. The first cousins in the consanguineous marriage in the third generation were also carriers, inheriting the recessive alleles from their carrier parents. The consanguineous marriage produced two affected children, one boy and one girl, and four unaffected children.

13. The trait represented in the following pedigree is expressed only in the males of the family. Is the trait Y-linked? Why or why not? If you believe the trait is not Y-linked, propose an alternate explanation for its inheritance.

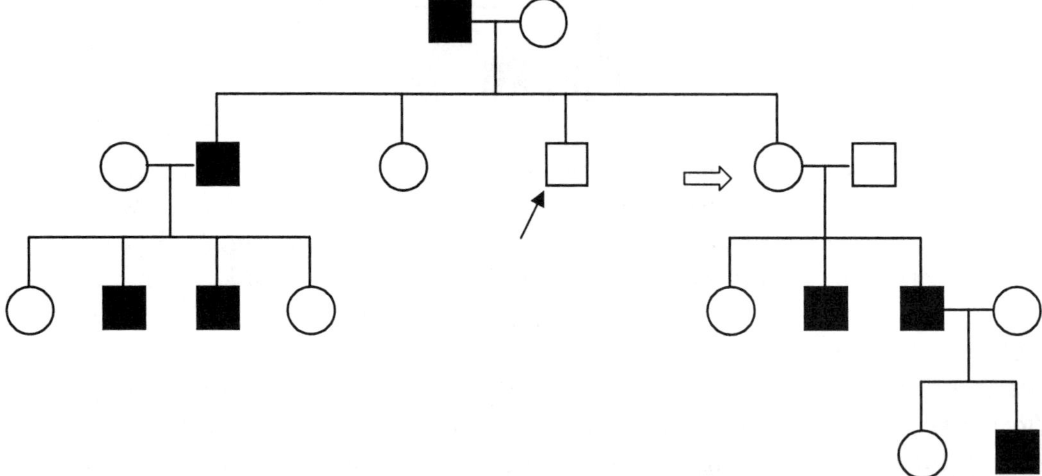

Y-linked traits are passed from father to son. This trait cannot be Y-linked because an affected father can have an unaffected son (indicated by a solid arrow) and also because we see sons inheriting the trait from their mother (indicated by an open arrow). Moreover, this trait cannot be X-linked because it is often passed from father to son, whereas X-linked traits are passed from father to daughter. The most probable mode of inheritance for this trait is sex-limited (only in males) autosomal dominant.

*14. A geneticist studies a series of characteristics in monozygotic twins and dizygotic twins, obtaining the following concordances. For each characteristic, indicate whether the rates of concordance suggest genetic influences, environmental influences, or both. Explain your reasoning.

Characteristic	Monozygotic concordance (%)	Dizygotic concordance (%)
Migraine headaches	60	30
Eye color	100	40
Measles	90	90
Clubfoot	30	10
High blood pressure	70	40
Handedness	70	70
Tuberculosis	5	5

Migraine headaches appear to be influenced by genetic and environmental factors. Markedly greater concordance in monozygotic twins, who share 100% genetic identity, than in dizygotic twins, who share 50% genetic identity, is indicative of a genetic influence. However, the fact that monozygotic twins show only 60% concordance despite sharing 100% genetic identity indicates that environmental factors also play a role.

Eye color appears to be purely genetically determined because the concordance is greater in monozygotic twins than in dizygotic twins. Moreover, the monozygotic twins have 100% concordance for this trait, indicating that environment has no detectable influence.

Measles appears to have no detectable genetic influence because there is no difference in concordance between monozygotic and dizygotic twins. Some environmental influence can be detected because monozygotic twins show less than 100% concordance.

Clubfoot appears to have genetic and environmental influences, by the same reasoning as for migraine headaches. A strong environmental influence is indicated by the high discordance in monozygotic twins.

High blood pressure has genetic and environmental influences, similar to clubfoot.

Handedness, like measles, appears to have no genetic influence because the concordance is the same in monozygotic and dizygotic twins. Environmental influence is indicated by the less than 100% concordance in monozygotic twins.

Tuberculosis similarly lacks indication of genetic influence, with the same degree of concordance in monozygotic and dizygotic twins. The primacy of environmental influence is indicated by the very low concordance in monozygotic twins.

15. In a study of schizophrenia (a mental disorder including disorganization of thought and withdrawal from reality), researchers looked at the prevalence of the disorder in the biological and adoptive parents of people who were adopted as children; they found the following results:

	Prevalence of schizophrenia (%)	
Adopted persons	Biological parents	Adoptive parents
With schizophrenia	12	2
Without schizophrenia	6	4

(Source: S. S. Kety et al., The biological and adoptive families of adopted individuals who become schizophrenic: prevalence of mental illness and other characteristics, *The Nature of Schizophrenia: New Approaches to Research and Treatment*, L. C. Wynne, R. L. Cromwell, and S. Matthysse, Eds. [New York: Wiley, 1978), pp. 25–37.]

What can you conclude from these results concerning the role of genetics in schizophrenia? Explain your reasoning.
These data suggest that schizophrenia has a strong genetic component. The biological parents of schizophrenic adoptees are far more likely to be schizophrenic than genetically unrelated individuals (the adoptive parents), despite the fact that the schizophrenic adoptees share the same environment as the adoptive parents. If environmental variables (such as chemicals in the water or food or power lines) were a major factor, then one would expect to see a higher frequency of schizophrenia in the adoptive parents. Another possibility is that this increased frequency of schizophrenia in the biological parents simply reflects a greater likelihood that schizophrenic parents give up their children for adoption. This latter possibility is ruled out by the data that the biological parents of nonschizophrenic adoptees do not show a similar increased frequency of schizophrenia compared to adoptive parents.

*16. The following pedigree illustrates the inheritance of Nance-Horan syndrome, a rare genetic condition in which affected persons have cataracts and abnormally shaped teeth.

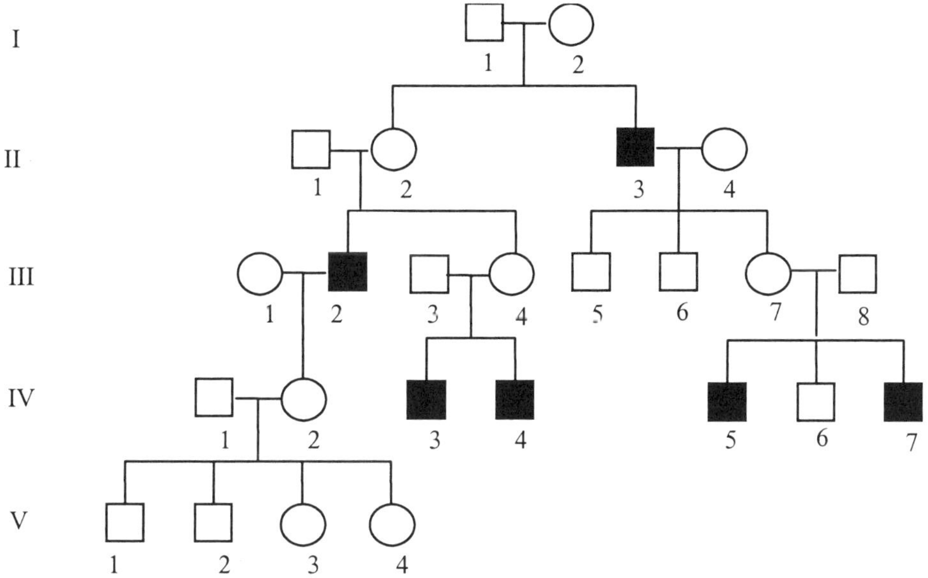

(Pedigree adapted from D. Stambolian, R. A. Lewis, K. Buetow, A. Bond, and R. Nussbaum. *American Journal of Human Genetics* 47[1990]:15.)

(a) On the basis of this pedigree, what do you think is the most likely mode of inheritance for Nance-Horan syndrome?
X-linked recessive. Only males have the condition, and unaffected female carriers have affected sons.

(b) If couple III-7 and III-8 have another child, what is the probability that the child will have Nance-Horan syndrome?
The probability is ¼. The female III-7 is a carrier, so there is a ½ probability that the child will inherit her X chromosome with the Nance-Horan allele and another ½ probability that the child will be a boy.

(c) If III-2 and III-7 mated, what is the probability that one of their children would have Nance-Horan syndrome?
The probability is ½ because half the boys will inherit the Nance-Horan allele from the III-7 carrier female. All the girls will inherit one Nance-Horan allele from the III-2 affected male, and half of them will get a second Nance-Horan allele from the III-2 female, so half the girls will also have Nance-Horan syndrome.

17. The following pedigree illustrates the inheritance of ringed hair, a condition in which each hair is differentiated into light and dark zones. What mode or modes of inheritance are possible for the ringed hair trait in this family?

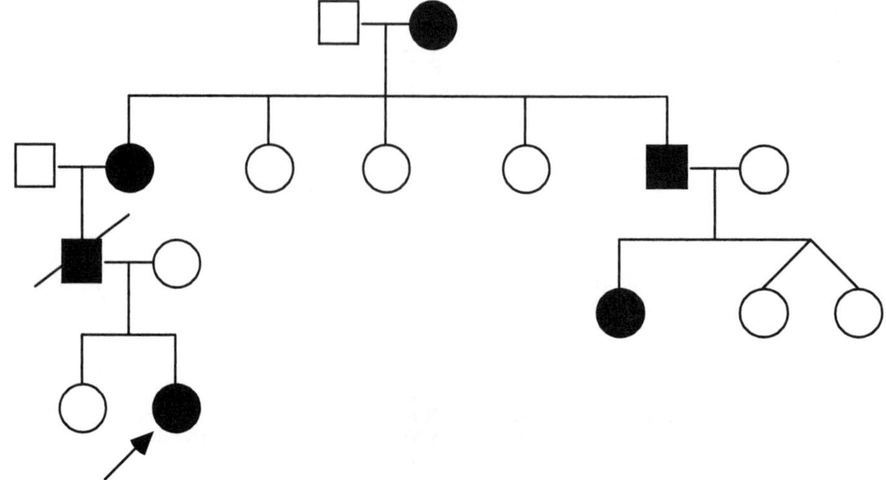

(Pedigree adapted from L. M. Ashley and R. S. Jacques, *Journal of Heredity* 41[1950]:83.)

This pedigree is consistent with autosomal dominant inheritance. Affected individuals marrying unaffected individuals have affected children, so the trait is dominant. Males do not pass the trait to all their daughters, so it cannot be X-linked dominant.

18. Ectodactyly is a rare condition in which the fingers are absent and the hand is split. This condition is usually inherited as an autosomal dominant trait. Ademar Freire-Maia reported the appearance of ectodactyly in a family in São Paulo, Brazil, whose pedigree is shown here. Is this pedigree consistent with autosomal dominant inheritance? If not, what mode of inheritance is most likely? Explain your reasoning.

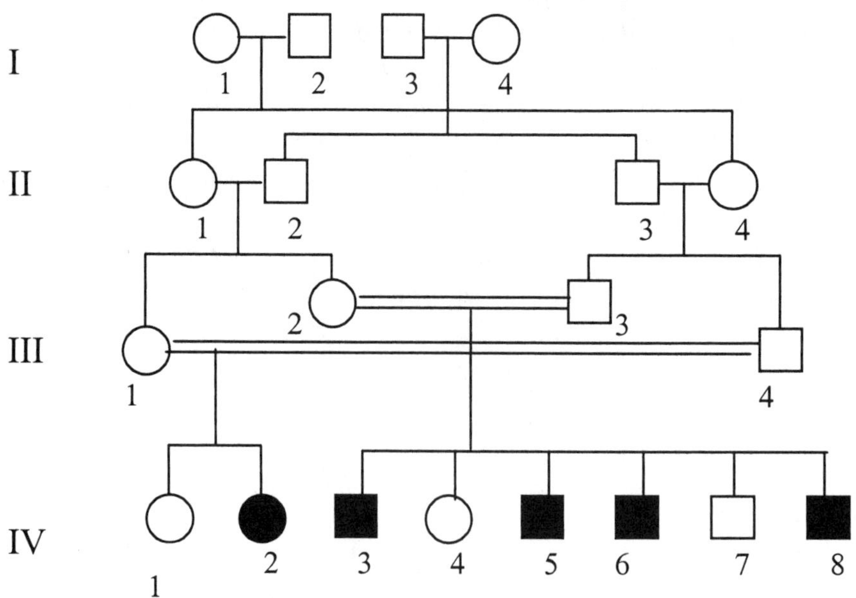

(Pedigree adapted from A.Freire-Maia, *Journal of Heredity* 62[1971]:53.)

This pedigree shows autosomal recessive inheritance, not autosomal dominant inheritance. It cannot be dominant because unaffected individuals have affected children. In generation II, two brothers married two sisters, so the members of generation III in the two families are as closely related as full siblings. A single recessive allele in one of the members of generation I was inherited by all four members of generation III. The consanguineous matings in generation III then produced children homozygous for the recessive ectodactyly allele. X-linkage is ruled out because the father of female IV-2 is unaffected; he has to be heterozygous.

CHALLENGE QUESTIONS

19. Draw a pedigree that represents an autosomal dominant trait, sex-limited to males, and that excludes the possibility that the trait is Y-linked.

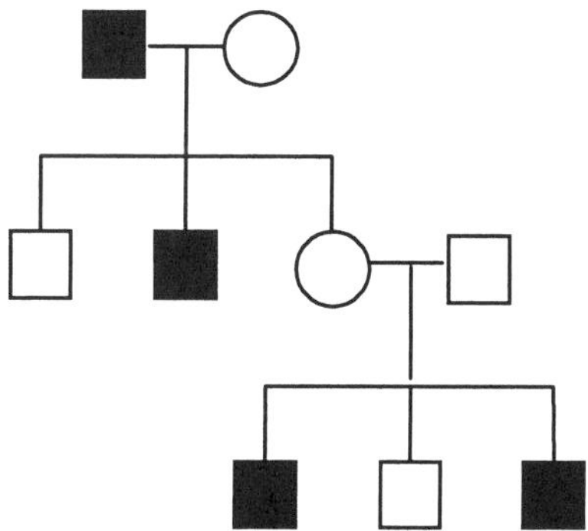

This pedigree excludes Y-linkage because not all the sons of an affected male are affected, an unaffected male has affected sons, and also because it is transmitted through an unaffected female to her sons.

20. Androgen insensitivity syndrome is a rare disorder of sexual development, in which people with an XY karyotype, genetically male, develop external female features. All persons with androgen insensitivity syndrome are infertile. In the past, some researchers proposed that androgen insensitivity syndrome is inherited as a sex-limited, autosomal dominant trait. (It is sex-limited because females cannot express the trait.) Other investigators suggested that this disorder is inherited as a X-linked recessive trait.

    Draw a pedigree that would show conclusively that androgen insensitivity syndrome is inherited as an X-linked recessive trait and that excludes the possibility that it is sex-limited, autosomal dominant. If you believe that no pedigree can conclusively differentiate between the two choices (sex-limited, X-linked recessive and sex-limited, autosomal dominant), explain why. Remember that all affected individuals are infertile.

    In this case, no pedigree can conclusively differentiate between sex-limited, X-linked recessive and sex-limited autosomal dominant. In both modes, only males will exhibit the trait, inherited from their mothers (because males with a single copy of the allele will be affected and sterile in either case). In both modes, mothers who have children with the trait will always be heterozygotes because these mothers will always have inherited a wild-type allele from their father (who cannot have the androgen insensitivity allele and still be fertile!). Therefore, both modes will exhibit inheritance of the trait from heterozygous female to heterozygous female, and half their XY children will have androgen insensitivity and be outwardly female and sterile.

CASE STUDY QUESTIONS

1. Dalia is a student in an introductory genetics class. On her first genetics exam, a question asks for the mode of inheritance of PKU. Dalia answers that PKU is a complex trait exhibiting multifactorial inheritance. Her teacher grades her answer wrong, saying that everyone knows that PKU is recessive. Dalia claims that her answer is correct and that she should receive credit for it. Who is correct and why?
   Dalia and her teacher are both correct. PKU acts as a simple recessive trait in most instances, but it is in fact multifactorial because the phenotype depends on environmental factors such as the level of phenylalanine in the diet. Moreover, PKU displays complexity because the expressivity of this phenotype varies depending on genetic background factors that determine blood concentrations of phenylalanine and influence the entry of phenylalanine into the brain. Therefore, Dalia deserves credit for her answer if she can explain her answer.

2. Does PKU exhibit variable expressivity? Explain your answer.
   Yes, the expressivity of PKU is dependent on both environmental factors (the amount of phenylalanine in the diet) and on the genetic background. People with the same PKU genotype can have different severity of symptoms because other factors determine the levels of phenylalanine in the blood and the accumulation of phenylalanine in the brain.

3. How is PKU treated? What are some of the difficulties with this treatment?
   PKU is treated primarily by carefully controlling the amount of phenylalanine in the diet. Because most foods contain phenylalanine, the diet is quite restrictive. The restrictive diet must begin immediately after birth, with weekly or biweekly monitoring of blood levels of phenylalanine throughout infancy and childhood. Moreover, phenylalanine is an essential amino acid, so it cannot be eliminated from the diet, but must be carefully controlled.

4. Why do some women with PKU produce children with mental retardation, even though the children do not have a genotype that causes PKU? Why don't men with PKU produce such children?
   PKU can have a genetic maternal effect if the mother has gone off the restricted phenylalanine diet during pregnancy. The fetus then becomes exposed to high levels of phenylalanine in the maternal blood circulation. Even though the fetus is heterozygous and would otherwise be phenotypically normal, maternal phenylalanine crosses the placental barrier to affect fetal brain development. Fathers with PKU cannot have any such effect on fetal phenylalanine levels.

5. Michael and Lauren have recently married and want to have children. Neither Michael nor Lauren has PKU, and both of their parents are similarly unaffected. However, Michael has a sister with PKU and Lauren has a brother with PKU. What is the probability that Michael and Lauren's first child will have PKU? Would the probability change if Michael's sister were shown to be a compound heterozygote?
Since Michael's sister has PKU, we can conclude that Michael's parents are both carriers. Similarly, Lauren's parents must both be carriers. Michael and Lauren can have a child with PKU only if both are carriers. If both are carriers, then the probability that a child of theirs will have PKU is ¼. Thus the probability of their first child having PKU = 1/4(probability that Michael is a carrier)(probability that Lauren is a carrier). The probability that Michael is a carrier is 2/3. The probability that Lauren is a carrier is also 2/3. Hence the probability of Michael and Lauren's first child having PKU is (1/4)(2/3)(2/3) = 1/9.

Chapter Seven: Linkage, Recombination, and Eukaryotic Gene Mapping

COMPREHENSION QUESTIONS

*1. What does the term recombination mean? What are two causes of recombination?
Recombination means that meiosis generates gametes with different allelic combinations than the original gametes the organism inherited. If the organism was created by the fusion of an egg bearing AB *and a sperm bearing* ab, *recombination generates gametes that are* Ab *and* aB. *Recombination may be caused by loci on different chromosomes that sort independently or by a physical crossing over between two loci on the same chromosome, with breakage and exchange of strands of homologous chromosomes paired in meiotic prophase I.*

*2. In a testcross for two genes, what types of gametes are produced with (a) complete linkage, (b) independent assortment, and (c) incomplete linkage?
(a) Complete linkage of two genes means that only nonrecombinant gametes will be produced; the recombination frequency is zero.
(b) Independent assortment of two genes will result in 50% of the gametes being recombinant and 50% being nonrecombinant, as would be observed for genes on two different chromosomes. Independent assortment may also be observed for genes on the same chromosome, if they are far enough apart that one or more crossovers occur between them in every meiosis.
(c) Incomplete linkage means that greater than 50% of the gametes produced are nonrecombinant and less than 50% of the gametes are recombinant; the recombination frequency is greater than 0 and less than 50%.

3. What effect does crossing over have on linkage?
Crossing over generates recombination between genes located on the same chromosome, and thus renders linkage incomplete.

4. Why is the frequency of recombinant gametes always half the frequency of crossing over?
Crossing over occurs at the four-strand stage, when two homologous chromosomes, each consisting of a pair of sister chromatids, are paired. Each crossover involves just two of the four strands and generates two recombinant strands. The remaining two strands that were not involved in the crossover generate two nonrecombinant strands. Therefore, the frequency of recombinant gametes is always half the frequency of crossovers.

*5. What is the difference between genes in coupling configuration and genes in repulsion? What effect does the arrangement of linked genes (whether they are in coupling configuration or in repulsion) have on the results of a cross?
Genes in coupling configuration have two wild-type alleles on the same chromosome and the two mutant alleles on the homologous chromosome. Genes in repulsion have a wild-type allele of one gene together with the mutant allele of the

second gene on the same chromosome, and vice versa on the homologous chromosome. The two arrangements have opposite effects on the results of a cross. For genes in coupling configuration, most of the progeny will be either wild type for both genes, or mutant for both genes, with relatively few that are wild type for one gene and mutant for the other. For genes in repulsion, most of the progeny will be mutant for only one gene and wild type for the other, with relatively few recombinants that are wild type for both or mutant for both.

6. How does one test to see if two genes are linked?
One first obtains individuals that are heterozygous for both genes. This may be achieved by crossing an individual homozygous dominant for both genes to one homozygous recessive for both genes, resulting in a heterozygote with genes in coupling configuration. Alternatively, an individual that is homozygous recessive for one gene may be crossed to an individual homozygous recessive for the second gene, resulting in a heterozygote with genes in repulsion. Then the heterozygote is mated to a homozygous recessive tester and the progeny of each phenotypic class are tallied. If the proportion of recombinant progeny is far less than 50%, the genes are linked. If the results are not so clear-cut, then they may be tested by chi-square, first for equal segregation at each locus, then for independent assortment of the two loci. Significant deviation from results expected for independent assortment indicates linkage of the two genes.

7. What is the difference between a genetic map and a physical map?
A genetic map gives the order of genes and relative distance between them based on recombination frequencies observed in genetic crosses. A physical map locates genes on the actual chromosome or DNA sequence, and thus represents the physical distance between genes.

*8. Why do calculated recombination frequencies between pairs of loci that are located relatively far apart underestimate the true genetic distances between loci?
The further apart two loci are, the more likely it is to get double crossovers between them. Unless there are marker genes between the loci, such double crossovers will be undetected because double crossovers give the same phenotypes as nonrecombinants. The calculated recombination frequency will underestimate the true crossover frequency because the double crossover progeny are not counted as recombinants.

9. Explain how one can determine which of three linked loci is the middle locus from the progeny of a three-point testcross.
Double crossovers always result in switching the middle gene with respect to the two nonrecombinant chromosomes. Hence, one can compare the two double crossover phenotypes with the two nonrecombinant phenotypes and see which gene is reversed. In the diagram on the facing page we see that the coupling relationship of the middle gene is flipped in the double crossovers with respect to the genes on either side. So whichever gene on the double crossover can be altered to make the double crossover resemble a nonrecombinant chromosome is the middle gene. If we

take either of the double crossover products l M r *or* L m R, *changing the* M *gene will make it resemble a nonrecombinant.*

*10. What does the interference tell us about the effect of one crossover on another?
A positive interference value results when the actual number of double crossovers observed is less than the number of double crossovers expected from the single crossover frequencies. Thus, positive interference indicates that a crossover inhibits or interferes with the occurrence of a second crossover nearby.

Conversely, a negative interference value, where more double crossovers occur than expected, suggests that a crossover event can stimulate additional crossover events in the same region of the chromosome.

11. List some of the methods for physically mapping genes and explain how they are used to position genes on chromosomes.
Deletion mapping: Recessive mutations are mapped by crossing mutants with strains containing various overlapping deletions that map to the same region as the recessive mutation. If the heterozygote with the mutation on one chromosome and the deletion on the homologous chromosome has a mutant phenotype, then the mutation must be located on the same physical portion of the chromosome that is deleted. If, on the other hand, the heterozygote has a wild-type phenotype (the mutation and the deletion complement), then the mutation lies outside the deleted region of the chromosome.

Somatic-cell hybridization: Human and mouse cells are fused. The resulting hybrid cell randomly loses human chromosomes and retains only a few. A panel of hybrids that retain different combinations of human chromosomes is tested for expression of a human gene. A correlation between the expression of the gene and the retention of a unique human chromosome in those cell lines indicates that the human gene must be located on that chromosome.

In situ hybridization: DNA probes that are labeled with either a radioactive or fluorescent tag are hybridized to chromosome spreads. Detection of the labeled hybridized probe by autoradiography or fluorescence imaging reveals which chromosome and where along that chromosome the homologous gene is located.

DNA sequencing: Overlapping DNA sequences are joined using computer programs to ultimately form chromosome-length sequence assemblies, or contigs. The locations of genes along the DNA sequence can be determined by searching for matches to known gene or protein amino acid sequences.

12. What is a lod score and how is it calculated?
The term lod means logarithm of odds. It is used to determine whether genes are linked, usually in the context of pedigree analysis. One first determines the probability of obtaining the observed progeny given a specified degree of linkage.

That probability is divided by the probability of obtaining the observed progeny if the genes are not linked and sort independently. The log of the ratio of these probabilities is the lod score. A lod score of 3 or greater, indicating that the specified degree of linkage results in at least a thousand fold greater likelihood of yielding the observed outcome than if the genes are unlinked, indicates linkage.

APPLICATION QUESTIONS AND PROBLEMS

*13. In the snail *Cepaea nemoralis,* an autosomal allele causing a banded shell (B^B) is recessive to the allele for unbanded shell (B^O). Genes at a different locus determine the background color of the shell; here, yellow (C^Y) is recessive to brown (C^{Bw}). A banded, yellow snail is crossed with a homozygous brown, unbanded snail. The F_1 are then crossed with banded, yellow snails (a testcross).

(a) What will be the results of the testcross if the loci that control banding and color are linked with no crossing over?

With absolute linkage, there will be no recombinant progeny. The F_1 inherited banded and yellow alleles (B^BC^Y) together on one chromosome from the banded yellow parent and unbanded and brown alleles (B^OC^{Bw}) together on the homologous chromosome from the unbanded brown parent. Without recombination, all the F_1 gametes will contain only these two allelic combinations, in equal proportions. Therefore, the F_2 testcross progeny will be ½ banded, yellow and ½ unbanded, brown.

(b) What will be the results of the testcross if the loci assort independently?

With independent assortment, the progeny will be:
¼ banded, yellow
¼ banded, brown
¼ unbanded, yellow
¼ unbanded, brown

(c) What will be the results of the testcross if the loci are linked and 20 map units apart?

The recombination frequency is 20%, so each of the two classes of recombinant progeny must be 10%. The recombinants are banded, brown and unbanded, yellow. The two classes of nonrecombinants are 80% of the progeny, so each must be 40%. The nonrecombinants are banded, yellow and unbanded, brown.
In summary:
40% banded, yellow
40% unbanded, brown
10% banded, brown
10% unbanded, yellow

*14. In silkmoths (*Bombyx mori*), red eyes (*re*) and white-banded wing (*wb*) are encoded by two mutant alleles that are recessive to those that produce wild-type traits (re^+ and wb^+); these two genes are on the same chromosome. A moth homozygous for red eyes and white-banded wings is crossed with a moth homozygous for the wild-type traits. The F_1 have normal eyes and normal wings. The F_1 are crossed with moths that have red eyes and white-banded wings in a testcross. The progeny of this testcross are:

wild-type eyes, wild-type wings	418
red eyes, wild-type wings	19
wild-type eyes, white-banded wings	16
red eyes, white-banded wings	426

(a) What phenotypic proportions would be expected if the genes for red eyes and white-banded wings were located on different chromosomes?
¼ wild-type eyes, wild-type wings
¼ red eyes, wild-type wings
¼ wild-type eyes, white-banded wings
¼ red eyes, white-banded wings

(b) What is the genetic distance between the genes for red eyes and white-banded wings?
The F_1 heterozygote inherited a chromosome with alleles for red eyes and white-banded wings (re wb) *from one parent and a chromosome with alleles for wild-type eyes and wild-type wings* (re^+ wb^+) *from the other parent. These are therefore the phenotypes of the nonrecombinant progeny, present in the highest numbers. The recombinants are the 19 with red eyes, wild-type wings and 16 with wild-type eyes, white-banded wings.*
RF = recombinants/total progeny × 100% = (19 + 16)/879 × 100% = 4.0%
The distance between the genes is 4 map units.

*15. A geneticist discovers a new mutation in *Drosophila melanogaster* that causes the flies to shake and quiver. She calls this mutation spastic (*sps*) and determines that spastic is due to an autosomal recessive gene. She wants to determine if the spastic gene is linked to the recessive gene for vestigial wings (*vg*). She crosses a fly homozygous for spastic and vestigial traits with a fly homozygous for the wild-type traits and then uses the resulting F_1 females in a testcross. She obtains the following flies from this testcross:

vg^+ sps^+	230
vg sps	224
vg sps^+	97
vg^+ sps	99
total	650

Are the genes that cause vestigial wings and the spastic mutation linked? Do a series of chi-square tests to determine if the genes have assorted independently.
To test for independent assortment, we first test for equal segregation at each locus, then test whether the two loci sort independently.
Test for vg:
Observed vg *= 224 + 97 = 321*
Observed vg^+ *= 230 + 99 = 329*
Expected vg *or* vg^+ *= ½ × 650 = 325*

$$\chi^2 = \Sigma\frac{(observed - expected)^2}{expected} = (321 - 325)^2/325 + (329 - 325)^2/325 = 16/325 + 16/325$$

$$= 0.098$$

We have n – *1 degrees of freedom, where* n *is the number of phenotypic classes = 2, so just 1 degree of freedom. From Table 3.4, we see that the P value is between 0.7 and 0.8. So these results do not deviate significantly from the expected 1:1 segregation.*

Similarly, testing for sps, *we observe 327* sps$^+$ *and 323* sps *and expect ½ × 650 = 325 of each:*

χ^2 *= 4/325 + 4/325 = .025, again with 1 degree of freedom. The P value is between 0.8 and 0.9, so these results do not deviate significantly from the expected 1:1 ratio.*

Finally, we test for independent assortment, where we expect 1:1:1:1 phenotypic ratios, or 162.5 of each.

Observed	*Expected*	*o – e*	*(o – e)^2*	*(o – e)^2/e*
230	*162.5*	*67.5*	*4556.25*	*28.0*
224	*162.5*	*61.5*	*3782.25*	*23.3*
97	*162.5*	*–65.5*	*4290.25*	*26.4*
99	*162.5*	*–63.5*	*4032.25*	*24.8*

We have four phenotypic classes, giving us three degrees of freedom. The chi-square value of 102.5 is off the chart, so we reject independent assortment.

Instead, the genes are linked, and the RF = (97 + 99)/650 × 100% = 30%, giving us 30 map units between them.

16. In cucumbers, heart-shaped leaves (*hl*) are recessive to normal leaves (*Hl*), and having numerous fruit spines (*ns*) is recessive to having few fruit spines (*Ns*). The genes for leaf shape and number of spines are located on the same chromosome; mapping experiments indicate that they are 32.6 map units apart. A cucumber plant having heart-shaped leaves and numerous spines is crossed with a plant that is homozygous for normal leaves and few spines. The F_1 are crossed with plants that have heart-shaped leaves and numerous spines. What phenotypes and proportions are expected in the progeny of this cross?
The recombinants should total 32.6%, so each recombinant phenotype will be 16.3% of the progeny. Because the F_1 inherited a chromosome with heart-shaped leaves and numerous spines (hl ns) *from one parent and a chromosome with normal leaves and few spines* (Hl Ns) *from the other parent, these are the nonrecombinant phenotypes, and together they total 67.4%, or 33.7% each. The two recombinant phenotypes are heart-shaped leaves with few spines* (hl Ns) *and normal-shaped leaves with numerous spines* (Hl ns).

Heart-shaped, numerous spines	*33.7%*
Normal-shaped, few spines	*33.7%*
Heart-shaped, few spines	*16.3%*
Normal-shaped, numerous spines	*16.3%*

*17. In tomatoes, tall (*D*) is dominant over dwarf (*d*), and smooth fruit (*P*) is dominant over pubescent (*p*) fruit, which is covered with fine hairs. A farmer has two tall and smooth tomato plants, which we will call plant A and plant B. The farmer crosses plants A and B with the same dwarf and pubescent plant and obtains the following numbers of progeny:

	Progeny of	
	Plant A	Plant B
Dd Pp	122	2
Dd pp	6	82
dd Pp	4	82
dd pp	124	4

(a) What are the genotypes of plant A and plant B?
The genotypes of both plants are DdPp.

(b) Are the loci that determine height of the plant and pubescence linked? If so, what is the map distance between them?
Yes. From the cross of plant A, the map distance is 10/256 = 3.9% or 3.9 m.u. The cross of plant B gives 6/170 = 3.5% or 3.5 m.u. If we pool the data from the two crosses, we get 16/426 = 3.8% or 3.8 m.u.

(c) Explain why different proportions of progeny are produced when plant A and plant B are crossed with the same dwarf pubescent plant.
The two plants have different coupling configurations. In plant A, the dominant alleles D *and* P *are coupled; one chromosome is* D P *and the other is* d p. *In plant B, they are in repulsion; its chromosomes have* D p *and* d P.

18. Alleles *A* and *a* occur at a locus that is located on the same chromosome as a locus with alleles *B* and *b*. *AaBb* is crossed with *aabb,* and the following progeny are produced:

AaBb	5
Aabb	45
aaBb	45
aabb	5

What conclusion can be made about the arrangement of the genes on the chromosome in the *AaBb* parent?
The results of this testcross reveal that Aabb and aaBb, with far greater numbers, are the progeny that received nonrecombinant chromatids from the AaBb parent. Given that all the progeny received ab from the aabb parent, the nonrecombinant progeny received either an Ab or an aB chromatid from the AaBb parent. Therefore, the A and B loci are in repulsion in the AaBb parent.

19. A cross between individuals with genotypes $a^+a\ b^+b \times aa\ bb$ produces the following progeny:

$a^+a\ b^+b$	83
$a^+a\ bb$	21
$aa\ b^+b$	19
$aa\ bb$	77

(a) Does the evidence indicate that the *a* and *b* loci are linked?

The ratio of a^+ *to* a *is 104/96, and the ratio of* b^+ *to* b *is 102/98, both close to 1:1 ratios. The four phenotypic classes are not present in 1:1:1:1 ratios (no need for chi-square test), so they are linked.*

(b) What is the map distance between *a* and *b*?

The recombinants are the two phenotypic classes with the fewest progeny: RF = (21 + 19)/200 = 40/200 = 0.2 = 20%; the two genes are 20 m.u. apart.

(c) Are the genes in the parent with genotype $a^+a\ b^+b$ in coupling configuration or in repulsion? How do you know?

They are in coupling configuration because the nonrecombinants are a^+b^+ *and* ab.

20. In tomatoes, dwarf (*d*) is recessive to tall (*D*) and opaque (light green) leaves (*op*) is recessive to green leaves (*Op*). The loci that determine the height and leaf color are linked and separated by a distance of 7 m.u. For each of the following crosses, determine the phenotypes and proportions of progeny produced.

(a) $\dfrac{D\quad Op}{d\quad op} \times \dfrac{d\quad op}{d\quad op}$

The recombinants in this cross would be D op *and* d Op, *and each would be 3.5% of the progeny, to total 7% recombinants. Each of the nonrecombinants would be 46.5%, to total the remaining 93%.*

Tall green	*46.5%*
Dwarf opaque	*46.5%*
Tall opaque	*3.5%*
Dwarf green	*3.5%*

(b) $\dfrac{D\quad op}{d\quad Op} \times \dfrac{d\quad op}{d\quad op}$

Here with the genes in repulsion, the recombinants are D Op *and* d op.

Tall green	*3.5%*
Dwarf opaque	*3.5%*
Tall opaque	*46.5%*
Dwarf green	*46.5%*

(c) $\dfrac{D\quad Op}{d\quad op} \times \dfrac{D\quad Op}{d\quad op}$

This is not a testcross, so we have to account for recombination in both parents. The most straightforward way is to do a Punnett square, including the types and proportions of gametes produced by meiosis in each parent. Because the genes are in coupling configuration in both parents, we can use the figures from part (a).

	D Op *0.465*	D op *0.035*	d Op *0.035*	d op *0.465*
D Op *0.465*	*Tall, green .216*	*Tall, green .016*	*Tall, green .016*	*Tall, green .216*
D op *0.035*	*Tall, green .016*	*Tall, opaque .001*	*Tall, green .001*	*Tall, opaque .016*
d Op *0.035*	*Tall, green .016*	*Tall, green .001*	*Dwarf, green .001*	*Dwarf, green .016*
d op *0.465*	*Tall, green .216*	*Tall, opaque .016*	*Dwarf, green .016*	*Dwarf, opaque .216*

In summary, we get

Tall green	*3(.216) + 4(.016) + 2(.001) = .714*
Dwarf opaque	*.216*
Tall opaque	*2(.016) + .001 = .033*
Dwarf green	*2(.016) + .001 = .033*

(d) $\dfrac{D \quad op}{d \quad Op} \times \dfrac{D \quad op}{d \quad Op}$

Again, this is not a testcross, and recombination in both parents must be taken into account. Both are in repulsion, so we use the proportions from part (b).

	D Op *0.035*	D op *0.465*	d Op *0.465*	d op *0.035*
D Op *0.035*	*Tall, green .001*	*Tall, green .016*	*Tall, green .016*	*Tall, green .001*
D op *0.465*	*Tall, green .016*	*Tall, opaque .216*	*Tall, green .216*	*Tall, opaque .016*
d Op *0.465*	*Tall, green .016*	*Tall, green .216*	*Dwarf, green .216*	*Dwarf, green .016*
d op *0.035*	*Tall, green .001*	*Tall, opaque .016*	*Dwarf, green .016*	*Dwarf, opaque .001*

In summary, we get

Tall, green	*3(.001) + 4(.016) + 2 (.216) = .499*
Dwarf, opaque	*.001*
Tall, opaque	*2(.016) + .216 = .248*
Dwarf, green	*2(.016) + .216 = .248*

21. In German cockroaches, bulging eyes (*bu*) are recessive to normal eyes (bu^+) and curved wings (*cv*) are recessive to straight wings (cv^+). Both traits are encoded by autosomal genes that are linked. A cockroach has genotype $bu^+bu\ cv^+cv$ and the

genes are in repulsion. Which of the following sets of genes will be found in the most common gametes produced by this cockroach?

a. $bu^+ cv^+$
b. *bu cv*
c. $bu^+ bu$
d. $cv^+ cv$
e. $bu\ cv^+$

Explain your answer.

The most common gametes will have (e) $bu\ cv^+$*. Equally common will be gametes that have* $bu^+ cv$*, not given among the choices. Since these genes are linked, in repulsion, the wild types alleles are on different chromosomes. Thus the cockroach has one chromosome with* $bu^+ cv$ *and the homologous chromosome with* $bu\ cv^+$*. Meiosis always produces nonrecombinant gametes at higher frequencies than recombinants, so gametes bearing* $bu\ cv^+$ *will be produced at higher frequencies than (a) or (b), which are the products of recombination. The choices (c) and (d) have two copies of one locus and no copy of the other locus. They violate Mendelian segregation: each gamete must contain one allele of each locus.*

*22. In *Drosophila melanogaster,* ebony body (*e*) and rough eyes (*ro*) are encoded by autosomal recessive genes found on chromosome 3; they are separated by 20 map units. The gene that encodes forked bristles (*f*) is X-linked recessive and assorts independently of *e* and *ro*. Give the phenotypes of progeny and their expected proportions when each of the following genotypes is test-crossed.

(a) $\dfrac{e^+ \quad ro^+}{e \quad ro} \quad \dfrac{f^+}{f}$

We can calculate the four phenotypic classes and their proportions for e *and* ro*, and then each of those classes will be split 1:1 for* f *because* f *sorts independently. The recombination frequency between* e *and* ro *is 20%, so each of the recombinants* (e^+ ro *and* e ro^+) *will be 10%, and each of the nonrecombinants* (e^+ ro^+ *and* e ro) *will be 40%. Each of these will then be split equally among* f^+ *and* f.

e^+ ro^+ f^+	*20%*
e^+ ro^+ f	*20%*
e ro f^+	*20%*
e ro f	*20%*
e^+ ro f^+	*5%*
e^+ ro f	*5%*
e ro^+ f^+	*5%*
e ro^+ f	*5%*

(b) $\dfrac{e^+ \quad ro}{e \quad ro^+} \quad \dfrac{f}{f}$

We can do the same calculations as in part (a), except the nonrecombinants are e^+ ro *and* e ro^+ *and the recombinants are* e^+ ro^+ *and* e ro.

$e^+ ro^+ f^+$	*5%*
$e^+ ro^+ f$	*5%*
$e\ ro\ f^+$	*5%*
e ro f	*5%*
$e^+ ro\ f^+$	*20%*
$e^+ ro\ f$	*20%*
$e\ ro^+ f^+$	*20%*
$e\ ro^+ f$	*20%*

*23. A series of two-point crosses were carried out among seven loci (*a, b, c, d, e, f,* and *g*), producing the following recombination frequencies. Map the seven loci, showing their linkage groups, the order of the loci in each linkage group, and distances between the loci of each linkage group.

Loci	% Recombination	Loci	% Recombination
a - b	50	*c - d*	50
a - c	50	*c - e*	26
a - d	12	*c - f*	50
a - e	50	*c - g*	50
a - f	50	*d - e*	50
a - g	4	*d - f*	50
b - c	10	*d - g*	8
b - d	50	*e - f*	50
b - e	18	*e - g*	50
b - f	50	*f - g*	50
b - g	50		

50% recombination indicates that the genes assort independently. Less than 50% recombination indicates linkage. Starting with the most tightly linked genes a *and* g, *we look for other genes linked to these and find only gene* d *has less than 50% recombination with* a *and* g. *So one linkage group consists of* a, g, *and* d. *We know that gene* g *is between* a *and* d *because the* a *to* d *distance is 12.*

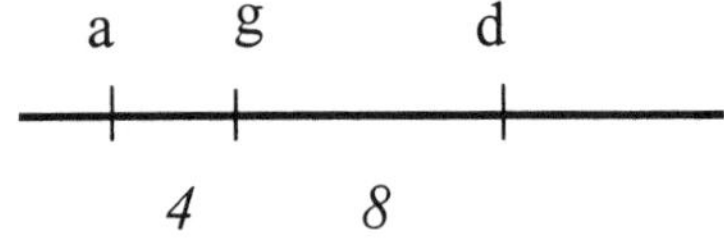

Similarly, we find a second linkage group of b, c, *and* e, *with* b *in the middle.*

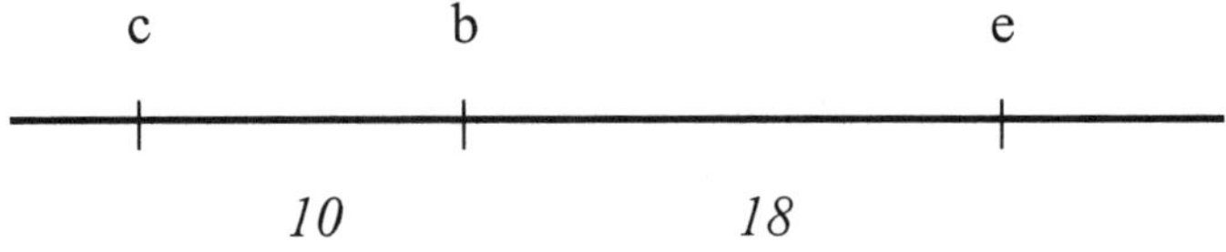

Gene f *is unlinked to either of these groups, on a third linkage group.*

*24. Waxy endosperm (*wx*), shrunken endosperm (*sh*), and yellow seedling (*v*) are encoded by three recessive genes in corn that are linked on chromosome 5. A corn plant homozygous for all three recessive alleles is crossed with a plant homozygous for all the dominant alleles. The resulting F_1 are then crossed with a plant homozygous for the recessive genes in a three-point testcross. The progeny of the testcross are given below:

wx	*sh*	*V*	87
Wx	*Sh*	*v*	94
Wx	*Sh*	*V*	3479
wx	*sh*	*v*	3478
Wx	*sh*	*V*	1515
wx	*Sh*	*v*	1531
wx	*Sh*	*V*	292
Wx	*sh*	*v*	280
total			10,756

(a) Determine order of these genes on the chromosome.
The nonrecombinants are Wx Sh V *and* wx sh v.
The double crossovers are wx sh V *and* Wx Sh v.
Comparing the two, we see that they differ only at the v *locus, so* v *must be the middle gene.*

(b) Calculate the map distances between the genes.
Wx-V *distance—recombinants are* wx V *and* Wx v:
RF = (292+280+87+94)/10,756 = 753/10,756 = .07 = 7% or 7 m.u.
Sh-V *distance—recombinants are* sh V *and* Sh v:
RF = (1515+1531+87+94)/10,756 = 3227/10,756 = 30 = 30% or 30 m.u.
The Wx-Sh *distance is the sum of these two distances: 7 + 30 = 37 m.u.l.*

(c) Determine the coefficient of coincidence and the interference among these genes.
Expected dcos = RF1 × RF2 × total progeny =.07(.30)(10,756) = 226
C.o.C. = observed dcos/expected dcos = (87+94)/226 = 0.80
Interference = 1 – C.o.C = 0.20

25. Fine spines (*s*), smooth fruit (*tu*), and uniform fruit color (*u*) are three recessive traits in cucumbers whose genes are linked on the same chromosome. A cucumber plant heterozygous for all three traits is used in a testcross and the progeny at the top of the following page are produced from this testcross.

S	*U*	*Tu*	2
s	*u*	*Tu*	70
S	*u*	*Tu*	21
s	*u*	*tu*	4
S	*U*	*tu*	82
s	*U*	*tu*	21
s	*U*	*Tu*	13
S	*u*	*tu*	17
total			230

(a) Determine the order of these genes on the chromosome.
Nonrecombinants are s u Tu *and* S U tu.
Double crossovers are s u tu *and* S U Tu.
Because Tu *differs between the nonrecombinants and the double crossovers,* Tu *is the middle gene.*

(b) Calculate the map distances between the genes.
S-Tu *distance: recombinants are* S Tu *and* s tu.
RF = (2 + 4 + 21 + 21)/230 = 48/230 = 21% or 21 m.u.
U-Tu *distance: recombinants are* u tu *and* U Tu.
RF = (2 + 4 + 13 + 17)/230 = 36/230 = 16% or 16 m.u.

(c) Determine the coefficient of coincidence and the interference among these genes.
Expected dcos = (48/230)(36/230)(230) = 7.5
C.o.C. = observed dcos/expected dcos = 6/7.5 = 0.8
I = 1 – C.o.C. = 0.2

(d) List the genes found on each chromosome in the parents used in the testcross.
In the correct gene order for the heterozygous parent: s Tu u *and* S tu U
For the testcross parent: s tu u *and* s tu u

*26. In *Drosophila melanogaster,* black body (*b*) is recessive to gray body (b^+), purple eyes (*pr*) are recessive to red eyes (pr^+), and vestigial wings (*vg*) are recessive to normal wings (vg^+). The loci coding for these traits are linked, with the map distances:

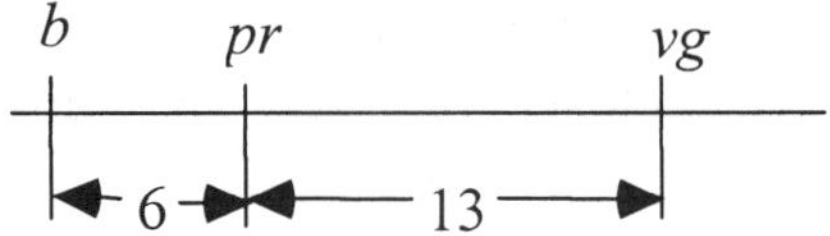

The interference among these genes is 0.5. A fly with black body, purple eyes, and vestigial wings is crossed with a fly homozygous for gray body, red eyes, and normal wings. The female progeny are then crossed with males that have black body, purple eyes, and vestigial wings. If 1000 progeny are produced from this testcross, what will be the phenotypes and proportions of the progeny?
Although we know what the recombination frequencies are between the pairs of genes, these recombination frequencies result from both single crossover (sco) and

double crossover (dco) progenies. So we must first calculate how many double crossover progeny we should get.

Working backward, given that interference = 0.5, the coefficient of coincidence = 1 – interference = 0.5.

We now use the C.o.C. to calculate the actual dco progeny:
C.o.C. = 0.5 = actual dcos/theoretical dcos = actual dcos/(.06)(.13)(1000)
The denominator calculates to 7.8, so actual dcos = 0.5(7.8) = 3.9.
We round 3.9 to 4 double crossover progeny.

Because the parents were either homozygous recessive for all three loci or homozygous dominant (wild type) for all three loci, the F_1 heterozygote fly has chromosomes with b pr vg *and* b^+ pr^+ vg^+. *These are therefore the nonrecombinant progeny phenotypes. The double crossover progeny will be* b pr^+ vg *and* b^+ pr vg^+. *We calculated above that there will be four double crossover progeny, so we should expect two progeny flies of each double crossover phenotype.*

Next, we know that the recombination frequency between b *and* pr *is 6% or 0.06. This recombination frequency arises from the sum of the single crossovers between* b *and* pr *and the double crossover progeny:*
scos(b-pr) + *dcos = .06(1,000) = 60. But we already calculated that dcos = 4, so substituting in the above equation, we get: scos*(b-pr) + *4 = 60; scos = 56. The single crossover phenotypes between* b *and* pr *are* b pr^+ vg^+ *and* b^+ pr vg. *These total 56, or 28 each.*

Similarly, scos(pr-vg) + *dcos = .13(1000) = 130; scos*(pr-vg) = *130 – dcos = 130 – 4 = 126. The single crossover phenotypes between* pr *and* vg *are* b pr vg^+ *and* b^+ pr^+ vg. *These total 126, or 63 each.*

The two remaining phenotypic classes are the nonrecombinants.
 # nonrecombinants = 1000 – scos(b-pr) – *cos*(pr-vg) – *dcos = 1000 – 56 – 126 – 4.*
So, # nonrecombinants = 1000 – 186 = 814. The nonrecombinant phenotypes are b^+ pr^+ vg^+ *and* b pr vg; *we expect 407 of each, to total 814.*
In summary, the expected numbers of all eight phenotypic classes are:

b^+ pr^+ vg^+	*407*
b pr vg	*407*
b^+ pr^+ vg	*63*
b pr vg^+	*63*
b^+ pr vg	*28*
b pr^+ vg^+	*28*
b^+ pr vg^+	*2*
b pr^+ vg	*2*

27. A group of geneticists are interested in identifying genes that may play a role in susceptibility to asthma. They study the inheritance of genetic markers in a series of families that have two or more children affected with asthma. They find an association between the presence or absence of asthma and a genetic marker on the short arm of chromosome 20 and calculate a lod score of 2 for this association. What does this lod score indicate about genes that may influence asthma?
The lod score of 2 indicates that the probability of observing this degree of association if the marker is linked to asthma is 100 times higher than if the marker

has no linkage to asthma. An lod score of 3, for an odds ratio of 1000-fold, is generally considered convincing evidence of linkage.

*28. The locations of six deletions have been mapped to the *Drosophila* chromosome shown on the following page. Recessive mutations *a, b, c, d, e,* and *f* are known to be located in the same region as the deletions, but the order of the mutations on the chromosome is not known. When flies homozygous for the recessive mutations are crossed with flies homozygous for the deletions, the following results are obtained, where "m" represents a mutant phenotype and a plus sign (+) represents the wild type. On the basis of these data, determine the relative order of the seven mutant genes on the chromosome.

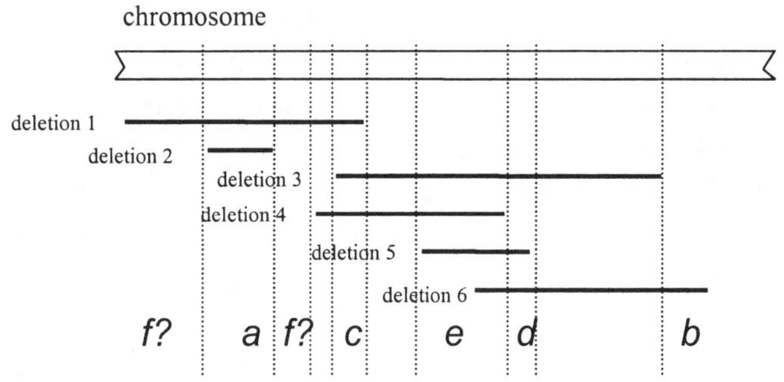

Deletion	a	b	c	d	e	f
1	*m*	+	*m*	+	+	*m*
2	*m*	+	+	+	+	+
3	+	*m*	*m*	*m*	*m*	+
4	+	+	*m*	*m*	*m*	+
5	+	+	+	*m*	*m*	+
6	+	*m*	+	*m*	+	+

The mutations are mapped to the intervals indicated on the figure above the table. The location of f *is ambiguous; it could be in either location shown above.*

29. A panel of cell lines was created from mouse-human somatic-cell fusions. Each line was examined for the presence of human chromosomes and for the production of an enzyme. The following results were obtained:

		Human chromosomes											
Cell line	Enzyme	1	2	3	4	5	6	7	8	9	10	17	22
A	-	+	-	-	-	+	-	-	-	-	-	+	-
B	+	+	+	-	-	-	-	-	+	-	-	+	+
C	-	+	-	-	-	+	-	-	-	-	-	-	+
D	-	-	-	-	+	-	-	-	-	-	-	-	-
E	+	+	-	-	-	-	-	-	+	-	+	+	-

On the basis of these results, which chromosome has the gene that codes for the enzyme?
The enzyme is produced only in cell lines B and E. Of all the chromosomes, only chromosome 8 is present in just these two cell lines and absent in all the other cell lines that do not produce the enzyme. Therefore, the gene for the enzyme is most likely on chromosome 8.

*30. A panel of cell lines was created from mouse-human somatic-cell fusions. Each line was examined for the presence of human chromosomes and for the production of three enzymes. The following results were obtained:

	Enzyme			Human chromosomes								
Cell line	1	2	3	4	8	9	12	15	16	17	22	X
A	+	-	+	-	-	+	-	+	+	-	-	+
B	+	-	-	-	-	+	-	-	+	+	-	-
C	-	+	+	+	-	-	-	-	-	+	-	+
D	-	+	+	+	+	-	-	-	+	-	-	+

On the basis of these results, give the chromosome location of enzyme 1, enzyme 2, and enzyme 3.
Enzyme 1 is located on chromosome 9. Chromosome 9 is the only chromosome that is present in the cell lines that produce enzyme 1 and absent in the cell lines that do not produce enzyme 1.

Enzyme 2 is located on chromosome 4. Chromosome 4 is the only chromosome that is present in cell lines that produce enzyme 2 (C & D) and absent in cell lines that do not produce enzyme 2 (A & B).

Enzyme 3 is located on the X chromosome. The X chromosome is the only chromosome present in the three cell lines that produce enzyme 3 and absent in the cell line that does not produce enzyme 3.

CHALLENGE QUESTION

31. In calculating map distances, we did not concern ourselves with whether double crossovers were two-stranded, three-stranded, or four-stranded; yet, these different types of double crossovers produce different types of gametes. Can you explain why we do not need to determine how many strands take part in double crossovers in diploid organisms? (Hint: Draw out the types of gametes produced by the different types of double crossovers and see how they contribute to the determination of map distances.)
The three-stranded double crossovers all generate two recombinant chromosomes and two nonrecombinant chromosomes.

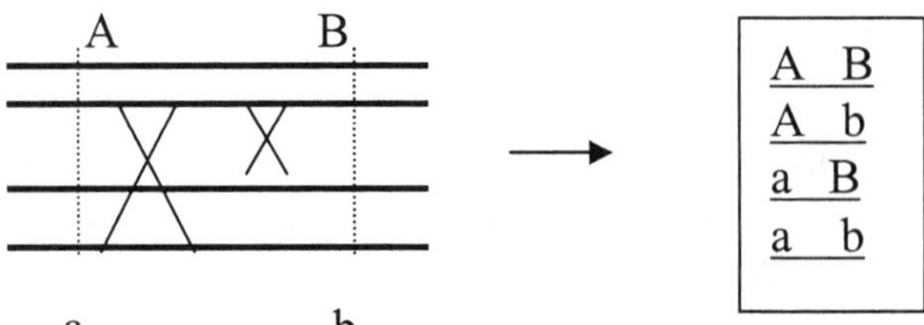

The four-stranded double crossovers always generate four recombinant chromosomes.

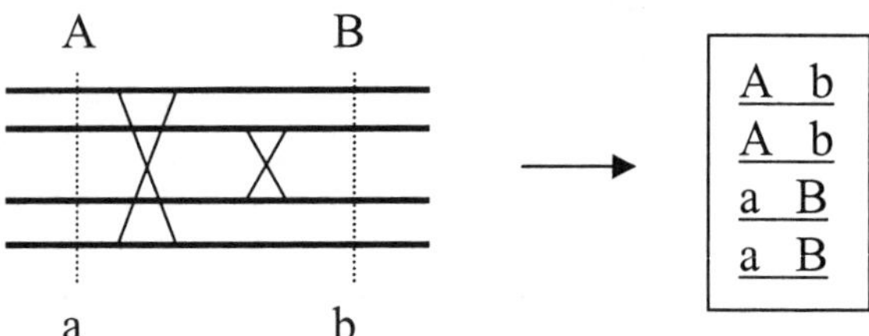

The two-stranded double crossovers always generate four nonrecombinant chromosomes.

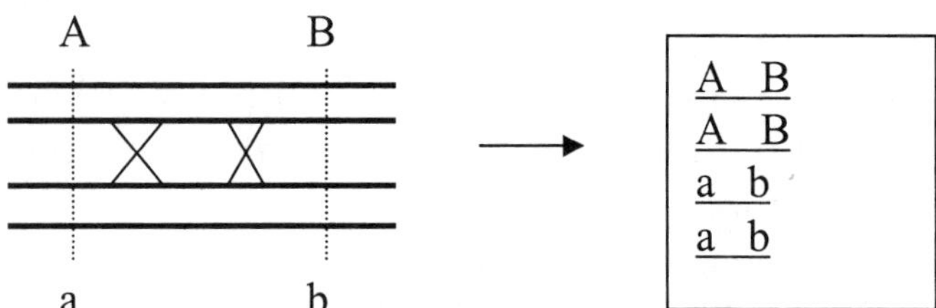

Because the four-stranded and two-stranded double crossovers are equally probable, the net result of all types of double crossovers averaged together is 50% recombinant and 50% nonrecombinant chromosomes.

Chapter Eight: Bacterial and Viral Genetic Systems

COMPREHENSION QUESTIONS

*1. List some of the characteristics that make bacteria and viruses ideal organisms for many types of genetic studies.
(1) Reproduction is rapid, asexual, and produces lots of progeny.
(2) Their genomes are small and haploid.
(3) They are easy to grow in the laboratory.
(4) Techniques are available for isolating and manipulating their genes.

2. Explain how auxotrophic bacteria are isolated.
Unlike prototrophic bacteria (wild-type) that can grow on minimal media, auxotrophic bacteria are mutant strains of bacteria that are unable to grow on minimal media. In other words, auxotrophs are not nutritionally self-sufficient. To isolate an auxotrophic bacterium from a culture of wild-type bacteria, first spread the bacterial culture out on a petri dish containing nutritionally complete growth medium allowing prototrophic and auxotrophic colonies to grow. Using the replica plating technique, transfer a few cells from each colony to replica plates. One of the replica plates should contain a selective medium that lacks a nutrient required by the auxotroph for growth, and the other replica plate should contain a nutritionally complete medium. The auxotrophic colonies should grow only on the nutritionally complete medium and not the selective medium. Prototrophic colonies should grow on both types of media.

3. Briefly explain the differences between F^+, F^-, Hfr, and F' cells.
An F^+ cell will contain the F factor as a circular plasmid separate from the chromosome. The Hfr cell has the F factor integrated into its chromosome. In F' strains, the F factor exists as a separate circular plasmid, but the plasmid carries bacterial genes that were originally part of the bacterial chromosome. The F^- strain does not contain the F Factor and can receive DNA from cells that contain the F Factor (F^+, Hfr, and F' cells).

*4. What types of matings are possible between F^+, F^-, Hfr, and F' cells? What outcomes do these matings produce? What is the role of F factor in conjugation?

Types of matings	*Outcomes*
$F^+ \times F^-$	*Two F^+ cells*
$Hfr \times F^-$	*One F^+ cell and one F^- cell*
$F' \times F^-$	*Two F' cells*

The F factor contains a number of genes involved in the conjugation process, including genes necessary for the synthesis of the sex pilus. The F factor also has an origin of replication that allows for the factor to be replicated during the conjugation process.

*5. Explain how interrupted conjugation, transformation, and transduction can be used to map bacterial genes. How are these methods similar and how are they different?

To map genes by conjugation, an interrupted mating procedure is used. During the conjugation process, an Hfr strain is mixed with an F^- strain. The two strains must have different genotypes and must remain in physical contact for the transfer to occur. At regular intervals, the conjugation process is interrupted. The chromosomal transfer from the Hfr strain always begins with a portion of the integrated F factor and proceeds in a linear fashion. To transfer the entire chromosome would require approximately 100 minutes. The time required for individual genes to be transferred is relative to their position on the chromosome and the direction of transfer initiated by the F factor. Gene distances are typically mapped in minutes. The genes that are transferred by conjugation to the recipient must be incorporated into the recipient's chromosome by recombination to be expressed.

In transformation, the relative frequency at which pairs of genes are transferred or cotransformed indicates the distance between the two genes. Closer gene pairs are cotransformed more frequently. As was the case with conjugation, the donor DNA must recombine into the recipient cell's chromosome. Physical contact of the donor and recipient cells is not needed. The recipient cell uptakes the DNA directly from the environment. So, the DNA from the donor strain has to be isolated and broken up before transformation can take place.

A viral vector is needed for the transfer of DNA by transduction. DNA from the donor cell is packaged into a viral protein coat. The viral particle containing the bacterial donor DNA then infects another bacterial cell or the recipient. The donor bacterial DNA is incorporated into the recipient cell's chromosome by recombination. Only genes that are close together on the bacterial chromosome can be cotransduced. Therefore, the rate of cotransduction, like the rate of cotransformation, gives an indication of the physical distances between genes on the chromosome.

6. What is horizontal gene transfer and how might it occur?

Horizontal gene transfer occurs when a bacterial cell acquires genes from another species. Genome analysis experiments have shown that bacterial species have even acquired DNA from eukaryotic organisms. Three mechanisms that could lead to horizontal gene transfer are transduction, transformation, and conjugation.

7. What types of genomes do viruses have?

Viral genomes can consist of either DNA or RNA molecules. The viral nucleic acids can be either double-stranded or single-stranded, depending on the type of virus.

8. Briefly describe the differences between the lytic cycle of virulent phages and the lysogenic cycle of temperate phages.

Virulent phages reproduce strictly by the lytic cycle and ultimately result in the death of the host bacterial cell. During the lytic cycle, a virus injects its genome into the host cell. The genome directs production and assembly of new viral particles. A viral enzyme is produced and breaks open the cell, releasing new viral particles into the environment.

Temperate phages can utilize either the lytic or lysogenic cycle. The infection cycle begins when a viral particle injects its genome into the host cell. In the lysogenic cycle, the viral genome integrates into the host chromosome as a prophage. The inactive prophage can remain part of the bacterial chromosome for an extended period and is replicated along with the bacterial chromosome prior to cell division. Certain environmental stimuli can trigger the prophage to exit the lysogenic cycle and enter the lytic cycle.

9. Briefly explain how genes in phages are mapped.
To map genes in phages, bacterial cells are doubly infected with phage particles that differ in two or more genes. During the production of new phage progeny, the phage DNAs can undergo recombination, thus resulting in the formation of recombinant plaques. The rate of recombination is used to determine the linear order and relative distances between genes. The farther apart two genes are on the chromosome, the more frequently they will recombine.

*10. How does specialized transduction differ from generalized transduction?
In generalized transduction, bacterial genes are transferred from one bacterial cell to another by a virus. In specialized transduction, only genes from a particular locus on the bacterial chromosome are transferred to another bacterium. The process of specialized transduction requires lysogenic phages that integrate into specific locations on the host cell's chromosome. When the phage DNA excises from the host chromosome and the excision process is imprecise, the phage DNA will carry a small part of the bacterial DNA. The hybrid DNA must be injected by the phage into another bacterial cell during another round of infection.

Transfer of DNA by generalized transduction requires that the host DNA be broken down into smaller pieces and that a piece of the host DNA is packaged into a phage coat instead of phage DNA. The defective phage cannot produce new phage particles upon a subsequent infection, but it can inject the bacterial DNA into another bacterium or recipient. Through a double crossover event, the donor DNA can become incorporated into the bacterial recipient's chromosome.

*11. Briefly explain the method used by Benzer to determine whether two different mutations occurred at the same locus.
To conduct the complementation test, Benzer infected cells of E. coli *K with large numbers of the two mutant phage types. For successful infection to occur on the* E. coli *K strains, each mutant phage needs to supply the gene product or protein missing in the other. Complementation will happen only if the mutations are at separate loci. If the two mutations are at the same locus, then complementation of gene products will not occur and no plaques will be produced on the* E. coli *K lawns.*

12. What is the difference between a positive-strand RNA virus and a negative-strand RNA virus?

A positive-strand RNA virus corresponds to messenger RNA. In other words, the genomic RNA codes directly for viral proteins. In negative-strand RNA viruses, a copy of the complementary RNA molecule is synthesized initially. From the complementary copy, translation of viral proteins can occur.

*13. Explain how a retrovirus, which has an RNA genome, is able to integrate its genetic material into that of a host having a DNA genome.
Retroviruses are able to integrate their genomes into the host cell's DNA genome through the action of the enzyme reverse transcriptase. Reverse transcriptase can synthesize complementary DNA from either a RNA or DNA template. The retrovirus enzyme synthesizes a double-stranded copy of DNA using the retroviral single-stranded RNA as the template. The newly synthesized DNA molecule can then integrate into the host chromosome to form a provirus.

14. Briefly describe the genetic structure of a typical retrovirus.
Retroviral genomes all have three genes in common: gag, pol, *and* env. *Proteins that make up the viral capsid are encoded by the* gag *gene. Reverse transcriptase and an enzyme called integrase are encoded by the* pol *gene. While reverse transcriptase synthesizes double-stranded viral DNA from an RNA template, integrase results in the insertion of the viral DNA into the host chromosome. Finally the* env *gene encodes for proteins found on the viral envelope.*

15. What is a prion? How can prions lack nucleic acids and be infectious?
Prions (or proteinaceous infectious particles) appear to be infectious proteins. The prion protein PrP^{sc} is similar to a normal cellular protein PrP^{c}. However, the abnormal and infectious PrP^{sc} can modify the structure of PrP^{c} converting it into the abnormal protein, which can then convert more PrP^{c} into PrP^{sc}. The normal cellular protein assumes a helical conformation while the PrP^{sc} folds into a flattened β-sheet. PrP^{sc} interacts with PrP^{c} resulting in the normal protein being refolded into the β-sheet conformation or abnormal form.

APPLICATION QUESTIONS AND PROBLEMS

*16. John Smith is a pig farmer. For the past five years, Smith has been adding vitamins and low doses of antibiotics to his pig food; he says that these supplements enhance the growth of the pigs. Within the past year, however, several of his pigs have died from infections of common bacteria, which failed to respond to large doses of antibiotics. Can you offer an explanation for the increased rate of mortality due to infection in Smith's pigs? What advice might you offer Smith to prevent this problem in the future?
Over the past five years, Farmer Smith, by using low doses of antibiotics, has been selecting for bacteria that are resistant to the antibiotics. The doses used killed sensitive bacteria, but not those bacteria that were moderately sensitive or slightly resistant. Over time, only resistant bacteria will be present in his pigs because any sensitive bacteria have been eliminated by the low doses of antibiotics.

In the future, Farmer Smith can continue to use the vitamins, but should use the antibiotics only when a sick pig requires them. In this manner, he will not be selecting for antibiotic resistant bacteria, and the chances of the antibiotic therapy successfully treating his sick pigs will be greater.

17. Rarely, conjugation of Hfr and F^- cells produces two Hfr cells. Explain how this occurs.
Hfr strains contain an F factor integrated into the bacterial chromosome. The F factor mediates transfer of the bacterial chromosome. During conjugation of an Hfr strain with an F^- strain, the transfer process begins within the F factor. So literally, part of the F factor is the first to arrive in F^- cell. However, the remaining part of the F factor is transferred last. Because nearly 100 minutes are required to completely transfer the donor chromosome, the two cells must remain in contact for the entire 100 minutes. So, if the donor and recipient cell are not disturbed and the transfer process is not interrupted, then the entire Hfr strain's chromosome, including the F factor, can be donated to the F^- cell. Following recombination between the donor and recipient chromosomes, this results in the production of a second Hfr strain.

18. A strain of Hfr cells that is sensitive to the antibiotic streptomycin (str^s) has the genotype gal^+ his^+ bio^+ pur^+ gly^+. These cells were mixed with an F^- strain that is resistant to streptomycin (str^r) and has genotype gal^- his^- bio^- pur^- gly^-. The cells were allowed to undergo conjugation. At regular intervals, a sample of cells was removed and conjugation was interrupted by placing the sample in a blender. The cells were then plated on medium that contains streptomycin. The cells that grew on this medium were then tested for the presence of genes transferred from the Hfr strain. Genes from the donor Hfr strain first appeared in the recipient F^- strain at the times listed here. On the basis of these data, give the order of the genes on the bacterial chromosome and indicate the minimum distances between them.

gly^+	3 minutes
his^+	14 minutes
bio^+	35 minutes
gal^+	36 minutes
pur^+	38 minutes

The closer genes are to the F factor, the more quickly they will be transferred. The transfer process will occur in a linear fashion. By interrupting the mating process using a blender, the transfer will stop and the F^- strain will have received only genes carried on the piece of the Hfr strain's chromosome that entered the F^- cell prior to the disruption.

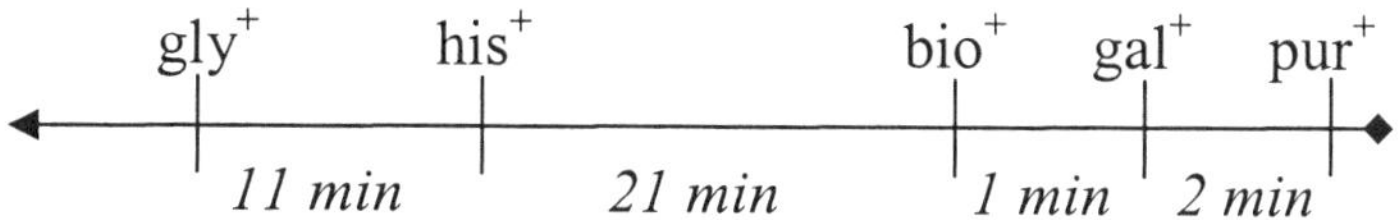

*19. A series of Hfr strains that have genotype m^+ n^+ o^+ p^+ q^+ r^+ are mixed with an F^- strain that has genotype m^- n^- o^- p^- q^- r^-. Conjugation is interrupted at regular intervals and the order of appearance of genes from the Hfr strain is determined in the recipient cells. The order of gene transfer for each Hfr strain is:

Hfr 5 m^+ q^+ p^+ n^+ r^+ o^+
Hfr 4 n^+ r^+ o^+ m^+ q^+ p^+
Hfr 1 o^+ m^+ q^+ p^+ n^+ r^+
Hfr 9 q^+ m^+ o^+ r^+ n^+ p^+

What is the order of genes on the circular bacterial chromosome? For each Hfr strain, give the location of the F factor in the chromosome and its polarity.

In each of the Hfr strains, the F factor has been inserted into a different location in the chromosome. The orientation of the F factor in the strains varies as well.

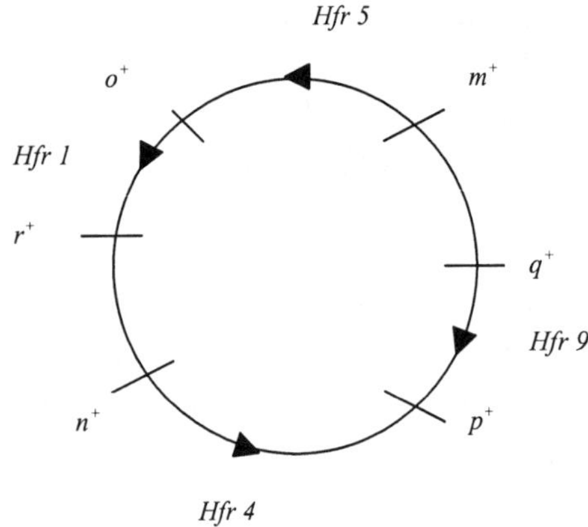

*20. Crosses of three different Hfr strains with separate samples of an F^- strain are carried out, and the following mapping data are provided from studies of interrupted conjugation:

		Appearance of genes in F^- cells				
Hfr1:	Genes	b^+	d^+	c^+	f^+	g^+
	Time	3	5	16	27	59
Hfr2:	Genes	e^+	f^+	c^+	d^+	b^+
	Time	6	24	35	46	48
Hfr3	Genes	d^+	c^+	f^+	e^+	g^+
	Time	4	15	26	44	58

Construct a genetic map for these genes, indicating their order on the bacterial chromosome and the distances between them.

The F factor for each Hfr strain has been inserted into a different location on the chromosome, and the orientation of the F factor varies in the different strains. Although most of the selective markers transferred from each Hfr strain to the F⁻ strain are the same, some of the markers for a given Hfr strain are not transferred due to the mating being disrupted prior to the transfer of that selective marker. The relative position of the genes to each other in minutes does not vary. So, for the different Hfr strains, the distance in minutes between each gene remains constant. The genes and their relative positions are shown below. Times are in minutes.

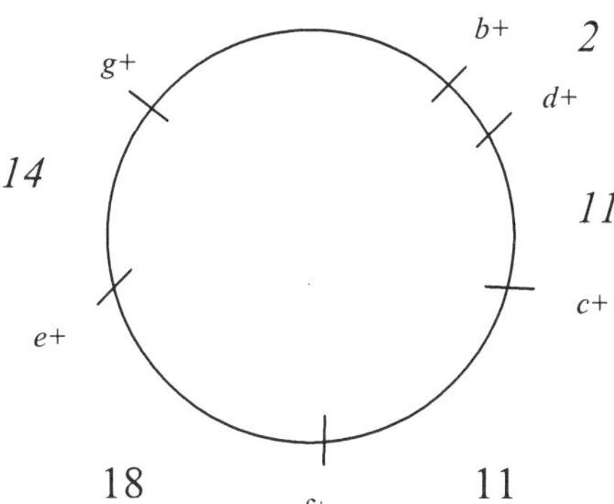

21. DNA from a strain of *Bacillus subtilis* with the genotype $trp^+\ tyr^+$ is used to transform a recipient strain with the genotype $trp^-\ tyr^-$. The following numbers of transformed cells were recovered:

Genotype	Number of transformed cells
$trp^+\ tyr^-$	154
$trp^-\ tyr^+$	312
$trp^+\ tyr^+$	354

What do these results suggest about the linkage of the *trp* and *tyr* genes?
During transformation, only genes that are closely linked or located near each other on the donor's chromosome will be transformed together. In other words, a higher cotransformation frequency indicates a shorter distance between the two genes on the donor's chromosome.

To calculate the cotransformation frequency of the trp$^+$ *and* tyr$^+$ *genes from* Bacillus subtilis, *divide the number of transformed cells with the genotype* trp$^+$ tyr$^+$ *by the total number of transformed cells (354/820). The frequency of cotransformation is 0.43 or 43%. The high level of cotransformation indicates that these two genes are closely linked.*

22. DNA from a strain of *Bacillus subtilis* with genotype $a^+\ b^+\ c^+\ d^+\ e^+$ is used to transform a strain with genotype $a^-\ b^-\ c^-\ d^-\ e^-$. Pairs of genes are checked for cotransformation and the following results are obtained:

Pair of genes	Cotransformation
a^+ and b^+	no
a^+ and c^+	no
a^+ and d^+	yes
a^+ and e^+	yes
b^+ and c^+	yes
b^+ and d^+	no
b^+ and e^+	yes
c^+ and d^+	no
c^+ and e^+	yes
d^+ and e^+	no

On the basis of these results, what is the order of the genes on the bacterial chromosome?

Only genes located near each other on the bacterial chromosome will be cotransformed together. However, by performing transformation experiments and screening for different pairs of cotransforming genes, a map of the gene order can be determined. Gene pairs that never result in cotransformation must be farther apart on the chromosome, while gene pairs that result in cotransformation are more closely linked. From the data, we see that gene a^+ *cotransforms with both* e^+ *and* d^+. *However, genes* d^+ *and* e^+ *do not exhibit cotransformation, indicating that* a^+ *and* e^+ *are more closely linked than* d^+ *and* e^+. *Gene* a^+ *does not exhibit cotransformation with either gene* b^+ *or* c^+, *yet gene* e^+ *does. This indicates that gene* e^+ *is more closely linked to genes* b^+ *and* c^+ *than is gene* a^+. *The orientation of genes* b^+ *and* c^+ *relative to* e^+ *cannot be determined from the data provided.*

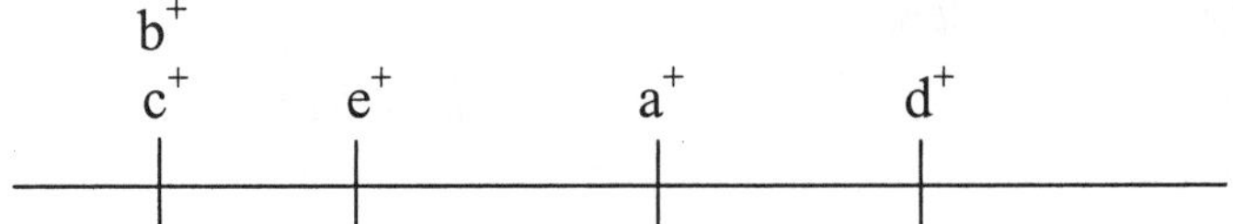

23. DNA from a bacterial strain that is *his⁺ leu⁺ lac⁺* is used to transform a strain that is *his⁻ leu⁻ lac⁻*. The following percentages of cells were transformed:

Donor strain	Recipient strain	Genotype of transformed cells	Percent
his⁺ leu⁺ lac⁺	*his⁻ leu⁻ lac⁻*	*his⁺ leu⁺ lac⁺*	0.02%
		his⁺ leu⁺ lac⁻	0.00%
		his⁺ leu⁻ lac⁺	2.00%
		his⁺ leu⁻ lac⁻	4.00%
		his⁻ leu⁺ lac⁺	0.10%
		his⁻ leu⁻ lac⁺	3.00%
		his⁻ leu⁺ lac⁻	1.50%

(a) What conclusions can you make about the order of these three genes on the chromosome?

The percentages of cotransformation between his⁺, leu⁺, *and* lac⁺ *loci must be examined. Genes that cotransform more frequently will be closer together on the donor chromosome. Cotransformation between* lac⁺ *and* his⁺ *occurs in 2.02 % of the transformed cells. By comparing this value with the cotransformation of* lac⁺ *and* leu⁺ *at 0.12% and the cotransformation of* his⁺ *and* leu⁺ *at .02%, we can see that* lac⁺ *and* his⁺ *cotransform more frequently. Therefore,* lac⁺ *and* his⁺ *must be more closely linked than the other gene pairs combinations. Because* lac⁺ *and* leu⁺ *cotransform more frequently than* his⁺ *and* leu⁺, leu⁺ *must be located closer to* lac⁺ *than it is to* his⁺.

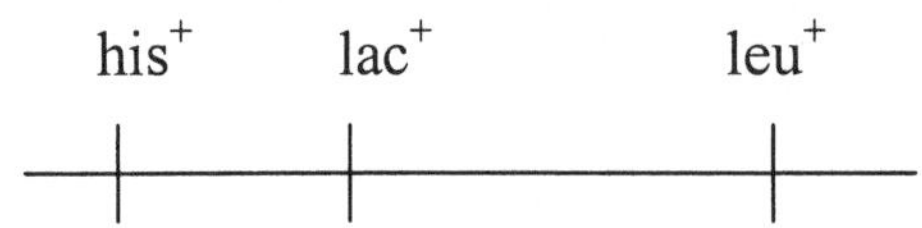

(b) Which two genes are closest?
From the cotransformation frequencies, we can predict that lac$^+$ *and* his$^+$ *are the closest two genes.*

24. Two mutations that affect plaque morphology in phages (a^- and b^-) have been isolated. Phages carrying both mutations (a^- b^-) are mixed with wild-type phages (a^+ b^+) and added to a culture of bacterial cells. Subsequent to infection and lysis, samples of the phage lysate are collected and cultured on bacterial cells. The following numbers of plaques are observed:

Plaque phenotype	Number
a^+ b^+	2043
a^+ b^-	320
a^- b^+	357
a^- b^-	2134

What is the frequency of recombination between the *a* and *b* genes?
First, we must identify the progeny phage whose plaque phenotype is different from either of the infecting phage. The original infecting phages were wild-type (a$^+$ b$^+$) *and doubly mutant* (a b). *Any phage that give rise to the* a$^+$ b$^-$ *plaque phenotype or the* a$^-$ b$^+$ *plaque phenotype were produced by recombination between the two types of infecting phage particles.*

Plaque phenotype	*Number*
a$^+$ b$^+$	*2043*
a$^+$ b$^-$	*320 (recombinant)*
a$^-$ b$^+$	*357 (recombinant)*
a$^-$ b$^-$	*2134*
Total plaques	*4854*

The frequency of recombination is calculated by dividing the total number of recombinant plaques by the total number of plaques (677/4854), which gives a frequency of 0.14, or 14%.

*25. A geneticist isolates two mutations in bacteriophage. One mutation causes the clear plaques (*c*) and the other produces minute plaques (*m*). Previous mapping experiments have established that the genes responsible for these two mutations are 8 map units apart. The geneticist mixes phages with genotype c^+ m^+ and genotype c^- m^- and uses the mixture to infect bacterial cells. She collects the progeny phages and cultures a sample of them on plated bacteria. A total of 1000 plaques are observed. What numbers of the different types of plaques (c^+ m^+, c^- m^-, c^+ m^-, c^- m^+) should she expect to see?
We know that the two genes are 8 map units apart. These 8 map units correspond to a percent recombination between the two genes of 8%. When the geneticist mixes the two phages (m$^+$ c$^+$ × m$^-$ c$^-$) *creating a double infection of the bacterial cell, she should expect the two types of recombinant plaque phenotypes,* m$^+$ c$^-$ *and* m$^-$ c$^+$, *to*

comprise 8% of the progeny phage. The remaining 92% will be a combination of the wild-type phage and the doubly mutant phage.

Plaque phenotype	*Expected number*
$c^+ m^+$	*460*
$c^- m^-$	*460*
$c^+ m^-$	*40 (recombinant)*
$c^+ m^-$	*40 (recombinant)*
Total plaques	*1000*

26. The geneticist carries out the same experiment described in Problem 23, but this time she mixes phages with genotypes $c^+ m^-$ and $c^- m^+$. What results are expected with *this* cross?
We know that the two genes are 8 map units apart, corresponding to 8 percent recombination. The phage used by the geneticist in this experiment will produce recombinants with different phenotypes from her previous experiment, but the number of recombinants will remain the same.

Plaque phenotype	*Expected number*
$c^+ m^+$	*40 (Recombinant)*
$c^- m^-$	*40 (Recombinant)*
$c^- m^+$	*460*
$c^+ m^-$	*460*
Total plaques	*1000*

*27. A geneticist isolates two r mutants (r_{13} and r_2) that cause rapid lysis. He carries out the following crosses and counts the number of plaques listed here:

Genotype of parental phage	Progeny	Number of plaques
$h^+ r_{13}^- \times h^- r_{13}^+$	$h^+ r_{13}^+$	1
	$h^- r_{13}^+$	104
	$h^+ r_{13}^-$	110
	$h^- r_{13}^-$	2
	total	216
$h^+ r_2^- \times h^- r_2^+$	$h^+ r_2^+$	6
	$h^- r_2^+$	86
	$h^+ r_2^-$	81
	$h^- r_2^-$	7
	total	180

(a) Calculate the recombination frequencies between r_2 and h and between r_{13} and h.
To determine the recombination frequencies, the recombinant offspring must be identified. The recombination frequency is calculated by dividing the total number of recombinant plaques by the total number of plaques.

Genotype of parents	*Progeny*	*Number of plaques*
$h^+r_{13}^- \times h^-r_{13}^+$	$h^+r_{13}^+$ *(recombinant)*	*1*
	$h^-r_{13}^-$	*104*
	$h^+r_{13}^-$	*110*
	$h^-r_{13}^-$ *(recombinant)*	*2*
	total	*216*
$h^+r_2^- \times h^-r_2^+$	$h^+r_2^+$ *(recombinant)*	*6*
	$h^-r_2^+$	*86*
	$h^+r_2^-$	*81*
	$h^-r_2^-$ *(recombinant)*	*7*
	total	*180*

The recombination frequency between r_2 *and* h *is 13/180 = .072 or 7.2%. The RF between* r_{13} *and* h *is 3/216 = 0.014 or 1.4%.*

(b) Draw all possible linkage maps for these three genes.

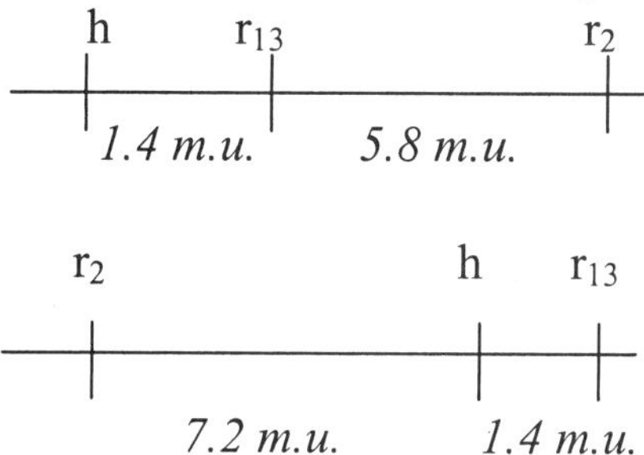

*28. *E. coli* cells are simultaneously infected with two strains of phage λ. One strain has a mutant host range, is temperature sensitive, and produces clear plaques (genotype = *h st c*); another strain carries the wild-type alleles (genotype = $h^+\ st^+\ c^+$). Progeny phage are collected from the lysed cells and are plated on bacteria. These genotypes of the progeny phage are:

Progeny phage genotype	Number of plaques
$h^+\ c^+\ t^+$	321
$h\ c\ t$	338
$h^+\ c\ t$	26
$h\ c^+\ t^+$	30
$h^+\ c\ t^+$	106
$h\ c^+\ t$	110
$h^+\ c^+\ t$	5
$h\ c\ t^+$	6

(a) Determine the order of the three genes on the phage chromosome.

First we need to identify the progeny phage that has genotypes similar to the parents and the progeny phage with genotypes that differ from the parents. The parental genotypes are $h^+c^+\ t^+$ *and* h c t. *Any genotype that differs from those two genotypes had to be generated by recombination. By comparing the genotype of the double recombinant phage progeny with the nonrecombinants, we can predict the gene order.*

Phage genotype	*Number of progeny*	*Type*
$h^+c^+t^+$	*321*	*Parental*
h c t	*338*	*Parental*
h^+c t	*26*	*Recombinant*
h c^+t^+	*30*	*Recombinant*
h^+ c t^+	*106*	*Recombinant*
h c^+t	*110*	*Recombinant*
h^+ c^+ t	*5*	*Double recombinant*
h c t^+	*6*	*Double recombinant*
Total	*942*	

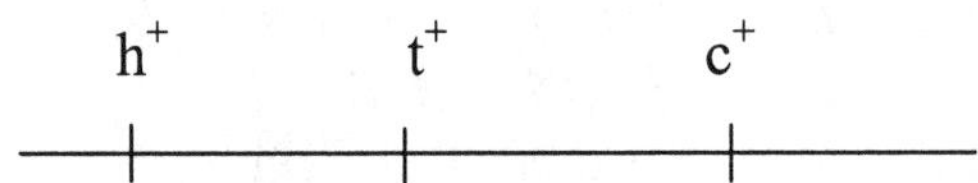

(b) Determine the map distances between the genes.

The map distances can be calculated by determining the percent recombination between each gene pair. The double recombinant progeny, h^+ c^+ *and* h c, *appear to be parentals. However, this genotype was generated by a double crossover event. To consider the double crossover events, multiply the number of double recombinant progeny by two.*

h^+ t^+: *[(26+30+5+6)/942] × 100 % = 7.1% or 7.1 map units*
h^+ c^+: *[(26+30+106+110+10+12)/942] × 100% = 31.2 % or 31.2 m. u.*
t^+ c^+: *[(106+110+5+6)/942] × 100 % = 24.1 % or 24.1 map units*

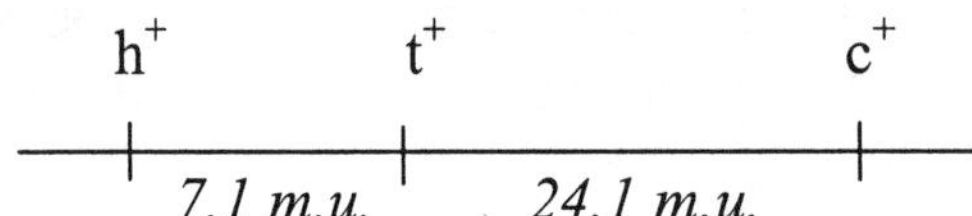

(c) Determine the coefficient of coincidence and the interference.

$$COC = \frac{\text{(observed number of double recombinants)}}{\text{(expected number of double recombinants)}}$$

COC = (6 + 5)/(.071 × .241 × 942) = 0.68
Interference = 1 – COC = 1 – 0.68 = 0.32

29. A donor strain of bacteria with genes a^+ b^+ c^+ is infected with phages to map the donor chromosome with generalized transduction. The phage lysate from the bacterial cells is collected and used to infect a second strain of bacteria that are a^- b^- c^-. Bacteria with the a^+ gene are selected and the percentage of cells with cotransduced b^+ and c^+ genes are recorded.

Donor	Recipient	Selected gene	Cells with cotransduced gene (%)
$a^+\ b^+\ c^+$	$a^-\ b^-\ c^-$	a^+	25 b^+
		a^+	3 c^+

Is the *b* or *c* gene closer to *a*? Explain your reasoning.
The gene b$^+$ *cotransduces more frequently with* a$^+$*, the selective marker, than does* c$^+$*. Because genes that are closer together on the donor bacterial chromosome cotransduce more frequently, we can see that* b$^+$ *is closer to* a$^+$.

30. A donor strain of bacteria with genotype *leu*$^+$ *gal*$^-$ *pro*$^+$ is infected with phages. The phage lysate from the bacterial cells is collected and used to infect a second strain of bacteria that are *leu*$^-$ *gal*$^+$ *pro*$^-$. The second strain is selected for *leu*$^+$, and the following cotransduction data obtained:

Donor	Recipient	Selected gene	Cells with cotransduced gene (%)
$leu^+\ gal^-\ pro^+$	$leu^-\ gal^+\ pro^-$	leu^+	47 pro^+
		leu^+	26 gal^-

Which genes are closest, *leu* and *gal* or *leu* and *pro*?
Because leu *and* pro *cotransduce together more frequently, they must be the closest.*

31. A geneticist isolates two new mutations from the *rII* region of bacteriophage T4, called rII_X and rII_Y. *E. coli* B cells are simultaneously infected with phages carrying the rII_X mutation *and* with phages carrying the rII_Y mutation. After the cells have lysed, samples of the phage lysate are collected. One sample is grown on *E. coli* K cells and a second sample on *E. coli* B cells. There are 8322 plaques on the *E coli* B and 3 plaques on *E. coli* K. What is the recombination frequency between these two mutations?

The recombination frequency $= \frac{(2 \times \text{plaques on K})}{(\text{Total number of plaques})}$

Recombination frequency $= (2 \times 3)/8322 = 7.2 \times 10^{-4}$ *or 0.072 %*

32. A geneticist is working with a new bacteriophage called phage Y3 that infects *E. coli*. He has isolated eight mutant phages that fail to produce plaques when grown on *E. coli* strain K. To determine whether these mutations occur at the same functional gene, he simultaneously infects *E. coli* K cells with paired combinations of the mutants and looks to see whether plaques formed. He obtains the results at the top of the following page. (A plus sign means that plaques were formed on *E. coli* K; a minus sign means no plaques were formed on *E. coli* K).

Mutant	1	2	3	4	5	6	7	8
1								
2	+							
3	+	+						
4	+	–	+					
5	–	+	+	+				
6	–	+	+	+	–			
7	+	–	+	–	+	+		
8	–	+	+	+	–	–	+	

(a) To how many functional genes (cistrons) do these mutations belong?

The geneticist is essentially conducting the "trans" portion of the cis/trans test. If complementation occurs between the different phage mutants that are infecting the E. coli *K cells, then plaques will form on the lawn of* E. coli *K. Complementation can occur only when the mutations of the different phages are located on different cistrons or functional genes. Phage mutants that do not complement each other have mutations that lie on the same cistrons.*

From the formation of plaques on E. coli *K, we can see three groups of phages that failed to complement with other phages within their group but did complement the phages in the other groups. Because there are three groups, we can infer the presence of three cistrons or functional genes.*

(b) Which mutations belong to the same functional gene?

We will identify the groups as group 1, group 2 and group 3.

Group 1: Mutants 1, 5, 6, and 8
Group 2: Mutants 2, 4, and 7
Group 3: Mutant 3

CHALLENGE QUESTIONS

33. As a summer project, a microbiology student independently isolates two mutations in *E. coli* that are auxotrophic for glycine (*gly*⁻). The student wants to know whether these two mutants occur at the same cistron. Outline a procedure that the student could use to determine whether these two *gly*⁻ mutations occur within the same cistron.

To determine if the two gly⁻ *are with the same cistrons, a strain of bacteria will have to be constructed that contains both genes, but on different DNA molecules within the strain. Only by both mutations being present in the same cell can complementation of the two mutations be tested. The student will need to create a merodiploid or partial diploid strain. A method for doing this is to create an F' that contains one of the* gly⁻ *markers. Because* gly⁻ *results in an auxotrophic mutant, it would be difficult to use as an initial screen for an F' that contains it. So, the student will need to use Hfr × F⁻ matings to map the location of* gly⁻ *marker and identify other protrophic markers that are nearby. By screening for F' strains that have the nearby protrophic markers, an F′ strain that contains the* gly$^-$ *marker should be identified.*

Next the F′ strain having the gly$^-$ *marker should be mated to a F$^-$ that contains the other* gly$^-$ *mutation. The identified exconjugants should contain both* gly$^-$ *mutations on separate DNA molecules within the cell. If the exconjugant can grow on minimal media not supplemented with glycine, then complementation has occurred and the two* gly$^-$ *mutations are located on different cistrons.*

34. A group of genetics students mixes two auxotrophic strains of bacteria: one is *leu*$^+$ *trp*$^+$ *his*$^-$ *met*$^-$ and the other is *leu*$^-$ *trp- his*$^+$ *met*$^+$. After mixing the two strains, they plate the bacteria on minimal medium and observe a few prototrophic colonies (*leu*$^+$ *trp*$^+$ *his*$^+$ *met*$^+$). They assume that some gene transfer has occurred between the two strains. How can they determine whether the transfer of genes is due to conjugation, transduction, or transformation?
Conjugation requires the direct contact of the donor bacterial strain and the recipient. If the transfer does not occur when the bacteria are kept physically separate, then conjugation is not the likely pathway. Another test would be to conduct interrupted mating experiments (assuming that one of the bacterial strains is an Hfr *strain) to see if the transfer of the different markers is time dependent, which is also indicative of conjugation.*

If the transfer occurs by transformation, then extraction of DNA from either strain and exposure of the other strain to the extracted DNA should result in the transfer of the DNA molecules. By selecting one of the mutations as a selective marker and measuring cotransformation frequencies between the selective marker and the other genes individually, the frequency of the transfer will hint toward the mechanism of transfer.

Finally, if the transfer is by transduction, then by exposing the one cell type to extracted DNA from the other cell type, transfer of the genes would not be expected. Potential cotransduction frequencies could be measured similarly to the cotransformation frequency. Also, the presence of plaques might be evident.

Chapter Nine: Chromosome Variation

COMPREHENSION QUESTIONS

*1. List the different types of chromosome mutations and define each.
Chromosome rearrangements:

Deletion: loss of a portion of a chromosome.

Duplication: addition of an extra copy of a portion of a chromosome.

Inversion: a portion of the chromosome is reversed in orientation.

Translocation: a portion of one chromosome becomes incorporated into a different (nonhomologous) chromosome.

Aneupoloidy: loss or gain of one or more chromosomes so that the chromosome number deviates from 2n or the normal euploid complement.

Polyploidy: Gain of entire sets of chromosomes so the chromosome number changes from 2n to 3n (triploid), 4n (tetraploid), and so on.

*2. Why do extra copies of genes sometimes cause drastic phenotypic effects?
The expression of some genes is balanced with the expression of other genes; the ratios of their gene products, usually proteins, must be maintained within a narrow range for proper cell function. Extra copies of one of these genes cause that gene to be expressed at proportionately higher levels, thereby upsetting the balance of gene products.

3. Draw a pair of chromosomes as they would appear during synapsis in prophase I of meiosis in an individual heterozygous for a chromosome duplication.
In the figure below, adapted from Figure 9.6b, the vertical dashed lines denote the locations of the genes labeled A, B, C,...G. One chromosome has duplicated a segment containing genes C, D, and E.

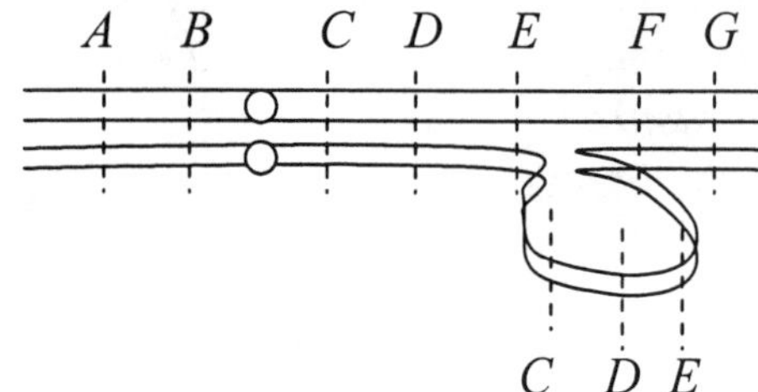

4. How does a deletion cause pseudodominance?
An individual heterozygous for a deletion has only one copy of the genes that are missing in the deleted portion of the chromosome. Any alleles, even recessive, of these genes are therefore expressed phenotypically.

*5. What is the difference between a paracentric and a pericentric inversion?
A paracentric inversion does not include the centromere; a pericentric inversion includes the centromere.

6. How do inversions cause phenotypic effects?
Although inversions do not result in loss or duplication of chromosomal material, inversions can have phenotypic consequences if the inversion disrupts a gene at one of its breakpoints or if a gene near a breakpoint is altered in its expression because of a change in its chromosomal environment, such as relocation to a heterochromatic region. Such effects on gene expression are called position effects.

*7. Draw a pair of chromosomes as they would appear during synapsis in prophase I of meiosis in an individual heterozygous for a paracentric inversion.
In the following figure, adapted from Figure 9.12, the inverted sequence with genes E, F, and G is in the looped region of the chromosomes.

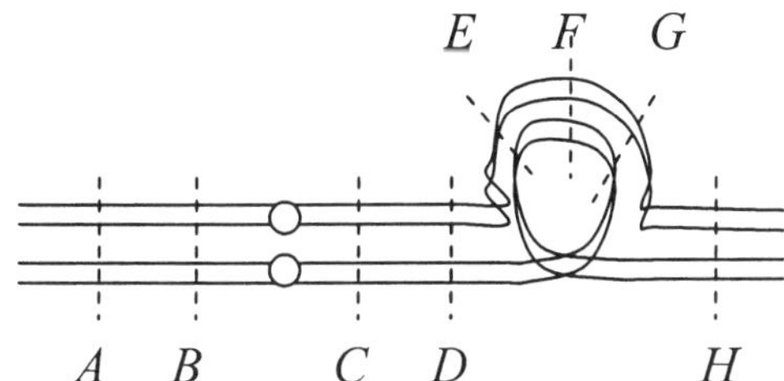

8. Explain why recombination is suppressed in individuals heterozygous for paracentric and pericentric inversions.
A crossover within a paracentric inversion produces a dicentric and an acentric recombinant chromatid. The acentric fragment is lost, and the dicentric fragment breaks, resulting in chromatids with large deletions that lead to nonviable gametes or embryonic lethality. A crossover within a pericentric inversion produces recombinant chromatids that have duplications or deletions. Again, gametes with these recombinant chromatids do not lead to viable progeny.

*9. How do translocations produce phenotypic effects?
Like inversions, translocations can produce phenotypic effects if the translocation breakpoint disrupts a gene or if a gene near the breakpoint is altered in its expression because of relocation to a different chromosomal environment (a position effect).

10. Sketch the chromosome pairing and the different segregation patterns that can arise in an individual heterozygous for a reciprocal translocation.

Chromosome pairing

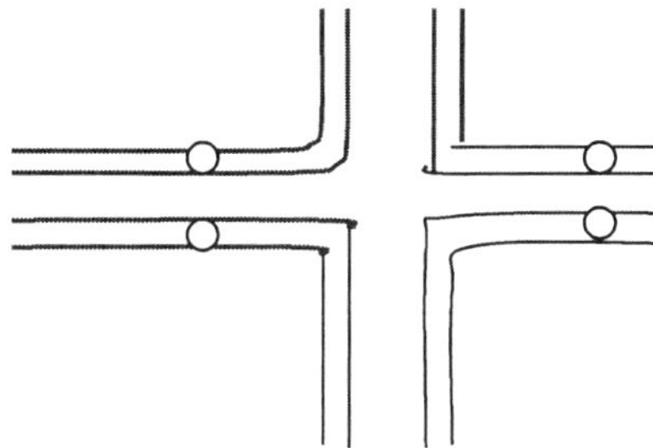

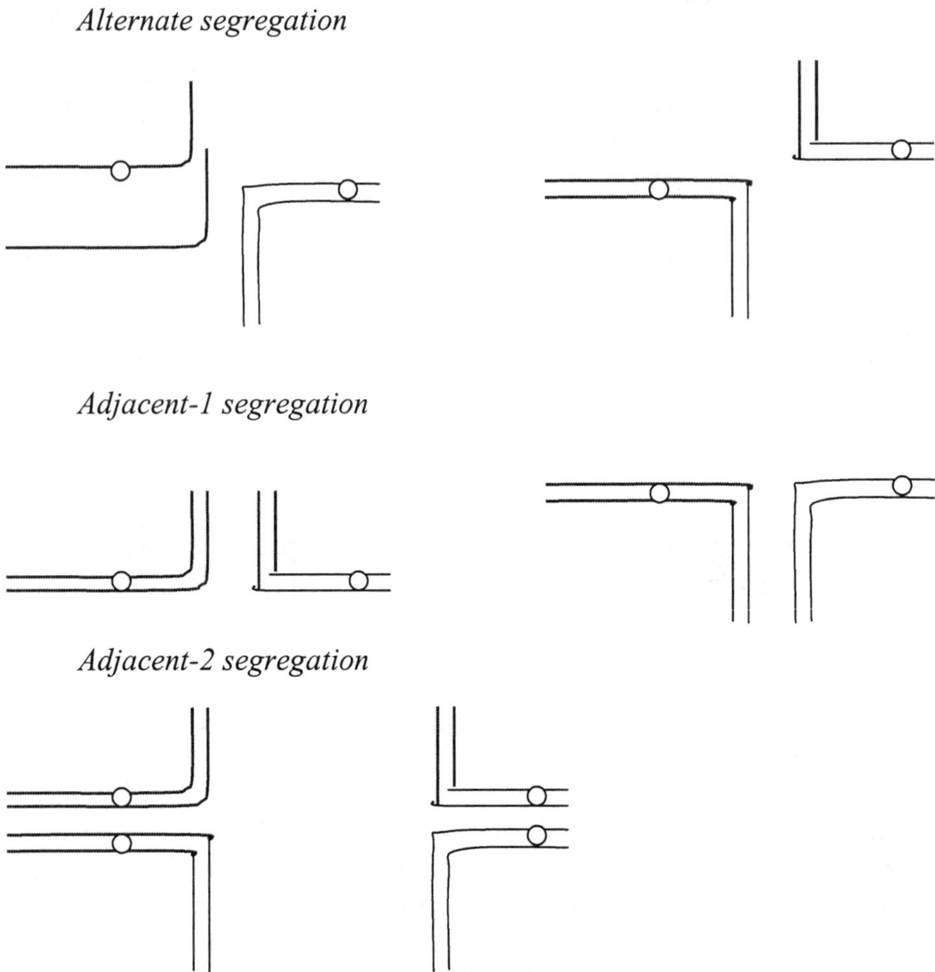

11. What is a Robertsonian translocation?
The long arms of two acrocentric chromosomes are joined to a common centromere through translocation, resulting in a large metacentric chromosome and a very small chromosome with two very short arms. The very small chromosome may be lost.

12. List four major types of aneuploidy.
Nullisomy: having no copies of a chromosome.
Monosomy: having only one copy of a chromosome.
Trisomy: having three copies of a chromosome.
Tetrasomy: having four copies of a chromosome.

*13. Why are sex-chromosome aneuploids more common in humans than autosomal aneuploids?
Individuals with more than one X chromosome maintain only one active X chromosome and inactivate all others in the form of Barr bodies. Thus, the extra copies of the X chromosome do not cause an imbalance in the dosage of functional

genes, and people with 1, 2, 3, or 4 X chromosomes all have similar levels of gene expression for most X-linked genes.

The Y chromosome is small and contains relatively few genes, most of which appear to be focused on male sex determination and male fertility. There are no genes on the Y chromosome that are essential for human development or viability (after all, half of all people appear to get along perfectly well without a Y chromosome). Thus, having extra copies of the Y chromosome does not affect human embryonic development.

*14. What is the difference between primary Down syndrome and familial Down syndrome? How does each arise?
Primary Down syndrome is caused by spontaneous, random nondisjunction of chromosome 21, leading to trisomy 21. Familial Down syndrome most frequently arises as a result of a Robertsonian translocation of chromosome 21 with another chromosome, usually chromosome 14. Translocation carriers do not have Down syndrome, but their children have an increased incidence of Down syndrome. If the translocated chromosome segregates with the normal chromosome 21, the gamete will have two copies of chromosome 21 and result in a child with familial Down syndrome.

*15. What is uniparental disomy and how does it arise?
Uniparental disomy refers to the inheritance of both copies of a chromosome from the same parent. This may arise originally from a trisomy condition in which the early embryo loses one of the three chromosomes, and the two remaining copies are from the same parent.

16. What is mosaicism and how does it arise?
Mosaicism is a condition in which an individual has patches of cells that have chromosomal abnormalities and patches of cells with normal chromosome content. Mosaicism may arise from mitotic nondisjunction during early embryonic divisions. Mosaicism can also arise from X-inactivation in a heterozygous female.

*17. What is the difference between autopolyploidy and allopolyploidy? How does each arise?
*In autopolyploidy, all sets of chromosomes are from the same species. Autopolyploids typically arise from mitotic nondisjunction of all the chromosomes in an early 2*n *embryo, resulting in an autotetraploid, or from meiotic nondisjunction that results in a 2*n *gamete fusing with a 1*n *gamete to form an autotriploid. In allopolyploidy, the chromosomes of two different species are contained in one individual through the hybridization of two related species followed by mitotic nondisjunction. Fusion of their gametes results in a functionally diploid hybrid with a haploid set of chromosomes from each parent. If an early embryonic cell then undergoes mitotic nondisjunction and doubles each chromosome, then a fertile 4*n *allotetraploid individual having two copies of each chromosome from each species may result.*

18. Explain why autopolyploids are usually sterile, while allopolyploids are often fertile.
Autopolyploids arise from duplication of their own chromosomes. During meiosis, the presence of more than two homologous chromosomes results in faulty alignment of homologues in prophase I, and subsequent faulty segregation of the homologues in anaphase I. The resulting gametes have an uneven distribution of chromosomes and are genetically unbalanced. These gametes usually produce lethal chromosome imbalances in the zygote. Allopolyploids, however, have chromosomes from different species. As long as they have a diploid set of chromosomes from each species, as in an allotetraploid or even an allohexaploid, the homologous chromosome pairs from each species can align and segregate properly during meiosis. Their gametes will be balanced and will produce viable zygotes when fused with other gametes from the same type of allopolyploid individual.

APPLICATION QUESTIONS AND PROBLEMS

*19. Which types of chromosome mutations:
(a) increase the amount of genetic material on a particular chromosome?
Duplications
(b) increase the amount of genetic material for all chromosomes?
Polyploidy
(c) decrease the amount of genetic material on a particular chromosome?
Deletions
(d) change the position of DNA sequences on a single chromosome without changing the amount of genetic material?
Inversions
(e) move DNA from one chromosome to a nonhomologous chromosome?
Translocations

*20. A chromosome has the following segments, where • represents the centromere.
A B • C D E F G

What types of chromosome mutations are required to change this chromosome into each of the following chromosomes? (In some cases, more than one chromosome mutation may be required.)
(a) A B A B • C D E F G: *Tandem duplication of AB.*
(b) A B • C D E A B F G: *Displaced duplication of AB.*
(c) A B • C F E D G: *Paracentric inversion of DEF.*
(d) A • C D E F G: *Deletion of B.*
(e) A B • C D E: *Deletion of FG.*
(f) A B • E D C F G: *Paracentric inversion of CDE.*
(g) C • B A D E F G: *Pericentric inversion of ABC.*
(h) A B • C F E D F E D G: *Duplication and inversion of DEF.*
(i) A B • C D E F C D F E G: *Duplication of CDEF, inversion of EF.*

*21. A chromosome initially has the following segments:
A B • C D E F G

Draw and label the chromosome that would result from each of the following mutations:
(a) Tandem duplication of DEF: *A B • C D E F D E F G.*
(b) Displaced duplication of DEF: *A B • C D E F G D E F.*
(c) Deletion of FG: *A B • C D E.*
(d) Deletion of CD: *A B • E F G.*
(e) Paracentric inversion that includes DEFG: *A B • C G F E D.*
(f) Pericentric inversion of BCDE: *A E D C • B F G.*

22. The following diagrams represent two nonhomologous chromosomes:
A B • C D E F G
R S • T U V W X

What type of chromosome mutation would produce the following chromosomes?
(a) A B • C D
R S • T U V W X E F G
Nonreciprocal translocation of E F G
(b) A U V B • C D E F G
R S • T W X
Nonreciprocal translocation of U V
(c) A B • T U V F G
R S • C D E W X
Reciprocal translocation of C D E and T U V
(d) A B • C W G
R S • T U V D E F X
Reciprocal translocation of D E F and W

*23. A species has $2n = 16$ chromosomes. How many chromosomes will be found per cell in each of the following mutants in this species?
(a) Monosomic: *15*
(b) Autotriploid: *24*
(c) Autotetraploid: *32*
(d) Trisomic: *17*
(e) Double monosomic: *14*
(f) Nullisomic: *14*
(g) Autopentaploid: *40*
(h) Tetrasomic: *18*

*24. The *Notch* mutation is a deletion on the X chromosome of *Drosophila melanogaster.* Female flies heterozygous for *Notch* have an indentation on the margin of their wings; *Notch* is lethal in the homozygous and hemizygous conditions. The *Notch* deletion covers the region of the X chromosome that contains the locus for white eyes, an X-linked recessive trait. Give the phenotypes and proportions of progeny produced in the following crosses:

(a) A red-eyed, Notch female is mated with a white-eyed male.

$X^NX^+ \times X^wY \rightarrow$ ¼ X^NX^w *Notch female with white eyes*
¼ X^NY *lethal*
¼ X^+X^w *female with red eyes (wild type)*
¼ X^+Y *wild-type male*

Overall, ⅓ of live progeny will be Notch females with white eyes, ⅓ will be wild-type females, and ⅓ will be wild-type males.

(b) A white-eyed, Notch female is mated with a red-eyed male.

$X^NX^w \times X^+Y \rightarrow$ ¼ X^NX^+ *Notch females*
¼ X^NY *lethal*
¼ X^+X^w *wild-type females*
¼ X^wY *white-eyed males*

The surviving progeny will therefore be ⅓ Notch red-eyed females, ⅓ wild-type females, and ⅓ white-eyed males.

(c) A white-eyed, Notch female is mated with a white-eyed male.

$X^NX^w \times X^wY \rightarrow$ ¼ X^NX^w *Notch, white-eyed females*
¼ X^NY *lethal*
¼ X^wX^w *white-eyed females*
¼ X^wY *white-eyed males*

Viable progeny are ⅓ Notch white-eyed females, ⅓ white-eyed females, and ⅓ white-eyed males.

25. A geneticist examines plant cells that are undergoing meiosis and notices that dicentric bridges occur in some of the cells. What does this observation tell you about the chromosomes of the plant from which the cells were taken?
Dicentric bridges occur as a result of crossing over within a paracentric inversion, in individuals heterozygous for the inversion. Therefore, this plant must be heterozygous for one or more paracentric inversions.

26. The green nose fly normally has six chromosomes, two metacentric and four acrocentric. A geneticist examines the chromosomes of an odd-looking greennose fly and discovers that it has only five chromosomes; three of them are metacentric and two are acrocentric. Explain how this change in chromosome number might have occurred.
A Robertsonian translocation between two of the acrocentric chromosomes would result in a new metacentric chromosome and a very small chromosome that may have been lost.

27. Species I is diploid ($2n = 8$) with chromosomes AABBCCDD; related species II is diploid ($2n = 8$) with chromosomes MMNNOOPP. Individuals with the following sets of chromosomes represent what types of chromosome mutations?
(a) AAABBCCDD: *Trisomy A*

(b) MMNNOOOOPP: *Tetrasomy O*
(c) AABBCDD: *Monosomy C*
(d) AAABBBCCCDDD: *Triploidy*
(e) AAABBCCDDD: *Ditrisomy A and D*
(f) AABBDD: *Nullisomy C*
(g) AABBCCDDMMNNOOPP: *Allotetraploidy*
(h) AABBCCDDMNOP: *Allotriploidy*

28. Species I has 2n = 8 chromosomes and species II has 2n = 14 chromosomes. What would be the expected chromosome numbers in individual organisms with the following chromosome mutations? Give all possible answers.
a. Allotriploid of species I and II—*Such allotriploids could have 1n from species I and 2n from species II for 3n = 18; alternatively they could have 2n from species I and 1n from species II for 3n = 15.*
b. Autotetraploid of species II—*4n = 28*
c. Trisomic of species I—*2n + 1 = 9*
d. Monosomic of species I—*2n – 1 = 13*
e. Tetrasomic of species I—*2n + 2 = 10*
f. Allotetraploid of species and II—*2n + 2n = 22; 1n + 3n = 25; 3n + 1n = 19*

*29. A wild-type chromosome has the following segments:
A B C • D E F G H I.
An individual is heterozygous for the following chromosome mutations. For each mutation, sketch how the wild-type and mutated chromosomes would pair in prophase I of meiosis, showing all chromosome strands.

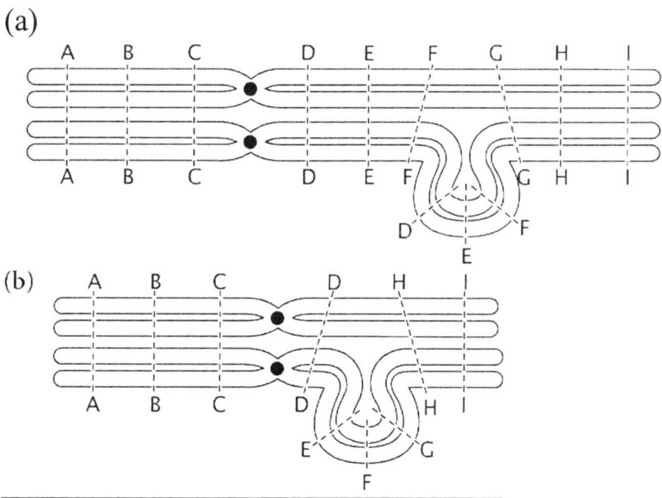

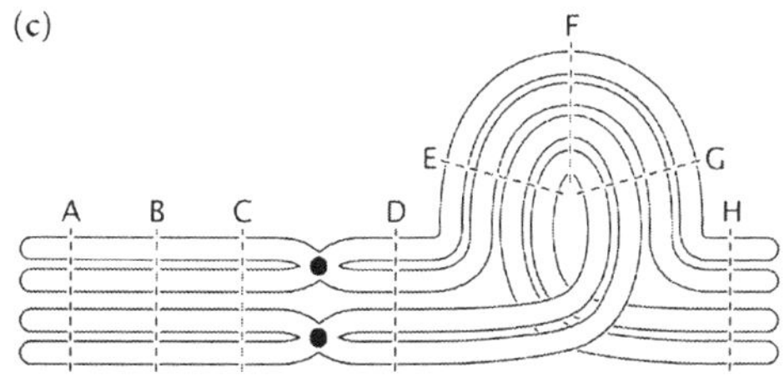

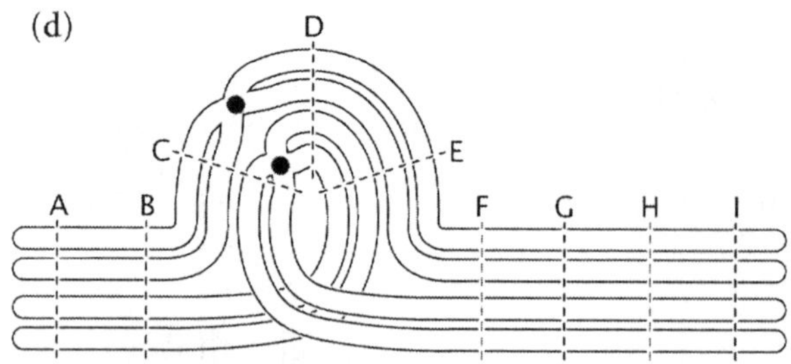

30. An individual that is heterozygous for a pericentric inversion has the following two chromosomes:

A B C D • E F G H I
A B C F E• D G H I

(a) Sketch the pairing of these two chromosomes in prophase I of meiosis, showing all four strands.

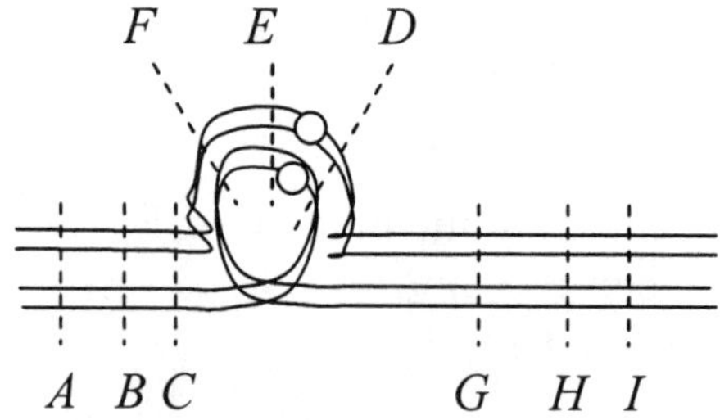

(b) Draw the chromatids that would result from a single crossover between the E and F segments.

A B C D •E F G H I
A B C D •E F C B A
I H G F E•D G H I
A B C F E•D G H I

(c) What will happen when the chromosomes separate in anaphase I of meiosis?

Each cell will have a pair of sister chromatids that consist of one functional, intact chromatid (containing one copy of all the genes, whether inverted or normal) joined to a nonfunctional chromatid that contains a large duplication and deletion.

31. Draw the chromatids that would result from a two-strand double crossover between E and F in problem 30.

A B C D • E F G H I
A B C D • E F G H I
A B C F E • D G H I
A B C F E • D G H I
Here all four chromatids will be functional.

*32. An individual heterozygous for a reciprocal translocation possesses the following chromosomes.

A B • C D E F G
A B • C D V W X
R S • T U E F G
R S • T U V W X

(a) Draw the pairing arrangement of these chromosomes in prophase I of meiosis.

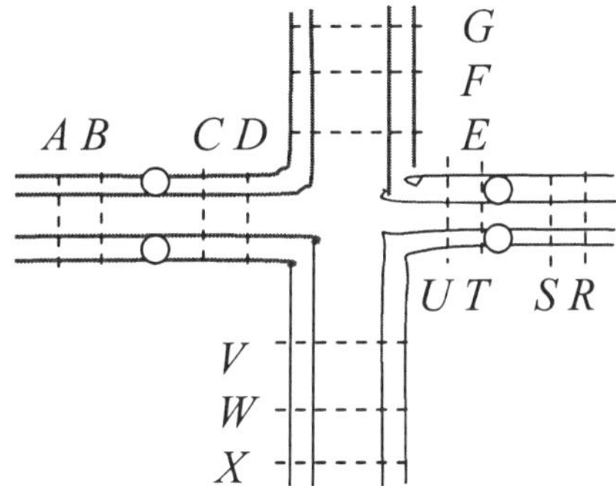

(b) Diagram the alternate, adjacent-1, and adjacent-2 segregation patterns in anaphase I of meiosis.

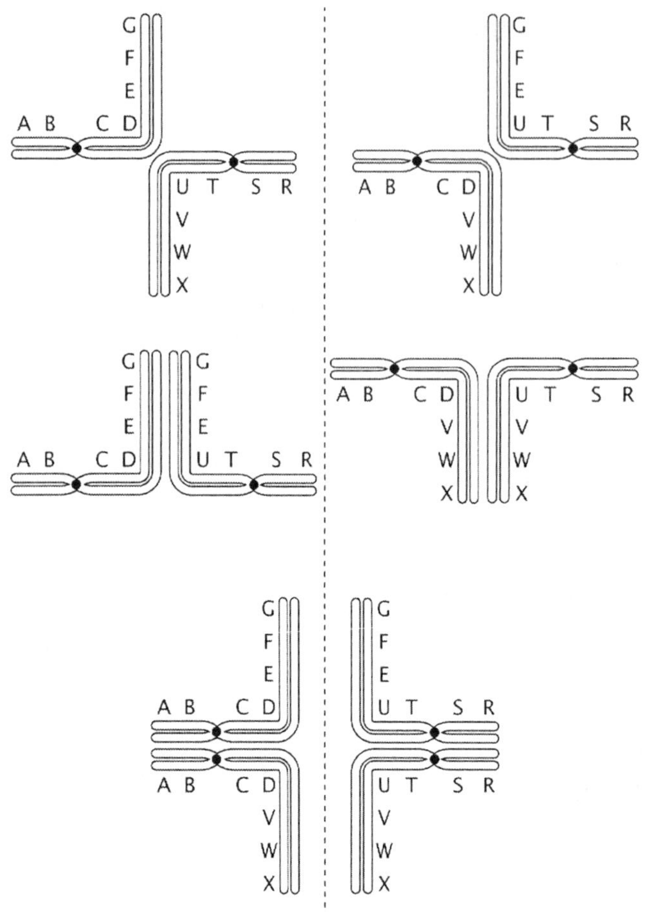

(c) Give the products that result from alternate, adjacent-1, and adjacent-2 segregation.

Alternate: Gametes contain either both normal or both translocation chromosomes, and are all viable.

AB•CDEFG + RS•TUVWX and
AB•CDVWX + RS•TUEFG

Adjacent-1: Gametes contain one normal and one translocation chromosome, resulting in duplication of some genes and deficiency for others.

AB•CDEFG + RS•TUEFG and
AB•CDVWX + RS•TUVWX

Adjacent-2 (rare): Gametes also contain one normal and one translocation chromosome, with duplication of some genes and deficiency for others.

AB•CDEFG + AB•CDVWX and
RS•TUVWX + RS•TUEFG

33. Red-green color blindness is an X-linked recessive disorder. A young man with a 47,XXY karyotype (Klinefelter syndrome) is color blind. His 46,XY brother also is color blind. Both parents have normal color vision. Where did the nondisjunction occur that gave rise to the young man with Klinefelter syndrome?

Because the father has normal color vision, the mother must be the carrier for color blindness. The color-blind young man with Klinefelter syndrome must have inherited two copies of the color-blind X chromosome from his mother. The nondisjunction event therefore most likely took place in meiosis II of the egg.

34. Some people with Turner syndrome are 45,X/46,XY mosaics. Explain how this mosaicism could arise.
Such mosaicism could arise from mitotic nondisjunction early in embryogenesis, in which the Y chromosome fails to segregate and is lost. All the mitotic descendents of the resulting 45,X embryonic cell will also be 45,X. The fetus will then consist of a mosaic of patches of 45,X cells and patches of normal 46,XY cells.

35. What would be the chromosome number of progeny resulting from the following crosses in wheat (see Figure 9.30)? What type of polyploidy (allotriploid, allotetraploid, etc.) would result from each cross?
a. Einkorn wheat and Emmer wheat.
Einkorn is T. monococcum ($2n = 14$); Emmer is T. turgidum ($4n = 28$). Gametes from Einkorn have $n = 7$ and gametes from Emmer have $2n = 14$ chromosomes. The progeny from this cross would be $3n = 21$ and allotriploid.
b. Bread wheat and Emmer wheat.
Bread wheat (T. aestivum) is allohexaploid ($6n = 42$) and produce gametes with $3n = 21$ chromosomes. Progeny from this cross would have ($3n = 21$) + ($2n = 14$) or $5n = 35$ chromosomes and would be allopentaploid.
c. Einkorn wheat and Bread wheat.
($n = 7$) + ($3n = 21$) produces progeny with $4n = 28$ chromosomes, and are allotetraploid.

*36. Bill and Betty have had two children with Down syndrome. Bill's brother has Down syndrome and his sister has two children with Down syndrome. On the basis of these observations, which of the following statements is most likely correct? Explain your reasoning.
(a) Bill has 47 chromosomes.
(b) Betty has 47 chromosomes.
(c) Bill and Betty's children each have 47 chromosomes.
(d) Bill's sister has 45 chromosomes.
(e) Bill has 46 chromosomes.
(f) Betty has 45 chromosomes.
(g) Bill's brother has 45 chromosomes.
The high incidence of Down syndrome in Bill's family and among Bill's relatives is consistent with familial Down syndrome, caused by a Robertsonian translocation involving chromosome 21. Bill and his sister, who are unaffected, are phenotypically normal carriers of the translocation and have 45 chromosomes. Their children and Bill's brother, who have Down syndrome, have 46 chromosomes, but one of these chromosomes is the translocation that has an extra copy of the long arm of chromosome 21. From the information

given, there is no reason to suspect that Bill's wife Betty has any chromosomal abnormalities. Therefore, statement (d) is most likely correct.

*37. Tay-Sachs disease is an autosomal recessive disease that causes blindness, deafness, brain enlargement, and premature death in children. It is possible to identify carriers for Tay-Sachs disease by means of a blood test. Mike and Sue have been tested for the Tay-Sachs gene; Mike is a heterozygous carrier for Tay-Sachs, but Sue is homozygous for the normal allele. Mike and Sue's baby boy is completely normal at birth, but at age 2 develops Tay-Sachs disease. Assuming that a new mutation has not occurred, how could Mike and Sue's baby have inherited Tay-Sachs disease?
Mike and Sue's baby could have inherited Tay-Sachs disease by uniparental disomy. A nondisjunction in meiosis II during spermatogenesis could have produced a sperm carrying two copies of the chromosome bearing the Tay-Sachs allele. Fertilization of a normal egg would then produce a trisomic zygote. Loss of the mother's normal chromosome during the first mitotic division would then produce an embryo in which both remaining copies of the chromosome bear the Tay-Sachs allele from Mike.

38. In mammals, sex-chromosome aneuploids are more common than autosomal aneuploids, but in fishes sex-chromosome aneuploids and autosomal aneuploids are found with equal frequency. Offer an explanation for these differences in mammals and fishes.
In mammals, the higher frequency of sex chromosome aneuploids compared to autosomal aneuploids is due to X-chromosome inactivation and the lack of essential genes on the Y chromosome. If fishes do not have chromosome inactivation, and both their sex chromosomes have numerous essential genes, then the frequency of aneuploids should be similar for both sex chromosomes and autosomes.

*39. A young couple is planning to have children. Knowing that there have been a substantial number of stillbirths, miscarriages, and fertility problems on the husband's side of the family, they see a genetic counselor. A chromosome analysis reveals that, whereas the woman has a normal karyotype, the man possesses only 45 chromosomes and is a carrier for a Robertsonian translocation between chromosomes 22 and 13.

(a) List all the different types of gametes that might be produced by the man.
If the translocation segregates away from the normal chromosomes 22 and 13, then the resulting gametes will have either (1) normal chromosome 13 and normal chromosome 22 or (2) translocated chromosome 13+22.
If the translocation and chromosome 22 segregate away from chromosome 13, then the resulting gametes will have either (3) translocated chromosome 13+22 and normal chromosome 22 or (4) normal chromosome 13.
If the translocation and chromosome 13 segregate away from chromosome 22 (probably rare), then the gametes will have either (5) normal chromosome 13 and translocated chromosome 13+22 or (6) normal chromosome 22.

(b) What types of zygotes will develop when each of the gametes produced by the man fuses with a normal gamete from the woman?

(1) 13, 13, 22, 22; normal
(2) 13, 13+22, 22; translocation carrier
(3) 3, 13+22, 22, 22; trisomy 22
(4) 13, 13, 22; monosomy 22
(5) 13, 13, 13+22, 22; trisomy 13
(6) 13, 22, 22; monosomy 13

(c) If trisomies and monosomies involving chromosome 13 and 22 are lethal, what proportion of the surviving offspring will be carriers of the translocation?
Half, or 50%, from the answer in part (b).

CHALLENGE QUESTIONS

40. Red-green color blindness is a human X-linked recessive disorder. Jill has normal color vision, but her father is color blind. Jill marries Tom, who also has normal color vision. Jill and Tom have a daughter who has Turner syndrome and is color blind.

(a) How did the daughter inherit color blindness?
The daughter, with Turner syndrome, is 45,XO. A normal egg cell with a color-blind X chromosome was fertilized by a sperm carrying no sex chromosome. Such a sperm could have been produced by a nondisjunction event during spermatogenesis.

(b) Did the daughter inherit her X chromosome from Jill or from Tom?
The color-blind daughter must have inherited her X chromosome from Jill. Tom is not color blind and therefore could not have a color-blind allele on his single X chromosome. Jill's father was color blind, so Jill must have inherited a color-blind X chromosome from him and passed it on to her daughter.

41. Humans and many complex organisms are diploid, possessing two sets of genes, one inherited from the mother and one from the father. However, there are a number of eukaryotic organisms that spend the majority of their life cycle in a haploid state. Many of these, such as *Neurospora* (a fungus) and yeast, still undergo meiosis and sexual reproduction, but most of the cells that make up the organism are haploid.

Considering that haploid organisms are fully capable of sexual reproduction and generating genetic variation, why are most complex organisms diploid? In other words, what might be the evolutionary advantage of existing in a diploid state instead of a haploid state? And why might a few organisms, such as *Neurospora* and yeast, exist as haploids?

The most obvious advantage to being a diploid is genetic redundancy. Most mutations reduce or eliminate gene function, and are therefore recessive. Having two copies means that the organism, and its component cells, are able to survive the vast majority of mutations that would be lethal or deleterious in a haploid state. This redundancy is especially important for organisms that have large, complex genomes, complex development and relatively lengthy life cycles. Another advantage to diploidy is that gene expression levels can be higher than in haploid cells, often leading to larger cell sizes and larger, more robust organisms. Indeed, many cultivated polyploid crop plants have higher growth rates and yields than their diploid relatives. A third advantage is that the ability to carry recessive mutations in a masked state allows diploid populations to accumulate and harbor much more genetic diversity. Variant forms of genes that may be harmful to the organism and selected against in the current environment, may prove advantageous when the environment changes.

These advantages are less important for organisms that have small genomes and simpler, shorter life cycles. Yeast and Neurospora genomes are highly adapted to their ecological niches, and haploid growth exposes and weeds out less favorable genetic variants. The abililty to replicate their haploid genomes more quickly may give them a selective advantage over diploids. Indeed these species become diploid and undergo meiosis only when conditions become less favorable for growth.

Chapter Ten: DNA: The Chemical Nature of the Gene

COMPREHENSION QUESTIONS

*1. What three general characteristics must the genetic material possess?

(1) The genetic material must contain complex information.
(2) The genetic material must replicate or be replicated faithfully.
(3) The genetic material must encode the phenotype.

2. Briefly outline the history of our knowledge of the structure of DNA until the time of Watson and Crick. Which do you think were the principal contributions and developments?

1869: Johann Friedrich Miescher isolates nuclei from white blood cells and extracts a substance that was slightly acidic and rich in phosphorous. He calls it nuclein.
Late 1800s: Albrecht Kossel determines that DNA contains the four nitrogenous bases: adenine, guanine, cytosine, and thymine.
1920s: Phoebus Aaron Levine discovers that DNA consists of repeating units, each consisting of a sugar, a phosphate, and a nitrogenous base.
1950: Erwin Chargaff formulates Chargaff's rules (A = T and G = C).
1947: William Ashbury begins studying DNA structure using X-ray diffraction.
1951–1953: Rosalind Franklin, working in Maurice Wilkins' lab, obtains higher resolution pictures of DNA structure using X-ray diffraction techniques.
1953: Watson and Crick propose the model of DNA structure.

All of these scientists contributed information that helped Watson and Crick determine the structure of the DNA double helix. Erwin Chargaff and Rosalind Franklin made two important contributions that directly led to the discoveries by Watson and Crick. By combining Chargaff's rules with Rosalind Franklin's X-ray diffraction data, Watson and Crick were able to predict accurately the structure of the DNA double helix.

*3. What experiments demonstrated that DNA is the genetic material?

Experiments by Hershey and Chase in the 1950s using the bacteriophage T2 and E. coli *cells demonstrated that DNA is the genetic material of the bacteriophage. Also, the experiments by Avery, Macleod, and McCarty demonstrated that the transforming material initially identified by Griffiths was DNA.*

4. What is transformation? How did Avery and his colleagues demonstrate that the transforming principle is DNA?

Transformation occurs when a transforming material (or DNA) genetically alters the bacterium that absorbs the transforming material. Avery and his colleagues demonstrated that DNA is the transforming material by using enzymes that destroyed the different classes of biological molecules. Enzymes that destroyed proteins or nucleic acids had no effect on the activity of the transforming material. However, enzymes that destroyed DNA eliminated the biological activity of the transforming material. Avery and his colleagues were also able to isolate the transforming material and demonstrate that it had chemical properties similar to DNA.

*5. How did Hershey and Chase show that DNA is passed to new phages in phage reproduction?
Hershey and Chase used the radioactive isotope ^{32}P to demonstrate that DNA is passed to new phage particles during phage reproduction. The progeny phage released from bacteria infected with ^{32}P-labeled phages emitted radioactivity from ^{32}P. The presence of the ^{32}P in the progeny phage indicated that the infecting phage had passed DNA on to the progeny phage.

6. Why was Watson and Crick's discovery so important?
By deciphering the structure of the DNA molecule, Watson and Crick provided the foundation for molecular studies of the genetic material or DNA, allowing scientists to discern how genes function to produce phenotypes. Their model also suggested a possible mechanism for the replication of DNA that would ensure the fidelity of the replicated copies.

*7. Draw and label the three parts of a DNA nucleotide.
The three parts of a DNA nucleotide are phosphate, deoxyribose sugar, and a nitrogenous base.

Nitrogen-containing base

Phosphate

Deoxyribose sugar

Deoxyguanosine 5′-phosphate (dGMP)

8. How does an RNA nucleotide differ from a DNA nucleotide?
DNA nucleotides, or deoxyribonucleotides, have a deoxyribose sugar that lacks an oxygen molecule at the 2' carbon of the sugar molecule. Ribonucleotides, or RNA nucleotides, have a ribose sugar with an oxygen linked to the 2' carbon of the sugar molecule. Ribonucleotides may contain the nitrogenous base uracil, but not thymine. DNA nucleotides contain thymine, but not uracil.

9. How does a purine differ from a pyrimidine? What purines and pyrimidines are found in DNA and RNA?

A purine consists of a six-sided ring attached to a five-sided ring. A pyrimidine consists of only a six-sided ring. In both DNA and RNA, the purines found are adenine and guanine. DNA and RNA differ in their pyrimidine content. The pyrimidine cytosine is found in both RNA and DNA. However, DNA contains the pyrimidine thymine, whereas RNA contains the pyrimidine uracil but not thymine.

*10. Draw a short segment of a single polynucleotide strand, including at least three nucleotides. Indicate the polarity of the strand by labeling the 5' end and the 3' end.

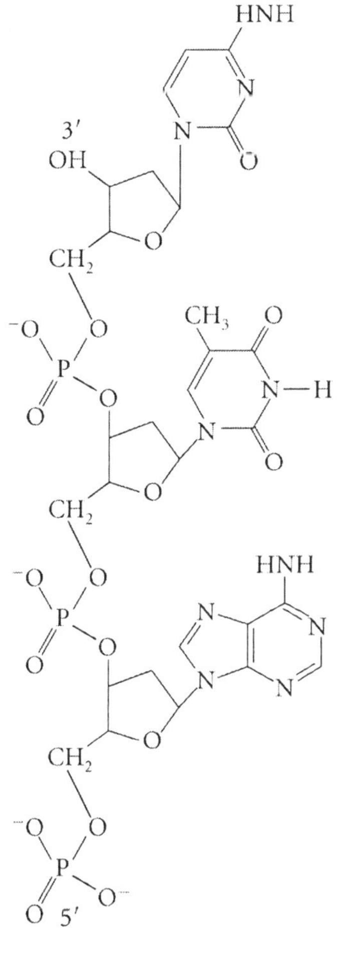

11. Which bases are capable of forming hydrogen bonds with each another?
Adenine is capable of forming two hydrogen bonds with thymine. Guanine is capable of forming three hydrogen bonds with cytosine.

12. What different types of chemical bonds are found in DNA and where do they occur?
The deoxyribonucleotides in a single chain or strand of DNA are held by covalent bonds called phosphodiester linkages between the 3′ end of the deoxyribose sugar of a nucleotide and the 5′ end of the deoxyribose sugar of the next nucleotide in the chain. Two chains of deoxyribonucleotides are held together by hydrogen bonds between the complementary nitrogenous bases of the nucleotides in each chain.

*13. What is local variation in DNA structure and what causes it?
Because DNA is not a static, rigid structure that is invariant, the local variation in DNA structure refers to the actual variations that exist in a DNA molecule. For instance, B-DNA is described as having an average of 10 bases per turn. However, the actual values may be less than or greater than 10, depending on the environmental conditions.

14. What are some of the important genetic implications of the DNA structure?
Referring back to question 1, the structure of DNA gives insight into the three fundamental genetic processes. The Watson and Crick model suggests that the genetic information or instructions are encoded in the nucleotide sequences. The complementary polynucleotide strands indicate how faithful replication of the genetic material is possible. Finally, the arrangement of the nucleotides is such that they specify the primary structure or amino acid sequence of protein molecules.

*15. What are the major transfers of genetic information?
The major transfers of genetic information are replication, transcription, and translation. These are the components of the central dogma of molecular biology.

16. What are hairpins and how do they form?
Hairpins are a type of secondary structure found in single strands of nucleotides. The formation of hairpins occurs when sequences of nucleotides on the single strand are inverted complementary repeats of one another.

17. What is DNA methylation?
DNA methylation is the addition of methyl groups (–CH_3) to certain positions on the nitrogenous bases on the nucleotide.

APPLICATION QUESTIONS AND PROBLEMS

18. A student mixes some heat-killed type IIS *Streptococcus pneumonia* bacteria with live type IIR bacteria and injects the mixture into a mouse. The mouse develops pneumonia and dies. The student recovers some type IIS bacteria from the dead mouse. It is the only experiment conducted by the student. Has the student demonstrated that transformation has taken place? What other explanations might explain the presence of the type IIS bacteria in the dead mouse?
No, the student has not demonstrated that transformation has taken place. Unlike Griffiths, who used strains IIR and IIIS to demonstrate transformation, the student is using strains IIR and IIS. A mutation in the IIR strain injected into the mouse could be sufficient to convert the IIR strain into the virulent IIS strain. By not conducting the appropriate control of injecting IIR bacteria only, the student cannot determine whether the conversion from IIR to IIS is due to transformation or to a mutation.

Although heat may have killed all the IIS bacteria, the student has not demonstrated that the heat was sufficient to kill all the IIS bacteria. A second useful control experiment would have been to inject the heat-killed IIS into mice and see if any of the IIS bacteria survived the heat treatment.

19. Explain how heat-killed type IIIS bacteria in Griffith's experiment genetically altered the live type IIR bacteria? (Hint: See the discussion of transformation in Chapter 8.)

    The IIR strain of Streptococcus pneumonia must have been naturally competent or in other words was capable of taking up DNA from the environment. The heat-killed strains of IIIS bacteria lysed releasing their DNA into the environment allowing for the IIIS chromosomal DNA fragments to come in contact with IIR cells. The IIIS DNA responsible for the virulence of the IIIS strain was taken up by a IIR cell and integrated into the IIR cell's chromosome, thus "transforming" the IIR cell into a virulent IIIS cell.

*20. (a) Why did Hershey and Chase choose ^{32}P and ^{35}S for use in their experiment?
Proteins contain sulfur in the amino acids cysteine and methionine. However, proteins do not typically contain phosphorous (or have very limited amounts due to the phosphorylation of certain proteins by protein kinases). DNA contains a lot of phosphorous due to its sugar-phosphate backbone but no sulfur. Hershey and Chase chose the isotopes ^{32}P and ^{35}S because these radioactive elements would allow them to distinguish between proteins and DNA molecules. Only DNA would contain the isotope ^{32}P, and only proteins would contain the isotope ^{35}S.

(b) Could they have used radioactive isotopes of carbon (C) and oxygen (O) instead? Why or why not?
No, they could not have used radioactive isotopes of oxygen and carbon. DNA and proteins contain significant amounts of carbon and oxygen. The molecules were labeled with the radioactive isotopes in vivo *by using phage to infect* E. coli *cells grown in media containing the radioactive isotopes. Because both proteins and DNA contain carbon and oxygen, both molecules in the phage progeny would have received the radioactive isotopes. Hershey and Chase would have been unable to isolate only DNA molecules or proteins that contain radioactive isotopes of these elements.*

21. What results would you expect if the Hershey and Chase experiment were conducted on tobacco mosaic virus?
Infection by TMV results in the protein coat and RNA genome entering the host cell. Inside the plant cell, the TMV protein coat unwinds, releasing the viral genome, which initiates infection. If Hershey and Chase had used ^{32}P and ^{35}S to label TMV particles, the RNA molecules would have been labeled with the ^{32}P and the viral proteins would have been labeled with ^{35}S. However, the protein coat and the RNA genome would have entered the cell, so "radioactive ghost proteins" would not have been located outside the cell. Newly synthesized viral RNAs would have contained measurable levels of ^{32}P.

22. DNA molecules of different size are often separated using a technique called electrophoresis (see Chapter 18). With this technique, DNA molecules are placed in a gel, an electrical current is applied to the gel, and the DNA molecules migrate toward the positive (+) pole of the current. What aspect of its structure causes DNA molecules to migrate toward to the positive pole?

The phosphate backbone of DNA molecules typically carries a negative charge thus making the DNA molecules attractive to the positive pole of the current.

*23. Each nucleotide pair of a DNA double helix weighs about 1×10^{-21} g. The human body contains approximately 0.5 g of DNA. How many nucleotide pairs of DNA are in the human body? If you assume that all the DNA in human cells is in the B-DNA form, how far would the DNA reach if stretched end to end?

If each nucleotide pair of a DNA double helix weighs approximately 1×10^{-21} g, and the human body contains 0.5 grams of DNA, then the number of nucleotide pairs can be estimated as: (0.5 g DNA / human)/(1×10^{-21} g / nucleotide) = 5×10^{20} nucleotides pairs/human.

DNA that is in B form has an average distance of .34 nm between each nucleotide pair. If a human possesses 5×10^{20} nucleotide pairs, then that DNA stretched end to end would reach: (5×10^{20} nucleotides / human) × (0.34 nm/nucleotide pair) = 1.7×10^{20} nm or 1.7×10^{8} km.

24. What aspects of its structure contribute to the stability of the DNA molecule? Why is RNA less stable than DNA?

Several aspects contribute to the stability of the DNA molecule. The relatively strong phosphodiester linkages connect the nucleotides of a given strand of DNA. The helical nature of the double-stranded DNA molecule results in the negatively charged phosphates of each strand being arranged to the outside and away from each other. The complementary nature of the nitrogenous bases of the nucleotides helps hold the two strands of polynucleotides together. The stacking interactions of the bases, which allow for any base to follow another in a given strand, also play a major role in holding the two strands together. Finally, the ability of DNA to have local variations in its secondary structure contributes to its stability.

RNA nucleotides or ribonucleotides contain an extra oxygen at the 2' carbon. This extra oxygen at each nucleotide makes RNA a less stable molecule.

*25. Which of the following relations will be found in the percentages of bases of a double-stranded DNA molecule?

A double-stranded DNA molecule will contain equal percentages of A and T nucleotides and equal percentages of G and C nucleotides. The combined percentage of A and T bases added to the combined percentage of the G and C bases should equal 100.

(a) A + T = G + C *No*
(b) A + G = T + C *Yes*
(c) A + C = G + T *Yes*

(d) $\frac{A+T}{C+G}=1.0$ *No*

(e) $\frac{A+G}{C+T}=1.0$ *Yes*

(f) $\frac{A}{C}=\frac{G}{T}$ *No*

(g) $\frac{A}{G}=\frac{T}{C}$ *Yes*

(h) $\frac{A}{T}=\frac{G}{C}$ *No*

*26. If a double-stranded DNA molecule has 15% thymine, what are the percentages of all the other bases?
The percentage of thymine (15%) should be approximately equal to the percentage of adenine (15%). The remaining percentage of DNA bases will consist of cytosine and guanine bases (100% – 15% – 15% = 70%); these should be in equal amounts (70%/2 = 35%). Therefore, the percentages of each of the other bases if the thymine content is 15% are adenine = 15%; guanine = 35%; and cytosine = 35%.

27. A virus contains 10% adenine, 24% thymine, 30% guanine, and 36% cytosine. Is the genetic material in this virus double-stranded DNA, single-stranded DNA, double-stranded RNA, or single-stranded RNA? Support your answer.
Most likely the viral genome is single-stranded DNA. The presence of thymine indicates that the viral genome is DNA. For the molecule to be double-stranded DNA, we would predict equal percentages of adenine and thymine bases and equal percentages of guanine and cytosine bases. Neither the percentages of adenine and guanine bases nor the percentages of guanine and cytosine bases are equal, indicating that the viral genome is single-stranded.

28. Suppose that each of the bases in DNA were capable of pairing with any other base. What effect would this have on DNA's capacity to serve as the source of genetic information?

DNA's ability to be replicated faithfully and to encode phenotypes would be destroyed. If each base could pair with any other base, the result during replication would be changes in the DNA sequences of the newly replicated strands. The two new molecules of DNA would not be identical to the original molecule or to each other because different bases would be inserted in each newly synthesized strand. If the DNA base sequence was constantly changing due to the random pairing of bases, then the no consistent "code" could be maintained. This lack of a code would inhibit the ability of a DNA molecule to faithfully code for any particular protein.

*29. A B-DNA molecule has 1 million nucleotide pairs.

(a) How many complete turns are there in this molecule?

B-form DNA contains approximately 10 nucleotides per turn of the helix. A B-DNA molecule of 1 million nucleotide pairs will have about the following number of complete turns: (1,000,000 nucleotides) / 10 nucleotides/turn) = 100,000 complete turns.

(b) If this same molecule were in the Z-DNA configuration, how many complete turns would it have?

If the same DNA molecule assumes a Z-DNA configuration, then each turn would consist of about 12 nucleotides. The determination of the number of complete turns in the 1 million nucleotide molecule is:
(1,000,000 nucleotides) / (12 nucleotides / turn) = 83333.3 or 83333 complete turns.

30. For entertainment on a Friday night, a genetics professor proposed that his children diagram a polynucleotide strand of DNA. Having learned about DNA in preschool, his 5-year-old daughter was able to draw a polynucleotide strand, but she made several mistakes. The daughter's diagram (represented here) contained at least 10 mistakes.

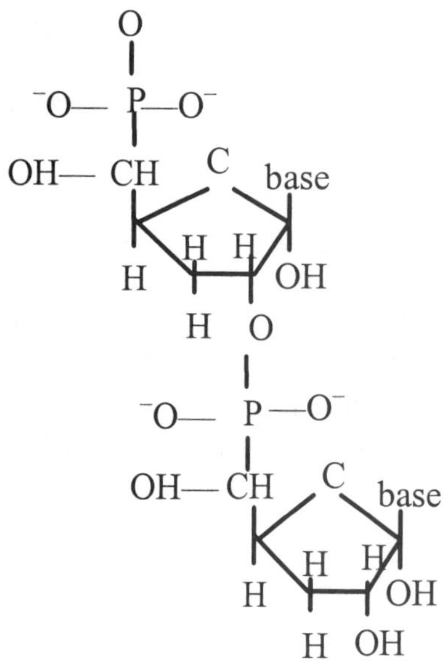

O
$^-$O— P—O$^-$
OH— CH C base
H H
H OH
H O
$^-$O— P—O$^-$
OH—CH C base
H H
H OH
H OH

(a) Make a list of all the mistakes in the structure of this DNA polynucleotide strand.

(1) Neither 5′ carbon of the two sugars is directly linked to phosphorous.

(2) Neither 5′ carbon of the two sugars has an OH group attached.

(3) Neither sugar molecule has oxygen in its ring structure between the 1′ and 4′ carbons.

(4) In both sugars, the 2′ carbon has an –OH group attached, which does not occur in deoxyribonucleotides.

(5) At the 3′ position in both sugars, only hydrogen is attached, as opposed to an –OH group.

(6) The 1′ carbon of both sugars has an –OH group, as opposed to just a hydrogen attached.

(b) Draw the correct structure for the polynucleotide strand.

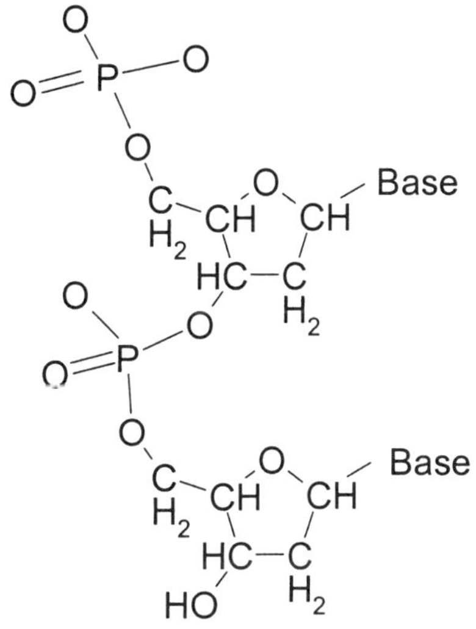

O
O=P—O
O
C—CH O CH— Base
H_2 HC—C
H_2
O
O=P—O
O
C—CH O CH— Base
H_2 HC—C
HO H_2

*31. Chapter 1 considered the theory of the inheritance of acquired characteristics and noted that this theory is no longer accepted. Is the central dogma consistent with the theory of the inheritance of acquired characteristics? Why or why not?

The central dogma of molecular biology is not consistent with the theory of inheritance of acquired characteristics. The flow of information predicted by the central dogma is:

DNA ⟶ *RNA* ⟶ *Protein*

One exception to the central dogma is reverse transcription, whereby RNA codes for DNA. However, biologists currently do not know of a process that will allow for the flow of information from proteins back to DNA. The theory of inheritance of acquired characteristics necessitates such a flow of information from proteins back to the DNA.

32. Write a sequence of bases in an RNA molecule that would produce a hairpin structure.

For a hairpin structure to form in a RNA molecule, an inverted complementary RNA sequence separated by a region of noncomplementary sequence is necessary. The inverted complements form the stem structure, and the loop of the hairpin is formed by the noncomplementary sequences.

5′—UGCAU—3′...unpaired nucleotides...5′—AUGCA—3′

```
          ∩
          UA
          AU
          CG
          GC
5'---------UA---------3'
```

*33. The following sequence is present in one strand of a DNA molecule:

5'—CATTGACCGA—3'

Write out the sequence on the same strand that produces an inverted repeat and the sequence on the complementary strand.

The presence of inverted complementary sequences on the same strand of a DNA molecule will result in a double-stranded DNA molecule containing a palindrome and could result in the formation of a cruciform structure.

5′—CATTGACCGA—3′..........................5′—TCGGTCAATG—3′
3′—GTAACTGGCT—5′..........................3′—AGCCAGTTAC—5′

34. Write out a sequence of nucleotides on a strand of DNA that will form a hairpin structure

For a hairpin structure to form in a strand of DNA, the DNA strand must contain inverted complementary DNA sequences separated by a region of noncomplementary sequence. The inverted complements form the stem structure, and the loop of the hairpin is formed by the noncomplementary sequences

5′—TGCATTACTCAATGCA—3′

or

```
   C T
  A   C
   T A
   G C
   C G
   A T
   T A
5' T A 3'
```

CHALLENGE QUESTIONS

35. Researchers have proposed that early life on Earth used RNA as its source of genetic information and that DNA eventually replaced RNA as the source of genetic information. What aspects of DNA structure might make it better suited than RNA to be the genetic material?

Due to the lack of an attached oxygen molecule at the 2' carbon position of the sugar molecule in deoxynucleotides, DNA molecules are more stable and less reactive than RNA molecules. The double-helical nature of the DNA molecule provides a greater opportunity for DNA repair and fidelity during replication. If mistakes occur in one strand, the complementary strand can serve as a template for corrections.

*36. Suppose that an automated, unmanned probe is sent into deep space to search for extraterrestrial life. After wandering for many lightyears among the far reaches of the universe, this probe arrives on a distance planet and detects life. The chemical composition of life on this planet is completely different from that of life on Earth, and its genetic material is not composed of nucleic acids. What predictions can you make about the chemical properties of the genetic material on this planet?
Although the chemical composition of the genetic material may be different DNA, it more than likely will have similar properties to DNA. As discussed earlier in the chapter, the genetic material must possess three general characteristics:
(1) It must contain complex information.
(2) It must replicate or be replicated faithfully.
(3) It must encode the phenotype.
Even if the material is not DNA, it must meet these criteria. For instance, if the material could not be replicated or duplicated faithfully, then life on that planet could not continue because ultimately no offspring could be produced. A lack of fidelity would result in the loss of information. Genetic material from any lifeform has to store the information necessary for the survival of that organism. Also, the genetic material will need to be stable. Unstable molecules will not allow for long-term storage of information, resulting in the loss of information and change in phenotype.

37. How might ^{32}P and ^{35}S be used to demonstrate that the transforming principle is DNA? Briefly outline an experiment that would show that DNA and not protein is the transforming principle.
The first step would be to label the DNA and proteins of the donor bacteria cells with ^{35}S and ^{32}P. The DNA could be labeled by growing a culture of bacteria in the presence of ^{32}P. The cells as they replicate ultimately will incorporate radioactive phosphorous into their DNA. A second culture of bacteria should be grown in the presence of ^{35}S, which ultimately will be incorporated into proteins.

Material from each culture should be used to transform bacteria cells that previously had not been exposed to the radioactive isotopes. Transformed cells (or colonies) that would be identified by the acquisition of a new phenotype should contain low levels of the radioactive material due to the uptake of the labeled molecules. If the transforming material were protein, then cells transformed by the material from the ^{35}S exposed bacterial cultures would also contain ^{35}S. If the transforming material were DNA, then the cells transformed by the material from the ^{32}P exposed bacterial cultures would also contain ^{32}P.

38. Scientists have reportedly isolated short fragments of DNA from fossilized dinosaur bones hundreds of millions of years old. The technique used to isolate this DNA is the polymerase chain reaction (PCR), which is capable of amplifying very small amounts of DNA a million fold (see Chapter 16). Critics have claimed that the DNA isolated from dinosaur bones is not of ancient origin but instead represents contamination of the samples with DNA from present-day organisms, such as bacteria, mold, or humans. What precautions, analyses, and control experiments could be carried out to ensure that DNA recovered from fossils is truly of ancient origin?
An initial precaution would be to handle all the material in the most sterile manner possible. People handling the samples should wear gloves and masks to help keep the area as devoid of extraneous DNA as possible. Instruments used in the sampling should be sterilized to eliminate any contamination by bacteria, viruses, fungi, and so on. They should also be treated to remove trace DNAs. In addition, the source material surrounding the bones should be treated to remove contaminating DNAs.

Controls also need to be conducted. The DNA from people involved in the procedure should be tested to see if amplification occurs. Material at various locations around the site and isolated bugs and microorganisms from the area should be sampled to see if similar amplification patterns emerge. The design of the primers used for amplification should be considered carefully and should be executed considering the sequences of potential dinosaur descendants, such as birds or reptiles, in an attempt to limit random amplifications. Furthermore, every experiment should be reproducible.

Chapter Eleven: Chromosome Structure and Transposable Elements

COMPREHENSION QUESTIONS

*1. How does supercoiling arise? What is the difference between positive and negative supercoiling?

Supercoiling arises from:

(1) Overwinding (positive supercoiling) or underwinding (negative supercoiling) the DNA double helix.

(2) When the DNA molecule does not have free ends, as in circular DNA molecules.

(3) When the ends of the DNA molecule are bound to proteins that prevent them from rotating about each other.

2. What functions does supercoiling serve for the cell?

Supercoiling compacts the DNA. Negative supercoiling helps to unwind the DNA duplex for replication and transcription.

*3. Describe the composition and structure of the nucleosome. How do core particles differ from chromatosomes?

The nucleosome core particle contains two molecules each of histones H2A, H2B, H3, and H4, which form a protein core with 145-147 bp of DNA wound around the core. Chromatosomes contain the nucleosome core with a molecule of histone H1.

4. Describe in steps how the double helix of DNA, which is 2 nm in width, gives rise to a chromosome that is 700 nm in width.

DNA is first packaged into nucleosomes; the nucleosomes are packed to form a 30 nm fiber. The 30 nm fiber forms a series of loops that pack to form a 250 nm fiber, which in turn coils to form a 700 nm chromatid.

5. What are polytene chromosomes and chromosomal puffs?

Polytene chromosomes are giant chromosomes formed by repeated rounds of DNA replication without nuclear division, found only in the larval salivary glands of Drosophila *and a few other species of flies. Certain regions of polytene chromosomes can become less condensed, resulting in localized swelling, or chromosomal puffs, because of intense transcriptional activity at the site.*

*6. Describe the function and molecular structure of the centromere.

Centromeres are the points of attachment for mitotic and meiotic spindle fibers and are required for the movement of chromatids to the poles in anaphase. Centromeres have distinct centromeric DNA sequences where the kinetochore proteins bind. For some species, such as yeast, the centromere is compact, consisting of only 125 bp. For other species, including Drosophila *and mammals, the centromere is larger, comprising several kilobasepairs of DNA sequence.*

*7. Describe the function and molecular structure of a telomere.
Telomeres are the ends of the linear chromosomes in eukaryotes. They cap and stabilize the ends of the chromosomes to prevent degradation by exonucleases or joining of the ends. Telomeres also enable replication of the ends of the chromosome. Telomeric DNA sequences consist of repeats of a simple sequence, usually in the form of $5'C_n(A/T)_m$.

8. What is the C value of an organism?
The C value is the amount of DNA per cell of an organism.

*9. Describe the different types of DNA sequences that exist in eukaryotes.
Unique-sequence DNA, present in only one or a few copies per haploid genome, represent most of the protein coding sequences, plus a great deal of sequences with unknown function.

Moderately repetitive sequences, between a few hundred to a few thousand base pairs long, are present in up to several thousand copies per haploid genome. Some moderately repetitive DNA consists of functional genes that code for rRNAs and tRNAs, but most is made up of transposable elements and remnants of transposable elements.

Highly repetitive DNA, or satellite DNA, consists of clusters of tandem repeats of short (often less than 10 base pairs) sequences present in hundreds of thousands to millions of copies per haploid genome.

10. What is the difference between euchromatin and heterochromatin?
Euchromatin undergoes regular cycles of condensation during mitosis and decondensation during interphase, whereas heterochromatin remains highly condensed throughout the cell cycle, except transiently during replication. Nearly all transcription takes place in euchromatic regions, with little or no transcription of heterochromatin.

*11. What general characteristics are found in many transposable elements? Describe the differences between replicative and nonreplicative transposition.
Most transposable elements have terminal inverted repeats and are flanked by short direct repeats. Replicative transposons use a copy-and-paste mechanism in which the transposon is replicated and inserted in a new location, leaving the original transposon in place. Nonreplicative transposons use a cut-and-paste mechanism in which the original transposon is excised and moved to a new location.

*12. What is a retrotransposon and how does it move?
A retrotransposon is a transposable element that relocates through an RNA intermediate. First it is transcribed into RNA. A reverse transcriptase encoded by the retrotransposon then reverse transcribes the RNA template into a DNA copy of the transposon, which then integrates into a new location in the host genome.

*13. Describe the process of replicative transposition through DNA intermediates. What enzymes are involved?

First, a transposase makes single-stranded nicks on either side of the transposon and on either side of the target sequence. Second, the free ends of the transposon are joined by a DNA ligase to the free ends of the DNA at the target site. Third, the free 3' ends of DNA on either side of the transposon are used to replicate the transposon sequence, forming the cointegrate. The enzymes normally required for DNA replication are required for this step, including DNA polymerase. The cointegrate has two copies of the transposon and the target site sequence on one side of each copy. Fourth, the cointegrate undergoes resolution, which involves a crossing over within the transposon, by resolvase enzymes such as those used in homologous recombination.

*14. Draw and label the structure of a typical insertion sequence.

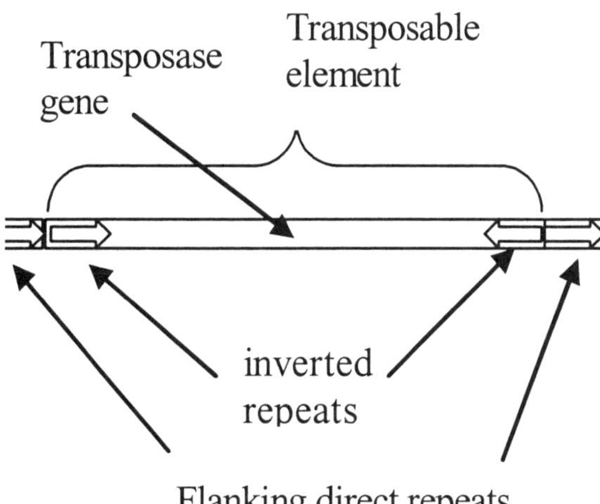

15. Draw and label the structure of a typical composite transposon in bacteria.

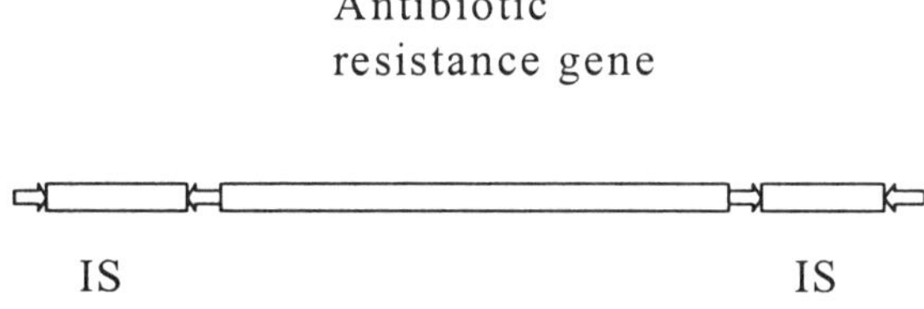

16. How are composite transposons and retrotransposons alike and how are they different?

Composite transposons and retrotransposons are similar in their complexity and some aspects of structure. Both have long flanking repeat sequences: IS elements for composite transposons, long terminal repeat sequences for retrotransposons. Both generate direct duplications of their target site sequences.

However, the IS elements in composite transposons may be either direct repeats or inverted, whereas retrotransposons have direct long terminal repeats. Furthermore, retrotransposons transpose through an RNA intermediate and depend on reverse transcriptase for transposition, whereas composite transposons transpose through DNA intermediates by the action of transposases encoded by one of their insertion sequences.

17. Explain how *Ac* and *Ds* elements produce variegated corn kernels.
A mutation caused by insertion of either an Ac *or* Ds *element in the pigment-producing gene causes the corn kernel to be colorless. During development of the kernel, the* Ac *or* Ds *element may transpose out of the pigment gene, and pigment production in the cell and its descendents will be restored. If the excision of the transposable element occurs early in kernel development, the kernel will have a relatively large sector of purple pigment. If the excision occurs later, the kernel will have relatively smaller sectors or specks of purple pigment.*

18. Briefly explain hybrid dysgenesis and how *P* elements lead to hybrid dysgenesis.
When a sperm containing P *elements from a* P+ *male fly fertilizes an egg from a female that does not contain* P *elements, the resulting zygote undergoes hybrid dysgenesis: transposition of* P *elements causes chromosomal abnormalities and disrupts gamete formation. The hybrid progeny fly is therefore sterile. Hybrid dysgenesis does not occur if the female fly is* P+ *because the* P *elements also encode a repressor that prevents* P *element transposition. The egg cells from a* P+ *female fly contain enough of this repressor protein in the cytoplasm to repress* P *element transposition in the progeny.*

19. What are some differences between LINEs and SINEs?
Although both LINEs (long interspersed elements) and SINEs (short interspersed elements) transpose through RNA intermediates, they have different origins. LINEs resemble retroviruses, whereas SINEs resemble the 7S RNA gene. SINEs, with a typical length of 200 - 400 bp, are also significantly smaller than LINEs, which range up to 6,000 bp for full length elements.

*20. Briefly summarize three hypotheses for the widespread occurrence of transposable elements.
The cellular function hypothesis proposes that transposable elements have a function in the cell or organism, such as regulation of gene expression.
The genetic variation hypothesis suggests that transposable elements serve to generate genomic variation. A larger pool of genomic variants would accelerate evolution by natural selection.
The selfish DNA hypothesis suggests that transposable elements are simply parasites, serving only to replicate and spread themselves.

APPLICATION QUESTIONS AND PROBLEMS

*21. Compare and contrast prokaryotic and eukaryotic chromosomes. How are they alike and how do they differ?
Prokaryotic chromosomes are usually circular, whereas eukaryotic chromosomes are linear. Prokaryotic chromosomes generally contain the entire genome, whereas

each eukaryotic chromosome has only a portion of the genome: the eukaryotic genome is divided into multiple chromosomes. Prokaryotic chromosomes are generally smaller and have only a single origin of DNA replication. Eukaryotic chromosomes are often many times larger than prokaryotic chromosomes and contain multiple origins of DNA replication. Prokaryotic chromosomes are typically condensed into nucleoids, which have loops of DNA compacted into a dense body. Eukaryotic chromosomes contain DNA packaged into nucleosomes, which are further coiled and packaged into successively higher-order structures. The condensation state of eukaryotic chromosomes varies with the cell cycle.

22. (a) In a typical eukaryotic cell, would you expect to find more molecules of the H1 histone or more molecules of the H2A histone? Explain your reasoning.
    Because each nucleosome contains two molecules of histone H2A and only one molecule of histone H1, eukaryotic cells will have more H2A than H1.

    (b) Would you expect to find more molecules of H2A or more molecules of H3? Explain your reasoning.
    Because each nucleosome contains two molecules of H2A and two molecules of H3, eukaryotic cells should have equal amounts of these two histones.

23. Suppose you examined polytene chromosomes from the salivary glands of fruit fly larvae and counted the number of chromosomal puffs observed in different regions of DNA.

    (a) Would you expect to observe more puffs from euchromatin or from heterochromatin? Explain your answer.
    Euchromatin is less condensed and capable of being transcribed, whereas heterochromatin is highly condensed and rarely transcribed. Because chromosomal puffs are sites of active transcription, they should occur primarily in euchromatin.

    (b) Would you expect to observe more puffs in unique-sequence DNA, moderately repetitive DNA, or repetitive DNA? Why?
    Highly repetitive DNA consists of simple tandem repeats usually found in heterochromatic regions and are rarely transcribed. Moderately repetitive DNA comprises transposons and remnants of transposons. Again, with the exception of the rDNA cluster, these sequences are rarely transcribed or transcribed at low levels. The most actively transcribed genes occur as single-copy sequences, or as small gene families. Therefore, more chromosomal puffs would be observed in unique-sequence DNA than in moderately or highly repetitive DNA.

*24. A diploid human cell contains approximately 6 billion base pairs of DNA.

    (a) How many nucleosomes are present in such a cell? (Assume that the linker DNA encompasses 40 bp.)
    Given that each nucleosome contains about 140 bp of DNA tightly associated with the core histone octamer, another 20 bp associated with histone H1, and 40 bp in the linker region, then one nucleosome occurs for every 200 bp of DNA.

6×10^9 bp divided by 2×10^2 bp/nucleosome = 3×10^7 nucleosomes (30 million).

(b) How many histone proteins are complexed to this DNA?

Each nucleosome contains two of each of the following histones: H2A, H2B, H3, and H4. A nucleosome plus one molecule of histone H1 constitute the chromatosome. Therefore, nine histone protein molecules occur for every nucleosome.

3×10^7 nucleosomes × 9 histones = 2.7×10^8 molecules of histones are complexed to 6 billion bp of DNA.

*25. Would you expect to see more or less acetylation in regions of DNA that are sensitive to digestion by DNase I? Why?

More acetylation. Regions of DNase I sensitivity are less condensed than DNA that is not sensitive to DNase I, the sensitive DNA is less tightly associated with nucleosomes, and it is in a more open state. Such a state is associated with acetylation of lysine residues in the N-terminal histone tails. Acetylation eliminates the positive charge of the lysine residue and reduces the affinity of the histone for the negatively charged phosphates of the DNA backbone.

26. Suppose a chemist develops a new drug that neutralizes the positive charges on the tails of histone proteins. What would be the most likely effect of this new drug on chromatin structure? Would you predict that this drug would have any effect on gene expression? Explain your answers.

Such a drug would disrupt the ionic interactions between the histone tails and the phosphate backbone of DNA, and thereby cause a loosening of the DNA from the nucleosome. The drug may mimic the effects of histone acetylation, which neutralizes the positively charged lysine residues. Changes in chromatin structure would result from the altered nucleosome-DNA packing and possible changes in interaction with other chromatin modifying enzymes and proteins. Changes in transcription would result because DNA may be more accessible to transcription factors.

27. A YAC that contains only highly repetitive, nonessential DNA is added to mouse cells that are growing culture. The cells are then divided into two groups, A and B. A laser is then used to damage the centromere on the YACs in cells of group A. The centromeres on the YACs of group B are not damaged. In spite of the fact that the YACs contain no essential DNA, the cells in group A divide more slowly than those in group B. Provide a possible explanation.

The onset of anaphase in mitosis is regulated in part by attachment of spindle fibers to the kinetochores of each chromosome. If the YACs in cells of group A suffered damage to the kinetochores as a result of laser treatment, then the attachment of spindle fibers may be impaired. The cell may be arrested in mitotic metaphase or initiate anaphase more slowly because it senses that the YAC centromere has not completed attachment to the spindle fibers.

*28. Which of the following two molecules of DNA has the lower melting temperature? Why?

AGTTACTAAAGCAATACATC	AGGCGGGTAGGCACCCTTA
TCAATGATTTCGTTATGTAG	TCCGCCCATCCGTGGGAAT

The molecule on the left, with a higher percentage of A–T base pairs, will have a lower melting temperature than the molecule on the right, which has mostly G–C base pairs. A–T base pairs have two hydrogen bonds, and thus less stability, than G–C base pairs, which have three hydrogen bonds.

29. Which of the following pairs of sequences might be found at the ends of an insertion sequence?
(a) 5'—GGGCCAATT—3' and 5'—CCCGGTTAA—3'.
(b) 5'—AAACCCTTT—3' and 5'—AAAGGGTTT—3'.
(c) 5'—TTTCGAC—3' and 5'—CAGCTTT—3'.
(d) 5'—ACGTACG—3' and 5'—CGTACGT—3'.
(e) 5'—GCCCCAT—3' and 5'—GCCCAT—3'.
The pairs of sequences in (b) and (d) are inverted repeats because they are both reversed and complementary and might be found at the ends of insertion sequences. The sequences in (a) are complementary, but not inverted. The sequences in (c) are reversed, but not complementary. The sequences in (e) are imperfect direct repeats.

*30. A particular transposable element generates flanking direct repeats that are 4 bp long. Give the sequence that will be found on both sides of the transposable element if this transposable element inserts at the position indicated on each of the following sequences:
(a) 5'—ATTCGAAC**TGAC**(transposable element)**TGAC**CGATCA—3'
(b) 5'—ATT**CGAA**(transposable element)**CGAA**CTGACCGATCA—3'
For (a) and (b) the target site duplication is indicated in bold.

*31. White eyes in *Drosophila melanogaster* result from an X-linked recessive mutation. Occasionally, white-eye mutants give rise to offspring that possess white eyes with small red spots. The number, distribution, and size of the red spots are variable. Explain how a transposable element could be responsible for this spotting phenomenon.
Such a fly may be homozygous (female) or hemizygous (male) for an allele of the white-eye locus that contains a transposon insertion. The eye cells in these flies cannot make red pigment. During eye development, the transposon may spontaneously transpose out of the white-eye locus, restoring function to this gene so the cell and its mitotic progeny can make red pigment. Depending on how early during eye development the transposition occurs, the number and size of red spots in the eyes will be variable.

32. Two different strains of *Drosophila melanogaster* are mated in reciprocal crosses. When strain A males are crossed with strain B females, the progeny are normal. However, when strain A females are crossed with strain B males, many mutations and chromosome rearrangements occur in the gametes of the F1 progeny and they are effectively sterile. Explain these results.
These results could be explained by hybrid dysgenesis, with strain B harboring P elements and strain A having no P elements. When sperm from strain A males fertilize eggs with P elements from strain B females, the progeny are normal because the strain B egg cytoplasm contains a repressor of P element transposition. However, when P^+ sperm cells from strain B fertilize P^- eggs from strain A, the P elements undergo a burst of transposition in the embryo because the P^- egg cytoplasm lacks the repressor.

*33. An insertion sequence contains a large deletion in its transposase gene. Under what circumstances would this insertion sequence be able to transpose?
Without a functional transposase gene of its own, the transposon would be able to transpose only if another transposon of the same type were in the cell and able to express a functional transposase enzyme. This transposase enzyme will recognize the inverted repeats and transpose its own element as well as other nonautonomous copies of the transposon with the same inverted repeats.

*34. What factor do you think determines the length of the flanking direct repeats that are produced in transposition?
The length of the flanking direct repeats that are generated depends on the number of base pairs between the staggered single-stranded nicks made at the target site by the transposase.

35. A transposable element is found to encode a transposase enzyme. On the basis of this information, what conclusions can you make about the likely structure and method of transposition of this element?
This element probably has short inverted terminal repeats, the bacterial IS elements, and transposes through a DNA intermediate, using either a cut-and-paste nonreplicative mechanism or a copy-and-paste replicative mechanism. Because it does not encode a reverse transcriptase, it is not likely to be a retrotransposon.

36. AZT (zidovudine) is a drug used to treat patients with AIDS. AZT works by blocking the reverse transcriptase enzyme used by human immunodeficiency virus (HIV), the causative agent of AIDS. Do you expect that AZT would have any effect on transposable elements? If so, what type of transposable elements would be affected and what would be the most likely effect?
AZT should affect retrotransposons because they transpose through an RNA intermediate that is reverse transcribed to DNA by reverse transcriptase. If endogenous reverse transcriptases in human cells have similar sensitivity to AZT as HIV reverse transcriptase, then AZT should inhibit retrotransposons.

37. A transposable element is found to encode a reverse transcriptase enzyme. On the basis of this information, what conclusions can you make about the likely structure and method of transposition of this element?
Like other retrotransposons, this element probably has long terminal direct repeats and transposes through an RNA intermediate that is reverse transcribed to DNA.

38. Transposition often produces chromosome rearrangements, such as deletions, inversions, and translocations. Can you suggest a reason why transposition leads to these chromosome mutations?
Transposons lead to multiple copies of the same DNA sequence dispersed on the same chromosome and among different chromosomes. Pairing between the dispersed copies of the transposon and homologous recombination will lead to DNA rearrangements, such as inversions, translocations, and deletions.

39. A geneticist examines an ear of corn in which most kernels are yellow, but she finds a few kernels with purple spots, as shown below. Give a possible explanation for the appearance of the purple spots in these otherwise yellow kernels, accounting for their different sizes. Hint: See section on *Ac* and *Ds* elements in maize in this chapter.
The appearance of purple spots of varying sizes in these few yellow corn kernels could be explained by transposition. The yellow kernels may be due to inactivation of a pigment gene by insertion of a Ds *element in the plant bearing this ear. Because the* Ds *element cannot transpose on its own, the mutant allele is stable in the absence of* Ac *and the plant produces yellow kernels when fertilized by pollen from the same strain (lacking* Ac*). However, a few kernels may have been fertilized by pollen from a different strain with an active* Ac *element. The* Ac *element can then mobilize transposition of the* Ds *element out of the pigment gene, restoring pigment gene function. Excision of the* Ds *element earlier in kernel development will produce larger clones of cells producing purple pigment. Excision later in kernel development will produce smaller clones of purple cells.*

40. A geneticist studying the DNA of the Japanese bottle fly finds many copies of a particular sequence that appears similar to the *copia* transposable element in *Drosophila*. Using recombinant DNA techniques, the geneticist places an intron into a copy of this DNA sequence and inserts it into the genome of a Japanese bottle fly. If the sequence is a transposable element similar to *copia*, what prediction would you make concerning the fate of the introduced sequence in the genomes of offspring of the fly receiving it?
Because copia *is a retrotransposon, the* copia*-like element probably transposes through an RNA intermediate. The recombinant transposon containing an intron will be transcribed into an RNA molecule containing the intron. However, the intron will be removed by splicing. Reverse transcription of the spliced RNA will generate copies of the transposon that lack the intron. Therefore, the daughter transposons will lack the intron sequence.*

CHALLENGE QUESTIONS

41. An explorer discovers a strange new species of plant and sends some of the plant tissue to a geneticist to study. The geneticist isolates chromatin from the plant and examines it with the electron microscope. She observes what appear to be beads on a string. She then adds a small amount of nuclease, which cleaves the string into individual beads that each contain 280 bp of DNA. After digestion with more nuclease, she finds that a 120 bp fragment of DNA remains attached to a core of histone proteins. Analysis of the histone core reveals histones in the following proportions:

H1	12.5%
H2A	25%
H2B	25%
H3	0%
H4	25%
H7 (a new histone)	12.5%

On the basis of these observations, what conclusions could the geneticist make about the probable structure of the nucleosome in the chromatin of this plant?
The 120 bp of DNA associated with the histone core is smaller than the 140 bp associated with typical nucleosomes. The new plant also is lacking histone H3. The new histone H7 apparently does not replace histone H3 in the nucleosome core because it is present in the same ratio as histone H1, or half of the ratios of nucleosomal core histones H2A, H2B, and H4. Finally, the 280 bp fragments with limited DNase digestion are larger than the 200 bp fragments seen with typical eukaryotic chromatin.

These observations suggest a model in which the nucleosome core consists of just six histones, two each of H2A, H2B, and H4, explaining the lack of H3 and the smaller amount of DNA. The longer DNA per nucleosome can be explained in part by a molecule of H7 either in the spacer between nucleosomes, or perhaps helping to cap nucleosomes in conjunction with H1.

42. Although highly repetitive DNA is common in eukaryotic chromosomes, it does not code for proteins; in fact, it is probably never transcribed into RNA. If highly repetitive DNA does not code for RNA or proteins, why is it present in eukaryotic genomes? Suggest some possible reasons for the widespread presence of highly repetitive DNA.
Highly repetitive DNA may have important structural roles for eukaryotic chromosomes. Highly repetitive DNA is present in regions of the chromosome that are heterochromatic, near the centromere and near the telomeres. Highly repetitive DNA clusters may play important roles in facilitating chromatin condensation in mitotic or meiotic prophase. They may serve to insulate critical regions from chromatin decondensation or from transcription. Near the telomeres, they may serve as buffers between the ends of the chromosome and essential protein coding genes, to minimize deleterious effects from loss of telomeric DNA. Heterochromatic

sequences may also regulate the expression of genes located in the heterochromatic region of the chromosome or near the heterochromatin/euchromatin border.

43. As discussed in the chapter, *Alu* sequences are retrotransposons that are common in the human genome. *Alu* sequences are thought to have evolved from the 7S RNA gene, which encodes an RNA molecule that takes part in transporting newly synthesized proteins across the endoplasmic reticulum. The 7S RNA gene is transcribed by RNA polymerase III, which uses an internal promoter (see Chapter 13). How might this observation explain the large number of copies of *Alu* sequences?
Retrotransposons must be transcribed in order to transpose. With an internal promoter, every copy of the Alu *sequence will have a promoter to drive transcription, and hence transposition.* Alu *sequences will then be able to proliferate exponentially.*

44. Houck and Kidwell proposed that *P* elements were carried from *Drosophila willistoni* to *D. melanogaster* by mites that fed on fruit flies. What evidence do you think would be required to demonstrate that *D. melanogaster* acquired *P* elements in this way? Propose a series of experiments to provide such evidence.
This hypothesis requires not only that mites pick up P *elements when they feed on* P+ *fruit flies, but also that they can transmit the* P *elements to a host that does not have them.*

Mites that have infected a laboratory colony of D. willistoni *should be isolated and tested for the presence of* P *elements. If* P *elements are present in these mites, these mites should then be allowed to infect a colony of* D. melanogaster *that is free of* P *elements. After several generations the colony of* D. melanogaster *should be tested for the presence of* P *elements after they have been disinfected of the mites. One way to test for the presence of* P *elements would be to mate females from this test colony with males that are* P+. *Fertile progeny, testifying to a lack of hybrid dysgenesis, would indicate that these females are* P+. *These experiments would show whether mites are capable of transmitting* P *elements from one species to another.*

Chapter Twelve: DNA Replication and Recombination

COMPREHENSION QUESTIONS

1. What is semiconservative replication?
In semiconservative replication, the original two strands of the double helix serve as templates for new strands of DNA. When replication is complete, two double-stranded DNA molecules will be present. Each will consist of one original template strand and one newly synthesized strand that is complementary to the template.

*2. How did Meselson and Stahl demonstrate that replication in *E. coli* takes place in a semiconservative manner?
Meselson and Stahl grew E. coli *cells in a medium containing the heavy isotope of nitrogen (^{15}N) for several generations. The ^{15}N was incorporated in the DNA of the* E. coli *cells. The* E. coli *cells were then switched to a medium containing the common form of nitrogen (^{14}N) and allowed to proceed through a few cycles of cellular generations. Samples of the bacteria were removed at each cellular generation. Using equilibrium density gradient centrifugation, Meselson and Stahl were able to distinguish DNAs that contained only ^{15}N from DNAs that contained only ^{14}N or a mixture of ^{15}N and ^{14}N because DNAs containing the ^{15}N isotope are "heavier." The more ^{15}N a DNA molecule contains, the further it will sediment during equilibrium density gradient centrifugation. DNA from cells grown in the ^{15}N medium produced only a single band at the expected position during centrifugation. After one round of replication in the ^{14}N medium, one band was present following centrifugation, but the band was located at a position intermediate to that of a DNA band containing only ^{15}N and a DNA band containing only ^{14}N. After two rounds of replication, two bands of DNA were present. One band was located at a position intermediate to that of a DNA band containing only ^{15}N and a DNA band containing only ^{14}N, while the other band was at a position expected for DNA containing only ^{14}N. These results were consistent with the predictions of semiconservative replication and incompatible with the predictions of conservative and dispersive replication.*

*3. Draw a molecule of DNA undergoing theta replication. On your drawing, identify (1) origin, (2) polarity (5' and 3' ends) of all template strands and newly synthesized strands, (3) leading and lagging strands, (4) Okazaki fragments, and (5) location of primers.

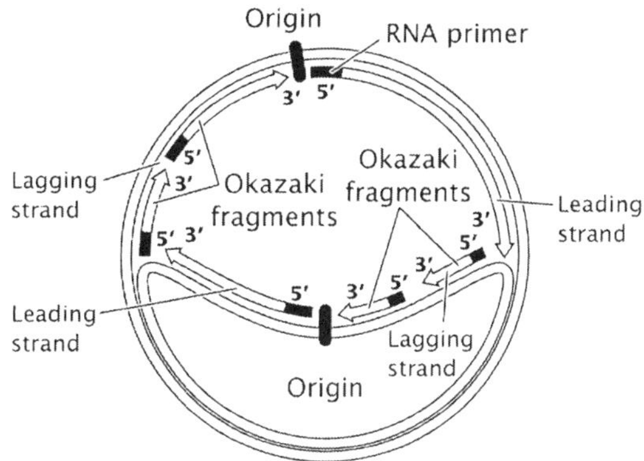

4. Draw a molecule of DNA undergoing rolling-circle replication. On your drawing, identify (1) origin, (2) polarity (5' and 3' ends) of all template strands and newly synthesized strands, (3) leading and lagging strands, (4) Okazaki fragments, and (5) location of primers.

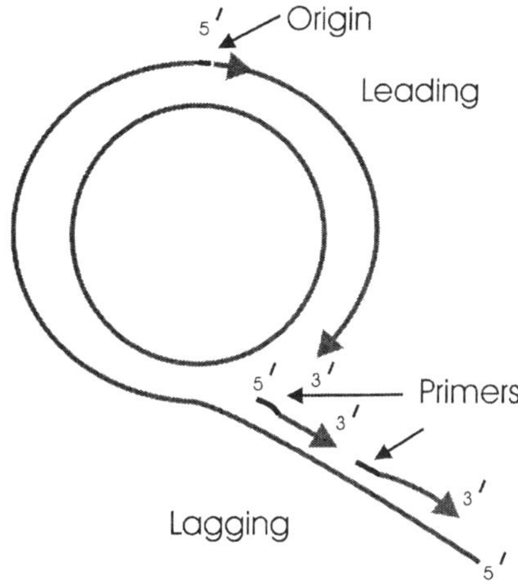

5. Draw a molecule of DNA undergoing eukaryotic linear replication. On your drawing, identify (1) origin, (2) polarity (5' and 3' ends) of all template strands and newly synthesized strands, (3) leading and lagging strands, (4) Okazaki fragments, and (5) location of primers.

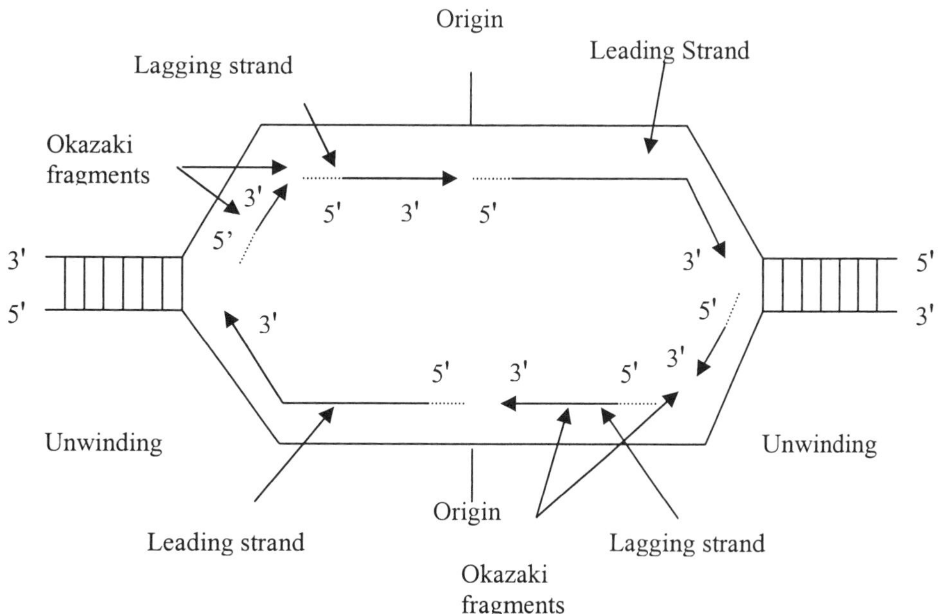

6. What are three major requirements of replication?
(1) A single-stranded DNA template.
(2) Nucleotide substrates for synthesis of the new polynucleotide strand.
(3) Enzymes and other proteins associated with replication to assemble the nucleotide substrates into a new DNA molecule.

*7. What substrates are used in the DNA synthesis reaction?
The substrates for DNA synthesis are the four types of deoxyribonucleoside triphosphates: deoxyadenosine triphosphate, deoxyguanosine triphosphate, deoxycytosine triphosphate, and deoxythymidine triphosphate.

8. List the different proteins and enzymes taking part in bacterial replication. Give the function of each in the replication process.
DNA polymerase III is the primary replication polymerase. It elongates a new nucleotide strand from the 3'–OH of the primer.

DNA polymerase I removes the RNA nucleotides of the primers and replaces them with DNA nucleotides.

DNA ligase connects Okazaki fragments by sealing nicks in the sugar phosphate backbone.

DNA primase synthesizes the RNA primers that provide the 3'–OH group needed for DNA polymerase III to initiate DNA synthesis.

DNA helicase unwinds the double-helix by breaking the hydrogen bonding between the two strands at the replication fork.

DNA gyrase reduces DNA supercoiling and torsional strain that is created ahead of the replication fork by making double-stranded breaks in the DNA and passing another segment of the helix through the break before resealing it. Gyrase is also called topoisomerase II.

Initiator proteins bind to the replication origin and unwind short regions of DNA.

Single-stranded binding protein (SSB protein) stabilizes single-stranded DNA prior to replication by binding to it, thus preventing the DNA from pairing with complementary sequences.

9. What similarities and differences exist in the enzymatic activities of DNA polymerases I, II, and III? What is the function of each type of DNA polymerase in bacterial cells?
Each of the three DNA polymerases has a 5' to 3' polymerase activity. They differ in their exonuclease activities. DNA polymerase I has a 3' to 5' as well as a 5' to 3' exonuclease activity. DNA polymerase II and DNA polymerase III have only a 3' to 5' exonuclease activity.
(1) DNA polymerase I carries out proofreading. It also removes and replaces the RNA primers used to initiate DNA synthesis.
(2) DNA polymerase II functions as a DNA repair polymerase. It restarts replication after DNA damage has halted replication. It has proofreading activity.

(3) DNA polymerase III is the primary replication enzyme and also has a proofreading function in replication.

*10. Why is primase required for replication?
Primase is a DNA-dependent RNA polymerase. Primase synthesizes the short RNA molecules, or primers, that provide a 3′–OH to which DNA polymerase can attach deoxyribonucleotides in replication initiation. The DNA polymerases require a 3′–OH to which they add nucleotides, and therefore they cannot initiate replication. Primase does not have this requirement.

11. What three mechanisms ensure the accuracy of replication in bacteria?
(1) Highly accurate nucleotide selection by the DNA polymerases when pairing bases.
(2) The proofreading function of DNA polymerase, which removes incorrectly inserted bases.
(3) A mismatch repair apparatus that repairs mistakes after replication is complete.

12. How does replication licensing ensure that DNA is replicated only once at each origin per cell cycle?
Only replication origins to which replication licensing factor (RPF) has bound can undergo initiation. Shortly after the completion of mitosis, RPF binds the origin during G_1 and is removed by the replication machinery during S phase.

*13. In what ways is eukaryotic replication similar to bacterial replication, and in what ways is it different?
Eukaryotic and bacterial replication of DNA replication share some basic principles:
(1) Semiconservative replication.
(2) Replication origins serve as starting points for replication.
(3) Short segments of RNA called primers provide a 3′–OH for DNA polymerases to begin synthesis of the new strands.
(4) Synthesis occurs in a 5′ to 3′ direction.
(5) The template strand is read in a 3′ to 5′ direction.
(6) Deoxyribonucleoside triphosphates are the substrates.
(7) Replication is continuous on the leading strand and discontinuous on the lagging strand.
Eukaryotic DNA replication differs from bacterial replication in that:
(1) It has multiple origins of replications per chromosome.
(2) It has several different DNA polymerases with different functions.
(3) Immediately following DNA replication, assembly of nucleosomes takes place.

14. Outline in words and pictures how telomeres at the end of eukaryotic chromosomes are replicated.
Telomeres are replicated by the enzyme telomerase. Telomerase, a

ribonucleoprotein, consists of protein and an RNA molecule that is complementary to the 3′ end of the DNA of a eukaryotic chromosome. The RNA molecule also serves as a template for the addition of nucleotides to the 3′ end. After the 3′ end has been extended, the 5′ end of the DNA can be extended as well, possibly by lagging strand synthesis of a DNA polymerase using the extended 3′ end as a template.

DNA replication of the linear eukaryotic chromosomes generates a 3′ overhang. Part of the RNA sequence within telomerase is complementary to the overhang.

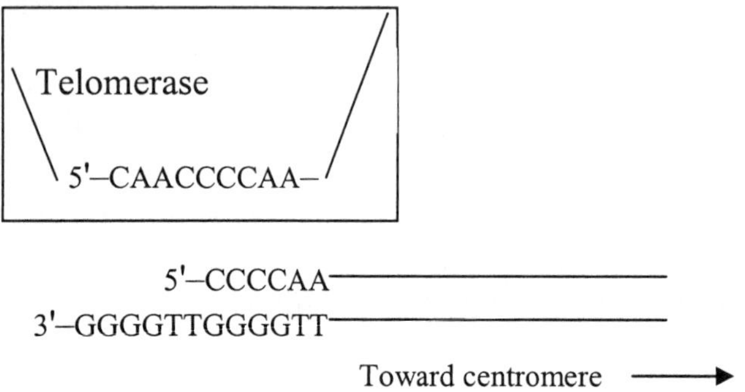

Telomerase RNA sequence pairs with the 3′ overhang and serves as a template for the addition of DNA nucleotides to the 3′ end of the DNA molecule, which serves to extend the 3′ end of the chromosome.

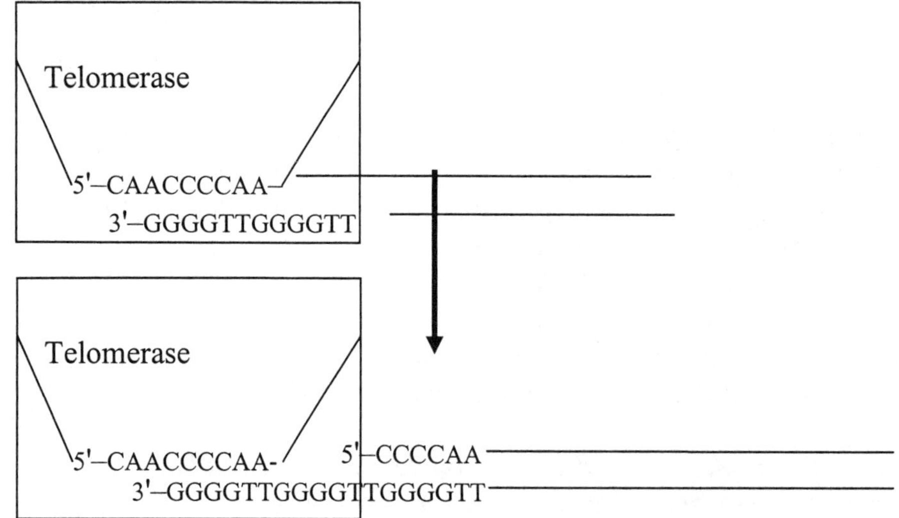

Additional nucleotides to the 5′ end are added by DNA synthesis using a DNA polymerase with priming by primase.

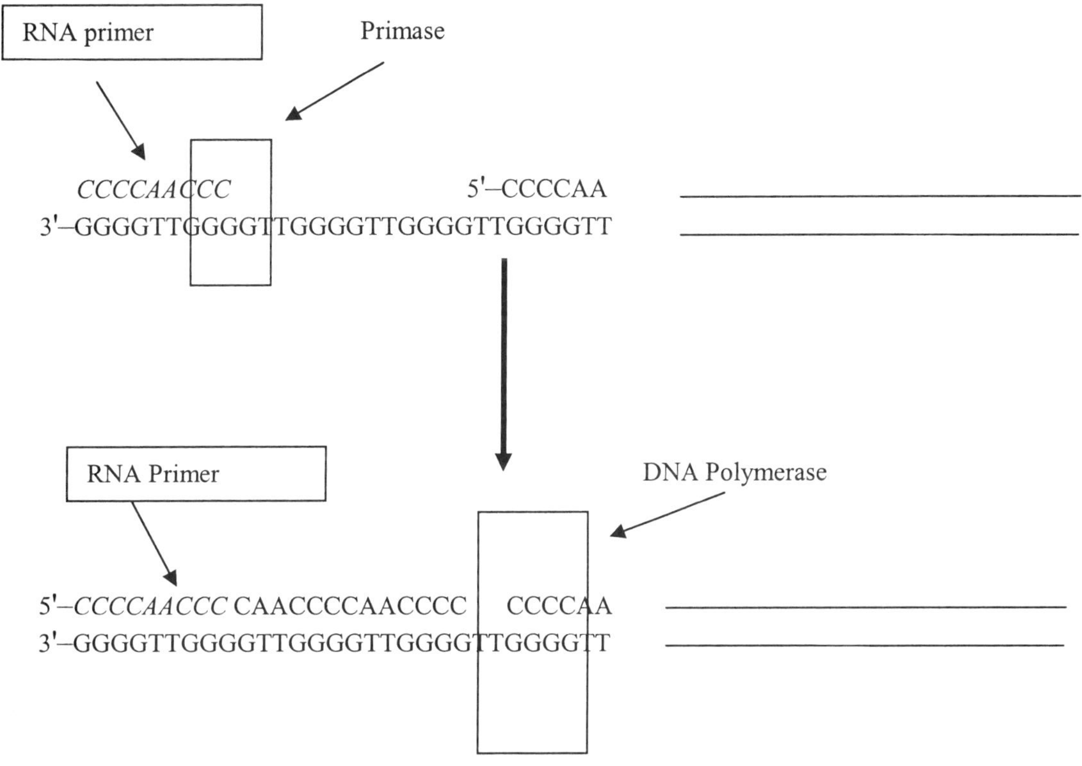

*15. What are some of the enzymes taking part in recombination in *E. coli* and what roles do they play?

(1) RecBCD protein unwinds double-stranded DNA and can cleave nucleotide strands.
(2) RecA protein allows a single strand to invade a double-stranded DNA.
(3) RuvA and RuvB proteins promote branch migration during homologous recombination.
(4) RuvC protein is resolvase, a protein that resolves the Holliday structure by cleavage of the DNA.
(5) DNA ligase repairs nicks or cuts in the DNA generated during recombination.

APPLICATION QUESTIONS AND PROBLEMS

*16. Suppose a future scientist explores a distant planet and discovers a novel form of double-stranded nucleic acid. When this nucleic acid is exposed to DNA polymerases from *E. coli,* replication takes place continuously on both strands. What conclusion can you make about the structure of this novel nucleic acid?

Each strand of the novel double-stranded nucleic acid must be oriented parallel to the other, as opposed to the antiparallel nature of earthly double-stranded DNA. Replication by E. coli *DNA polymerases can proceed continuously only in a 5' to 3' direction, which requires the template to be read in a 3' to 5' direction. If replication is continuous on both strands, the two strands must have the same direction and be parallel.*

*17. Phosphorous is required to synthesize the deoxyribonucleoside triphosphates used in DNA replication. A geneticist grows some *E. coli* in a medium containing nonradioactive phosphorous for many generations. A sample of the bacteria is then transferred to a medium that contains a radioactive isotope of phosphorus (^{32}P). Samples of the bacteria are removed immediately after the transfer and after one and two rounds of replication. What will be the distribution of radioactivity in the DNA of the bacteria in each sample? Will radioactivity be detected in neither, one, or both strands of the DNA?
In the initial sample removed immediately after transfer, very little if any ^{32}P should be incorporated into the DNA because replication in the medium containing ^{32}P has not yet occurred. After one round of replication in the ^{32}P containing medium, one strand of each newly synthesized DNA molecule will contain ^{32}P, while the other strand will contain only nonradioactive phosphorous. After two rounds of replication in the ^{32}P containing medium, 50% of the DNA molecules will have ^{32}P in both strands, while the remaining 50% will contain ^{32}P in one strand and nonradioactive phosphorous in the other strand.

18. A line of mouse cells is grown for many generations in a medium with ^{15}N. Cells in G_1 are then switched to a new medium that contains ^{14}N. Draw a pair of homologous chromosomes from these cells at the following stages, showing the two strands of DNA molecules found in the chromosomes. Use different colors to represent strands with ^{14}N and ^{15}N.
(a) Cells in G_1, before switching to medium with ^{14}N.

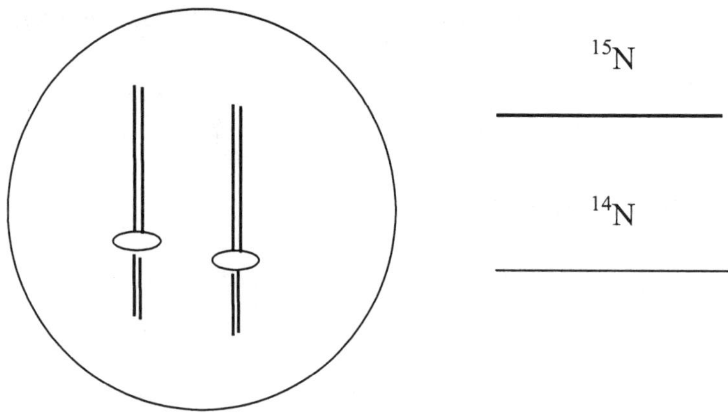

(b) Cells in G_2, after switching to medium with ^{14}N.

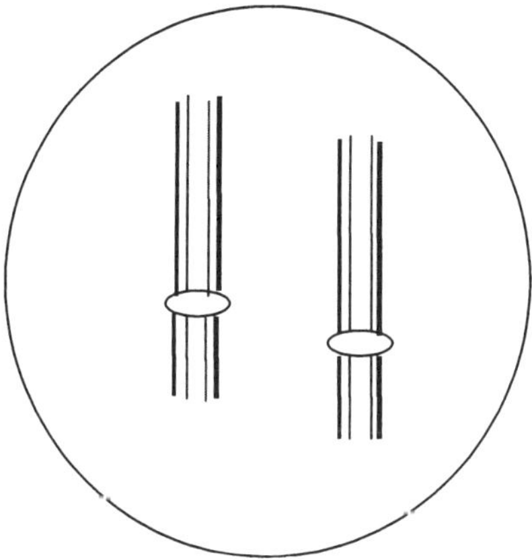

(c) Cells in anaphase of mitosis, after switching to medium with ^{14}N.

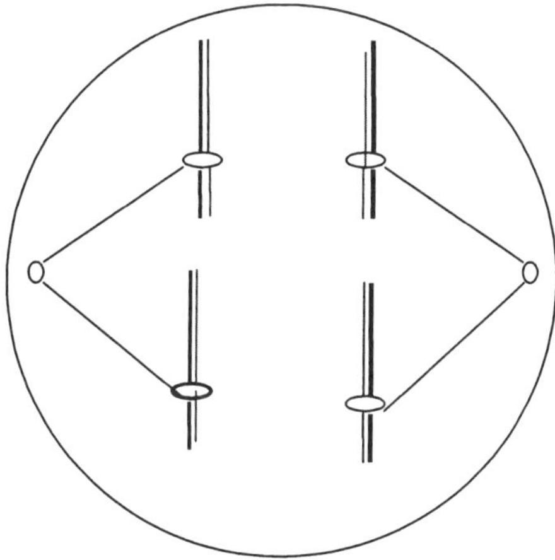

(d) Cells in metaphase I of meiosis, after switching to medium with ^{14}N.

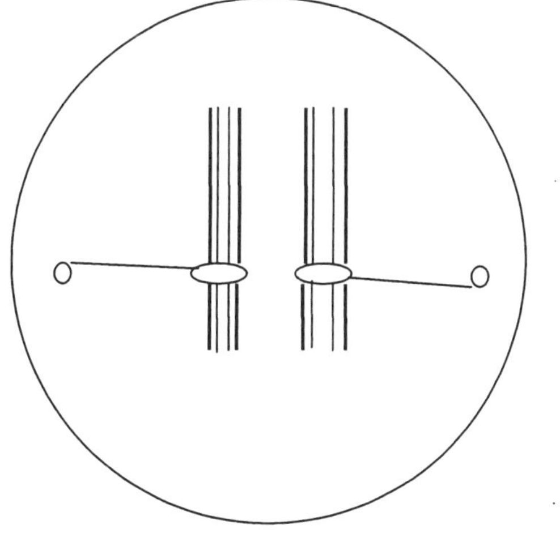

(e) Cells in anaphase II of meiosis, after switching to medium with ^{14}N.

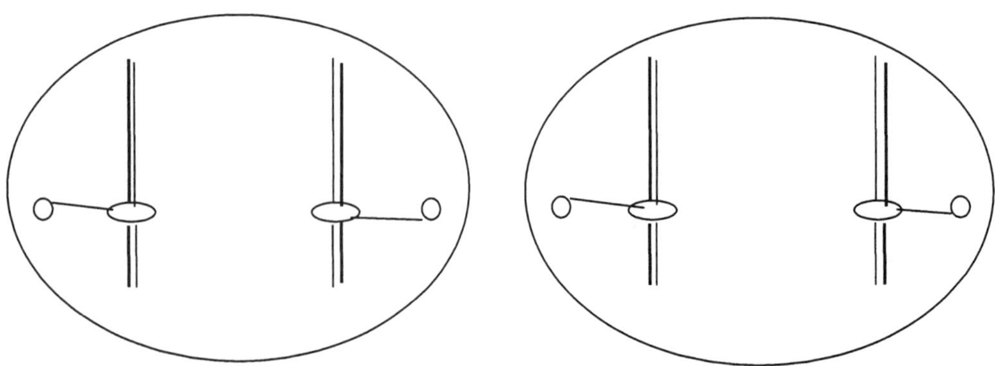

*19. A circular molecule of DNA contains 1 million base pairs. If DNA synthesis at a replication fork occurs at a rate of 100,000 nucleotides per minute, how long will theta replication require to completely replicate the molecule, assuming that theta replication is bidirectional? How long will replication of this circular chromosome take by rolling-circle replication? Ignore replication of the displaced strand in rolling-circle replication.
In bidirectional replication there are two replication forks, each proceeding at a rate of 100,000 nucleotides per minute. So, it would require 5 minutes for the circular DNA molecule to be replicated by bidirectional replication because each fork could synthesize 500,000 nucleotides (5 minutes × 100,000 nucleotides per minute) within the time period. Because rolling-circle replication is unidirectional and thus has only one replication fork, 10 minutes will be required to replicate the entire circular molecule.

20. A bacterium synthesizes DNA at each replication fork at a rate of 1000 nucleotides per second. If this bacterium completely replicates its circular chromosome by theta replication in 30 minutes, how many base pairs of DNA will its chromosome contain?
Each replication complex is synthesizing DNA at each fork at a rate of 1000 nucleotides per second. So for each second, 2000 nucleotides are being synthesized by both forks (1000 nucleotides / second × 2 forks = 2000 nucleotides / second) or 120,000 nucleotides per minute. If the bacterium requires 30 minutes to replicate its chromosome, then the size of the chromosome is 3,600,00 nucleotides (120,000 nucleotides / minute × 30 minutes = 3,600,000).

*21. The following diagram represents a DNA molecule that is undergoing replication. Draw in the strands of newly synthesized DNA and identify the following:
(a) Polarity of newly synthesized strands
(b) Leading and lagging strands
(c) Okazaki fragments
(d) RNA primers

Primer

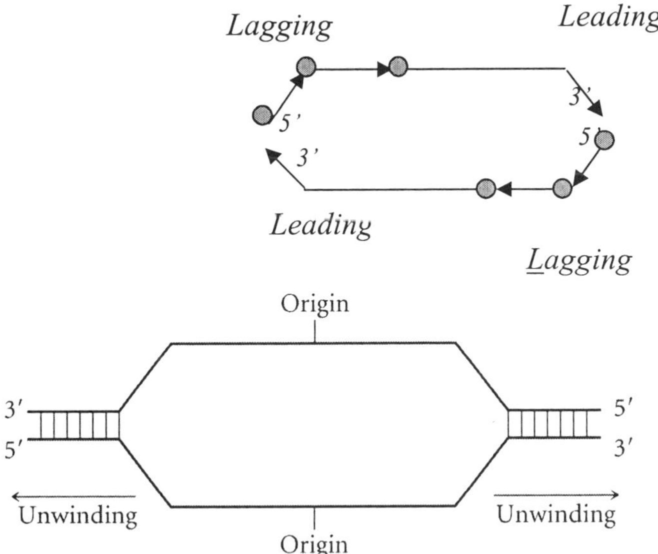

*22. What would be the effect on DNA replication of mutations that destroyed each of the following activities in DNA polymerase I?
(a) 3' → 5' exonuclease activity
The 3' → 5' exonuclease activity is important for proofreading newly synthesized DNA. If the activity is nonfunctional, then the fidelity of replication by DNA polymerase I will decrease, resulting in more misincorporated bases in the DNA.
(b) 5' → 3' exonuclease activity
Loss of the 5' → 3' exonuclease activity would result in the RNA primers used to initiate replication not being removed by DNA polymerase I.
(c) 5' → 3' polymerase activity
DNA polymerase I would be unable to synthesize new DNA strands if the 5' → 3' polymerase activity was destroyed. RNA primers could be removed by DNA polymerase I using the 5' → 3' exonuclease activity, but could not be replaced by DNA polymerase I.

23. How would DNA replication be affected in a cell that is lacking topoisomerase?

Topoisomerase II or gyrase reduces the positive supercoiling or torsional strain that develops ahead of the replication fork due to the unwinding of the double helix. If the topoisomerase activity was lacking, then the torsional strain would continue to increase, making it more difficult to unwind the double helix. Ultimately, the increasing strain would lead to an inhibition of the replication fork movement.

*24. A line of mammalian cells possesses a mutation that destroys the replication licensing factor. What would be the effect of this mutation on the cells? What would be the effect if the cells possessed a mutation that caused the licensing factor to remain attached to origins even after replication?

Replication licensing at the origins by replication licensing factor is necessary for the functioning of initiator proteins at the origins of replication. If replication licensing factor is destroyed then the initiator proteins will not act at the origins to separate the DNA strands, thus replication will not begin. The cell lines would be unable to replicate their chromosomes.

If replication licensing factor remains on the origins of replication even after replication, then the initiator proteins will bind to the licensed origins and begin a new round of replication. The end result would be cell lines with multiple copies of each chromosome.

25. A number of scientists who study ways to treat cancer have become interested in telomerase. Why would they be interested in telomerase? How might cancer drug therapies that target telomerase work?

Telomerase is an enzyme that functions in cells that undergo continuous cell divison and may play a role in the lack of cellular aging in these cells. Cells that lack telomerase exhibit progressive shortening of the chromosomal ends or telomeres. This shortening leads to unstable chromosomes and ultimately to programmed cell death. Many tumor cells also express telomerase, which may assist these cells in becoming immortal. If the telomerase activity in these cancer cells could be inhibited, then cell division might be halted in the cancer cells thus controlling cancer cell growth. Chemically modified antisense RNAs or DNA oligonucleotides complementary to the telomerase RNA sequence might block the telomerase activity by base pairing with the telomerase RNA, making it unavailable as a template. A second strategy would be target the DNA synthesis activity of the telomerase protein preventing the telomere DNA from being synthesized.

CHALLENGE QUESTIONS

26. Conditional mutations express their mutant phenotype only under certain conditions (the restrictive conditions) and express the normal phenotype under other conditions (the permissive conditions). One type of conditional mutation is a temperature-sensitive mutation, which expresses the mutant phenotype only at

certain temperatures.

Strains of *E. coli* have been isolated that contain temperature-sensitive mutations in the genes encoding different components of the replication machinery. In each of these strains, the protein produced by the mutated gene is nonfunctional under the restrictive conditions. These strains are grown under permissive conditions and then abruptly switched to the restrictive condition. After one round of replication under the restrictive condition, the DNA from each strain is isolated and analyzed. What would you predict to see in the DNA isolated from each strain in the following list?

<u>Temperature-sensitive mutation in gene encoding:</u>

(a) DNA ligase

DNA ligase is required to seal the nicks left after DNA polymerase I removes the RNA primers used to begin DNA synthesis. If DNA ligase is not functioning, multiple nicks in the lagging strand will be expected and the Okazaki fragments will not be joined. However, DNA replication will take place.

(b) DNA polymerase I

RNA primers are removed by DNA polymerase I. After one round of replication, the DNA molecule would still contain the RNA nucleotide primers since DNA polymerase I is not functioning. However, replication of the DNA will take place.

(c) DNA polymerase III

No replication would be expected. DNA polymerase III is the primary replication enzyme. If it is not functioning, then neither lagging- nor leading-strand DNA synthesis will take place.

(d) Primase

No replication would be expected. Primase synthesizes the short RNA molecules that act as primers for DNA synthesis. If the RNA primers are not synthesized, then no free 3′–OH will be available for DNA polymerase III to attach DNA nucleotides. Thus, DNA synthesis will not take place.

(e) Initiator protein

Initiator proteins bind to the oriC *and unwind the DNA, allowing for the binding of DNA helicase and single-stranded binding proteins to the DNA. If these proteins are unable to bind, then DNA replication initiation will not occur, and so DNA synthesis will not take place.*

*27. Regulation of replication is essential to genomic stability, and normally the DNA is replicated just once every cell cycle (during S phase). Normal cells produce protein A, which increases in concentration during S phase. In cells that have a mutated copy of the gene for protein A, however, replication occurs continuously throughout the cell cycle, with the result that cells may have 50 times the normal amount of DNA. Protein B is normally present in G_1 but disappears from the cell nucleus during S phase. In cells with a mutated copy of the gene for protein A, the levels of protein B fail to decrease during S phase and, instead, remain high throughout the cell cycle. When the gene for protein B is mutated, no replication occurs.

Propose a mechanism for how protein A and protein B might normally regulate replication so that each cell gets the proper amount of DNA. Explain how mutation of these genes produces the effects described above.

Protein B may be needed for replication to successfully initiate at replication origins. Protein B is present at the beginning of S phase but disappears during the stage. Protein A may be responsible for removing or inactivating protein B. As levels of protein A increase, the levels of protein B decrease preventing extra initiation events. When protein A is mutated, it can no longer inactivate protein B, thus successive rounds of replication can begin due to the high levels of protein B. When protein B is mutated, it cannot assist initiation and replication ceases.

Chapter Thirteen: Transcription

COMPREHENSION QUESTIONS

*1. Draw an RNA nucleotide and a DNA nucleotide, highlighting the differences. How is the structure of RNA similar to that of DNA? How is it different?
RNA and DNA are polymers of nucleotides that are held together by phosphodiester bonds. An RNA nucleotide contains ribose sugar, whereas a DNA nucleotide contains deoxyribose sugar. Also, the pyrimidine base uracil is found in RNA but thymine is not. DNA, however, contains thymine but not uracil. Finally, an RNA polynucleotide is typically single-stranded even though RNA molecules can pair with other complementary sequences. DNA molecules are almost always double-stranded.

Thymine (DNA only) | Uracil (RNA only)

Deoxyribonucleotide | Ribonucleotide

2. What are the major classes of cellular RNA? Where would you expect to find each class of RNA within eukaryotic cells?
Cellular RNA molecules are made up of six classes:
(1) Ribosomal RNA or rRNA is found in the cytoplasm.
(2) Transfer RNA or tRNA is found in the cytoplasm.
(3) Messenger RNA is found in the cytoplasm (however, pre-mRNA is found only in the nucleus).
(4) Small nuclear RNA (snRNA) is found in the nucleus as part of riboproteins called snrps.
(5) Small nucleolar RNA (snoRNA) is found in the nucleus.
(6) Small cytoplasmic RNA (scRNA) is found in the cytoplasm.

*3. What parts of DNA make up a transcription unit? Draw and label a typical transcription unit in a bacterial cell.

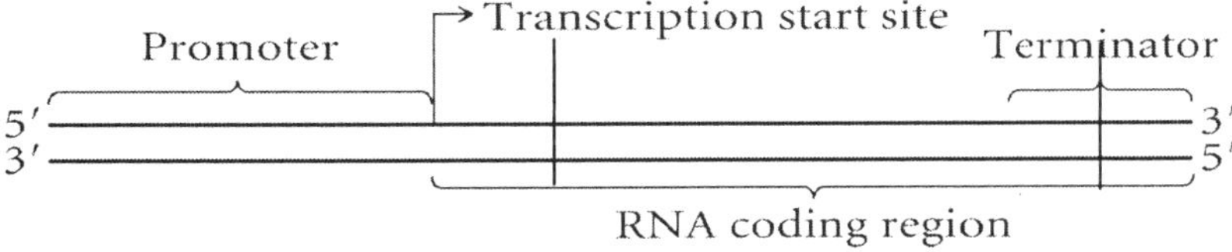

4. What is the substrate for RNA synthesis? How is this substrate modified and joined together to produce an RNA molecule?
Four ribonucleoside triphosphates serve as the substrate for RNA synthesis: adenosine triphosphate, guanosine triphosphate, cytosine triphosphate, and uridine monophosphate. The enzyme RNA polymerase uses a DNA polynucleotide strand as a template to synthesize a complementary RNA polynucleotide strand. The nucleotides are added to the RNA polynucleotide strand, one at time, at the 3′–OH of the RNA molecule. As each nucleoside triphosphate is added to the growing polynucleotide chain, two phosphates are removed from the 5′ end of the nucleotide. The remaining phosphate is linked to the 3′–OH of the RNA molecule to form the phosphodiester bond.

5. Describe the structure of bacterial RNA polymerase.
Bacterial RNA polymerase consists of several polypeptides. The RNA polymerase core enzyme is composed of four polypeptide subunits: two copies of the alpha subunit, the beta subunit, and the beta prime subunit. The addition of a sigma factor to the core enzyme forms the RNA polymerase holoenzyme.

*6. Give the names of the three RNA polymerases found in eukaryotic cells and the types of RNA they transcribe.
RNA polymerase I transcribes rRNA.
RNA polymerase II transcribes pre-mRNA and some snRNAs.
RNA polymerase III transcribes small RNA molecules such as 5S rRNA, tRNAs, and some small nuclear RNAs.

7. What are the three basic stages of transcription? Describe what happens at each stage.
(1) Initiation: Transcription proteins assemble at the promoter to form the basal transcription apparatus and begin synthesis of RNA.
(2) Elongation: RNA polymerase moves along the DNA template in a 3′ to 5′ direction unwinding the DNA and synthesizing RNA in a 5′ to 3′ direction.
(3) Termination: Synthesis of RNA is terminated, and the RNA molecule separates from the DNA template.

*8. Draw and label a typical bacterial promoter. Include any common consensus sequences.

```
                                              Transcription start site
5'—TTGACA---------------------TATAAT---|---3'
3'—AACTGT---------------------ATATTA-------5'
     –35                       –10
    Region                    Region
```

An upstream element rich in AT sequences is found only in some bacterial promoters and is located upstream of the –35 consensus sequence. However, the typical bacterial promoter consists of the –35 and –10 consensus sequences.

9. What are the two basic types of terminators found in bacterial cells? Describe the structure of each.
The two basic types of terminators in bacterial cells are rho-independent and rho-dependent terminators. Rho-independent terminators consist of inverted repeats that can form a hairpin structure. Immediately following the inverted repeats is a string of six adenine nucleotides. Rho-dependent terminators require the interaction of the protein rho with RNA polymerase. Two features are typical for rho-dependent termination: (1) variable DNA sequences that cause RNA polymerase to pause during transcription and (2) upstream from variable region lies DNA sequence that encodes RNA devoid of secondary structure, which also serves as the rho binding site.

10. How is the process of transcription in eukaryotic cells different from that in bacterial cells?
Eukaryotic transcription requires the action of three RNA polymerases. Each type of polymerase recognizes and transcribes from different types of promoters. Binding to the promoter and initiation from the promoter requires the action of many protein factors; different promoters require different sets of protein factors. The RNA molecule produced by transcription in eukaryotic cells usually requires extensive processing, such as the addition of a 5' cap, a 3' poly(A) tail, and the removal of introns prior to becoming functional. Bacterial promoters tend to be more uniform in composition, and only one RNA polymerase does transcription. Bacterial RNAs are typically functional once transcription has taken place.

*11. How are promoters and enhancers similar? How are they different?
Promoters and enhancers are DNA sequences that affect transcription. Transcriptional activator proteins bind to sequences of regulatory promoters and enhancers and stimulate transcription by interacting with the basal transcriptional apparatus. Promoters are essential for the binding of general transcriptional factors and RNA. Enhancers increase or stimulate transcription. Promoters are typically adjacent to the transcriptional start site and are highly position dependent, whereas enhancers function at a distance from the gene and function independently of position and direction.

12. How can an enhancer affect the transcription of a gene that is thousands of nucleotides away?
Enhancers are DNA sequences bound by transcriptional activator proteins. The DNA sequence located between the promoter and the enhancer is looped out, allowing for the interaction of the enhancer-bound proteins with proteins needed at the promoter, which stimulates transcription.

13. Compare the roles of general transcription factors and transcriptional activator proteins.
General transcription factors form the basal transcription apparatus together with RNA polymerase and are needed to initiate minimal levels of transcription.

Transcriptional activator proteins bring about higher levels of transcription by stimulating the assembly of the basal transcriptional apparatus at the start site.

14. What are some of the common consensus sequences found in RNA polymerase II promoters?
Four common consensus sequences can be found at RNA polymerase II promoters, although each sequence is not found at all the promoters.
(1) TATA box is the most common of the core promoter elements and has a consensus sequence of TATAAA. The TATA box is located –25 to –35 base pairs upstream of the transcriptional start site.
(2) TFIIB recognition element (BRE) is located –32 to –38 base pairs upstream of the transcriptional start site. Its consensus sequence is G/C G/C G/C CGCC.
(3) Initiator element (Inr) is a less common consensus sequence in RNA polymerase II promoters than the TATA box or BRE. The Inr overlaps the start site and has the consensus sequence YYAN T/A YY.
(4) Downstream core promoter element (DPE) is found in many promoters that have the initiator element. The DPE is located +30 base pairs downstream of the start site and has the consensus sequence RG A/T CGTG.

*15. What protein associated with a transcription factor is common to all eukaryotic promoters? What is its function in transcription?
The TATA binding protein (TBP) binds most eukaryotic promoters at the TATA box and positions the active site of RNA polymerase over the transcriptional start site.

*16. Compare and contrast transcription and replication. How are these processes similar and how are they different?
Characteristics of transcription and replication:
(1) Utilize a DNA template.
(2) Synthesize molecules in a 5′ to 3′ direction.
(3) Synthesize molecules that are antiparallel and complementary to the template.
(4) Uses nucleotide triphosphates as substrates.
(5) Are complexes of proteins and enzymes necessary for catalysis.
Characteristics of transcription:
(1) Unidirectional synthesis of only a single strand of nucleic acid.
(2) Initiation does not require a primer.
(3) Subject to numerous regulatory mechanisms.
(4) Each gene is transcribed separately.
Characteristics of replication:
(1) Bidirectional synthesis of two strands of nucleic acid.
(2) *Initiates from replication origins.*

APPLICATION QUESTIONS AND PROBLEMS

17. RNA polymerases carry out transcription at a much slower rate than DNA polymerases carry out replication. Why is speed more important in replication than in transcription?

DNA polymerases are required to replicate much larger regions of DNA such as entire chromosomes. Speed is essential to complete the replication process in a timely manner. RNA polymerases typically transcribe only small areas of the chromosomes. The speed required for replication by DNA polymerases is not needed by the RNA polymerases to transcribe these smaller regions.

18. Write the consensus sequence for the following set of nucleotide sequences:
    A G G A G T T
    A G C T A T T
    T G C A A T A
    A C G A A A A
    T C C T A A T
    T G C A A T T

    The consensus sequence is identified by determining which nucleotide is used most frequently at each position. For the two nucleotides that occur at an equal frequency at the first position, both are listed at that position in the sequence and identified by a slash mark:
    T/A G C A A T T

*19. List at least five properties that DNA polymerases and RNA polymerases have in common. List at least three differences.
    Similarities: (1) Both use DNA templates, (2) DNA template is read in 3′ to 5′ direction, (3) complementary strand is synthesized in a 5′ to 3′ direction that is antiparallel to the template, (4) both use triphosphates as substrates, and (5) actions are enhanced by accessory proteins.

    Differences: (1) RNA polymerases use ribonucleoside triphosphates as substrates, whereas DNA polymerases use deoxyribonucleoside triphophates; (2) DNA polymerases require a primer that provides an available 3′–OH group so that synthesis can begin, whereas RNA polymerases do not require primers to begin synthesis; and (3) RNA polymerases synthesize a copy off only one of the DNA strands, whereas DNA polymerases can synthesize copies off both strands.

20. RNA molecules have *three* phosphates at their 5' end, but DNA molecules never do. Explain this difference.
    During initiation of DNA replication, DNA nucleotide triphosphates must be attached to a 3′–OH of a RNA molecule by DNA polymerase. This process removes the terminal two phosphates of the nucleotides. If the RNA molecule is subsequently removed, then a single phosphate would remain at the 5′ end of the DNA molecule. RNA polymerase does not require the 3′–OH to initiate synthesis of RNA molecules. So, the 5′ end of a RNA molecule will retain all three of the phosphates from the original nucleotide triphosphate substrate.

21. An RNA molecule has the following percentages of bases: A = 23%, U = 42%, C = 21%, G = 14%.
    (a) Is this RNA single-stranded or double-stranded? How can you tell?

The RNA molecule is likely to be single-stranded. If the molecule was double-stranded, we would expect nearly equal percentages of adenine and uracil, as well as equal percentages of guanine and cytosine. In this RNA molecule, the percentages of these potential base pairs are not equal, so the molecule is single-stranded.

(b) What would be the percentages of bases in the template strand of the DNA that contains the gene for this RNA?
Because the DNA template strand is complementary to the RNA molecule, we would expect equal percentages for bases in the DNA complementary to the RNA bases. So, in the DNA we would expect A = 42%, T = 23%, C = 14%, and G = 21%.

*22. The following diagram represents DNA that is part of the RNA-coding sequence of a transcription unit. The bottom strand is the template strand. Give the sequence found on the RNA molecule transcribed from this DNA and label the 5' and 3' ends of the RNA.

5'–A T A G G C G A T G C C A–3'
3'–T A T C C G C T A C G G T–5' ← template strand

Because the RNA molecule would be complementary to the template strand and synthesized in an antiparallel fashion, the sequence would be:
5'–A U A G G C G A U G C C A–3'.

The RNA strand contains the same sequence as the nontemplate DNA strand except that the RNA strand contains uracil in place of thymine.

23. The following sequence of nucleotides is found in a single-stranded DNA template:
A T T G C C A G A T C A T C C C A A T A G A T
Assume that RNA polymerase proceeds along this template from left to right.

(a) Which end of the DNA template is 5' and which end is 3'?
RNA is synthesized in a 5' to 3' direction by RNA polymerase, which reads the DNA template in a 3' to 5' direction. So, if the polymerase is moving from left to right on the template then the 3' end must be on the left and the 5' end on the right.

3'–A T T G C C A G A T C A T C C C A A T A G A T–5'

(b) Give the sequence and label the 5' and 3' ends of the RNA copied from this template.

5'–U A A C G G U C U A G U A G G G U U A U C U A–3'

24. Write out a hypothetical sequence of bases that might be found in the first 20 nucleotides of a promoter of a bacterial gene. Include both strands of DNA and label the 5' and 3' ends of both strands. Be sure to include the start site for transcription and any consensus sequences found in the promoter.

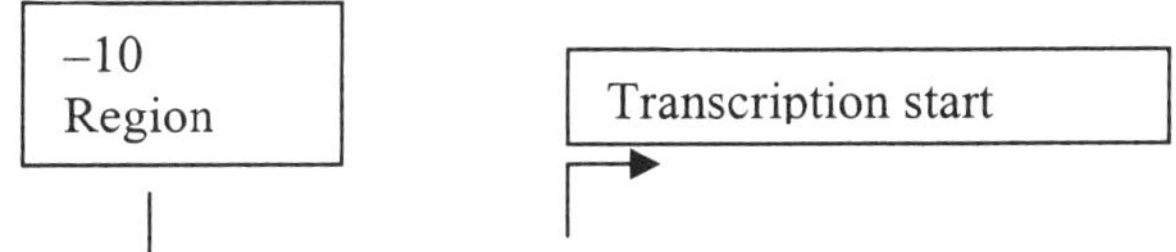

5'–GGACTATATGATGCGGCCCAT–3'

3'–CCTGATATACTACGCCGGGTA–5'

The –10 region, or Pribnow Box, has the consensus sequence of TATAAT. However, few bacterial promoters actually contain the exact consensus sequence. A common sequence at the transcription start site is 5'–CAT–3' with transcription beginning at the "A."

25. The following diagram represents a transcription unit in a hypothetical DNA molecule:

5'...TTGACA...TATAAT...3'
3'...AACTGT...ATATTA...5'

(a) On the basis of the information given, is this DNA from a bacterium or from a eukaryotic organism?
The DNA is from a bacterium as evidenced by the TATAAT sequence that corresponds to the –10 consensus sequence of a bacterial promoter and the TTGACA sequence that is identical to the –35 consensus sequence of a bacterial promoter.

(b) If this DNA molecule is transcribed, which strand will be the template strand and which will be the nontemplate strand?
5'...TTGACA...TATAAT...3' Nontemplate
3'...AACTGT...ATATTA...5' Template

(c) Where, approximately, will the start site of transcription be?

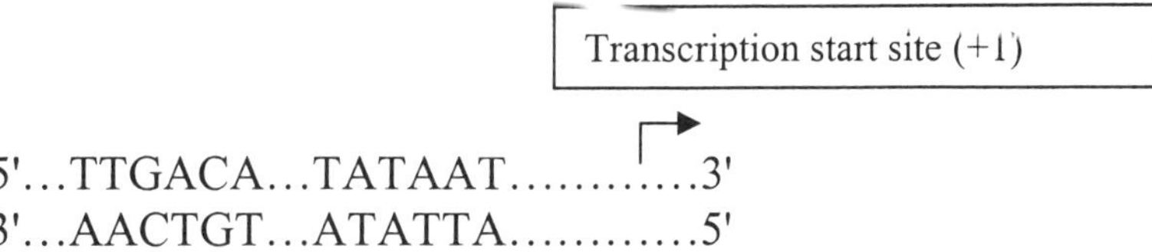

5'...TTGACA...TATAAT............3'
3'...AACTGT...ATATTA............5'

The start site should be located approximately 9 nucleotides downstream from the last nucleotide of the –10 consensus sequence (TATAAT) or the +1 nucleotide.

*26. What would be the most likely effect of a mutation at the following locations in *E. coli* gene?

(a) –8
A mutation at the –8 position would probably affect the –10 consensus sequence (TATAAT), which is centered on position –10. This consensus sequence is

necessary for binding of RNA polymerase. A mutation in there would most likely decrease transcription.

(b) –35

A mutation in the –35 region could affect the binding of the sigma factor to the promoter. Deviations away from the consensus typically down regulate transcription, so transcription is likely to be reduced or inhibited.

(c) –20

The –20 region is located between the consensus sequences of an E. coli *promoter. Although the holoenzyme may cover the site, it is unlikely that a mutation will have any effect on transcription.*

(d) Start site

A mutation in the start site would have little effect on transcription. The position of the start site relative to the promoter is more important than the sequence at the start site.

27. A strain of bacteria possesses a temperature-sensitive mutation in the gene that encodes the sigma factor. At elevated temperatures, the mutant bacteria produce a sigma factor that is unable to bind to RNA polymerase. What effect will this mutation have on the process of transcription when the bacteria are raised at elevated temperatures?

Binding of the sigma factor to the RNA polymerase core enzyme forms the RNA polymerase holoenzyme. Only the holoenzyme binds to the promoter. Without the sigma factor, RNA polymerase will be unable to bind the promoter and transcription initiation will not occur. Any RNA polymerase that has completed transcription inititation and has begun elongation will complete transcription because the sigma factor is not needed for elongation. However, no further initiation will be possible at the elevated temperature.

28. Computer programmers, working with molecular geneticists, have developed computer programs that can identify genes within long stretches of DNA sequences. Imagine that you are working with a computer programmer on such a project. Based on what you know about the process of transcription, what sequences might you suggest the computer program look for to identify the beginning and end of a gene?

The sequences recommended to the programmer will depend on the type of organism whose DNA is being studied. However, both bacteria and eukaryotic organisms have both promoter and termination consensus DNA sequences that could be recognized by the programmer's computer program as potential sites of interest for genes. The program should be able to recognize the consensus sequences as well as deviations from the consensus since the consensus sequence may only rarely be present. The identification by the program of both promoter sequences, terminator sequences and/or potential cleavage sites in a region of the DNA should provide evidence of a gene's presence in that region. For bacterial promoters, the –35 sequences and –10 sequences would be good starting points. At the end of bacterial genes are terminators. The program should be able to find inverted repeats and strings of A's which are indicative of Rho-independent termination. For Rho-dependent termination sequences may prove more difficult to identify. The presence of both the promoter sequences and a termination sequence in a region of DNA would be evidence of a bacterial gene. Similar promoter and termination sequences as well as 3' cleavage site sequences *in eukaryotic cells could also be used. For RNA polymerase II genes, identifying sequences of the core promoter such as the TFIIB recognition element(BRE) sequences, the TATA box, the initiator element sequence, and the downstream core promoter element sequences might indicate the beginning of a gene. Although not every RNA polymerase II promoter contains each sequence, the core promoter will contain at least one of these sequences. Transcription terminators at the ends of RNA polymerase II transcribed genes could prove difficult to recognize, however another potential sequence for identification would be the cleavage site consensus sequences at the 3' regions of the gene.*

*29. The following diagram represents a transcription unit on a DNA molecule.
(a) Assume that this DNA molecule is from a bacterial cell. Draw in the approximate location of the promoter and terminator for this transcription unit.

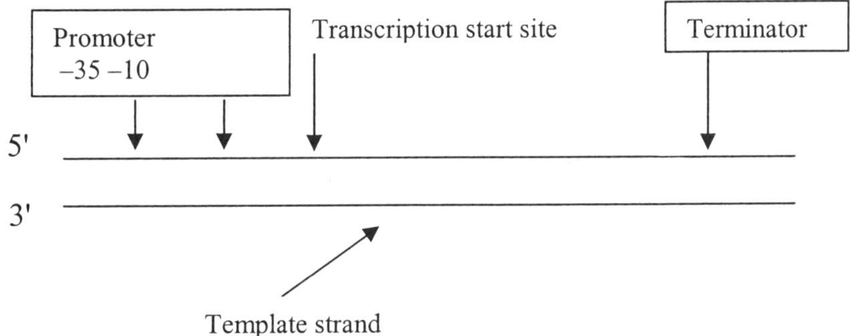

(b) Assume that this DNA molecule is from a eukaryotic cell. Draw in the approximate location of an RNA polymerase II promoter.

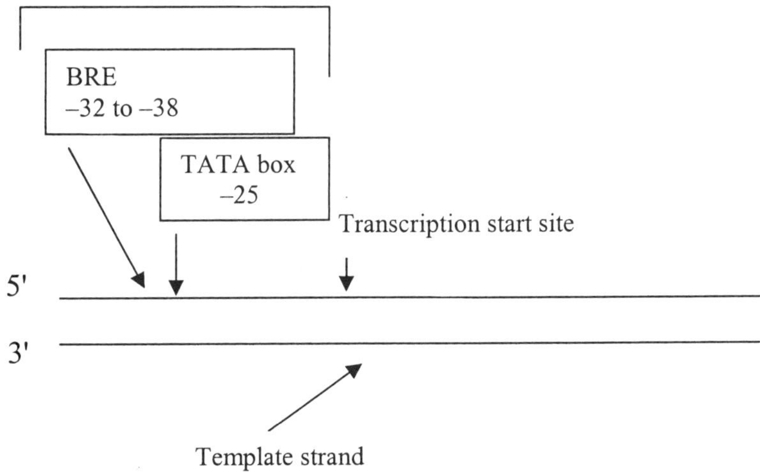

(c) Assume that this DNA molecule is from a eukaryotic cell. Draw in the approximate location of an internal RNA polymerase III promoter.

Two examples of RNA polymerase III promoters are shown.

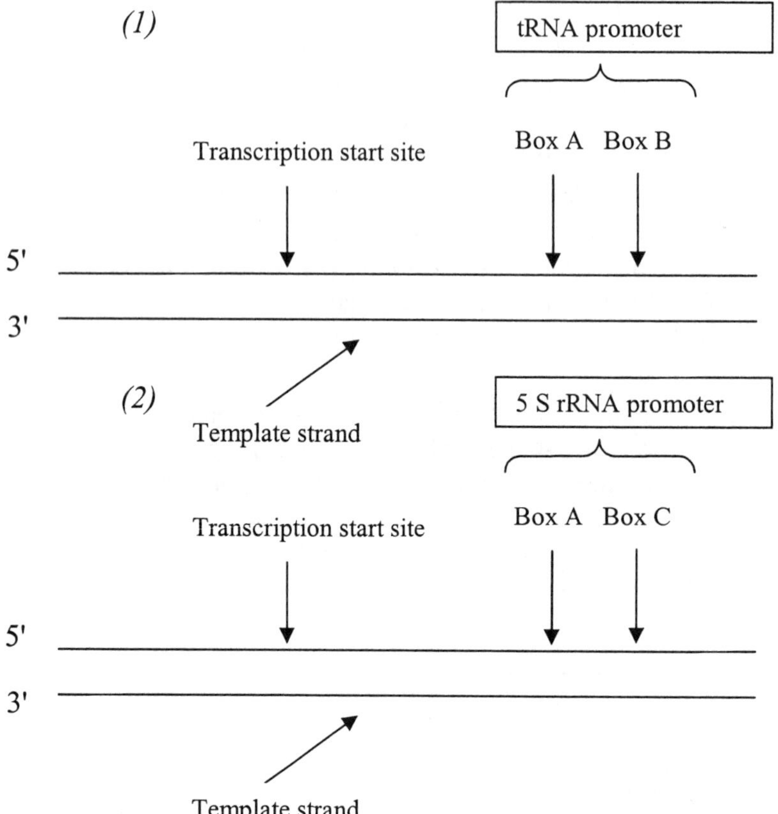

30. The following DNA nucleotides are found near the end of a bacterial transcription unit. Find the terminator in this sequence:
3'–AGCATACAGCAGACCGTTGGTCTGAAAAAAGCATACA–5'

(a) Mark the point at which transcription will terminate.

(b) Is this terminator intrinsic or rho-dependent?
Based on the potential hairpin structure that can form and the run of U's that will be synthesized in the RNA, this terminator is an intrinsic terminator.

(c) Draw a diagram of the RNA that will be transcribed from this DNA, including its nucleotide sequence and any secondary structures that form.

```
                 A
               C   A
                GC
                GC
                UA
                CG
                UA
5'–UCGUAUGUCG CUUUU–3'
```

*31. A strain of bacteria possesses a temperature-sensitive mutation in the gene that encodes the rho subunit of RNA polymerase. At high temperatures, rho is not functional. When these bacteria are raised at elevated temperatures, which of the following effects would you expect to see? Explain your reasoning for accepting or rejecting each of these five options.

(a) Transcription does not take place.

Because the rho protein is involved in transcription termination, it should not affect transcription initiation or elongation. So you would expect to see transcription.

(b) All RNA molecules are shorter than normal.

Without the rho protein, transcription would be expected to continue past the normal termination site of rho-dependent terminators producing some longer molecules than expected.

(c) All RNA molecules are longer than normal.

Only RNA molecules produced from genes using rho-dependent termination should be longer. Genes that are terminated through rho-independent termination should remain unaffected.

(d) Some RNA molecules are longer than normal.

You would expect to see some RNA molecules that are longer than normal. Only genes that use rho-dependent termination would be expected to not terminate at the normal termination site, thus producing some RNA molecules that are longer than normal.

(e) RNA is copied from both DNA strands.

RNA will be copied from only a single strand because rho protein does not affect transcription initiation or elongation.

32. Suppose that the string of A's following the inverted repeat in a rho-independent terminator was deleted, but the inverted repeat was left intact. How would this deletion affect termination? What would happen when RNA polymerase reached this region?

Termination would not take place if the string of A's following the inverted repeat were deleted. Although RNA polymerase may stall briefly at the hairpin, the presence of the A–U base pairs is needed to destabilize the DNA-RNA interaction and to end transcription. If the string of A's is not present, then transcription by RNA polymerase would continue.

*33. Through genetic engineering, a geneticist mutates the gene that encodes TBP in cultured human cells. This mutation destroys the ability of TBP to bind to the TATA box. Predict the effect of this mutation on cells that possess it.

If TBP cannot bind the TATA box, then genes with these promoters will be transcribed at very low levels or not at all. Because the TATA box is the most common promoter element for RNA polymerase II transcription units and is found in some RNA polymerase III promoters, transcription will decline significantly. The lack of proteins encoded by these genes will most likely result in cell death.

34. Elaborate repair mechanisms are associated with replication to prevent permanent mutations in DNA, yet no similar repair is associated with transcription. Can you think of a reason for these differences in replication and transcription? (Hint: Think about the relative effects of a permanent mutation in a DNA molecule compared with one in an RNA molecule.)

RNA molecules are constantly being synthesized and subsequently degraded. Typically, RNA is produced regularly by new transcription. A defective RNA molecule would soon be replaced and would only briefly affect a cell. Furthermore, the defective RNA would not be passed to offspring of that cell. DNA mutations, however, are permanent. All RNAs transcribed from that gene would be affected by a change in the DNA sequence. The mutation in the DNA molecule would be passed to progeny cells and propagated. Finally, replication occurs only once during the cell cycle. Once the mutation occurs, there is no subsequent replication that could produce the correct sequence.

CHALLENGE QUESTIONS

35. Enhancers are sequences that affect the initiation of the transcription of genes that are hundreds or thousands of nucleotides away. Enhancer-binding proteins usually interact directly with transcription factors at promoters by causing the intervening DNA to loop out. An enhancer of the bacteriophage T4 does not function by looping of the DNA (D. R. Herendeen, et al., 1992, *Science* 256:1298–1303). Propose some additional mechanisms (other than DNA looping) by which this enhancer might affect transcription at a gene thousands of nucleotides away.

For T4 late genes, the replication fork serves as a mobile enhancer. Three T4 replication proteins also serve as transcriptional activators from the enhancers. Two potential mechanisms by which these proteins can activate transcription are (1) DNA tracking by the enhancer proteins and (2) topological changes in the DNA induced by the binding of the T4 enhancer by the transcriptional activator proteins. In the DNA tracking model, the enhancer binding proteins move along the DNA and activate the transcriptional apparatus either at the promoter, or, perhaps, the activator proteins deliver the transcriptional apparatus to the promoter. In the second model, binding of the enhancer by the activator proteins leads to topological changes in the DNA structure that makes the promoter more accessible to the transcriptional apparatus.

36. Many genes in bacteria and eukaryotes contain numerous sequences that potentially cause pauses or premature terminations of transcription. Nevertheless, transcription of these genes within a cell normally produces multiple RNA molecules thousands of nucleotides long without pausing or terminating prematurely. However, when a single round of transcription takes place on such templates in a test tube, RNA synthesis is frequently interrupted by pauses and premature terminations, which slow the rate at which transcription occurs and frequently reduces the length of mRNA molecules produced. The majority of these pauses and premature terminations occur when RNA polymerase temporarily backtracks (i.e., backs up) for one or two nucleotides along the DNA. Experimental studies have demonstrated that most transcriptional delays and premature terminations disappear if several RNA polymerases are simultaneously transcribing the DNA molecule. Propose an explanation for why transcription is more rapid and results in longer mRNA when the template DNA is being transcribed by multiple RNA polymerases.
When a single RNA polymerase is transcribing a template in a test tube, the backtracking often may lead to the RNA polymerase and the 3′ end of the RNA transcript losing contact with each other resulting in premature termination. If other RNA polymerases are present on the template, they may act cooperatively with the leading RNA polymerase that has backtracked by pushing the leading RNA polymerase foward thus continuing transcription before termination can occur.

*37. The location of the TATA box in two species of yeast, *Saccharomyces pombe* and *S. cerevisiae,* differs dramatically. The TATA box of *S. pombe* is about 30 nucleotides upstream of the start site, similar to the location for most other eukaryotic cells. However, the TATA box of *S. cerevisiae* can be as many as 120 nucleotides upstream of the start site. To understand how the TATA box functions in these two species, a series of experiments was conducted to determine which components of the transcription apparatus of these two species could be interchanged. In these experiments, different components of the transcription apparatus were switched in *S. pombe* and *S. cerevisiae,* and the effects of the switch on the level of RNA synthesis and on the start point of transcription were observed. TFIID from *S. pombe* could be used in *S. cerevisiae* cells and vice versa, without any effect on the transcription start site in either cell type. Switching TFIIB, TFIIE, or RNA polymerase did alter the level of transcription. However, the following pairs of components could be exchanged without affecting transcription: TFIIE together with TFIIH; and TFIIB together with RNA polymerase. Exchange of TFIIE-TFIIH did not alter the start point, but exchange of TFIIB-RNA polymerase did shift it (Y. Li, P. M. Flanagan, H. Tschochner, and R. D. Kornberg, 1994, *Science* 263:805–807).

On the basis of these results, what conclusions can you draw about how the different components of the transcription apparatus interact and which components are responsible for setting the start site? Propose a mechanism for the determination of the start site in eukaryotic RNA polymerase II promoters.

When TFIIE and TFIIH were individually exchanged, transcriptional levels were altered. However, the paired exchange of TFIIE and TFIIH had no effect on the level of transcription, suggesting that TFIIE and TFIIH must interact to promote transcription, or that the absence of their interaction will inhibit transcription. Similar results were obtained by the paired exchange of TFIIB and RNA polymerase, which suggests interactions between TFIIB and RNA polymerase are needed for transcription. Furthermore, the exchange of TFIIB and RNA polymerase shifted the start point of transcription indicating that TFIIB plays a role in positioning RNA polymerase at the proper start site for transcription.

38. The relationship between chromatin structure and transcription activity has been the focus of recent research. In one set of experiments, the level of *in vitro* transcription of a *Drosophila* gene by RNA polymerase II was studied with the use of DNA and various combinations of histone proteins.

First, the level of transcription was measured for naked DNA with no associated histone proteins. Next, the level of transcription was measured after nucleosome octamers (without H1) were added to the DNA. The addition of the octamers caused the level of transcription to drop by 50%. When both nucleosome octamers and H1 proteins were added to the DNA, transcription was greatly repressed, dropping to less than 1% of that obtained with naked DNA (see the adjoining table). GAL4-VP16 is a protein that binds to the DNA of certain eukaryotic genes. When GAL4-VP16 is added to DNA, the level of RNA polymerase II transcription is greatly elevated. Even in the presence of the H1 protein, GAL4-VP16 stimulates high levels of transcription.

Propose a mechanism for how the H1 protein represses transcription and how GAL4-VP16 overcomes this repression. Explain how your proposed mechanism would produce the results obtained in these experiments.

Treatment	Relative Amount of Transcription
Naked DNA	100
DNA + octamers	50
DNA + octamers + H1	<1
DNA + GAL4-VP16	1000
DNA + octamers + GAL4-VP16	1000
DNA + octamers + H1 + GAL4-VP16	1000

(Based on experiments reported in an article by G. E. Croston et al., 1991, *Science* 251:643–649.)

The addition of H1 in the presence of nucleosome octamers almost completely inhibits transcription, suggesting that the promoter is not available to the transcriptional apparatus. If GAL4-VP16 could remove and prevent H1 from blocking the promoter, then transcription initiation would occur. Also, the addition of GAL4-VP16 to the DNA increases transcription tenfold over naked DNA. So GAL4-VP16 is likely a transcriptional activator protein as well.

Chapter Fourteen: RNA Molecules and RNA Processing

COMPREHENSION QUESTIONS

*1. What is the concept of colinearity? In what way is this concept fulfilled in bacterial and eukaryotic cells?
According to the concept of colinearity, the number of amino acids in a protein should be proportional to the number of nucleotides in the gene encoding the protein. In bacteria, the concept is nearly fulfilled. However, in eukaryotes, long regions of noncoding DNA sequence split coding regions. So, the concept of colinearity at the DNA level is not fulfilled.

2. What are some characteristics of introns?
Eukaryotic genes commonly contain introns. However, introns are rare in bacterial genes. The number of introns found in an organism's genome is typically related to complexity—more complex organisms possess more introns. Introns typically do not encode proteins and are usually larger in size than exons.

*3. What are the four basic types of introns? In which genes are they found?
(1) Group I introns are found in rRNA genes and some bacteriophage genes.
(2) Group II introns are found in protein-encoding genes of mitochondria, chloroplasts, and a few eubacteria.
(3) Nuclear pre-mRNA introns are found in the protein-encoding genes of the nucleus.
(4) Transfer RNA introns are found in tRNA genes.

*4. What are the three principal elements in mRNA sequences in bacterial cells?
(1) The 5′ untranslated region, which contains the Shine-Dalgarno sequence.
(2) The protein-encoding region.
(3) The 3′ untranslated region.

5. What is the function of the Shine-Dalgarno consensus sequence?
The Shine-Dalgarno consensus sequence functions as the ribosome binding site on the mRNA molecule.

*6. (a) What is the 5' cap?
The 5′ end of eukaryotic mRNA is modified by the addition of the 5′ cap. The cap consists of an extra guanine nucleotide linked 5′ to 5′ to the mRNA molecule that is methylated at position 7 of the base and of sugars that are methylated at the 2′–OH of the ribose sugar of adjacent bases.
(b) How is the 5' cap added to eukaryotic pre-mRNA?
Initially, the terminal phosphate of the three 5′ phosphates linked to the end of the mRNA molecule is removed. Subsequently, a guanine nucleotide is attached to the 5′ end of the mRNA using a 5′ to 5′ phosphate linkage. Next, a methyl group is attached to position 7 of the guanine base. Ribose sugars of adjacent nucleotides may also be methylated, but at the 2′–OH.

(c) What is the function of the 5' cap?

CAP binding proteins recognize the 5' cap and stimulate binding of the ribosome to the 5' cap and to the mRNA molecule. The 5' cap may also increase mRNA stability in the cytoplasm. Finally, the 5' cap is needed for efficient splicing of the intron that is nearest the 5' end of the pre-mRNA molecule.

7. How is the poly(A) tail added to pre-mRNA? What is the purpose of the poly(A) tail?

Initially a complex consisting of several proteins forms on the 3' UTR of the pre-mRNA molecule. Cleavage and polyadenylation specificity factor (CPSF) bind to the AAUAAA consensus sequence, which is located upstream of the 3' cleavage site. Another protein, cleavage stimulation factor (CsTF), binds downstream of the cleavage site. Two cleavage factors (CFI and CFII) and polyadenylate polymerase (PAP) also become part of the complex. Once the complex has formed, the pre-mRNA is cleaved. CsTF and the two cleavage factors leave the complex. PAP adds approximately 10 adenine nucleotides to the 3' end of the pre-mRNA molecule. The addition of the short poly(A) tail allows for the binding of the poly(A) binding protein (PABII) to the tail. PABII increases the rate of polyadenylation, which subsequently allows for more PABII protein to bind the tail.

The presence of the poly(A) tail increases the stability of the mRNA molecule through the interaction of proteins at the poly(A) tail.

*8. What makes up the spliceosome? What is the function of the spliceosome?

The spliceosome consists of five small ribonucleoproteins (snRNPs). Each snRNP is composed of multiple proteins and a single small nuclear RNA molecule or snRNA. The snRNPs are identified by which snRNA (U1, U2, U3, U4, U5, or U6) each contains. Splicing of pre-mRNA nuclear introns takes place within the spliceosome.

9. Explain the process of pre-mRNA splicing in nuclear genes.

Removal of an intron from the pre-mRNA requires the assembly of the spliceosome complex on the pre-mRNA, cleavage at both the 5' and 3' splice sites of the intron, and two transesterification reactions ultimately leading to the joining of the two exons. Initially, snRNP U1 binds to the 5' splice site through complementary base pairing of the U1 snRNA. Next, snRNP U2 binds to the branch point within the intron. The U5 and U4–U6 complex joins the spliceosome, resulting in the looping of the intron so that the branch point and 5' splice site of the intron are now adjacent to each other. U1 and U4 now disassociate from the spliceosome, and the spliceosome is activated. The pre-mRNA is then cleaved at the 5' splice site, producing an exon with a 3'–OH. The 5' end of the intron folds back and forms 5'–2' phosphodiester linkage through the first transesterification reaction with the adenine nucleotide at the branch point of the intron. This looped structure is called the lariat. Next, the 3' splice site is cleaved and then immediately ligated to the 3'–OH of the first exon through the second transesterification reaction. Thus, the exons are now joined and the intron has been excised.

10. Describe two types of alternative processing pathways. How do they lead to the production of multiple proteins from a single gene?
Alternative processing of pre-mRNA can take the form of either alternative splicing of pre-mRNA introns or the alternative cleavage of 3' cleavage sites in a pre-mRNA molecule containing two or more cleavage sites for polyadenylation. Alternative splicing results in different exons of the pre-mRNA being ligated to form mature mRNA. Each mRNA formed by an alternative splicing process will yield a different protein. In pre-mRNA molecules with multiple 3' cleavage sites, cleavage at the different sites will generate mRNA molecules that differ in size. Each alternatively cleaved RNA will code for a different protein. A single pre-mRNA transcript can undergo both alternative processing steps, thus potentially producing multiple proteins.

*11. What is RNA editing? Explain the role of guide RNAs in RNA editing.
RNA editing alters the sequence of a RNA molecule after transcription either by the insertion, deletion, or modification of nucleotides within the transcript. The guide RNAs (gRNAs) provide templates for the alteration of nucleotides in RNA molecules undergoing editing and are complementary to regions within the pre-edited RNA molecule. At these complementary regions, the gRNAs base pair to the pre-edited RNA molecule. The alteration of nucleotides at the paired region follows.

*12. Summarize the different types of processing that can take place in pre-mRNA.
Several modifications to pre-mRNA take place to produce mature mRNA.
(1) Addition of the 5' cap to the 5' end of the pre-mRNA.
(2) Cleavage of the 3' end of a site downstream of the AAUAAA consensus sequence of the last exon.
(3) Addition of the poly(A) tail to the 3' end of the mRNA immediately following cleavage.
(4) Removal of the introns (splicing).

*13. What are some of the modifications in tRNA processing?
(1) Cleavage of the precursor RNA into smaller molecules.
(2) Nucleotides at both the 5' and 3' ends of the tRNAs maybe be removed or trimmed.
(3) Standard bases can be altered by base-modifying enzymes.

14. Describe the basic structure of ribosomes in bacterial and eukaryotic cells.
Ribosomes in both eukaryotes and bacteria consist of a complex of protein and RNA molecules. A functional ribosome is composed of a large and a small subunit. The bacterial 70S ribosome consists of a 30S small subunit and a 50S large subunit. Within the small subunit are a single 16S RNA molecule and 21 proteins. The 23S and the 5S RNA molecules, along with 31 proteins, are found in the large bacterial subunit. The eukaryotic 80S ribosome is comprised of a 60S large subunit and a 40S small subunit. Three RNA molecules, the 28S RNA, the 5.8S RNA, and the 5S RNA, are located in the large subunit as well as 49 proteins. The eukaryotic small subunit contains only a single 18S RNA molecule and 33 proteins.

*15. Explain how rRNA is processed.

Most rRNAs are synthesized as large precursor RNAs that are processed by methylation, cleavage, and trimming to produce the mature mRNA molecules. In E. coli, *methylation occurs to specific bases and the 2'–OH of the ribose sugars of the 30S rRNA precursor. The 30S precursor is cleaved and trimmed to produce the 16S rRNA, 23S rRNA, and the 5S rRNA. In eukaryotes, a similar process occurs. However, small nucleolar RNAs help to cleave and modify the precursor rRNAs.*

16. What is the origin of siRNAs and microRNAs? What do these RNA molecules do in the cell?

The siRNAs originate from the cleavage of mRNAs, RNA transposons, and RNA viruses by the enzyme Dicer. Dicer may produce multiple siRNAs from a single double-stranded RNA molecule. The double-stranded RNA molecule may occur due to the formation of hairpins or by duplexes between different RNA molecules. The miRNAs arise from the cleavage of individual RNA molecules that are distinct from other genes. The enzyme Dicer cleaves these RNA molecules that have formed small hairpins. A single miRNA is produced from a single RNA molecule.

Both siRNAs and microRNAs silence gene expression through a process called RNA interference. Both function by shutting off gene expression of a cell's own genes or to shut off expression of genes from the invading foreign genes of viruses or tranposons. The microRNAs typically silence genes that are different from those from which the microRNAs are transcribed. However, the siRNAs usually silence genes from which they are transcribed.

17. What are some similarities and differences between siRNAs and miRNAs?

Some similarities:

- *Both silence gene expression of foreign genes or a cell's own genes after combining with proteins to form a RNA-Induced-Silencing Complex (RISC)*
- *Both are dsRNA molecules 21 and 22 nucleotides in length.*
- *Both are produced by the action of the enzyme Dicer on larger RNA molecules.*

Some differences:

- *The miRNAs originate from transcription products of distinct genes while siRNAs originate from mRNA, RNA transposons, or RNA viruses.*
- *For miRNAs DICER cleaves single-stranded RNAs that form short hairpins, while for siRNAs DICER cleaves single-stranded RNAs that form long hairpins or RNA duplexes.*
- *The miRNAs silence genes that are different from the ones from which they were transcribed, while siRNAs silence genes from which they were transcribed.*
- *The miRNAs typically act by inhibiting translation although some do trigger the degradation of the mRNA, while the siRNAs work by stimulating degradation of the mRNA target molecule.*

APPLICATION QUESTIONS AND PROBLEMS

*18. At the beginning of the chapter, we considered Duchenne muscular dystrophy and the dystrophin gene. We learned that the gene causing Duchenne muscular dystrophy encompasses more than 2 million nucleotides, but less than 1% of the gene encodes the protein dystrophin. On the basis of what you now know about gene structure and RNA processing in eukaryotic cells, provide a possible explanation for the large size of the dystrophin gene.
The large size of the dystrophin gene is likely due to the presence of many intervening sequences or introns within the coding region of the gene. Excision of the introns through RNA splicing yields the mature mRNA that encodes the dystrophin protein.

19. How do the mRNA of bacterial cells and the pre-mRNA of eukaryotic cells differ? How do the mature mRNAs of bacterial and eukaryotic cells differ?
Bacterial mRNA is translated immediately upon being transcribed. Eukaryotic pre-mRNA must be processed. Bacterial mRNA and eukaryotic pre-mRNA have similarities in structure. Each has a 5' untranslated region as well as a 3' untranslated region. Both also have protein-coding regions. However, the protein-coding region of the pre-mRNA is disrupted by introns. The eukaryotic pre-mRNA must be processed to produce the mature mRNA. Eukaryotic mRNA has a 5' cap and a poly(A) tail, unlike bacterial mRNAs. Bacterial mRNA also contains the Shine-Dalgarno consensus sequence. Eukaryotic mRNA does not have the equivalent.

*20. Draw a typical eukaryotic gene and the pre-mRNA and mRNA derived from it. Assume that the gene contains three exons. Identify the following items and, for each item, give a brief description of its function.

(a) 5' untranslated region
The 5' untranslated region lies upstream of the translation start site. In bacteria, the ribosome binding site or Shine-Dalgarno sequence is found within the 5' untranslated region. However, eukaryotic mRNA does not have the equivalent sequence, and a eukaryotic ribosome binds at the 5' cap of the mRNA molecule.

(b) Promoter
The promoter is the DNA sequence that the transcription apparatus recognizes and binds to initiate transcription.

(c) AAUAAA consensus sequence
The AAUAAA consensus sequence lies downstream of the coding region of the gene. It determines the location of the 3' cleavage site in the pre-mRNA molecule.

(d) Transcription start site
The transcription start site begins the coding region of the gene and is located 25 to 30 nucleotides downstream of the TATA box.

(e) 3' untranslated region
The 3' untranslated region is a sequence of nucleotides at the 3' end of the mRNA that is not translated into proteins. However, it does affect the translation of the mRNA molecule as well as the stability of the mRNA.

(f) Introns

Introns are noncoding sequences of DNA that intervene within coding regions of a gene.

(g) Exons

Exons are transcribed regions that are not removed in intron processing. They include the 5'UTR, coding regions that are translated into amino acid sequences, and the 3'UTR.

(h) Poly(A) tail

A poly(A) tail is added to the 3' end of the pre-mRNA. It affects mRNA stability.

(i) 5' cap

The 5' cap functions in the initiation of translation and mRNA stability.

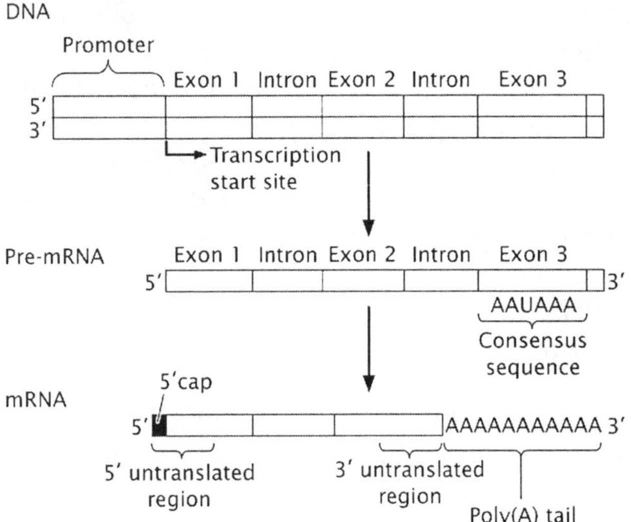

21. How would the deletion of the Shine-Dalgarno sequence affect a bacterial mRNA?
In bacteria, the small ribosomal subunit binds to the Shine-Dalgarno sequence to initiate translation. If the Shine-Dalgarno sequence is deleted, then translation initiation cannot take place, preventing protein synthesis.

*22. How would the deletion of the following sequences or features most likely affect a eukaryotic pre-mRNA?

(a) AAUAAA consensus sequence

The deletion of the AAUAA consensus sequence would prevent binding of the cleavage and polydenylation factor (CPSF), thus resulting in no polyadenylation of the pre-mRNA. This would affect the stability and translation of the mRNA.

(b) 5' cap

The deletion of the 5' cap would most likely prevent splicing of the intron that is nearest to the 5' cap. Ultimately, elimination of the cap will affect the stability of the pre-mRNA as well as its ability to be translated.

(c) Poly(A) tail

Polyadenylation increases the stability of the mRNA. If eliminated from the pre-mRNA, then the mRNA would be degraded quickly by nucleases in the cytoplasm.

23. What would be the most likely effect on the amino acid sequence of a protein of a mutation that occurred in an intron of the gene encoding the protein? Explain your answer.
Because introns are removed prior to translation, an intron mutation would have little effect on a protein's amino acid sequence unless the mutation occurred within the 5′ splice site, the 3′ splice site, or the branch point. If mutations within these sequences altered splicing, then the mature mRNA would be altered, thus altering the amino acid sequence of the protein. The result could be a protein with additional amino acid sequence. Or, possibly, the altered splicing could introduce a stop codon that stops translation prematurely.

24. A geneticist induces a mutation in the gene that codes for cleavage and polyadenylation specificity factor (CPSF) in a line of cells growing in the laboratory. What would be the immediate effect of this mutation on RNA molecules in the cultured cells?
CPSF is needed for cleavage of the 3′ UTR and for polyadenylation. Nonfunctional CPSF would result in mRNA lacking a poly(A) tail, and the mRNA would be degraded more quickly in the cytoplasm by nucleases.

*25. A geneticist mutates the gene for proteins that bind to the poly(A) tail in a line of cells growing in the laboratory. What would be the immediate effect of this mutation in the cultured cells?
The stability of the mRNA is dependent on the proteins that bind to the poly(A) tail. If the proteins are unable to bind to the tail, then the mRNA that contain poly(A) tails will be degraded at a much more rapid rate within the cells.

*26. An in vitro (within a test tube) splicing system has been developed that contains all the components (snRNAs, proteins, splicing factors) necessary for the splicing of nuclear pre-mRNA genes. When a piece of RNA containing an intron and two exons is added to the system, the intron is removed as a lariat and the exons are spliced together. What intermediate products of the splicing reaction would accumulate if the following components were omitted from the splicing system? Explain your reasoning.

(a) U1
RNAs will be unspliced and intact. U1 binding to the 5′ splice site is the first step in the splicing reaction, and is necessary for all subsequent steps.

(b) U2
RNAs will be unspliced and intact, but with U1 bound to the 5′ splice site. U2 binding to the branch site is the second step in the splicing reaction; without U2, no further reactions can take place.

(c) U6
RNAs will be unspliced and intact, but with U1 and U2 bound. The U6/U4 dimer and U5 form a catalytic complex. A mixture missing any one of these snRNPs will not be catalytically active.

(d) U5
Same as (c).

(e) U4
Same as (c).

27. The splicing system introduced in Problem 26 is used to splice an RNA molecule containing two exons and one intron. This time, however, the U2 snRNA used in the splicing reaction contains several mutations in the sequence that pairs with the U6 snRNA. What would be the effect of these mutations on the splicing process?
Disruption of the base pairing between U2 and U6 would result in splicing not taking place. Assembly of the spliceosome at the 5′ splice site and branch point will occur on the pre-mRNA. However, if U2 cannot pair with U6, then the spliceosome will not be activated. Thus, splicing will not take place.

28. A geneticist isolates a gene that contains five exons. He then isolates the mature mRNA produced by this gene. After making the DNA single-stranded, he mixes the single-stranded DNA and RNA. Some of the single-stranded DNA hybridizes (pairs) with the complementary mRNA. Draw a picture of what the DNA-RNA hybrids would look like under the electron microscope.

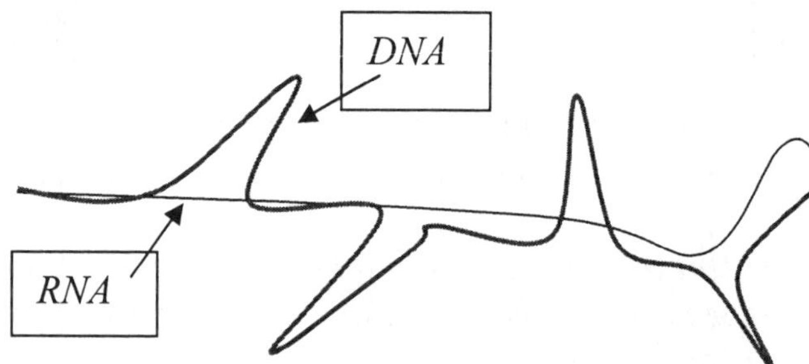

29. A geneticist discovers that two different proteins are encoded by the same gene. One protein has 56 amino acids and the other 82 amino acids. Provide a possible explanation for how the same gene could encode both of these proteins.
The premRNA molecules transcribed from the gene are likely processed by alternative processing pathways. Two possible mechanisms that could have produced the two different proteins from the same premRNA are alternative splicing or multiple 3′ cleavage sites in the premRNA. The cleavage of the premRNA molecule at different 3′ cleavage sites would produce alternatively processed mRNA molecules that differ in size. Translation from each of the alternative mRNAs would produce proteins containing different numbers of amino acids.

Alternative splicing of the premRNA could produce different mature mRNAs each containing a different number of exons. Again translation from each alternatively spliced mRNA would generate proteins that differ in the number of amino acids contained.

30. What would be the likely effects of deleting an exonic splicing enhancer from the exon of a gene?
If the exonic splicing enhancer (ESE) sequence is deleted from the exon of a gene, it is likely that the intron adjacent to the exon will not be removed. The ESE sequence is bound by SR proteins, which are necessary to recruit the spliceosome machinery to the splice site. If the spliceosome machinery does not assemble at the 5′ or 3′ splice site,

then it is likely that inappropriate splicing will occur. The result is that the exon will likely be removed during splicing events and not be part of the mRNA.

31. The chemical reagent psoralen can be used to elucidate nucleic acid structure. This chemical attaches itself to nucleic acids and, on exposure to UV light, forms covalent bonds between closely associated nucleotide sequences. Such cross-links provide information about the proximity of RNA molecules to one another in complex structures.

Psoralen cross-linking has been used to examine the structure of the spliceosome. In one study, the following cross-linked structures were obtained during splicing. U1, U2, U5, and U6 became cross-linked to pre-mRNA. U2 was cross-linked to U6 and to pre-mRNA. The U1, U5, and U6 cross-links with pre-mRNA were mapped to sequences near the 5' splice site, whereas the U2 snRNA cross-links with pre-mRNA were mapped to the branch site. After splicing, U2, U5, and U6 were cross-linked to the excised lariat.

Explain these results in regard to what is known about the structure of the spliceosome and how it functions in RNA splicing. (Based on D. A. Wassarman, and J. A. Steitz, 1992, Interactions of small nuclear RNAs with precursor messenger RNA during in vitro splicing, *Science* 257:1918–1925.)

Psoralen-induced cross-links are likely to form between RNA molecules that base pair with each other. In the spliceosome, the snRNAs of U1, U2, U5, and U6 form base pairs with the pre-mRNA. As might be expected, each of these snRNAs was covalently cross-linked to the pre-mRNA after the psoralen treatment. The snRNAs of U2 and U6 also share complementary sequences and form base pairs with each other within the spliceosome. Psoralen cross-linking of U2 and U6 indicates their close interaction in the spliceosome.

The location of the cross-links induced by psoralen treatment also matches the model of how the intron is spliced. U1 attaches at the 5' splice site of the pre-mRNA, which begins the assembly of the spliceosome on the intron. U6 aligns the 5' splice site with the branch point, while U5 helps to position both the 3' and 5' ends of the intron for splicing. Psoralen cross-links of U1, U5, and U6 map to the 5' splice site, which correlates to the suspected action of these molecules. U2 forms base pairs with the RNA sequence at the branch site, which is confirmed by the cross-linking of U2 to the branch site. Finally, before splicing occurs, U1 is released from the spliceosome and is not part of the active complex. However, U2, U5, and U6 are part of the active complex and help stabilize the intron secondary structure so that the two exons are in close proximity. The interaction of U2 and U6 orients the intron for splicing. When the intron is excised and released as a lariat, U2, U5, and U6 are still bound to the lariat, as indicated by the psoralen cross-linking of these snRNAs to the excised lariat. Only U4 is not linked to the pre-mRNA intron or exon. The lack of cross-linked U4 would be expected because U4 does not form base pairs with the pre-mRNA during the splicing reactions and functions to deliver U6 to the intron.

32. Explain how each of the following processes complicates the concept of colinearity.

(a) Trans-splicing

In trans-splicing, exons from different genes are spliced together during RNA processing events. Essentially, the mature mRNA product is not produced by DNA sequences that are contiguous or even necessarily on the same chromosome. This results in an amino acid sequence of the translated protein from trans-spliced mRNA being encoded by two or more different genes. According to the principle of colinearity, we would have expected the DNA sequence of a single gene to correspond to the amino acid sequence of the protein.

(b) Alternative splicing

Different mature mRNAs from a single gene can be produce by alternative splicing. Different arrangements of the gene's exons can occur in the mature mRNAs. Thus, different proteins can be encoded within the same gene as opposed to one gene corresponding to one protein as is predicted by the concept of colinearity.

(c) RNA editing

In RNA editing, genetic information is added to the premRNA after it is transcribed. In other words, the mature mRNA will contain information that was not part of the DNA from which it was transcribed. The result is that the nucleotide sequence of the gene does not correspond to the amino acid sequence of the protein—a clear violation of the concept of colinearity.

CHALLENGE QUESTIONS

33. In addition to snRNAs, the spliceosome contains a number of proteins. Some of these proteins are associated with the snRNAs to form snRNPs. Other proteins are associated with the spliceosome but are not associated with any specific snRNA.

One group of spliceosomal proteins comprises the precursor RNA-processing (PRP) proteins. Three PRP proteins that directly take part in splicing are PRP2, PRP16, and PRP22. The results of studies have shown that PRP2 is required for the first step of the splicing reaction, PRP16 acts at the second step, and PRP22 is required for the release of the mRNA from the spliceosome. Other studies have found that these PRP proteins have amino acid sequences similar to the sequences found in RNA helicase enzymes—enzymes that are capable of unwinding two paired RNA molecules. On the basis of this information, propose a functional role for PRP2, PRP16, and PRP22 in RNA splicing.

Formation of the spliceosome depends on the formation of base pairs between the nRNA of the snRNPs and the pre-mRNA. Also, U2 and U6 share complementary sequences, which allow them to pair together. Rearrangement of the spliceosome on the pre-mRNA will require the unwinding of these paired regions. The PRP proteins facilitate this unwinding and aid in conformational adjustments of the spliceosome.

Functioning PRP2 is required for the spliceosome to become active. It is likely that PRP2 has a role in altering the conformation of the spliceosome, converting it from an inactive form to an active one over the splice sites. Because PRP16 is needed for the second splicing reaction, its helicase function may be necessary to alter the conformation of the spliceosome over the 3' splice site so that splicing can occur.

Finally, PRP22 unwinds the pairing holding the spliceosome to the mRNA. This unwinding allows the spliceosome to be released from the mRNA.

34. In eukaryotic cells, a poly(A) tail is normally added to pre-mRNA molecules but not to rRNA or tRNA. Using recombinant DNA techniques, it is possible to connect a protein encoding gene (which is normally transcribed by RNA polymerase II) to a promoter for RNA polymerase I. This hybrid gene is subsequently transcribed by RNA polymerase I and the appropriate pre-mRNA is produced, but this pre-mRNA is not cleaved at the 3' end and a poly (A) tail is not be added.

 Propose a mechanism to explain how the type of promoter found at the 5 ' end of a gene can affect whether or not a poly (A) tail is added to the 3' end.
It is likely that the promoter influence on polyadenylation is due to the RNA polymerase II molecule that binds and initiates transcription from that promoter. Because only pre-mRNAs are cleaved and polyadenylated at the 3' end, RNA polymerase II could play an important role in stimulating 3' cleavage and polyadenylation. If a protein encoding gene is fused to a RNA polymerase I promoter, then RNA polymerase I will transcribe the protein encoding gene. Because RNA polymerase I will be responsible for the transciption events, RNA polymerase II will not be present and would be unable to stimulate the 3' cleavage reactions or the addition of the poly (A) tail. Recent research has indicated that RNA polymerase II is a critical component of the 3' cleavage reaction, which is necessary for the addition of the poly (A) tail.

35. SR proteins are essential to proper spliceosome assembly and are known to be involved in the regulation of alternative splicing. Surprisingly, the role of SR proteins in splice-site selection and alternative splicing is affected by the promoter that is used for transcription of the pre-mRNA. For example, it is possible through genetic engineering to create RNA polymerase II promoters that have somewhat different sequences. When pre-mRNAs with exactly the same sequences are transcribed from two different RNA polymerase II promoters that differ slightly in sequence, which promoter is used can affect how the pre-mRNA is spliced.

 Propose a mechanism for how the DNA sequence of an RNA polymerase II promoter could affect alternative splicing that takes place in the pre-mRNA.
Different promoters could attract different components of the RNA polymerase II transcriptional complex or even affect how transcription moves from the initiation stage to the elongation stage. Components of the transcription complex may interact with SR proteins and help locate these proteins at the ESEs within the exons. If different promoters have an affect on the transcription complex and its function, then the changes to the transcriptional complex may also affect their interactions with the SR proteins. The result might be that the SR proteins act at alternative splice sites when different promoters are present. Recent research has indicated that RNA polymerase II does interact with SR proteins.

Chapter Fifteen: The Genetic Code and Translation

COMPREHENSION QUESTIONS

1. What is the one gene, one enzyme hypothesis? Why was this hypothesis an important advance in our understanding of genetics?
The one gene, one enzyme hypothesis proposed by Beadle and Tatum states that each gene encodes a single, separate protein. Now that we know more about the nature of enzymes and genes, it has been modified to the one gene, one polypeptide hypothesis because many enzymes consist of multiple polypeptides. The original hypothesis helped establish a linear link between genes (DNA) and proteins.

*2. What three different methods were used to help break the genetic code? What did each reveal and what were the advantages and disadvantages of each?
Marshall Nirenberg and Johann Heinrich Matthaei used the enzyme polynucleotide kinase to create homopolymers of synthetic RNAs. Using a cell-free protein synthesizing system they were able to determine the amino acid coded by each homopolymer. By this method, the meanings for the amino acids specified by the codons UUU, AAA, CCC, and GGG were determined. The disadvantage is that the meanings for only four codons could be determined. The same system was also used to create copolymers that contained random mixtures of two nucleotides in a known ratio. Different amino acids in the protein depended on the ratio of the two nucleotides. To determine or predict the composition of the codons, the frequency of amino acids produced using the copolymer was compared with the theoretical frequencies expected for the codons. A disadvantage of this procedure is that it depended on random incorporation of the nucleotides, which did not always happen. A further problem was that the base sequence of the codon could not be determined—only the bases contained within the codon. The redundancy of the code also provided difficulties because several different codons could specify the same amino acid.

To solve these problems, Nirenberg and Leder mixed ribosomes bound to short RNAs of known sequences with charged tRNAs. The mixture was passed through a nitrocellulose filter to which the tRNAs paired to ribosome-mRNA stuck. They next determined the amino acids attached to the bound tRNAs. More than 50 codons were identified by this method. The difficulty is that not all tRNAs and codons could be identified with this method.

Gobind Khorana and his colleagues used a third method. They synthesized RNA molecules of known repeating sequences. Using a cell-free protein synthesizing system they produced proteins of alternating amino acids. However, this procedure could not specify which codon encodes which amino acid.

3. What are isoaccepting tRNAs?
Isoaccepting tRNAs are tRNA molecules that have different anticodon sequences but accept the same amino acids.

*4. What is the significance of the fact that many synonymous codons differ only in the third nucleotide position?

Synonymous codons code for the same amino acid, or, in other words, have the same meaning. A nucleotide at the third position of a codon pairs with a nucleotide in the first position of the anticodon. Unlike the other nucleotide positions involved in the codon-anticodon pairing, this pairing is often weak or "wobbles," and nonstandard pairings can occur. Because many synonymous codons differ at only the third nucleotide position, it is likely that in these codons the "wobble" and nonstandard base-pairing with the anticodons will result in the correct amino acid being inserted in the protein even if a nonstandard pairing occurs.

*5. Define the following terms as they apply to the genetic code:

(a) Reading frame

The reading frame refers to each different way that the groups of three nucleotides or codons can be read in a sequence of nucleotides. For any sequence of nucleotides, there are potentially three sets of codons that could specify the amino acid sequence of a polypeptide.

(b) Overlapping code

If an overlapping code is present, then a single nucleotide is included in more than one codon. The result for a sequence of nucleotides is that more than one type of polypeptide can be encoded within that sequence.

(c) Nonoverlapping code

In a nonoverlapping code, a single nucleotide is part of only one codon. For a sequence of RNA, this results in the production of a single type of polypeptide.

(d) Initiation codon

An initiation codon establishes the appropriate reading frame and specifies the first amino acid of the protein chain. Typically, the initiation codon is AUG; however, GUG and UUG can also serve as initiation codons.

(e) Termination codon

The termination codon signals the termination or end of translation and the end of the protein molecule. There are three types of termination codons—UAA, UAG, and UGA—which can also be referred to as stop codons or nonsense codons. These codons do not code for amino acids.

(f) Sense codon

A sense codon is a group of three nucleotides that code for an amino acid. There are 61 sense codons that code for the 20 amino acids commonly found in proteins.

(g) Nonsense codon

Nonsense codons or termination codons signal the end of translation. These codons do not code for amino acids.

(h) Universal code

In a universal code, each codon specifies or codes for the same amino acid in all organisms. The genetic code is nearly universal but not completely. Most of the exceptions occur in mitochondrial genes.

(i) Nonuniversal codons

Most codons are universal (or nearly universal) in that they specify the same amino acids in almost all organisms. However, there are exceptions where a codon has different meanings in different organisms. Most of the known exceptions are the termination codons, which in some organisms do code for amino acids. Occasionally, a sense codon is substituted for another sense codon.

6. How is the reading frame of a nucleotide sequence set?
The initiation codon on the mRNA sets the reading frame.

*7. How are tRNAs linked to their corresponding amino acids?
For each of the 20 different amino acids that are commonly found in proteins, there is a corresponding aminoacyl-tRNA synthetase that covalently links the amino acid to the tRNA molecule.

8. What role do the initiation factors play in protein synthesis?
Initiation factors are proteins that are required for the initiation of translation. In bacteria, there are three initiation factors (IF1, IF2, and IF3). Each one has a different role. IF1 promotes the disassociation of the large and small ribosomal subunits. IF3 binds to the small ribosomal subunit and prevents it from associating with the large ribosomal subunit. IF2 is responsible for binding GTP and delivering the fMet-tRNA$_f^{met}$ to the initiator codon on the mRNA. In eukaryotes, there are more initiation factors, but many have similar roles. Some of the eukaryotic initiation factors are necessary for recognition of the 5′ cap on the mRNA. Others possess a RNA helicase activity, which is necessary to resolve secondary structures.

9. How does the process of initiation differ in bacterial and eukaryotic cells?
Bacterial initiation of translation requires that sequences in the 16S rRNA of the small ribosomal subunit bind to the mRNA at the ribosome binding site or Shine-Dalgarno sequence. The Shine-Dalgarno sequence is essential in placing the ribosome over the start codon (typically AUG). In eukaryotes, there is no Shine-Dalgarno sequence. The small ribosomal subunit recognizes the 5′ cap of the eukaryotic mRNA with the assistance of initiation factors. Next, the ribosomal small subunit migrates along the mRNA scanning for the AUG start codon. In eukaryotes, the start codon is located with a consensus sequence called the Kozak sequence (5′–ACCAUGG–3′). Transcription in eukaryotes also requires more initiation factors.

*10. Give the elongation factors used in bacterial translation and explain the role played by each factor in translation.
Three elongation factors have been identified in bacteria: EF-TU, EF-TS, and EF-G. EF-TU joins with GTP followed by a tRNA charged with an amino acid. The charged tRNA is delivered to the ribosome at the "A" site. During the process of delivery, the GTP joined to EF-TU is cleaved to form a EF-TU-GDP complex. EF-TS is necessary to regenerate EF-TU-GTP. The elongation factor EF-G binds GTP and is necessary for the translocation or movement of the ribosome along the mRNA during translation.

11. What events bring about the termination of translation?
The process of termination begins when a ribosome encounters a termination codon. Because the termination codon would be located at the "A" site, no corresponding

tRNA will enter the ribosome. This allows for the release factors (RF_1, RF_2, and RF_3) to bind the ribosome. RF_1 recognizes and interacts with the stop codons UAA and UAG, while RF_2 can interact with UAA and UGA. A RF_3-GTP complex binds to the ribosome. Termination of protein synthesis is complete when the polypeptide chain is cleaved from the tRNA located at the "P" site. During this process, the GTP is hydrolyzed to GDP.

12. Give several examples of RNA-RNA interactions that take place in protein synthesis.
Several RNA-RNA interactions that take place during protein synthesis are important. The tRNA molecules form base pairs with codons on the mRNA. The 3′ end of the 16S rRNA within the small ribosomal subunit forms base pairs with Shine-Dalgarno sequence at the 5′ end of the mRNA. Ribosomal RNAs on the large and small subunit interact with tRNAs at the "A" and the "P" sites. The association of the large and small subunits of the ribosome is potentially the result of interactions between the 16S rRNA of the small subunit and the 23S rRNA of the large subunit.

13. How do prokaryotic cells overcome the problem of a stalled ribosome on a mRNA that has no termination codon? What about eukaryotic cells?
When a ribosome has stalled on a mRNA, a bacteria *cell uses transfer-messenger (tm)RNA to "move" the ribosome. The tmRNA consists of a tRNA component charged with alanine and a mRNA component. At the empty A-site of the stalled ribosome, elongation factor Tu delivers the charged tmRNA. The tmRNA acts as a tRNA and allows for the formation of a peptide bond between the alanine of the tmRNA and the peptide attached to the tRNA at the P site. The now uncharged tRNA from the P site is released and translation continues. However, the ribosome now uses the tmRNA as the template. Ten amino acids are added before a stop codon is reached. The result is the termination of translation and the release of the ribosome. The added amino acids target the peptide for degradation.*
Eukaryotic cells use nonstop mRNA decay. When a ribosome is stalled, the codon free A-site of the ribosome is recognized and bound by a protein that recruits other proteins to degrade the mRNA starting at the 3′ end.

14. What are some types of posttranslational modification of proteins?
Several different modifications can occur to a protein following translation. Frequently the amino terminal methionine may be removed. Sometimes in bacteria only the formyl group is cleaved from the N-formyl methionine, leaving a methionine at the amino terminal. More extensive modification occurs in some proteins that are originally synthesized as precursor proteins. These precursor proteins are cleaved and trimmed by protease enzymes to produce a functional protein. Glycoproteins are produced by the attachment of carbohydrates to newly synthesized proteins. Molecular chaperones are needed by many proteins to ensure that the proteins are folded correctly. Secreted proteins that are targeted for the membrane or other cellular locations frequently have 15 to 30 amino acids, called the signal sequence, removed from the amino terminal. Finally, acetylation of amino acids in the amino terminal of some eukaryotic proteins also occurs.

*15. Explain how some antibiotics work by affecting the process of protein synthesis.
A number of antibiotics bind the ribosome and inhibit protein synthesis at different steps in translation. Some antibiotics, such as streptomycin, bind to the small subunit and inhibit translation initiation. Other antibiotics, such as chloramphenicol bind to the large subunit and block elongation of the peptide by preventing peptide bond formation. Antibiotics such as tetracycline and neomycin, bind the ribosome near the "A" site yet have different effects. Tetracyclines block entry of charged tRNAs to the "A" site, while neomycin induces translational errors. Finally, some antibiotics such as erythromycin, block the translocation of the ribosome along the mRNA.

16. Compare and contrast the process of protein synthesis in bacterial and eukaryotic cells, giving similarities and differences in the process of translation in these two types of cells.
Bacterial and eukaryotic cells share several similarities as well as have several differences in protein synthesis. Initially, bacteria and eukaryotes share the universal genetic code. However, the initiation codon, AUG, in eukaryotic cells codes for methionine, whereas in bacteria the AUG codon codes for N-formyl methionine. In eukaryotes, transcription takes place within the nucleus, whereas most translation takes place in the cytoplasm (although some translation does take place within the nucleus). So, transcription and translation in eukaryotes are kept temporally and spatially separate. However, in bacterial cells transcription and translation occur nearly simultaneously.

Stability of mRNA in eukaryotic cells and bacterial cells is also different. Bacterial mRNA is typically short lived, lasting only a few minutes. Eukaryotic mRNA may last hours or even days. Charging of the tRNAs with amino acids is essentially the same in both bacteria and eukaryotes. The ribosomes of bacteria and eukaryotes are different as well. Bacteria and eukaryotes have large and small ribosomal subunits, but they differ in size and composition. The bacterial large ribosomal consists of two ribosomal RNAs, while the eukaryotic large ribosomal subunit consists of three.

During translation initiation, the bacterial small ribosomal subunit recognizes the Shine-Dalgarno consensus sequence in the 5′ UTR of the mRNA and to regions of the 16S rRNA. In most eukaryotic mRNAs, the small subunit binds the 5′ cap of the mRNA and scans downstream until it encounters the first AUG codon. Finally, elongation and termination in bacterial and eukaryotic cells are functionally similar, although different elongation and termination factors are used.

APPLICATION QUESTIONS AND PROBLEMS

*17. Sydney Brenner isolated *Salmonella typhimurium* mutants that were implicated in the biosynthesis of tryptophan and would not grow on minimal medium. When these mutants were tested on minimal medium to which one of four compounds (indole glycerol phosphate, indole, anthranilic acid, and tryptophan) had been added, the growth responses shown in the table on the facing page were obtained.

Give the order of indole glycerol phosphate, indole, anthranilic acid, and tryptophan in a biochemical pathway leading to the synthesis of tryptophan. Indicate which step in the pathway is affected by each of the mutations.

Mutant	Minimal medium	Anthranilic acid	Indole glycerol phosphate	Indole	Tryptophan
trp-1	–	–	–	–	+
trp-2	–	–	+	+	+
trp-3	–	–	–	+	+
trp-4	–	–	+	+	+
trp-6	–	–	–	–	+
trp-7	–	–	–	–	+
trp-8	-	+	+	+	+
trp-9	–	–	–	–	+
trp-10	–	–	–	–	+
trp-11	–	–	–	–	+

Based on the mutant strain's ability to grow on the above substrates, we can group the mutations into four groups that we will call group 1, group 2, group 3, and group 4.

Group 1 mutants can grow only on the minimal medium supplemented with trpytophan. Group 1: trp *1,* trp *10,* trp *11,* trp *9,* trp *6, and* trp *7.*

Group 2 mutants can grow on the minimal medium supplemented with either trpytophan or indole. Group 2: trp *3.*

Group 3 mutants can grow on the minimal medium supplemented with tryptophan, indole, or indole glycerol phosphate. Group 3: trp *2 and* trp *4.*

Group 4 mutants can grow on minimal medium supplemented with the addition of tryptophan, indole, indole glycerol phosphate, or anthranilic acid. Group 4: trp *8.*

By examining the compounds needed for growth by the different groups of mutants, we can identify the step in the pathway that is blocked by each mutation. For each group, the pathway step blocked will correspond to the step proceeding the last compound on which a mutant strain can grow. Any compound added to the minimal media that proceeds the block will not allow for growth of the mutant strain.

Group 1 mutants can only grow when tryptophan is added to the growth medium. So group 1 mutants are blocked at the last step in the biosynthesis process before tryptophan is synthesized. Because group 2 mutants can grow with either tryptophan or indole added to the growth medium, this suggests that in the pathway indole is the immediate precursor of tryptophan and that group 2 mutants are blocked in the step proceeding the synthesis of indole. Using the same type of analysis, we can create the pathway on the following page for synthesis of tryptophan.

Group 4 Group 3 Group 2 Group 1
Precursor → anthranilic acid → Indole glyerol phosphate → Indole →Tryptophan

18. The addition of a series of compounds yielded the following biochemical pathway:

precursor → compound I → compound II → compound III
enzyme A enzyme B enzyme C

Mutation *a* inactivates enzyme A, mutation *b* inactivates enzyme B, and mutation *c* inactivates enzyme C. Mutants, each having one of these defects, were tested on minimal medium to which compound I, II, and III was added. Fill in the results expected of these tests by placing a plus sign (+) for growth or a minus sign (–) for no growth in the following table:

Strain with mutation	Minimal medium to which is added: Compound I	Compound II	Compound III
a	+	+	+
b	–	+	+
c	–	–	+

To determine whether growth will occur on the minimal medium with the added compound, the step in the pathway where the mutation occurs must be considered. Because mutation a *affects enzyme A, then any strain with mutation* a *can only grow on minimal media with the addition of compound I, compound II or compound III. Strains containing mutation* b *can grow only with the addition of either compound II or compound III because enzyme B, which converts compound I to compound II, has been affected. Strains with mutation* c *can only grow with the addition of compound III since the enzyme needed to synthesize compound III from compound II has been mutated.*

*19. Assume that the number of different types of bases in RNA is four. What would be the minimum codon size (number of nucleotides) required if the number of different types of amino acids in proteins were:
(The number of codons possible must be equal to or greater than the number of different types of amino acids because the codons encode for the different amino acids. To calculate how many possible codons that are possible for a given codon size with four different types of bases in the RNA, the following formula can be used: 4^n, where n *is the number of nucleotides within the codon.)*

(a) 2
1, because 4^1 = 4 codons, which is more than enough to specify 2 different amino acids.

(b) 8
2.

(c) 17
3.

(d) 45
3.
(e) 75
4.

20. How many codons would be possible in a triplet code if only three bases (A, C, and U) were used?
To calculate the number of possible codons of a triplet code if only three bases are used, the following equation can be used: 3^n, *where* n *is the number of nucleotides within the codon. So, the number of possible codons is equal to* 3^3, *or 27 possible codons.*

*21. Using the genetic code given in Figure 15.14, give the amino acids specified by the bacterial mRNA sequences and indicate the amino and carboxyl ends of the polypeptide produced.
Each of the mRNA sequences begins with the three nucleotides AUG. This indicates the start point for translation and allows for a reading frame to be set. In bacteria, the AUG initiation codon codes for N-formyl-methionine. Also, for each of these mRNA sequences, a stop codon is present either at the end of the sequence or within the interior of the sequence.

The amino terminal refers to the end of the protein with a free amino group and will be the first peptide in the chain. The carboxyl terminal refers to the end of the protein with a free carboxyl group and is the last amino acid in the chain. For the following peptide chains reading from left to right, the first amino acid is located at the amino end, while the last amino acid is located at the carboxyl end.

(a) 5'–AUGUUUAAAUUUAAAUUUUGA–3'
Amino fMet–Phe–Lys–Phe–Lys–Phe Carboxyl
(b) 5'–AUGUAUAUAUAUAUAUGA--3'
Amino fMet–Tyr–Ile--Tyr--Ile Carboxyl
(c) 5'–AUGGAUGAAAGAUUUCUCGCUUGA–3'
Amino fMet–Asp–Glu–Arg–Phe–Leu–Ala Carboxyl
(d) 5'–AUGGGUUAGGGGACAUCAUUUUGA–3'
Amino fMet–Gly Carboxyl (The stop codon UAG occurs after the codon for glycine.)

22. A nontemplate strand on DNA has the following base sequence. What amino acid sequence would be encoded by this sequence?
5'–ATGATACTAAGGCCC–3'
To determine the amino acid sequence, we need to know the mRNA sequence and the codons present. The nontemplate strand of the DNA has the same sequence as the mRNA except that thymine containing nucleotides are substituted for the uracil containing nucleotides. So the mRNA sequence would be as follows: 5'–AUGAUACUAAGGCCC–3'.

Assuming that the AUG indicates a start codon, then the amino acid sequence would be starting from the amino end of the peptide and ending with the carboxyl end: fMet–Met–Leu–Arg–Pro.

*23. The following amino acid sequence is found in a tripeptide: Met–Trp–His. Give all possible nucleotide sequences on the mRNA, on the template strand of DNA, and on the nontemplate strand of DNA that could encode this tripeptide.
The potential mRNA nucleotide sequences encoding for the tripeptide Met–Trp–His can be determined by using the codon table found in Figure 15.14. From the table, we can see that the amino acid His has two potential codons, while the amino acids Met and Trp each have only one potential codon. So, there are two different mRNA nucleotide sequences that could encode for the tripeptide. Once the potential mRNA nucleotide sequences have been determined, the template and nontemplate DNA strands can be derived from these potential mRNA sequences.

(1) 5′–AUGUGGCAU–3′
DNA template: 3′–TACACCGTA–5′
DNA nontemplate: 5′–ATGUGGCAT–3′
(2) *5′–AUGUGGCAC–3′*
DNA template: 3′–TACACCGTG–5′
DNA nontemplate: 5′–ATGTGGCAC–3′

24. How many different mRNA sequences can code for a polypeptide chain with the amino acid sequence Met–Leu–Arg? (Be sure to include the stop codon.)
From Figure 15.14, we can determine that leucine and arginine each has six different potential codons. There are also three potential stop codons. As for methionine, only one codon, AUG, is typically found as the initiation codon. (However, UUG and GUG have been shown to serve as start codons on occasion. For this problem, we will ignore these rare cases.) So, the number of potential sequences is the product of the number of different potential codons for this tripeptide, which gives us a total of $(1 \times 6 \times 6 \times 3) =$ 108 different mRNA sequences that can code for the tripeptide Met–Leu–Arg.

*25. A series of tRNAs have the following anticodons. Consider the wobble rules given in Table 15.2, and give all possible codons with which each tRNA can pair.
From the wobble rules outlined in Table 15.2, we can see that when "A" occurs at the 5′ of the anticodon it can pair only with "U" in the 3′ end of the codon. When "C" is present at the 5′ of the anticodon, it can only pair with "G" at the 3′ of the codon. However, both "U" and "G" when present at the 5′ end of the anticodon can pair with two different nucleotides at the 3′ end of the codon (U with A or G; and G with U or C). The rare base iosine (I) is also found at the 5′ of the anticodon of tRNA on occasion. Iosine can pair with "A," "U," or "C" at the 3′ end of the codon.

(a) 5'–GGC–3'
Codons: 3′–CCG–5′ or 3′–UCG–5′.
(b) 5'–AAG–3'
Codon: 3′–UUC–5′.
(c) 5'–IAA–3'
Codons: 3′–AUU–5′ or 3′–UUU–5′ or 3′–CUU–5′.

(d) 5'–UGG–3'
Codons: 3'–ACC–5' or 3'–GCC–5'.

(e) 5'–CAG–3'
Codon: 3'–GUC–5'.

26. An anticodon on a tRNA has the sequence 5'–GCA–3'.
(a) What amino acid is carried by this tRNA?
The anticodon 5'–GCA–3' would pair with the codon 5'–CGU–3'. Based on the codon table in Figure 15.14, the amino acid encoded by this codon is cysteine. So, this tRNA is most likely carrying cysteine.
(b) What would be the effect if the G in the anticodon were mutated to a U?
The anticodon would now be 3'–ACU–5' and could pair to the codon 5'–UGA–3', a stop codon. The result would be that amino acid cysteine would be placed where the stop codon 5'–UGA–3' was located in the mRNA. Essentially, the stop codon would be suppressed and translation could continue.

27. Which of the following amino acid changes could result from a mutation that changed a single base? For each change that could result from the alteration of a single base, determine which position of the codon (first, second or third nucleotide) in the mRNA must be altered for the change to occur.
(a) Leu → Gln
Of the six codons that encode for Leu, only two could be mutated by the alteration of a single base to produce the codons for Gln:
CUA (Leu)—Change the second position to A to produce CAA (Gln).
CUG (Leu)—Change the second position to A to produce CAG (Gln).
(b) Phe → Ser
Both Phe codons (UUU and UUC) could be mutated at the second position to produce Ser codons:
UUU (Phe)—Change the second position to C to produce UCU (Ser).
UUC (Phe)—Change the second postion to C to produce UCC (Ser).
(c) Phe → Ile
Both Phe codons (UUU and UUC) could be mutated at the first position to produce Ile codons:
UUU (Phe)—Change the first position to A to produce AUU (Ile).
UUC (Phe)—Change the first position to A to produce AUC (Ile).
(d) Pro → Ala
All four codons for Pro can be mutated at the first position to produce Ala codons:
CCU (Pro)—Change the first position to G to produce GCU (Ala).
CCC (Pro)—Change the first position to G to produce GCC (Ala).
CCA (Pro)—Change the first position to G to produce GCA (Ala).
CCG (Pro)—Change the first position to G to produce GCG (Ala).
(e) Asn → Lys
Both codons for Asn can be mutated at a single position to produce Lys codons:
AAU (Asn)—Change the third position to A to produce AAA (Lys).
AAU (Asn)—Change the third position to G to produce AAG (Lys).

AAC (Asn)—Change the third postion to A to produce AAA (Lys).
AAC (Asn)—Change the third position to G to produce AAG (Lys).

(f) Ile → Asn

Only two of the three Ile codons can be mutated at a single position to produce Asn codons:

AUU (Ile)—Change the second position to A to produce AAU (Asn).
AUC (Ile)—Change the second position to A to produce AAC (Asn).

28. Arrange the following components of translation in the approximate order in which they would appear or be used during protein synthesis:

The components are in order according to when they are used or play a key role in translation. The potential exception is initiation factor 3. Initiation factor 3 could possibly be listed first because it is necessary to prevent the 30s ribosome from associating with the 50s ribosome. It binds to the 30s subunit prior to the formation of the 30s initiation complex. However, during translation events the release of initiation factor 3, allows the 70s initiation complex to form, a key step in translation.

fMet-tRNA^{fMet}

30S initiation complex

initiation factor 3

70S initiation complex

elongation factor Tu

peptidyl transferase

elongation factor G

release factor 1

29. Examine the following figure that displays a step in the process of translation.

Label the following on this figure.

a. 5′ and 3′ ends of the mRNA

b. A, P, and E sites

c. Start codon

d. Stop codon

e. Amino and carboxyl ends of the newly synthesized polypeptide chain

f. Approximate location of the next peptide bond that will be formed

g. Place on the ribosome where release factor 1 will bind

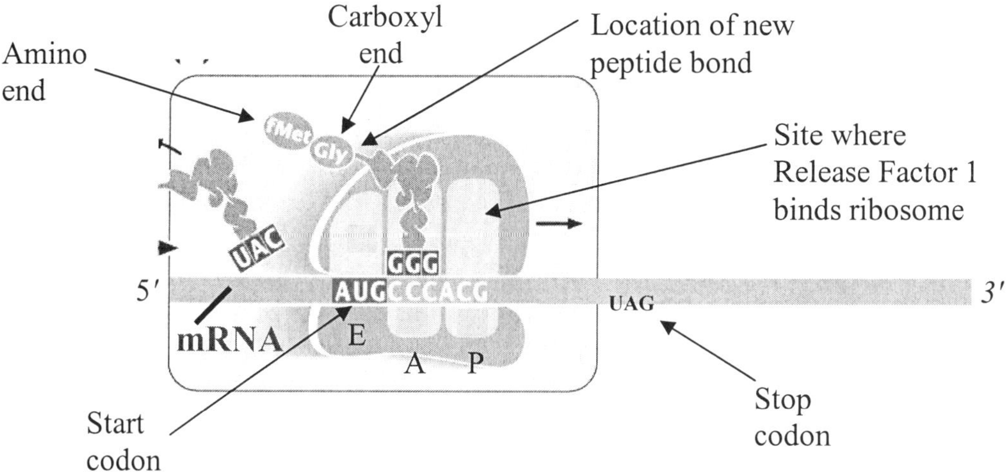

30. Answer the following questions for the figure in problem 29.

 a. What will be the anticodon of the next tRNA added to the A site of the ribosome?
 The anticodon 5′ CGU 3′ is complementary to the codon 5′ ACG 3′, which is located at the A site of the ribosome.

 b. What will be the next amino acid added to the growing polypeptide chain?
 The codon 5′ ACG 3′ encodes the amino acid threonine.

*31. A synthetic mRNA added to a cell-free protein-synthesizing system produces a peptide with the following amino acid sequence: Met–Pro–Ile–Ser–Ala. What would be the effect on translation if the following components were omitted from the cell-free protein-synthesizing system? What, if any, type of protein would be produced? Explain your reasoning.

 (a) Initiation factor 1
 The lack of IF1 would decrease the amount of protein synthesized. IF1 promotes the disassociation of the large and small ribosomal subunits. Translation initiation would occur, but at a slower rate because more of the small ribosomal subunits would be bound to the large ribosomal subunits.

 (b) Initiation factor 2
 No translation would occur. IF2 is necessary for translation initiation. The lack of IF2 would prevent fMet-$tRNA_f^{met}$ from being delivered to the small ribosomal subunit, thus blocking translation.

 (c) Elongation factor Tu
 Although translation initiation or delivery of the Met to the ribosome-mRNA complex would occur, no further amino acids would be delivered to the ribosome.

EF-TU is necessary for elongation where it binds GTP and the charged tRNA. This three-part complex enters the "A" site of the ribosome. If EF-TU is not present, then the charged tRNA will not enter the "A" site, thus stopping translation.

(d) Elongation factor G

EF-G is necessary for the translocation of the ribosome along the mRNA in a 5′ to 3′ direction. Once the formation of the peptide bond occurs between the Met and Pro, the lack of EF-G would prevent the movement of the ribosome along the mRNA, so no new codons would be read. However, the dipeptide Met-Pro would be formed because its formation would not require EF-G.

(f) Release factors R_1, R_2, and R_3

The release factors recognize the stop codons and promote cleavage of the peptide from the tRNA at the "P" site. The absence of the release factors prevents termination of translation at the stop codon, resulting in a larger peptide.

(g) ATP

ATP is required for the charging of the tRNAs with amino acids by the aminoacyl-tRNA synthetases. Without ATP, the charging would not take place, and no amino acids will be available for protein synthesis. So no protein synthesis will occur.

(h) GTP

GTP is required for initiation, elongation, and termination of translation. If GTP is absent, no protein synthesis will occur.

32. For each of the following sequences, place a check mark in the appropriate space to indicate the process most immediately affected by deleting the sequence. You should chose only one process for each sequence (one check mark per sequence).

Process Most Immediately Affected by Deletion

Sequence Deleted	*Replication*	*Transcription*	*RNA Processing*	*Translation*
a. ori site	✓			

The ori site or origin of replication is necessary for the initiation of replication.

Sequence Deleted	*Replication*	*Transcription*	*RNA Processing*	*Translation*
b. 3′ splice site consensus sequence			✓	

The 3′ splice site is necessary for proper excision of the intron. So, RNA processing events will be affected.

Sequence Deleted	*Replication*	*Transcription*	*RNA Processing*	*Translation*
c. poly(A) tail				✓

The poly(A) tail is involved in mRNA stability. If the tail is missing, then the mRNA will be degraded more rapidly thus affecting translation.

Sequence Deleted	*Replication*	*Transcription*	*RNA Processing*	*Translation*
d. terminator		✓		

The terminator is necessary for transcription termination. Deletion of the terminator will result in the production of an abnormally long RNA transcript.

Sequence Deleted	*Replication*	*Transcription*	*RNA Processing*	*Translation*
e. start codon				✓

The start codon is necessary for translation initiation.

Sequence Deleted	*Replication*	*Transcription*	*RNA Processing*	*Translation*
f. -10 consensus seq		✓		

In bacteria, the -10 sequence is an important component of the promoter. Deletion of the -10 sequence will prevent transcription initiation from occurring.

Sequence Deleted	*Replication*	*Transcription*	*RNA Processing*	*Translation*
g. Shine Dalgarno				✓

The Shine Dalgarno sequence or ribosome binding site is bound by the 30s subunit during the initiation of translation. If the sequence is deleted, then the ribosome will not bind to the mRNA molecule and translation will not occur.

CHALLENGE QUESTIONS

33. In what ways are spliceosomes and ribosomes similar? In what ways are they different? Can you suggest some possible reasons for their similarities?
Spliceosomes and ribosomes are both large complexes that are composed of several different RNA and protein molecules. In essence, the spliceosome and ribosome are RNA-based enzymes or ribozymes. The RNA molecules in both structures are necessary for catalysis. The 23S RNA molecule in the ribosome catalyzes the formation of peptide bonds between amino acids. Other rRNAs are also important for protein synthesis. In

spliceosomes, the snRNA molecules catalyze the cutting and splicing of pre-mRNA molecules to produce mature mRNA. The catalytic RNAs of the ribosome and the spliceosome may have originated during the RNA world when RNA molecules served to store information and to catalyze reactions that sustained life.

*34. Several experiments were conducted to obtain information about how the eukaryotic ribosome recognizes the AUG start codon. In one experiment, the gene that codes for methionine initiator tRNA ($tRNA_i^{Met}$) was located and changed. The nucleotides that specify the anticodon on $tRNA_i^{Met}$ were mutated so that the anticodon in the tRNA was 5'–CCA–3' instead of 5'–CAU–3'. When this mutated gene was placed into a eukaryotic cell, protein synthesis took place, but the proteins produced were abnormal. Some of the proteins produced contained extra amino acids, and others contained fewer amino acids.

(a) What do these results indicate about how the ribosome recognizes the starting point for translation in eukaryotic cells? Explain your reasoning.

By mutating the anticodon to 5'–CCA–3' from 5'–CAU–3' on $tRNA_i^{Met}$, the initiator tRNA will now recognize the codon 5'–UGG–3', which normally would code only for Trp. If translation initiation by the ribosome in eukaroytes occurs by binding the 5' cap of the mRNA followed by scanning, then the first 5'–UGG–3' codon recognized by the mutated $tRNA_i^{Met}$ will be the start site for translation. If the first 5'–UGG–3' codon occurs prior to the normal 5'–AUG–3' codon, then a protein containing extra amino acids could be produced. If the first 5'–UGG–3' codon occurs after the normal 5'–AUG–3', then a shorter protein will be produced. Finally, truncated proteins could also be produced by the first 5'–UGG–3' being out of frame of the normal coding sequence. If this happens, then most likely a stop codon will be encountered before the end of the normal coding sequence and will terminate translation. The data suggest that translation initiation takes place by scanning of the ribosome for the appropriate start sequence.

(b) If the same experiment had been conducted on bacterial cells, what results would you expect?

Very little or no protein synthesis would be expected. Translation initiation in bacteria requires the 16S RNA of the small ribosomal subunit to interact with the Shine-Dalgarno sequence. This interaction serves to line up the ribosome over the start codon. If the anticodon has been changed such that the start codon cannot be recognized, then protein synthesis is not likely to take place.

35. The redundancy of the genetic code means that some amino acids are specified by more than one codon. For example, the amino acid leucine is encoded by six different codons. Within a genome, synonymous codons do not occur equally; some synonymous codons occur much more frequently than others, and the preferred codons differ among different species. For example, in one species the codon UUA might be used most often to code for leucine, while in another species the codon CUU might be used most often. Speculate on a reason for this bias in codon usage and why the preferred codons are not the same in all organisms.

Synonymous codon usage patterns may depend on a variety of factors. Two potential factors that could affect usage patterns are the GC content of the organism and the relative amounts of isoaccepting tRNA molecules. The GC content of an organism reflects the relative proportions of nucleotides found in the DNA. In a given organism the bias for particular synonymous codons may reflect the overall GC content of that organism. For example, in organisms that have a high GC content, you might expect to find that the synonymous codon usage pattern reflects this bias resulting a preference for codons with more Gs and Cs. So, in two organisms that differ in GC content, the synonymous codon usage bias should reflect their differences in base composition.

Isoaccepting tRNAs are those that carry the same amino acid but have different anticodons. These isoaccepting tRNAs act at synonymous codons. For a given organism, the more prevalent synonymous codons may depend on the frequency of its tRNA and complementary anticodon. In different organisms the concentrations of the various isoaccepting tRNAs will vary leading to different usage patterns of synonymous codons in these organisms.

Chapter Sixteen: Control of Gene Expression

COMPREHENSION QUESTIONS

*1. Name six different levels at which gene expression might be controlled.
(1) Alteration or modification of the gene structure at the DNA level
(2) Transcriptional regulation
(3) Regulation at the level of mRNA processing
(4) Regulation of mRNA stability
(5) Regulation of translation
(6) Regulation by pos -translational modification of the synthesized protein

*2. Draw a picture illustrating the general structure of an operon and identify its parts.

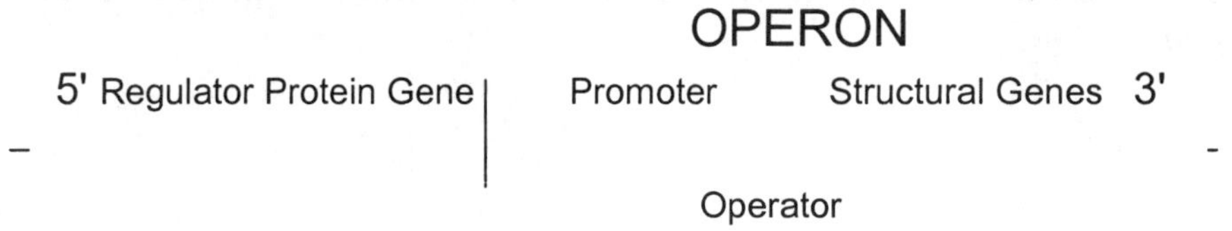

OPERON

5' Regulator Protein Gene | Promoter Structural Genes 3'

Operator

3. What is difference between positive and negative control? What is the difference between inducible and repressible operons?
Positive transcriptional control requires an activator protein to stimulate transcription at the operon. In negative control, a repressor protein inhibits or turns off transcription at the operon.

An inducible operon normally is not transcribed. It requires an inducer molecule to stimulate transcription either by inactivating a repressor protein in a negative inducible operon or by stimulating the activator protein in a positive inducible operon.

Transcription normally occurs in a repressible operon. In a repressible operon, transcription is turned off either by the repressor becoming active in a negative repressible operon or by the activator becoming inactive in a positive repressible operon.

*4. Briefly describe the *lac* operon and how it controls the metabolism of lactose.
The lac operon consists of three structural genes involved in lactose metabolism, the lacZ gene, the lacY gene, and the lacA gene. Each of these three genes has a different role in the metabolism of lactose. The lacZ gene codes for the enzyme β-galactosidase, which breaks the disaccharide lactose into galactose and glucose, and converts lactose into allolactose. The lacY gene, located downstream of the lacZ gene, codes for lactose permease. Permease is necessary for the passage of lactose through the E. coli cell membrane. The lacA gene, located downstream of lacY, encodes the enzyme thiogalactoside transacetylase whose function in lactose metabolism has not yet been determined. All of these genes share a common overlapping promoter and operator region. Upstream from the lactose operon is the lacI gene that encodes the lac operon repressor. The repressor binds at the

operator region and inhibits transcription of the lac operon by preventing RNA polymerase from successfully initiating transcription.

When lactose is present in the cell, the enzyme β-galactosidase converts some of it into allolactose. Allolactose binds to the lac repressor, altering its shape and reducing the repressor's affinity for the operator. If the repressor does not occupy the operator, RNA polymerase can initiate transcription of the lac structural genes from the lac promoter.

5. What is catabolite repression? How does it allow a bacterial cell to use glucose in preference to other sugars?
In catabolite repression, the presence of glucose inhibits or represses the transcription of genes involved in the metabolism of other sugars. Because the gene expression necessary for utilizing other sugars is turned off, only enzymes involved in the metabolism of glucose will be synthesized. Operons that exhibit catabolite repression are under the positive control of catabolic activator protein (CAP). For CAP to be active, it must form a complex with cAMP. Glucose affects the level of cAMP. The levels of glucose and cAMP are inversely proportional—as glucose levels increase, the level of cAMP decreases. Thus, CAP is not activated.

*6. What is attenuation? What is the mechanism by which the attenuator forms when tryptophan levels are high and the antiterminator forms when tryptophan levels are low?
Attenuation is the termination of transcription prior to the structural genes of an operon. It is a result of the formation of a termination hairpin or attenuator.

Two types of secondary structures can be formed by the mRNA 5′ UTR of the trp operon. If the 5′ UTR forms two hairpin structures from the base pairing of region 1 with region 2 and the pairing of region 3 with region 4, then transcription of the structural genes will not occur. The hairpin structure formed by the pairing of region 3 with region 4 results in a terminator being formed that stops transcription. When region 2 pairs with region 3, the resulting hairpin acts as an antiterminator allowing for transcription to proceed. Region 1 of the 5′ UTR also encodes a small protein and has two adjacent tryptophan codons (UGG).

Tryptophan levels affect transcription due to the coupling of translation with transcription in bacterial cells. When tryptophan levels are high, the ribosome quickly moves through region 1 and into region 2, thus preventing region 2 from pairing with region 3. Therefore, region 3 is available to form the attenuator hairpin structure with region 4, stopping transcription. When tryptophan levels are low, the ribosome stalls or stutters at the adjacent tryptophan codons in region 1. Region 2 now becomes available to base pair with region 3, forming the antiterminator hairpin. Transcription can now proceed through the structural genes.

*7. What is antisense RNA? How does it control gene expression?
Antisense RNA molecules are complementary to other DNA or RNA sequences. In bacterial cells, antisense RNA molecules can bind to a complementary region in the

5′ UTR of a mRNA molecule, blocking the attachment of the ribosome to the mRNA and thus stopping translation.

8. What are riboswitches? How do they control gene expression. How do riboswitches differ from RNA-mediated repression?
Riboswitches are regulatory sequences in RNA molecules. Most can fold into compact secondary structures consisting of a base stem and several branching hairpins. At riboswitches, regulatory molecules bind and influence gene expression by affecting the formation of secondary structures within the mRNA molecule. The binding of the regulatory molecule to a riboswitch sequence may result in repression or induction. Some regulatory molecules bind the riboswitch sequence and stabilize a terminator structure in the mRNA, which results in premature termination of the mRNA molecule. Other regulatory molecules bind riboswitch sequences resulting in the formation of secondary structures that block the ribosome binding sites of the mRNA molecules thus preventing translation initiation. In induction, the regulatory molecule acts as an inducer, stimulating the formation of a secondary structure in the mRNA that allows for transciption or translation to occur.
RNA-mediated repression occurs through the action of a ribozyme. In RNA-mediated repression, a RNA sequence within the 5′ untranslated region can act as a ribozyme that when stimulated by the presence of a regulatory molecule can induce self-cleavage of the mRNA molecule, which prevents translation of the molecule. When bound by a regulatory molecule, RNA mediated repression results in the self-cleavage of the mRNA molecule. When bound by a regulatory molecule, riboswitch sequences stimulate changes in the secondary structure of the mRNA molecule that affect gene expression.

9. What general features of transcriptional control are found in bacteriophage λ?
Three general features of lambda gene regulation have been identified. First, the four lambda operons exhibit both positive and negative control of transcription. Second, a cascade of reactions results in transcription. In other words, gene products from one operon affect the transcription of a second operon whose products regulate gene expression at a third operon. Finally, transcriptional antiterminator proteins are involved in lambda gene regulation. These proteins allow RNA polymerase to read through terminators.

*10. What changes take place in chromatin structure and what role do these changes play in eukaryotic gene regulation?
Changes in chromatin structure can result in repression or stimulation of gene expression. As genes become more transcriptionally active, chromatin shows increased sensitivity to DNase I digestion, suggesting that the chromatin structure is more open. Acetylation of histone proteins by acteyltransferase proteins results in the destabilization of the nucleosome structure and increases transcription as well as hypersensitivity to DNase I. The reverse reaction by deacetylases stabilizes nucleosome structure and lessens DNase I sensitivity. Other transcriptional factors and regulatory proteins, called chromatin remodeling complexes, bind directly to the DNA altering chromatin structure without acetylating histone proteins. The

chromatin remodeling complexes allow for transcription to be initiated by increasing accessibility to the promoters by transcriptional factors.

DNA methylation is also associated with decreased transcription. Methylated DNA sequences stimulate histone deacetylases to remove acetyl groups from the histone proteins, thus stabilizing the nucleosome and repressing transcription. Demethylation of DNA sequences is often followed by increased transcription, which may be related to the deacetylation of the histone proteins.

11. Briefly explain how transcriptional activator proteins and repressors affect the level of transcription of eukaryotic genes.
Transcriptional activator proteins stimulate transcription by binding DNA at specific base sequences such as an enhancer or regulatory promoter and attracting or stabilizing the basal transcriptional factor apparatus. Repressor proteins bind to silencer sequences or promoter regulator sequences. These proteins may inhibit transcription by either blocking access to the enhancer sequence by the activator protein, by preventing the activator from interacting with the basal transcription apparatus, or by preventing the basal transcription factor from being assembled.

12. What is an insulator?
An insulator or boundary element is a sequence of DNA that inhibits the action of regulatory elements called enhancers in a position dependent manner.

13. What is a response element? How do response elements bring about coordinated expression of eukaryotic genes?
Response elements are regulatory DNA sequences consisting of short consensus sequences located at various distances from the genes that they regulate. Under conditions of stress, a transcription activator protein binds to the response element and stimulates transcription. If the same response element sequence is located in the control regions of different genes, then these genes will be activated by the same stimuli, thus producing a coordinated response.

14. Outline the role of alternative splicing in the control of sex differentiation in *Drosophila.*
Sex development in fruit flies depends on alternative splicing as well as a cascade of genetic regulation. Early in the development of female fruit flies, a female-specific promoter is activated stimulating transcription at the sex-lethal (Sxl) *gene. Splicing of the pre-mRNA of the transformer (tra) gene is regulated by the* Sxl *protein. The mature mRNA produces the Tra protein. In conjugation with another protein, the Tra protein stimulates splicing of the pre-mRNA from the doublesex* (dsx) *gene. The resulting Dsx protein is required for the embryo to develop female characteristics. Male fruit flies do not produce the Sxl protein, which results in the Tra pre-mRNA in male fruit flies being spliced at an alternate location. The alternate Tra protein is not functional, resulting in the Dsx pre-mRNA splicing at a different location as well. Protein synthesis from this mRNA produces a male-specific doublesex protein, which causes development of male-specific traits.*

*15. What role does RNA stability play in gene regulation? What controls RNA stability in eukaryotic cells?

The total amount of protein synthesized is dependent on how much mRNA is available for translation. The amount of mRNA present is dependent on the rates of mRNA synthesis and degradation. Less-stable mRNAs will be degraded and become unavailable as templates for translation.

The presence of the 5′ cap, 3′ poly(A) tail, the 5′ UTR, 3′ UTR, and the coding region in the mRNA molecule affects stability. Poly(A) binding proteins (PABP) bind at the 3′ poly(A) tail. These proteins contribute to the stability of the tail, and protect the 5′ cap through direct interaction. Once a critical number of adenine nucleotides have been removed from the tail, the protection is lost and the 5′ cap is removed. The removal of the 5′ cap allows for 5′ to 3′ nucleases to degrade the mRNA.

16. Define RNA silencing. Explain how siRNAs arise and how they potentially affect gene expression.

RNA silencing, or RNA interference, occurs when double-stranded RNA molecules are cleaved and processed to produce small single-stranded interfering RNAs (siRNAs). These siRNAs bind to complementary sequences in mRNA molecules, stimulating cleavage and degradation of the mRNA. The siRNAs may also stimulate DNA methylation at DNA sequences complementary to the siRNAs.

The paired mRNA-siRNA attracts a protein-RNA complex that cleaves the mRNA in an area bound by the siRNA. Following the initial cleavage, the mRNA is further degraded. The cleavage and subsequent degradation of the mRNA makes it unavailable for translation. DNA methylation in the nucleus stimulated by siRNAs also affects transcription.

*17. What are some of the characteristics of *Arabidopsis thaliana* that make it a good model genetic organism?

<u>*Useful characteristics of Arabidopsis thaliana*</u>

- *Is an angiosperm and thus shares characteristics and life cycle similarities with other flowering plants.*
- *Is capable of self-fertilization or cross-fertilization.*
- *Is small in size, which is useful in laboratory environments.*
- *Can grow under low illumination levels, which also is useful in laboratory environments.*
- *Is a prolific reproducer with each plant capable of producing 10,000 to 40,000 seeds.*
- *Has a high germination frequency from its seeds.*
- *Has a small genome (~125 million base pairs) that has been completely sequenced.*
- *Has a number of ecotypes or variants that are available, which differ in genotypes and phenotypic characteristics.*
- *Can uptake genes from other organisms by means of a Ti plasmid.*

*18. Compare and contrast bacterial and eukaryotic gene regulation. How are they similar? How are they different?

Bacterial and eukaryotic gene regulation involves the action of protein repressors and protein activators. Cascades of gene regulation in which the activation of one set of genes affects another set of genes takes place in both eukaryotes and bacteria. Regulation of gene expression at the transcriptional level is also common in both types of cells.

Bacterial genes are often clustered in operons and are coordinately expressed through the synthesis of a single polygenic mRNA. Eukaryotic genes are typically separate, with each containing its own promoter and transcribed on individual mRNAs. Coordinate expression of multiple genes is accomplished through the presence of response elements. Genes sharing the same response element will be regulated by the same regulatory factors.

In eukaryotic cells, chromatin structure plays a role in gene regulation. Chromatin that is condensed inhibits transcription. So, for expression to occur, the chromatin must be altered to allow for changes in structure. Acetylation of histone proteins and DNA methylation are important in these changes.

At the level of transcription initiation, the process is more complex in eukaryotic cells. In eukaryotes, initiation requires a complex machine involving RNA polymerase, general transcription factors, and transcriptional activators. Bacterial RNA polymerase is either blocked or stimulated by the actions of regulatory proteins.

Finally, in eukaryotes the action of activator proteins binding to enhancers may take place at a great distance from the promoter and structural gene. These distant enhancers occur much less frequently in bacterial cells.

APPLICATION QUESTIONS AND PROBLEMS

*19. For each of the following types of transcriptional control, indicate whether the protein produced by the regulator gene will be synthesized initially as an active repressor, inactive repressor, active activator, or inactive activator.

(a) Negative control in a repressible operon

Inactive repressor

(b) Positive control in a repressible operon

Active activator

(c) Negative control in an inducible operon

Active repressor

(d) Positive control in an inducible operon

Inactive activator

*20. A mutation occurs at the operator site that prevents the regulator protein from binding. What effect will this mutation have in the following types of operons?

(a) Regulator protein is a repressor in a repressible operon.

The regulator protein-corepressor complex would normally bind to the operator and inhibit transcription. If a mutation prevented the repressor protein from binding at the operator, then the operon would never be turned off and transcription would occur all the time.

(b) Regulator protein is a repressor in an inducible operon.
In an inducible operon, a mutation at the operator site that blocks binding of the repressor would result in constitutive expression and transcription would occur all the time.

21. The *blob* operon produces enzymes that convert compound A into compound B. The operon is controlled by a regulatory gene *S*. Normally, the enzymes are synthesized only in the absence of compound B. If gene *S* is mutated, the enzymes are synthesized in the presence *and* in the absence of compound B. Does gene *S* produce a repressor or an activator? Is this operon inducible or repressible?
Because the blob operon is transcriptionally inactive in the presence of B, gene S most likely codes for a repressor protein that requires compound B as a corepressor. The data suggest that the blob operon is repressible because it is inactive in the presence of compound B, but active when compound B is absent.

*22. A mutation prevents the catabolite activator protein (CAP) from binding to the promoter in the *lac* operon. What will be the effect of this mutation on transcription of the operon?
Catabolite activator protein binds the CAP site of the lac operon and stimulates RNA polymerase to bind the lac promoter, thus resulting in increased levels of transcription from the lac operon. If a mutation prevents CAP from binding to the site, then RNA polymerase will bind the lac promoter poorly. This will result in significantly lower levels of transcription of the lac structural genes.

23. Under which of the following conditions would a *lac* operon produce the greatest amount of β-galactosidase? The least? Explain your reasoning.

	Lactose present	Glucose present
Condition 1	Yes	No
Condition 2	No	Yes
Condition 3	Yes	Yes
Condition 4	No	No

Condition 1 will result in the production of the maximum amount of β-galactosidase. For maximum transcription, the presence of lactose and the absence of glucose are required. Lactose (or allolactose) binds to the lac repressor reducing the affinity of the lac repressor to the operator. This decreased affinity results in the promoter being accessible to RNA polymerase. The lack of glucose allows for increased synthesis of cAMP, which can complex with CAP. The formation of CAP-cAMP complexes improves the efficiency of RNA polymerase binding to the promoter, which results in higher levels of transcription from the lac operon.

Condition 2 will result in the production of the least amount of β-galactosidase. With no lactose present, the lac repressor is active and binds to the operator, inhibiting transcription. The presence of glucose results in a decrease of cAMP levels. A CAP-cAMP complex does not form, and RNA polymerase will not be stimulated to transcribe the lac operon.

24. A mutant strain of *E. coli* produces β-galactosidase in the presence *and* in the absence of lactose. Where in the operon might the mutation in this strain occur?
Within the operon, the operator region is the most probable location of the mutation. If the mutation prevents the lac repressor protein from binding to the operator, then transcription of the lac structural genes will not be inhibited. Expression will be constitutive.

*25. For *E. coli* strains with the following *lac* genotypes, use a plus sign (+) to indicate the synthesis of β-galactosidase and permease or a minus sign (−) to indicate no synthesis of the enzymes.
In determining if expression of the β-galactosidase and the permease gene will occur, you should consider several factors. The presence of $lacZ^+$ *and* $lacY^+$ *on the same DNA molecule as a functional promoter (*$lacP^+$*) is required because the promoter is a cis-acting regulatory element. However the* $lacI^+$ *gene product or lac repressor is trans-acting and does not have to be located on the same DNA molecule as β-galactosidase and permease genes to inhibit expression. For the repressor to function, it does require that the cis-acting lac operator be on the same DNA molecule as the functional β-galactosidase and permease genes. Finally, the dominant* $lacI^s$ *gene product is also trans-acting and can inhibit transcription at any functional lac operator region.*

	Lactose absent		Lactose present	
Genotype of strain	β-Galactosidase	Permease	β-Galactosidase	Permease
$lacI^+lacP^+lacO^+lacZ^+lacY^+$	−	−	+	+
$lacI^-lacP^+lacO^+lacZ^+lacY^+$	+	+	+	+
$lacI^+lacP^+lacO^clacZ^+lacY^+$	+	+	+	+
$lacI^-lacP^+lacO^+lacZ^+lacY^-$	+	−	+	−
$lacI^-lacP^-lacO^+lacZ^+lacY^+$	−	−	−	−
$lacI^+lacP^+lacO^+lacZ^-lacY^+/$ $lacI^-lacP^+lacO^+lacZ^+lacY^-$		−	+	+
$lacI^-lacP^+lacO^clacZ^+lacY^+/$ $lacI^+lacP^+lacO^+lacZ^-lacY^-$	+	+	+	+
$lacI^-lacP^+lacO^+lacZ^+lacY^-/$ $lacI^+lacP^-lacO^+lacZ^-lacY^+$	−	−	+	−
$lacI^+lacP^-lacO^clacZ^-lacY^+/$ $lacI^-lacP^+lacO^+lacZ^+lacY^-$	−	−	+	−

$lacI^+ lacP^+ lacO^+ lacZ^+ lacY^+ /$ $lacI^+ lacP^+ lacO^+ lacZ^+ lacY^+$	–	–	+	+
$lacI^S lacP^+ lacO^+ lacZ^+ lacY^- /$ $lacI^+ lacP^+ lacO^+ lacZ^- lacY^+$	–	–	–	–
$lacI^S lacP^- lacO^+ lacZ^- lacY^+ /$ $lacI^+ lacP^+ lacO^+ lacZ^+ lacY^+$	–	–	–	–

26. Give all possible genotypes of a *lac* operon that produces β-galactosidase and permease under the following conditions. Do not give partial diploid genotypes.

	Lactose absent		Lactose present		
	β-Galactosidase	Permease	β-Galactosidase	Permease	Genotype
(a)	–	–	+	+	$lacI^+ lacP^+ lacO^+ lacZ^+ lacY^+$
(b)	–	–	–	+	$lacI^+ lacP^+ lacO^+ lacZ^- lacY^+$
(c)	–	–	+	–	$lacI^+ lacP^+ lacO^+ lacZ^+ lacY^-$
(d)	+	+	+	+	$lacI^- lacP^+ lacO^+ lacZ^+ lacY^+$ *or* $lacI^+ lacP^+ lacO^c lacZ^+ lacY^+$
(e)	–	–	–	–	$lacI^S lacP^+ lacO^+ lacZ^+ lacY^+$ *or* $lacI^+ lacP^- lacO^+ lacZ^+ lacY^+$
(f)	+	–	+	–	$lacI^- lacP^+ lacO^+ lacZ^+ lacY^-$ *or* $lacI^+ lacP^+ lacO^c lacZ^+ lacY^-$
(g)	–	+	–	+	$lacI^- lacP^+ lacO^+ lacZ^- lacY^+$ *or* $lacI^+ lacP^+ lacO^c lacZ^- lacY^+$

*27. Explain why mutations at the *lacI* gene are trans in their effects, but mutations in the *lacO* gene are cis in their effects.
The lacI gene encodes the lac repressor protein, which can diffuse within the cell and attach to any operator. It can therefore affect the expression of genes on the same or different molecules of DNA. The lacO gene encodes the operator. It affects the binding of DNA polymerase to the DNA, and therefore affects only the expression of genes on the same molecule of DNA.

*28. The *mmm* operon, which has sequences A, B, C, and D, encodes enzymes 1 and 2. Mutations in sequences A, B, C, and D have the following effects, where a plus sign (+) = enzyme synthesized, and a minus sign (–) = enzyme not synthesized.

	mmm absent		*mmm* present	
Mutation in sequence	Enzyme 1	Enzyme 2	Enzyme 1	Enzyme 2
No mutation	+	+	–	–
A	–	+	–	–
B	+	+	+	+
C	+	–	–	–
D	–	–	–	–

(a) Is the *mmm* operon inducible or repressible?
The data from the strain with no mutation indicate that the mmm operon is repressible. The operon is expressed in the absence of mmm, but inactive in the presence of mmm. This is typical of a repressible operon.

(b) Indicate which sequence (A, B, C, or D) is part of the following components of the operon:
Regulator gene __B__
When sequence B is mutated, gene expression is not repressed by the presence of mmm.
Promoter __D__
When sequence D is mutated, no gene expression occurs either in the presence or absence of mmm.
Structural gene for enzyme 1 __A__
When sequence A is mutated, enzyme 1 is not produced.
Structural gene for enzyme 2 __C__
When sequence C is mutated, enzyme 2 is not produce.

*29. Listed in parts *a* through *g* are some mutations that were found in the 5' UTR region of the *trp* operon of *E. coli*. What would the most likely effect of each of these mutations be on transcription of the *trp* structural genes?

(a) A mutation that prevented the binding of the ribosome to the 5' end of the mRNA 5' UTR
If the ribosome does not bind to the 5' end of the mRNA, then region 1 of the mRNA 5' UTR will be free to pair with region 2, thus preventing region 2 from pairing with region 3 of mRNA 5' UTR. Region 3 will be free to pair with region 4, forming the attenuator or termination hairpin. Transcription of the trp structural genes will be terminated. Essentially, no gene expression will occur.

(b) A mutation that changed the tryptophan codons in region 1 of the mRNA 5' UTR into codons for alanine
In the wild-type trp operon, low levels of tryptophan result in the ribosome pausing in region 1 of the mRNA 5' UTR. The pause permits regions 2 and 3 of the mRNA 5' UTR to form the antiterminator hairpin, allowing transcription of the structural genes to continue. If alanine codons have replaced tryptophan codons, then under conditions of low tryptophan, the stalling of the ribosome will not occur. The attenuator will form, stopping transcription. The ribosome will stall when alanine is low, so transcription of the structural genes will occur only when alanine is low.

(c) A mutation that created a stop codon early in region 1 of the mRNA 5' UTR
If region 1 of the mRNA 5' UTR is free to pair with region 2, then regions 3 and 4 of the mRNA 5' UTR can form the attenuator. An early stop codon will result in the ribosome "falling-off" region 1, allowing it to form a hairpin structure with region 2. Transcription will not occur because regions 3 and 4 are now free to form the attenuator.

(d) Deletions in region 2 of the mRNA 5' UTR
If region 2 of the mRNA 5' UTR is deleted, then the antiterminator cannot be formed. The attenuator will form and transcription will not occur.

(e) Deletions in region 3 of the mRNA 5' UTR
The trp operon mRNA 5' UTR will be unable to form the attenuator if region 3 contains a deletion. Attentuation or termination of transcription will not occur, resulting in continued transcription of the trp structural genes.

(f) Deletions in region 4 of the mRNA 5' UTR
Deletions in region 4 will prevent formation of the attenuator by the 5' UTR mRNA. Transcription will proceed.

(g) Deletion of the string of adenine nucleotides that follows region 4 in the 5' UTR
For the attenuator hairpin to function as a terminator, the presence of a string of uracil nucleotides following region 4 in the mRNA 5' UTR is required. The deletion of the string of adenine nucleotides in the DNA will result in no string of uracil nucleotides following region 4 of the mRNA 5' UTR. No termination will occur, and transcription will proceed.

30. Some mutations in the *trp* 5' UTR region increase termination by the attenuator. Where might these mutations occur and how might they affect the attenuator?
Mutations that disrupt the formation of the antiterminator will increase termination by the attenuator. Such disruptions could be caused by a deletion in region 2 that prevents region 2 from pairing with region 3. Mutations in region 1 could also affect the antiterminator if the mutations prevented the ribosome from stalling at the adjacent trpytophan codons within region 1. For example, any mutation that blocks translation initiation or stops translation early within region 1 would not allow the ribosome to migrate on the trp operon mRNA. Another type of mutation affecting antiterminator formation in region 1 is one that eliminates or replaces the two adjacent tryptophan codons in the small protein. Elimination of these codons would prevent the ribosome from stalling in region 1, thus increasing the rate of the terminator formation.

31. Some of the mutations mentioned in the Question 28 have an interesting property. They prevent the formation of the antiterminator that normally takes place when the tryptophan level is low. In one of the mutations, the AUG start codon for the 5' UTR peptide has been deleted. How might this mutation prevent antitermination from occurring?
The AUG start codon is necessary for the translation initiation of the 5' UTR peptide. If translation does not initiate, then the mRNA 5' UTR region 1 will be available to pair with region 2. The resulting hairpin will prevent the formation of the antiterminator.

32. Several examples of antisense RNA regulating translation in bacterial cells have been discovered. Molecular geneticists have also used antisense RNA to artificially control transcription in both bacterial and eukaryotic genes. If you wanted to inhibit the transcription of a bacterial gene with antisense RNA, what sequences might the antisense RNA contain?
To block transcription, you will need to disrupt the action of RNA polymerase either directly or indirectly. Antisense RNA containing sequences complementary to the gene's promoter should inhibit the binding of RNA polymerase. If transcription initiation by RNA polymerase requires the assistance of an activator protein, then antisense RNA complementary to the activator protein-binding site of the gene could also disrupt transcription. By binding the activator site, the antisense RNA would block access to the site by the activator and prevent RNA polymerase from being assisted by the activator to initiate transcription.

*33. What would be the effect of deleting the *Sxl* gene in a newly fertilized *Drosophila* embryo?
The Sxl *gene is necessary for proper splicing of the tra pre-mRNA and the production of the Tra protein, a protein needed for the development of female fruit flies. If the* Sxl *gene is absent, then no Tra protein will be produced and the embryo can develop only into a male.*

34. What would be the effect of a mutation that destroyed the ability of poly(A) binding protein (PABP) to attach to a poly(A) tail?
Poly(A) binding protein is necessary for the stability of the mRNA molecules in eukaryotic cells. The protein contributes to the stability of both the poly(A) tail and the 5' cap. If the PABP protein cannot bind the poly(A) tail, then the 5' cap will not be protected and thus will be removed, resulting in the mRNA being degraded more rapidly.

CHALLENGE QUESTIONS

35. Would you expect to see attenuation in the *lac* operon and other operons that control the metabolism of sugars? Why or why not?
No, attenuation of the lac operon and other operons that control the metabolism of sugars is not likely to occur. Most operons involved in the breakdown of sugar already exhibit two levels of control: a repressor and activation by CAP protein. The presence of the sugar alone is not sufficient to elicit high levels of expression from the operon. If present, glucose prevents CAP from being activated and thus glucose will be the primary sugar to be metabolized. If glucose levels are low or absent, then CAP will be activated by cAMP. So, the activation of the sugar operons will require the presence of the sugar and the absence of glucose. With glucose absent, the cell will quickly express the sugar metabolism operon so that energy production can occur. Finally, attenuation of the trp operon works because the ultimate result of gene expression in the operon is the amino tryptophan. The

presence of the two codons for tryptophan in the mRNA 5' UTR means that the translational rate will be affected by the levels of tryptophan in the cell.

36. A common feature of many eukaryotic mRNAs is the presence of a rather long 3' UTR, which often contains consensus sequences. Creatine kinase B (CK-B) is an enzyme important in cellular metabolism. Certain cells—termed U937D cells—have lots of CK-B mRNA, but no CK-B enzyme is present. In these cells, the 5' end of the CK-B mRNA is bound to ribosomes, but the mRNA is apparently not translated. Something inhibits the translation of the CK-B mRNA in these cells.
In recent experiments, numerous short segments of RNA containing only 3' UTR sequences were introduced into U937D cells. As a result, the U937D cells began to synthesize the CK-B enzyme, but the total amount of CK-B mRNA did not increase. Short segments of other RNA sequences did not stimulate synthesis of CK-B; only the 3' UTR sequences turned on translation of the enzyme.
On the basis of these experiments, propose a mechanism for how CK-B translation is inhibited in the U937D cells. Explain how the introduction of short segments of RNA containing the 3' UTR sequences might remove the inhibition.
From the above experimental data, translation of the CK-B protein is inhibited in the U937D cells—the CK-B mRNA is present and bound to the ribosome, but no protein is synthesized. A possible mechanism for the inhibition of translation could be the binding of translational repressors to the 3' UTR region of the CK-B mRNA. The action of soluble proteins inhibiting translation seems to be suggested by the response of the U937 cells to the short RNA sequences containing the 3' UTR. When these sequences are introduced to the U937D cells, the synthesis of CK-B occurs. Possibly exogenously applied 3' UTR sequences bind to the translational repressor proteins, making them unavailable to bind to the CK-B mRNA. If these factors are not present on the CK-B mRNA, then synthesis of the CK-B protein can take place.

37. In the fungus *Neurospora*, about 2–3% of cytosine bases are methylated. A recent study isolated those DNA sequences in *Neurospora* that contained 5-methylcytosine and found that almost all methylated sequences were located in the relict copies of transposable genetic elements. Based on these observations, propose a possible explanation for why *Neurospora* methylates its DNA and why DNA methylation in this species is associated with transposable genetic elements.
Heavy DNA methylation of cytosine to yield 5-methylcytosine containing nucleotides is associated with transcriptional repression in other organisms such as vertebrate animals and plants. Because almost all of the methylated sequences were associated with relic copies of transposons, this suggests that the methylation may be part of a process for inactivating these transposable elements. The Neurospora methylation could be a mechanism for inactivating the expression of genes found in these invading nucleic acid sequences and thus preventing the transposon sequences from spreading within the Neurospora genome.

Chapter Seventeen: Gene Mutations and DNA Repair

COMPREHENSION QUESTIONS

*1. What is the difference between somatic mutations and germ-line mutations?
Germ-line mutations are found in the DNA of germ (reproductive) cells and may be passed to offspring. Somatic mutations are found in the DNA of an organism's somatic tissue cells and cannot be passed to offspring.

*2. What is the difference between a transition and a transversion? Which type of base substitution is usually more common?
Transition mutations are base substitutions in which one purine (A or G) is changed to the other purine, or a pyrimidine (T or C) is changed to the other pyrimidine. Transversions are base substitutions in which a purine is changed to a pyrimidine or vice versa. Although transversions would seem to be statistically favored because there are eight possible transversions and only four possible transitions, about twice as many transition mutations are actually observed in the human genome.

*3. Briefly describe expanding trinucleotide repeats. How do they account for the phenomenon of anticipation?
Expanding trinucleotide repeats occur when DNA insertion mutations result in an increasing number of copies of a trinucleotide repeat sequence. Within a given family, a particular type of trinucleotide repeat may increase in number from generation to subsequent generation, increasing the severity of the mutation in a process called anticipation.

4. What is the difference between a missense mutation and a nonsense mutation? A silent mutation and a neutral mutation?
A base substitution that changes the sequence and the meaning of a mRNA codon, resulting in a different amino acid being inserted into a protein, is called a missense mutation. Nonsense mutations occur when a mutation replaces a sense codon with a stop (or nonsense) codon.

A nucleotide substitution that changes the sequence of a mRNA codon but not the meaning is called a silent mutation. In neutral mutations, the sequence and the meaning of a mRNA codon are changed. However, the amino acid substitution has little or no effect on protein function.

5. Briefly describe two different ways that intragenic suppressors may reverse the effects of mutations.
Intragenic suppression is the result of second mutations within a gene that restore a wild-type phenotype. The suppressor mutations are located at different sites within the gene from the original mutation. One type of suppressor mutation restores the original phenotype by reverting the meaning of a previously mutated codon to that of the original codon. The suppressor mutation occurs at a different position than the first mutation, which is still present within the codon. Intragenic suppression

may also occur at two different locations within the same protein. If two regions of a protein interact, a mutation in one of these regions could disrupt that interaction. The suppressor mutation in the other region would restore the interaction. Finally, a frameshift mutation due to an insertion or deletion could be suppressed by a second insertion or deletion that restores the proper reading frame.

*6. How do intergenic suppressors work?
Intergenic suppressor mutations restore the wild-type phenotype. However, they do not revert the original mutation. The suppression is a result of mutation in a gene other than the gene containing the original mutation. Because many proteins interact with other proteins, the original mutation may have disrupted the protein-protein interaction, while the second mutation restores the interaction. A second type of intergenic suppression occurs when a mutation within an anticodon region of a tRNA molecule allows for pairing at the codon containing the original mutation and the substitution of a functional amino acid in the protein.

*7. What is the difference between mutation frequency and mutation rate?
Mutation frequency is defined as the occurrence or frequency of mutation in a population of cells or individuals. The mutation rate is typically expressed as the number of mutations per biological unit such as per replication or cell division.

*8. What is the cause of errors in DNA replication?
Two types of events have been proposed that could lead to DNA replication errors: mispairing due to tautomeric shifts in nucleotides and mispairing through wobble or flexibility of the DNA molecule. Current evidence suggests that mispairing through wobble caused by flexibility in the DNA helix is the most likely cause.

9. How do insertions and deletions arise?
Strand slippage that occurs during DNA replication and unequal crossover events due to misalignment at repetitive sequences have been shown to cause deletions and additions of nucleotides to DNA molecules. Strand slippage results from the formation of small loops on either the template or the newly synthesized strand. If the loop forms on the template strand, then a deletion occurs. Loops formed on the newly synthesized strand result in insertions. If, during crossing over, a misalignment of the two strands at repetitive sequence occurs, then the resolution of the cross over will result in one DNA molecule containing an insertion and the other molecule containing a deletion.

*10. How do base analogs lead to mutations?
Base analogs have structures similar to the nucleotides and if present, may be incorporated into the DNA during replication. Many analogs have an increased tendency for mispairing, which can lead to mutations. DNA replication is required for the base analog-induced mutations to be incorporated into the DNA.

11. How do alkylating agents, nitrous acid, and hydroxylamine produce mutations?

Alkylating agents donate alkyl groups (either methyl or ethyl)) to the nucleotide bases. The addition of the alkyl group results in mispairing of the alkylated base and typically leads to transition mutations. Nitrous acid treatment results in the deamination of cytosine, producing uracil, which pairs with adenine. During the next round of replication, a CG to AT transition will occur. The deamination of guanine by nitrous acid produces xanthine. Xanthine can pair with either cytosine or thymine. If paired with thymine, then a CG to TA transition can occur. Hydroxylamine works by adding a hydroxyl group to cytosine, producing hydroxylaminocytosine. The hydroxylaminocytosine has an increased tendency to undergo tautomeric shifts, which allow pairings with adenine, resulting in GC to AT transitions.

12. What types of mutations are produced by ionizing and UV radiation?
Ionizing radiation promotes the formation of radicals and reactive ions that result in the breakage of phosphodiester linkages within the DNA molecule. Single- and double-strand breaks can occur. Double-strand breaks are difficult to repair accurately and may result in the deletion of genetic information. UV radiation promotes the formation of pyrimidine dimers between adjacent pyrimidines in a DNA strand. Inefficient repair of the dimers by error-prone DNA repair systems results in an increased mutation rate.

*13. What is the SOS system and how does it lead to an increase in mutations?
The SOS system is an error-prone DNA repair system consisting of at least 25 genes. Induction of the SOS system results in a bypass of damaged DNA regions, which allows for DNA replication across the damaged regions. However the bypass of damaged DNA results in a less accurate replication process, and thus more mutations will occur.

14. What is the purpose of the Ames test? How are *his*$^-$ bacteria used in this test?
The Ames test allows for rapid and inexpensive detection of potentially carcinogenic compounds using bacteria. The majority of carcinogenic compounds result in damage to DNA and are mutagens. The reversion of his$^-$ *bacteria to his*$^+$ *is used to detect the mutagenic potential of the compound being tested.*

*15. List at least three different types of DNA repair and briefly explain how each is carried out.
(1) *Mismatch repair. Replication errors that are the result of base-pair mismatches are repaired. Mismatch repair enzymes recognize distortions in the DNA structure due to mispairing and detect the newly synthesized strand by the lack of methylation on the new strand. The bulge is excised and DNA polymerase and DNA ligase fill in the gap.*
(2) *Direct repair. DNA damage is repaired by directly changing the damaged nucleotide back to its original structure.*
(3) *Base excision repair. The damaged base is excised, and then the entire nucleotide is replaced.*

(4) Nucleotide excision repair. Repair enzymes recognize distortions of the DNA double-helix. Damaged regions are excised by enzymes, which cut phosphodiester bonds on either side of the damaged region. The gap generated by the excision step is filled in by DNA polymerase.

16. What features do mismatch repair, base-excision repair, and nucleotide-excision repair have in common?
Mismatch repair, base excision repair, and nucleotide excision repair all result in the removal of nucleotides from DNA. All repair mechanisms that excise nucleotides share a common four-step pathway:
(1) DNA damage is detected;
(2) the damage is excised by DNA repair endonucleases;
(3) following excision, DNA polymerase adds nucleotides to the free 3′ OH group, using the remaining strand as a template; and
(4) ligation of nicks in the sugar phosphate backbone by DNA ligase.

APPLICATION QUESTIONS AND PROBLEMS

*17. A codon that specifies the amino acid Gly undergoes a single-base substitution to become a nonsense mutation. In accord with the genetic code given in Figure 15.12, is this mutation a transition or a transversion? At which position of the codon does the mutation occur?
By examining the four codons that encode for Gly, GGU, GGC, GGA, and GGG, and the three nonsense codons, UGA, UAA, and UAG, we can determine that only one of the Gly codons, GGA, could be mutated to a nonsense codon by the single substitution of a U for a G at the first position:

***G**GA → **U**GA*

Because uracil is a pyrimidine and guanine is a purine, the mutation is a transversion.

*18. (a) If a single transition occurs in a codon that specifies Phe, what amino acids could be specified by the mutated sequence?
Two codons can encode for Phe, UUU, and UUC. A single transition could occur at each of the positions of the codon resulting in different meanings.

Original codon	*Mutated codon (amino acid encoded)*
UUU	*CUU (Leu), UCU (Ser), UUC (Phe)*
UUC	*CUC (Ser), UCU (Ser), UUU (Ser)*

(b) If a single transversion occurs in a codon that specifies Phe, what amino acids could be specified by the mutated sequence?

Original codon	*Mutated codon (amino acid encoded)*
UUU	*AUU (Ile), UAU (Tyr), UUA (Leu), GUU (Val), UGU (Cys), UUG (Leu)*
UUC	*AUC (Ile), UAC (Tyr), UUA (Leu), GUC (Val), UGC (Cys), UUG (Leu)*

(c) If a single transition occurs in a codon that specifies Leu, what amino acids could be specified by the mutated sequence?

Original codon	*Mutated codon (amino acid encoded)*
CUU	*UUU (Phe), CCU (Pro), CUC (Leu)*
CUC	*UUC (Phe), CCC (Pro), CUG (Leu)*
CUA	*UUA (Leu), CCA (Pro), CUG (Leu)*
CUG	*UUG (Leu), CCG (Pro), CUA (Leu)*
UUG	*CUG (Leu), UCG (Ser), UUA (Ser)*
UUA	*CUA (Leu), UCG (Ser), UUG (Leu)*

(d) If a single transversion occurs in a codon that specifies Leu, what amino acids could be specified by the mutated sequence?

Original codon	*Mutated codon (amino acid encoded)*
UUA	*AUA (Met), UAA (Stop), UUU (Phe), GUA (Val), UGA (Stop), UUC (Phe)*
UUG	*AUG (Met), UAG (Stop), UUU (Phe), GUG (Val), UGG (Trp), UUC (Phe)*
CUU	*GUU (Val), CGU (Arg), CUG (Leu), AUU (Ile), UAU (Tyr), UUA (Leu)*
CUC	*AUC (Ile), CAC (His), CUA (Leu), GUC (Val), CGC (Arg), CUG (Leu)*
CUA	*AUA (Ile), CAA (Gln), CUC (Leu), GUA (Val), CGA (Arg), CUG (Leu)*
CUG	*AUG (Met), CAG (Gln), CUC (Leu), GUG (Val), CGG (Arg), CUU (Leu)*

19. Hemoglobin is a complex protein that contains four polypeptide chains. The normal hemoglobin found in adults—called adult hemoglobin—consists of two α and two β polypeptide chains, which are encoded by different loci. Sickle cell hemoglobin, which causes sickle cell anemia, arises from a mutation in the β chain of adult hemoglobin. Adult hemoglobin and sickle cell hemoglobin differ in a single amino acid: the sixth amino acid from one end in adult hemoglobin is glutamic acid, whereas sickle cell hemoglobin has valine at this position. After consulting the genetic code provided in Figure 15.12, indicate the type and location of the mutation that gave rise to sickle-cell anemia.

There are two possible codons for glutamic acid, GAA and GAG. Single base substitutions at the second position in both codons can produce codons that encode valine:

GAA--------> GUA (Val)
GAG--------> GUG (Val).

Both substitutions are transversions. However, in the gene encoding the β chain of hemoglobin, the GAG codon is the wild-type codon and the mutated GUG codon results in the sickle-cell phenotype.

*20. The following nucleotide sequence is found on the template strand of DNA. First, determine the amino acids of the protein encoded by this sequence by using the genetic code provided in Figure 15.12. Then give the altered amino acid sequence of the protein that will be found in each of the following mutations.

Sequence of DNA template: 3′–TAC TGG CCG TTA GTT GAT ATA ACT–5′
Nucleotide number → 1 24
mRNA sequence: 5′–AUG ACC GGC AAU CAA CUA UAU UGA–3′
amino acid sequence: Amino–Met Thr Gly Asn Gln Leu Tyr Stop–Carboxyl

(a) Mutant 1: A transition at nucleotide 11.
The transition results in the substitution of Ser for Asn.
original sequence: 3′–TAC TGG CCG TTA GTT GAT ATA ACT–5′
*mutated sequence: 3′–TAC TGG CCG T**C**A GTT GAT ATA ACT–5′*
*mRNA sequence: 5′–AUG ACC GGC A**G**U CAA CUA UAU UGA–3′*
*amino acids: Amino–Met Thr Gly **Ser** Gln Leu Tyr Stop–Carboxyl*

(b) Mutant 2: A transition at nucleotide 13.
The transition results in the formation of a UAA nonsense codon.
original sequence: 3′–TAC TGG CCG TTA GTT GAT ATA ACT–5′
*mutated sequence: 3′–TAC TGG CCG TTA **A**TT GAT ATA ACT–5′*
*mRNA sequence: 5′–AUG ACC GGC AAU **U**AA CUA UAU UGA–3′*
*amino acid sequence: Amino–Met Thr Gly Asn **STOP**–Carboxyl*

(c) Mutant 3: A one-nucleotide deletion at nucleotide 7.
The one-nucleotide deletion results in a frameshift mutation.
original sequence: 3′–TAC TGG CCG TTA GTT GAT ATA ACT–5′
*mutated sequence: 3′–TAC TGG **CGT TAG TTG ATA TAA CT**–5′*
*mRNA sequence: 5′–AUG ACC **GCA GUC AAC UAU AUU GA**–3′*
*amino acids: Amino–Met Thr **Ala Ile Asn Tyr Ile** –Carboxyl*

(d) Mutant 4: A T→ A transversion at nucleotide 15.
The transversion results in the substitution of His for Gln in the protein.
original sequence: 3′–TAC TGG CCG TTA GTT GAT ATA ACT–5′
*mutated sequence: 3′–TAC TGG CCG TTA GT**A** GAT ATA ACT–5′*
*mRNA sequence: 5′–AUG ACC GGC AAU CA**U** CUA UAU UGA–3′*
*amino acids: Amino–Met Thr Gly Asn **His** Leu Tyr Stop–Carboxyl*
or
*mutated sequence: 3′–TAC TGG CCG TTA GT**G** GAT ATA ACT–5′*
*mRNA sequence: 5′–AUG ACC GGC AAU CA**C** CUA UAU UGA–3′*
*amino acids: Amino–Met Thr Gly Asn **His** Leu Tyr Stop–Carboxyl*

(e) Mutant 5: An addition of TGG after nucleotide 6.
The addition of the three nucleotides results in the addition of Thr to the amino acid sequence of the protein.
original sequence: 3′–TAC TGG CCG TTA GTT GAT ATA ACT–5′
mutated sequence:3′–TAC TGG TGG CCG TTA GTT GAT ATA ACT–5′
*mRNA sequence: 5′–AUG ACC **ACC** GGC AAU CAA CUA UAU UGA–3′*
*amino acids: Amino–Met Thr **Thr** Gly Asn Gln Leu Tyr Stop–Carboxyl*

(f) Mutant 6: A transition at nucleotide 9.

The protein retains the original amino acid sequence.
original sequence: 3'–TAC TGG CCG TTA GTT GAT ATA ACT–5'
*mutated sequence: 3'–TAC TGG CC**A** TTA GTT GAT ATA ACT–5'*
*mRNA Sequence: 5'–AUG ACC GG**U** AAU CAA CUA UAU UGA –3'*
*amino acids: Amino–Met Thr **Gly** Asn Gln Leu Tyr Stop–Carboxyl*

21. A polypeptide has the following amino acid sequence:

Met-Ser-Pro-Arg-Leu-Glu-Gly

The amino acid sequence of this polypeptide was determined in series of mutants listed in parts *a* through *e*. For each mutant, indicate the type of change that occurred in the DNA (single-base substitution, insertion, deletion) and the phenotypic effect of the mutation (nonsense mutation, missense mutation, frameshift, etc.).

(a) Mutant 1: Met-Ser-Ser-Arg-Leu-Glu-Gly

A missense mutation has occurred resulting in the substitution of Ser for Pro in the protein. The change is most likely due to a single-base substitution in the Ser codon resulting in the production of a Pro codon. Four of the Ser codons can be changed to Pro codons by a single transition mutation.

Pro	*Ser*
***C**CU*	***U**CU*
***C**CC*	***U**CC*
***C**CA*	***U**CA*
***C**CG*	***U**CG*

(b) Mutant 2: Met-Ser-Pro

A single-base substitution has occurred in the Arg codon resulting in the formation of a stop codon. Two of the potential codons for Arg can be changed by single substitutions to stop codons. The phenotypic effect is a nonsense mutation.

Arg	*Stop*	
***C**GA*	***U**GA*	*transition mutation*
***A**GA*	***U**GA*	*transversion mutation*

(c) Mutant 3: Met-Ser-Pro-Asp-Trp-Arg-Asp-Lys

The deletion of a single nucleotide at the first position in the Arg codon (most likely CGA) has resulted in a frameshift mutation in which the mRNA is read in a different frame, producing a different amino acid sequence for the protein.

(d) Mutant 4: Met-Ser-Pro-Glu-Gly

A six base pair deletion has occurred, resulting in the elimination of two amino acids (Arg and Leu) from the protein. The result is a truncated polypeptide chain.

(e) Mutant 5: Met-Ser-Pro-Arg-Leu-Leu-Glu-Gly

The addition or insertion of three nucleotides into the DNA sequence has resulted in the addition of a Leu codon to the polypeptide chain.

*22. A gene encodes a protein with the following amino acid sequence:

Met-Trp-His-Arg-Ala-Ser-Phe.

A mutation occurs in the gene. The mutant protein has the following amino acid sequence:

Met-Trp-His-Ser-Ala-Ser-Phe.

An intragenic suppressor restores the amino acid sequence to that of the original protein:

Met-Trp-His-Arg-Ala-Ser-Phe.

Give at least one example of base changes that could produce the original mutation and the intragenic suppressor. (Consult the genetic code in Figure 15.12.)

Four of the six Arg codons could be mutated by a single-base substitution to produce a Ser codon. However, only two of the Arg codons mutated to form Ser codons could subsequently be mutated at a second position by a single-base substitution to regenerate the Arg codon. In both events, the mutations are transversions.

Original Arg Codon	*Ser Codon*	*Restored Arg Codon*
***C**GU*	*AGU*	*AG**G** or AG**A***
***C**GC*	*AGC*	*AG**G** or AG**A***

23. A gene encodes a protein with the following amino acid sequence:

Met-Lys-Ser-Pro-Ala-Thr-Pro

A nonsense mutation from a single-base-pair substitution occurs in this gene, resulting in a protein with the amino acid sequence Met-Lys. An intergenic suppressor mutation allows the gene to produce the full-length protein. With the original mutation and the intergenic suppressor present, the gene now produces a protein with the following amino acid sequence:

Met-Lys-Cys-Pro-Ala-Thr-Pro

Give the location and nature of the original mutation and the intergenic suppressor.

The original mutation is located in the Ser codon. Two of the six potential Ser codons (UCA and UCG) can be changed to stop codons by a single-base substitution.

*UCA → U**G**A (tranversion)*

*UCG → U**A**G (transversion)*

However, only the UCA codon is likely to be suppressed by Cys-tRNA containing a single-base substitution (a transversion mutation) in the anticodon (5′–ACA–3′ to 5′–UCA–3′), allowing for pairing with the UGA nonsense codon and suppression of the nonsense phenotype. The suppression is due to the insertion of Cys for the codon UGA.

*24. Can nonsense mutations be reversed by hydroxylamine? Why or why not?

No, hydroxylamine cannot reverse nonsense mutations. Hydroxylamine modifies cytosine-containing nucleotides and can only result in GC to AT transition mutations. In a stop codon, the GC to AT transition will result only in a different stop codon.

For 5′–UGA–3′:	*Template DNA:*	*3′–ACT–5′*
	Coding DNA:	*5′–TGA–3′*

Transition results:	*Template DNA*	*3′–ATT–5′*
GC to AT	*Coding DNA:*	*3′–TAA–5′*
	mRNA codon:	*5′–UAA–3′*
For 5′–UAG–3′:	*Template DNA:*	*3′–AT**C**–5′*
	Coding DNA:	*5′–TA**G**–3′*
Transition results:	*Template DNA:*	*3′–AT**T**–5′*
In GC to AT	*Coding DNA:*	*3′–TA**A**–5′*
	mRNA codon:	*5′–UA**A**–3′*

25. XG syndrome is a rare genetic disease that is due to an autosomal dominant gene. A complete census of a small European country reveals that 77,536 babies were born in 2000, of whom three had XG syndrome. In the same year, this country had a population of 5,964,321 people, and there were 35 living persons with XG syndrome. What are the mutation rate and mutation frequency of XG syndrome for this country?
To determine the incidence of XG syndrome or the mutation frequency in the country, divide the number of afflicted individuals by the total population of the country: 35/5,964,321 = 5.8×10^{-6}*, or approximately 1 out of 172,414 people is affected.*
The mutation rate for XG syndrome during the year 2000 is expressed as the number of mutations per babies born: 3/77,536 = 3.9×10^{-5}*.*

*26. The following nucleotide sequence is found in a short stretch of DNA:
5'–ATGT–3'
3'–TACA–5'
If this sequence is treated with hydroxylamine, what sequences will result after replication?
Hydroxylamine adds hydroxyl groups to cytosine, enabling the modified cytosine to occasionally pair with adenine, which ultimately can result in a GC to AT transition. So, only one base pair in the sequence will be affected. Ultimately, after replication, one of the dsDNA molecules will have the transition but not the other dsDNA molecule.

Original sequence	*Mutated sequence*
*5′–AT**G**T–3′*	*5′–AT**A**T–3′*
*3′–TA**C**A–5′*	*3′–TA**T**A–5′*

27. The following nucleotide sequence is found in a short stretch of DNA:
5'–AG–3'
3'–TC–5'
(a) Give all the mutant sequences that may result from spontaneous depurination occurring in this stretch of DNA.
The strand contains two purines, adenine and guanine. Because repair of depurination typically results in adenine being substituted for the missing purine, only the loss of the guanine by depurination will result in a mutant sequence.

*5′–A**G**–3′* *to* *5′–A**A**–3′*
*3′–T**C**–5′* *3′–T**T**–5′*

(b) Give all the mutant sequences that may result from spontaneous deamination occurring in this stretch of DNA.

Deamination of guanine, cytosine, and adenine can occur. However, the deamination of only cytosine and adenine are likely to result in mutant sequences because the deamination products can form improper base pairs. The deamination of guanine does not pair with thymine but can still form two hydrogen bonds with cytosine, thus no change will occur.

*5′–**A**G–3′* *if A is deaminated, then* *5′–**G**G–3′*
*3′–**T**C–5′* *3′–**C**C–5′*

*5′–A**G**–3′* *if C is deaminated then* *5′–A**A**–3′*
*3′–T**C**–5′* *3′–T**T**–5′*

28. In many eukaryotic organisms, a significant proportion of cytosine bases are naturally methylated to 5-methylcytosine. Through evolutionary time, the proportion of AT base pairs in the DNA of these organisms increases. Can you suggest a possible mechanism by which this increase occurs?

Spontaneous deamination of 5-methylcytosine produces thymine. If the subsequent repair of the GT mispairing is repaired incorrectly, then a GC to AT transition will result. Over time, the incorrect repairs will lead to an increase in the number of AT base pairs.

*29. A chemist synthesizes four new chemical compounds in the laboratory and names them PFI1, PFI2, PFI3, and PFI4. He gives the PFI compounds to a geneticist friend and asks her to determine their mutagenic potential. The geneticist finds that all four are highly mutagenic. She also tests the capacity of mutations produced by the PFI compounds to be reversed by other known mutagens and obtains the following results. What conclusions can you make about the nature of the mutations produced by these compounds?

	Reversed by				
Mutations produced by	2-Aminopurine	Nitrous acid	Hydroxylamine	Acridine orange	
PFI1		Yes	Yes	Some	No
PFI 2	No	No	No	No	
PFI3		Yes	Yes	No	No
PFI4		No	No	No	Yes

First consider the mutagenic actions of the reversion agents: 2-Aminopurine causes GC to AT and AT to GC transitions; Nitrous acid causes GC to AT and AT to GC transitions; Hydroxylamine causes GC to AT transitions; Acridine orange causes single-base insertions or deletions, resulting in frameshift mutations.

PFI1 causes both types of transitions, GC to AT and AT to GC. PFI1 mutations can be reversed by 2-aminopurine, nitrous acid, and occasionally by hydroxylamine, which suggests that it acts as a transition mutagen. The lack of reversion of all PFI1 mutations by hydroxylamine suggests that some of the mutations were caused by GC to AT substitutions because they are not reverted by hydroxylamine.

PFI2 causes transversions, or large deletions, because mutations caused by PF12 are not reverted by any of the agents.

PFI3 causes GC to AT transitions. Only GC to AT transitions could be reverted by 2-aminopurine and nitrous oxide but not by hydroxylamine.

PFI4 causes single-base insertions or deletions. PF14 is only reverted by acridine orange, an intercalating agent that only reverts single-base insertions or deletions.

*30. A plant breeder wants to isolate mutants in tomatoes that are defective in DNA repair. However, this breeder does not have the expertise or equipment to study enzymes in DNA repair systems. How could the breeder identify tomato plants that are deficient in DNA repair? What are the traits to look for?
The plant breeder should look for plants that have increased levels of mutations either in their germ-line or somatic tissues. Tomato plants with defective DNA repair systems may also be sensitive to high levels of sunlight and may need to be grown in a reduced light environment.

31. A genetics instructor designs a laboratory experiment to study the effects of UV radiation on mutation in bacteria. In the experiment, the students expose bacteria plated on petri plates to UV light for different lengths of time, place the plates in an incubator for 48 hours, and then count number of colonies that appear on each plate. The plates that have received more UV radiation should have more pyrimidine dimers, which block replication; thus, fewer colonies should appear on the plates exposed to UV light for longer periods of time. Before the students carry out the experiment, the instructor warns them that, while the bacteria are in the incubator, the students must not open the incubator door unless the room is darkened. Why should the bacteria not be exposed to light?
Exposure of DNA to UV light results in the formation of pyrimidine dimers in the DNA molecule. Often the repair of these dimers leads to mutations. Because the SOS repair system is error-prone and leads to an increased accumulation of mutations, UV light produces more mutations in bacteria when the SOS repair system is activated to repair the damage caused by the UV light. However, many species of bacteria have a direct DNA repair system that can repair pyrimidine dimers by breaking the covalent linkages between the pyrimidines that form the dimer. The enzyme that repairs the DNA is called photolyase, and is activated and energized by light. The photolyase is a very efficient repair enzyme and typically makes accurate repairs of the damage. If the bacteria in the UV radiation experiment are exposed to light, then the photolyase will be activated to repair the damage, resulting in fewer mutations in the irradiated bacteria.

CHALLENGE QUESTIONS

32. Tay Sachs disease is a severe, autosomal recessive genetic disease that produces deafness, blindness, seizures, and eventually death. The disease results from a defect in the HEXA gene, which codes for hexosaminidase A. This enzyme normally degrades GM2 gangliosides. In the absence of hexosaminidase A, GM2 gangliosides accumulate in the brain. Recent molecular studies have shown that the most common mutation causing Tay Sachs disease is a 4 bp insertion that produces a downstream premature stop codon. Further studies reveal that normal transcription of the HEXA gene occurs in individuals with Tay Sachs disease, but the HEXA mRNA is unstable. Propose a mechanism to account for how a premature stop codon could cause mRNA instability.
The stability of the mRNA molecule may be dependent on the translational machinery. If the ribosome terminates prematurely, then the mRNA molecule may become a target of ribonucleases. Essentially the premature stop codon has marked the HEXA mRNA for destruction in Tay Sachs individuals.

    Recent data has shown that the premature stop codon within the HEXA mRNA molecule leads to the rapid destruction of the HEXA mRNA within the cell. The destruction is likely accomplished through a process called nonsense-mediated mRNA decay (NMD). In the current model of NMD function, the NMD pathway is stimulated by the premature termination of protein synthesis at a nonsense codon located 50 or more nucleotides upstream of the final exon-exon junction. A exon junction protein complex (EJC) forms approximately 20 to 24 nucleotides upstream of the final exon-exon junction. The EJC stimulates decay of the mRNA molecule unless it is removed by the ribosome during translation. In the normal HEXA mRNA translation, the first ribosome removes the EJC because the termination codon is located downstream of the EJC. In individuals with Tay Sachs, the premature stop codon is located upstream of the EJC resulting in termination before the first ribosome can remove the EJC. Because the EJC is not removed, stimulation of NMD occurs and the mutated HEXA mRNA is degraded.

33. *Ochre* and *amber* are two types of nonsense mutations. Before the genetic code was worked out, Sydney Brenner, Anthony O. Stretton, and Samuel Kaplan applied different types of mutagens to bacteriophages in an attempt to determine the bases present in the codons responsible for *amber* and *ochre* mutations. They knew that *ochre* and *amber* mutants were suppressed by different types of mutations, demonstrating that each was a different termination codon. They obtained the following results.
    (1) A single-base substitution could convert an *ochre* mutation into an *amber* mutation.
    (2) Hydroxylamine induced *ochre* and *amber* mutations in wild-type phages.
    (3) 2-Aminopurine caused *ochre* to mutate to *amber*.
    (4) Hydroxylamine did not cause *ochre* to mutate to *amber*.
    These data do not allow the complete nucleotide sequence of the *amber* and *ochre* codons to be worked out, but they do provide some information about the bases found in the nonsense mutations.

(a) What conclusions about the bases found in the codons of *amber* and *ochre* mutations can be made from these observations?
In considering the data, it is important to remember the mutagenic actions of hydroxylamine and 2-aminopurine. Hydroxylamine produces only GC to AT transition mutations. However, 2-aminopurine can produce both types of transitions, GC to AT and AT to GC.

Because hydroxylamine can be used to produce amber and ochre mutations from wild-type phages, then amber and ochre codons must contain uracil and/or adenine. The production of amber mutations from ochre codons by 2-aminopurine but not by hydroxylamine suggests that ochre mutations do contain adenine and uracil, while amber mutations contain guanine and also contain either adenine and/or uracil.

(b) Of the three nonsense codons (UAA, UAG, UGA), which represents the *ochre* mutation?
Based on the mutagenesis data, only the UAA codon matches the results for the ochre mutation. It is the only stop codon that does not contain guanine.

34. To determine whether radiation associated with the atomic bombings of Hiroshima and Nagasaki produced recessive germ-line mutations, scientists examined the sex ratio of the children of the survivors of the blasts. Can you explain why an increase in germ-line mutations might be expected to alter the sex ratio?
Recessive germ-line mutations on the X chromosome have the potential to alter the sex ratios. Female offspring who have a single recessive mutation on the X chromosome will also possess a nonmutated allele on their other X chromosome. Therefore, females will be heterozygous for the pair of alleles. Male offspring, however, have only one X chromosome, and a mutation within an allele on the X chromosome will result in a phenotypic change. If the allele mutated is essential, then the mutation for males will be lethal. Essentially, recessive mutations on the X chromosome in the germ-line could lead to fewer males being born.

35. Trichothiodystrophy is an inherited disorder in humans that is characterized by premature aging, including osteoporosis, osteosclerosis, early graying, infertility, and reduced life span. Recent studies have shown that the mutation that causes this disorder occurs in a gene that encodes a DNA helicase. Propose a mechanism for how a mutation in a DNA helicase might cause premature aging. Be sure to relate the symptoms of the disorder to possible functions of the helicase enzyme.
The DNA helicase protein (XPD) is involved in DNA repair process, specifically in nucleotide excision repair to displace damaged nucleotides. If the helicase is not functioning properly, then the damaged nucleotides may not be removed, which may result in the accumulation of DNA damage within the chromosomes of these cells. The accumulation of DNA damage in these cells could lead either to apoptosis (programmed cell death) or to cellular senescence followed by cellular death. In either case, the cells affected will not divide and reproduce through cellular division processes. Cells responsible for bone formation, hair color, and fertility if affected by the DNA damage and subsequent lack of repair would ultimately cease dividing. This lack of new cell production would lead to bone

weakening and loss, lack of production of new melanocytes needed for hair color, and a lack of production of reproductive cells.

Because this DNA helicase is also associated with a transcription factor (TFIIH) needed in transcription by RNA polymerase II, a second possibility may involve its role in transcription leading to a decline in levels of transcription. Current data suggest that the DNA helicase may be involved in transcriptionally coupled DNA repair where its function may be to remove the stalled RNA polymerase from DNA that has been damaged. The removal of the RNA polymerase would allow for repair of the damaged section. If the RNA polymerase is not removed, it may prevent repair. A decline in transcriptional function due to malfunctioning transcription factors or a lack of repair because of a stalled RNA polymerase could lead to a decline in cellular functions and cellular death as well.

36. Achondroplasia is an autosomal dominant disorder characterized by disproportionate short stature—the legs and arms are short compared to the head and trunk. The disorder is due to a base substitution in the gene for fibroblast growth factor receptor 3 (FGFR3), which is located on the short arm of chromosome 4.

Although achondroplasia is clearly inherited as an autosomal dominant trait, more than 80% of individuals with achondroplasia are born to parents with normal stature. This indicates that most cases are caused by newly arising mutations; these newly arising cases (not inherited from an affected parent) are referred to as sporadic. Molecular studies have demonstrated that sporadic cases of achondroplasia are almost always caused by mutations inherited from the father (paternal mutations). In addition, the occurrence of achondroplasia is higher among older fathers; indeed, approximately 50% of children with achondroplasia are born to fathers older than age 35. There is no association with maternal age. The mutation rate for achondroplasia (about 4×10^{-5} mutations per gamete) is high compared to other genetic disorders.

Propose an explanation for why most spontaneous mutations for achondroplasia are paternal in origin and why the occurrence of achondroplasia is higher among older fathers.

In men, sperm cells are produced throughout much of their life. The cells responsible for the sperm production are called spermatogonia. These spermatogonia divide by mitosis to produce more spermatogonia and produce spermatocytes, cells that eventually will divide by meiosis to produce sperm cells. These continued cell divisions by the spermatogonia particularly to produce more spermatogonia could lead to an increased chance of mutations within the DNA of the spermatogonia cell. Essentially, the more cell divisions the greater the risk for mutation. Also some DNA sequences may be more susceptible to mutations than others. These locations are called hot spots. Potentially the base substitution occurs at a hot spot in the FGFR3 gene as more cell divisions occur. In addition, as men age, their exposure to environmental factors over time may increase mutation rates including that of the FGFR3 gene.

However, recent data does not support that the increase in mutations is due only to an increase in mutations as a result of spermatogonia mitosis. A second possibility is that the mutation may confer a positive benefit to the spermatogonia or ultimately to the sperm cells produced. Potentially, the mutated sperm have a higher survival rate than normal sperm cells leading to an increased risk of fertilization by the FGFR3 mutation containing sperm cells.

CASE STUDY QUESTIONS

1. Explain how the traditional view of PKU resulting from an autosomal recessive mutation is an oversimplification of the true situation.
In the most simple case, PKU can result from an autosomal mutation at the PAH locus. However, other genetic factors and environmental influences can affect the overall phenotype. Individuals (such as siblings) that have an identical PAH genotype can have different mental abilities suggesting that factors such as other genes and the environment have an effect on phenylalanine levels in the brain. Most individuals with PKU are compound heterozygotes in that they contain two different disease-causing PAH alleles. Also even within a single chromosome the PAH locus may have more than one mutation. Finally about 2% of PKU individuals do not have a mutation in the PAH locus responsible for the phenotype. Mutations at BH_4 locus have also been implicated. The BH_4 locus codes for tetrahydrobioperin (BH_4) a cofactor needed for the proper metabolic function of PAH. Also mutations at other genetic loci encoding for proteins that are involved in the production or recycling of BH_4 can result in the PKU phenotype.

2. Suppose that you isolated some DNA from the *PAH* locus and separated it into single strands. You then isolated mature mRNA encoded by *PAH* locus and mixed it with the single-stranded DNA. When the single-stranded DNA hybridizes (pairs) with the complementary mRNA, there are large regions of the DNA that loop out from the RNA. What is the cause of these loops?
Isolated genomic DNA will contain both the exons and introns of a gene. The mature mRNA encoded by the PAH locus has undergone RNA processing and as part of the processing the introns are removed leaving only exon encoded sequences. When the single-stranded DNA hybridizes with the mRNA from the PAH locus the exonic regions are complementary and base pair. However, the single-stranded DNA also contains the introns from the PAH locus and there is no corresponding complementary region on the mRNA. So, these intron containing DNA regions loop out from the RNA.

3. Upstream of the transcription start site for the PAH locus are several DNA sequences that contain several GC boxes and a CAAT box. What role do these sequences play in the expression of phenylalanine hydroxylase? What would be the most likely phenotypic effect if some of these sequences were deleted?
GC boxes and the CAAT box are likely part of the regulatory promoter of the phenylalanine hydroxylase gene. Both the GC boxes and the CAAT box may serve as binding sites for transcriptional activator proteins that stimulate transcription from the PAH locus. If some of the sequences were deleted then there would be a decrease

or complete absence of transcription from the PAH locus. The result would be a reduction or complete absence of phenylalanine hydroxylase expression. Such an individual containing these deletions would likely possess the phenylketonuric phenotype.

4. What would be the most likely effect on the structure of phenylalanine hydroxylase of the following types of mutations occurring at the PAH locus?
   (a) Missense mutation.
   A missense mutation would occur when a single base pair substitution changes a codon such that a different amino acid would be inserted in the phenylalanine hydroxylase protein. The effect would depend on the chemical properties of the original amino acid and on the chemical properties of the substituted amino acid. If chemically similar, then little change in the protein structure might be anticipated (a neutral mutation). If different, then significant alterations in the protein structure could occur. The result could be a change in the active site that reduces the enzyme activity or more likely as evidenced in other missense mutations in the phenylalanine hydroxylase protein, the change will lead to misfolding of the protein and aggregation. The misfolded protein will then be rapidly degraded.
   (b) Nonsense mutation.
   A nonsense mutation occurs when a sequence change results in a premature stop codon in the mRNA. The result will be a truncated protein. The likely effect is that truncated phenylalanine hydroxylase protein cannot assume its proper secondary and tertiary structures and will not have metabolic activity. Ultimately the defective protein will be rapidly degraded.
   (c) Small deletion or insertion.
   A deletion or insertion of a nucleotide could result in the reading frame of translation being shifted (or a frameshift mutation). The shift can have a drastic effect on the protein in that any subsequent codons will be affected thus changing many amino acids in the protein. The likely effect is a significant change in the structure of phenylalanine hydroxylase. However, if the insertion or deletion is a multiple of 3 nucleotides the effect may not be as drastic because the reading frame will be maintained. Essentially only a single codon may be affected.

Chapter Eighteen: Recombinant DNA Technology

COMPREHENSION QUESTIONS

1. List some of the effects and applications of recombinant DNA technology.
Recombinant DNA technology has had profound effects on all fields of biology. Whole genomes have been sequenced, structures of genes elucidated, the patterns of molecular evolution studied. Recombinant DNA technology is now used to diagnose and screen for genetic diseases, and gene therapy is being explored. Recombinant DNA is used to make pharmaceutical products, such as recombinant insulin and clotting factors. Genetically modified organisms will change the lives of farmers and improve agricultural productivity and the quality of food and fiber.

2. What common feature is seen in the sequences recognized by type II restriction enzymes?
The recognition sequences are palindromic, and 4–8 base pairs long.

3. What role do restriction enzymes play in bacteria? How do bacteria protect their own DNA from the action of restriction enzymes?
Restriction enzymes cut foreign DNA, such as viral DNA, into fragments. Bacteria protect their own DNA by modifying bases, usually by methylation, at the recognition sites.

*4. Explain how gel electrophoresis is used to separate DNA fragments of different lengths.
Gel electrophoresis acts as a molecular sieve. The gel is an aqueous matrix of agarose or polyacrylamide. DNA molecules are loaded into a slot or well at one end of the gel. When an electric field is applied, the negatively-charged DNA molecules migrate towards the positive electrode. Shorter DNA molecules are less hindered by the agarose or polyacrylamide matrix and migrate faster than longer DNA molecules, which must wind their way around obstacles and through the pores in the gel matrix.

*5. After DNA fragments are separated by gel electrophoresis, how can they be visualized?
DNA molecules can be visualized by staining with a fluorescent dye. Ethidium bromide intercalates between the stacked bases of the DNA double helix, and the ethidium bromide-DNA complex fluoresces orange when irradiated with an ultraviolet light source. Alternatively, they can be visualized by attaching radioactive or chemical labels to the DNA before it is placed in the gel.

6. What is the purpose of Southern blotting? How is it carried out?
Southern blotting is used to detect and visualize specific DNA fragments that have a sequence complementary to a labeled DNA probe. DNA is first cleaved into fragments with restriction endonucleases. The fragments are separated by size via gel electrophoresis. These fragments are then denatured and transferred by

blotting onto the surface of a membrane filter. The membrane filter now has single-stranded DNA fragments bound to its surface, separated by size as in the gel. The filter is then incubated with a solution containing a denatured, labeled probe DNA. The probe DNA hybridizes to its complementary DNA on the filter. After washing excess unbound probes, the labeled probe hybridized to the DNA on the filter can be detected using the appropriate methods to visualize the label. For radioactively labeled probes, the bound probe is detected by exposure to X-ray film.

*7. What are the differences between Southern, Northern, and Western blotting?
In all three techniques, macromolecules are separated by size via gel electrophoresis and are transferred to the surface of a membrane filter. Southern blotting transfers DNA, Northern blotting transfers RNA, and Western blotting transfers protein.

*8. Give three important characteristics of cloning vectors.
Cloning vectors should have:
(1) an origin of DNA replication so they can be maintained in a cell.
(2) a gene, such as antibiotic resistance, to select for cells that carry the vector.
(3) a unique restriction site or series of sites to cut and ligate a foreign DNA molecule.

9. Briefly describe four different methods for inserting foreign DNA into plasmids, giving the strengths and weaknesses of each.
Restriction cloning: Vector and foreign DNA are cut with the same restriction enzyme, then ligated together with DNA ligase. This is the most straightforward, simplest method of cloning. The disadvantage is that matching restriction sites may not be available. Ligation also produces undesirable products, such as the vector ligating to itself, without the foreign DNA insert.

Tailing: Vector and foreign DNA fragments are blunt-ended either by filling in or trimming overhanging sequences after restriction digestion, or by cutting with a blunt-cutting restriction endonuclease. Terminal deoxynucleotidyl transferase is used to add complementary homopolymeric DNA tails to the 3' ends of the vector and insert. After annealing of the insert to the vector by the complementary tails, the molecules are covalently joined with DNA ligase. This method has the advantage that the vector cannot self-ligate. The disadvantage is that the procedure destroys the restriction site in the vector, so the insert cannot be cut out. Also, the added tail sequences may interfere with the function of either the foreign DNA insert or the vector.

Ligation of blunt ends with T4 DNA ligase: Any two blunt ends of DNA fragments can be ligated, so the method does not depend on the presence of suitable restriction sites. The disadvantage is that blunt-end ligations are inefficient, and vector self-ligation is difficult to control.

Addition of linkers: Small synthetic DNA sequences containing restriction sites are ligated to blunt ends of DNA fragments. Restriction digestion then generates sticky ends that can be ligated as in restriction cloning. The advantage is that again, the method does not depend on the availability of suitable restriction sites; any restriction site can be added. The disadvantage is that this involves extra steps and is more difficult than simple restriction cloning.

10. How are plasmids transferred into bacterial cells?
Plasmids are transferred into bacterial cells by transformation. Most laboratory strains of E. coli *require chemical treatment to render them competent to take up DNA by transformation. Another common alternative is electroporation, using an electric discharge to drive DNA into cells.*

*11. Briefly explain how an antibiotic-resistance gene and the *lacZ* gene can be used as markers to determine which cells contain a particular plasmid.
Many plasmids designed as cloning vectors carry a gene for antibiotic resistance and the lacZ gene. The lacZ gene on the plasmid has been engineered to contain multiple unique restriction sites. Foreign DNAs are inserted into one of the unique restriction sites in the lacZ gene. After transformation, E. coli *cells carrying the plasmid are plated on a medium containing the appropriate antibiotic to select for cells that carry the plasmid. The medium also contains an inducer for the lac operon, so the cells express the lacZ gene, and X-gal, a substrate for beta-galactosidase that will turn blue when cleaved by β-galactosidase. The colonies that carry plasmid without foreign DNA inserts will have intact lacZ genes, make functional β-galactosidase, cleave X-gal, and turn blue. Colonies that carry plasmid with foreign DNA inserts will not make functional β-galactosidase because the lacZ gene is disrupted by the foreign DNA insert. They will remain white. Thus, cells carrying plasmids with inserts will form white colonies.*

12. How are genes inserted into bacteriophage λ vectors? What advantages do λ vectors have over plasmids?
λ vectors are prepared by removing the non essential third of the genome from the center, leaving two "arms." The two arms are ligated to either side of a foreign DNA molecule and then packaged into phage particles. λ vectors are very efficient at delivering DNA into E. coli *cells. Moreover, they have the capacity to carry foreign DNA fragments up to 23 kb in length. Finally, because only DNA molecules 40 to 50 kb can be packaged into a λ phage particle, self-ligated vectors or inserts by themselves will not be packaged or delivered into cells.*

*13. What is a cosmid? What are the advantages of using cosmids as gene vectors?
A cosmid is a plasmid vector with a plasmid origin of DNA replication, unique restriction sites for cloning, and selectable marker genes, that also has the lambda cos site so it can be packaged into λ phage particles for efficient delivery into E. coli cells. It can accommodate DNA fragments up to 44 kb.

14. What are yeast artificial chromosomes and shuttle vectors? When are these cloning vectors used?
Yeast artificial chromosomes (YACs) and shuttle vectors are used as cloning vectors to replicate and maintain foreign DNA sequences in eukaryotic cells. Yeast

artificial chromosomes have a yeast origin of DNA replication, a centromeric sequence, and telomeres. They are used to clone fragments of DNA on the order of 1 million bp in length. Shuttle vectors are plasmids that carry both bacterial and eukaryotic origins of DNA replication, so they can be replicated in either E. coli *or eukaryotic host cells.*

*15. How does a genomic library differ from a cDNA library? How is each created?
A genomic library is generated by cloning fragments of chromosomal DNA into a cloning vector. Chromosomal DNA is randomly fragmented by shearing or by partial digestion with a restriction enzyme. A cDNA library is made from mRNA sequences. Cellular mRNAs are isolated and then reverse transcriptase is used to copy the mRNA sequences to cDNA, which are cloned into plasmid or phage vectors.

16. How are probes used to screen DNA libraries? Explain how a synthetic probe can be prepared when the protein product of a gene is known.
The DNA library must first be plated out, either as colonies for plasmid libraries or phage plaques on a bacterial lawn for phage libraries. The colonies or plaques are transferred to membrane filters. A nucleic acid probe can be used to identify colonies or phage plaques that contain identical or similar sequences by hybridization. If no cloned DNA probe is available, but the amino acid sequence of the protein is known, then degenerate synthetic oligonucleotides can be synthesized that represent all possible coding sequences for a sequence of 7 to 10 amino acids. The synthetic oligonucleotides can be end-labeled and used as hybridization probes to screen a DNA library.

17. Explain how chromosome walking can be used to find a gene.
Chromosome walking is used to isolate DNA clones encoding genes defined solely by mutations, for which no DNA or amino acid sequence is known. Mapping experiments locate the gene to a chromosome and to a region of the chromosome. The chromosome walk begins with a neighboring gene that has been previously cloned. Clones that overlap this initial gene are isolated, then the overlapping clones are used to isolate additional overlapping clones, until the set of overlapping clones spans the chromosomal distance between the previously cloned gene and the unknown gene.

18. Discuss some of the considerations that must go into developing an appropriate cloning strategy.
In developing a cloning strategy, one must consider what is known about the gene to be cloned, the size and nature of the gene, and the ultimate purpose for cloning the gene. What is known about the gene will determine whether a nucleic acid or antibody probe can be used, or whether chromosome walking must be used. Large genes with many introns will require vectors capable of handling large inserts, whereas small genes can be cloned in plasmids. The ultimate purpose, whether for sequencing or expression, and in what host cell, will determine the choice of cloning vector.

*19. Briefly explain how the polymerase chain reaction is used to amplify a specific DNA sequence. What are some of the limitations of PCR?

First the double-stranded template DNA is denatured by high temperature. Then synthetic oligonucleotide primers corresponding to the ends of the DNA sequence to be amplified are annealed to the single-stranded DNA template strands. These primers are extended by a thermostable DNA polymerase so that the target DNA sequence is duplicated. These steps are repeated 30 times or more. Each cycle of denaturation, primer annealing, and extension results in doubling the number of copies of the target sequence between the primers.

PCR amplification is limited by several factors. One is that sequence of the gene to be amplified must be known, at least at the ends of the region to be amplified, in order to synthesize the PCR primers. Another is that the extreme sensitivity of the technique renders it susceptible to contamination. A third limitation is that the most common thermostable DNA polymerase used for PCR, Taq DNA polymerase, has a relatively high error rate. A fourth limitation is that PCR amplification is usually limited to DNA fragments of up to a few thousand base pairs; optimized DNA polymerase mixtures and reaction conditions extend the amplifiable length to around 20 kb.

*20. Briefly explain in situ hybridization, giving some applications of this technique.

In situ hybridization involves hybridization of radiolabeled or fluorescently labeled DNA or RNA probes to DNA or RNA molecules that are still in the cell. This technique can be used to visualize the expression of specific mRNAs in different cells and tissues, and the location of genes on metaphase or polytene chromosomes.

21. What is DNA footprinting?

DNA footprinting is a technique used to determine the binding sites of proteins on DNA sequences. DNA sequences containing a protein binding site are end-labeled, and then partially digested with DNase in the presence or absence of the DNA binding protein. In the absence of the protein, DNase generates a continuous ladder of fragments generated by DNase cleavage at all possible sites. This ladder of fragments is visualized by gel electrophoresis and autoradiography. In the presence of the protein, the binding sites are protected from DNase cleavage, resulting in a gap in the ladder of DNA fragments. The position of this gap identifies the protein binding site in the DNA sequence.

22. Briefly explain how site-directed mutagenesis is carried out.

In oligonucleotide-directed mutagenesis, an oligonucleotide containing the desired mutation in the sequence is synthesized. This mutant oligonucleotide is annealed to denatured target DNA template and used to direct DNA synthesis. The result is a double-stranded DNA molecule with a mismatch at the site to be mutated. When transformed into bacterial cells, bacterial repair enzymes will convert the molecule to the mutant form about 50% of the time.

*23. What are knockout mice, how are they produced, and for what are they used?
Knockout mice have a target gene disrupted or deleted. First, the target gene is cloned. The middle portion of the gene is replaced with a selectable marker, typically the neo gene that confers resistance to G418. This construct is then introduced back into mouse embryonic stem cells and cells with G418 resistance are selected. The surviving cells are screened for cells where the chromosomal copy of the target gene has been replaced with the neo-containing construct by homologous recombination of the flanking sequences. These embryonic stem cells are then injected into mouse blastocyst-stage embryos and these chimeric embryos are transferred to the uterus of a pseudopregnant female mouse. The knockout cells will participate in the formation of many tissues in the mouse fetus, including germ-line cells. The chimeric offspring are interbred to produce offspring that are homozygous for the knockout allele.

The phenotypes of the knockout mice provide information about the function of the gene.

24. Give some advantages that mice possess as model genetic organisms.
Mice are small, have a relatively short generation time with multiple progeny per litter, and are easily maintained and bred in the laboratory. Highly inbred strains are available; these provide animals with essentially identical genetic backgrounds. The long history of mouse genetics means that many mutations have been identified and characterized. As a mammal, the mouse is closer to humans in physiology than most other genetic model organisms, and the mouse genome is similar to the human genome in size and content of protein-coding genes. Finally, transgenic techniques allow creation of transgenic mice with either foreign genes added or targeted genes knocked out or replaced.

25. Describe how RFLPs can be used in gene mapping.
RFLPs provide molecular genetic markers that segregate in crosses or pedigrees like traditional genetic markers, and therefore can be used to map genes. The major advantage is that RFLPs are highly variable, so that most individuals in the population are heterozygous. For example, the inheritance of a disease gene in a human pedigree can be tested for co-inheritance with a set of RFLP markers spaced several centiMorgans apart on each human chromosome. The genotype of each individual in the pedigree for the disease gene can be deduced from its phenotypes and the phenotypes of its progeny, and the RFLP genotypes can be visualized by Southern blot analysis, probing with each RFLP marker probe. Statistically significant co-inheritance of the disease allele with an RFLP marker indicates linkage.

*26. What is DNA fingerprinting? What types of sequences are examined in DNA fingerprinting?
DNA fingerprinting is the typing of an individual for genetic markers at highly variable loci. This is useful for forensic investigations, to determine whether the suspect could have contributed to the evidentiary DNA obtained from blood or

other bodily fluids found at the scene of a crime. Other applications include paternity testing and the identification of bodily remains.

The RFLP loci traditionally used for DNA fingerprinting are called variable number of tandem repeat (VNTR) loci; these consist of short tandem repeat sequences located in introns or spacer regions between genes. The number of repeat sequences at the locus does not affect the phenotype of the individual in any way, so these loci are highly variable in the population. More recently, tandem repeat loci with smaller repeat sequences of just a few nucleotides, called short tandem repeats (STRs), have been adopted because they can be amplified by PCR.

27. What is gene therapy?
Gene therapy is the correction of a defective gene by either gene replacement or the addition of a wild-type copy of the gene. For this to work, enough of the cells of the critically affected tissues or organs must be transformed with the functional copy of the gene to restore normal physiology.

28. As the first recombinant DNA experiments were being carried out, there was concern among some scientists about this research. What were these concerns and how were they addressed?
The initial concerns were that recombinant DNA technology would lead to the release of novel pathogens that could attack humans or other species. Even the release of non pathogenic organisms could have adverse effects on ecological systems if they outcompete native species. NIH established a set of guidelines for experiments involving recombinant DNA technology, specifying both biological and physical containment levels for experiments at different risk levels. Stringent approval processes were required for the release of any genetically modified organism into the field.

APPLICATION QUESTIONS AND PROBLEMS

*29. Suppose that a geneticist discovers a new restriction enzyme in the bacterium *Aeromonas ranidae*. This restriction enzyme is the first to be isolated from this bacterial species. Using the standard convention for abbreviating restriction enzymes, give this new restriction enzyme a name (for help, see footnote to Table 18.2).
The first three letters are taken from the genus and species name, and the Roman numeral indicates the order in which the enzyme was isolated. Therefore, the enzyme should be named AraI.

30. How often, on average, would you expect a type II restriction endonuclease to cut a DNA molecule if the recognition sequence for the enzyme had five bp? (Assume that the four types of bases are equally likely to be found in the DNA and that the bases in a recognition sequence are independent.) How often would the endonuclease cut the DNA if the recognition sequence had 8 bp?
Because DNA has four different bases, the frequency of any sequence of n bases is equal to $1/(4^n)$. A five-bp recognition sequence will occur with a frequency of

$1/(4^5)$, or once every 1024 bp. An 8-bp recognition sequence will occur with a frequency of 1 per 4^8, or 65,536 bp.

*31. A microbiologist discovers a new type II restriction endonuclease. When DNA is digested by this enzyme, fragments that average 1,048,500 bp in length are produced. What is the most likely number of base pairs in the recognition sequence of this enzyme?
Here, 4^n = 1,048,500. So n = 10; a 10-bp recognition sequence is most likely.

32. Will restriction sites for an enzyme that has 4 bp in its restriction site be closer together, farther apart, or similarly spaced, on average, compared with those of an enzyme that has 6 bp in its restriction site? Explain your reasoning.
The restriction sites for an enzyme with a 4-bp recognition sequence should be spaced closer together than the sites for an enzyme with a 6-bp recognition sequence. The 4-bp recognition sequence will occur with an average frequency of once every 4^4 = 256 bp, whereas the 6-bp recognition sequence will occur with an average frequency of once every 4^6 = 4096 bp.

*33. About 60% of the base pairs in a human DNA molecule are AT. If the human genome has 3 billion base pairs of DNA, about how many times will the following restriction sites be present?
(a) *Bam*HI (restriction site = 5'—GGATCC—3')
(b) *Eco*RI (restriction site = 5'—GAATTC—3')
(c) *Hae*III (restriction site = 5'—GGCC—3')

We must first calculate the frequency of each base. Given that AT base pairs consist 60% of the DNA, we deduce that the frequency of A is 0.3 and frequency of T is 0.3. The GC base pairs must consist of 40% of the DNA; therefore, the frequency of G is 0.2 and the frequency of C is 0.2.

(a) *Bam*H1 GGATCC *is then (0.2)(0.2)(0.3)(0.3)(0.2)(0.2) = 0.000144*
3,000,000,000(0.000144) = 432,000 times
(b) *Eco*RI GAATTC = *(0.2)(0.3)(0.3)(0.3)(0.3)(0.2) = 0.000324*
3,000,000,000(0.000324) = 972,000 times
(c) *Hae*III GGCC = *(0.2)(0.2)(0.2)(0.2) = 0.0016*
3,000,000,000(0.0016) = 4,800,000 times

*34. Restriction mapping of a linear piece reveals the following *Eco*RI restriction sites.

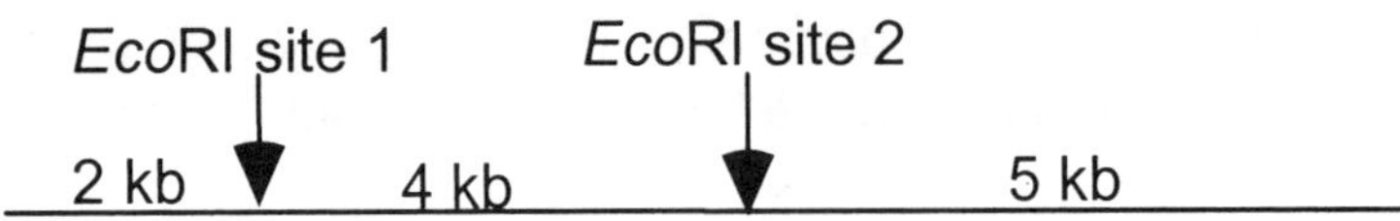

(a) This piece of DNA is cut with *Eco*RI, the resulting fragments are separated by gel electrophoresis, and the gel is stained with ethidium bromide. Draw a picture of the bands that will appear on the gel.

(b) If a mutation that alters *Eco*RI site 1 occurs in this piece of DNA, how will the banding pattern on the gel differ from the one you drew in part *a*?
(c) If mutations that alter *Eco*RI site 1 and 2 occur in this piece of DNA, how will the banding pattern on the gel differ from the one you drew in part *a*?
(d) If a 1000-bp insertion occurred between the two restriction sites, how would the banding pattern on the gel differ from the one you drew in part *a*?
(e) If a 500-bp deletion occurred between the two restriction sites, how would the banding pattern on the gel differ from the one you drew in part *a*?

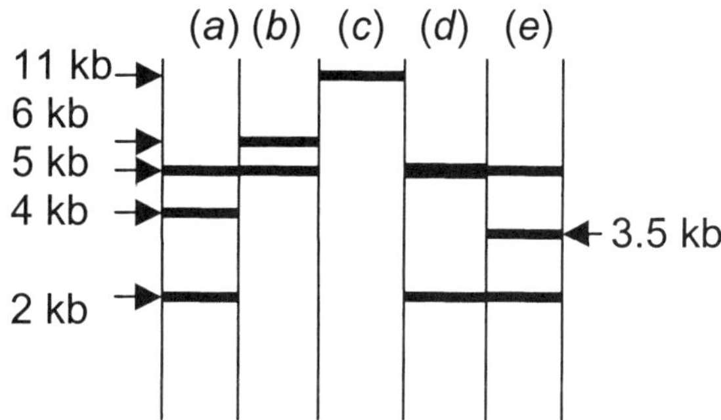

*35. Which vectors (plasmid, λ phage, cosmid) can be used to clone a continuous fragment of DNA with the following lengths?
(a) 4 kb – *plasmid*
(b) 20 kb – *λphage*
(c) 35 kb – *cosmid*

36. A geneticist uses a plasmid for cloning that has a gene that confers resistance to penicillin and the *lacZ* gene. The geneticist inserts a piece of foreign DNA into a restriction site that is located within the *lacZ* gene and transforms bacteria with the plasmid. Explain how the geneticist could identify bacteria that contain a copy of a plasmid with the foreign DNA.
The geneticist should plate the bacteria on agar medium containing penicillin to select for cells that have taken up the plasmid. The medium should also have X-gal and an inducer of the lac operon, such as IPTG or even lactose. Cells that have taken up a plasmid without foreign DNA will have an intact lacZ gene, produce functional β-galactosidase, and cleave X-gal to make a blue dye. These colonies will turn blue. In contrast, cells that have taken up a plasmid containing a foreign DNA inserted into the lacZ gene will be unable to make functional β-galactosidase. These colonies will be white.

*37. Suppose that you have just graduated from college and have started working at a biotechnology firm. Your first job assignment is to clone the pig gene for the hormone prolactin. Assume that the pig gene for prolactin has not yet been isolated, sequenced, or mapped; however, the mouse gene for prolactin has been cloned and the amino acid sequence of mouse prolactin is known. Briefly explain two different strategies that you might use to find and clone the pig gene for prolactin.

One strategy would be to use the mouse gene for prolactin as a probe to find the homologous pig gene from a pig genomic or cDNA library.

A second strategy would be to use the amino acid sequence of mouse prolactin to design degenerate oligonucleotides as hybridization probes to screen a pig DNA library.

Yet a third strategy would be to use the amino acid sequence of mouse prolactin to design a pair of degenerate oligonucleotide PCR primers to PCR amplify the pig prolactin gene.

38. A genetic engineer wants to isolate a gene from a scorpion that encodes the deadly toxin found in its stinger, with the ultimate purpose of transferring this gene to bacteria and producing the toxin for use as a commercial pesticide. Isolating the gene requires a DNA library. Should the genetic engineer create a genomic library or a cDNA library? Explain your reasoning.
A cDNA library. Bacteria cannot splice introns. If the engineer wants to express the toxin in bacteria, then he needs a cDNA sequence that has been reverse transcribed from mRNA, and therefore has no intron sequences.

*39. A protein has the following amino acid sequence:
Met-Tyr-Asn-Val-Arg-Val-Tyr-Lys-Ala-Lys-Trp-Leu-Ile-His-Thr-Pro
You wish to make a set of probes to screen a cDNA library for the sequence that encodes this protein. Your probes should be at least 18 nucleotides in length.
(a) Which amino acids in the protein should be used to construct the probes so that the least degeneracy results? (Consult the genetic code in Figure 15.12)
A probe of 18 nucleotides must be based on six amino acids. The six amino acid stretch with the least degeneracy is Val-Tyr-Lys-Ala-Lys-Trp. This sequence avoids the amino acids arg and leu, which have six codons each.
(b) How many different probes must be synthesized to be certain that you will find the correct cDNA sequence that specifies the protein?
Val and Ala have four codons each, Tyr and Lys have two codons each, and Trp has one codon. Therefore, there are $4 \times 2 \times 2 \times 4 \times 2 \times 1$ *possible sequences, or 128.*

*40. A gene in mice is discovered that is similar to a gene in yeast. How might it be determined whether this gene is essential for development in mice?
This gene must first be cloned, possibly by using the yeast gene as a probe to screen a mouse genomic DNA library. The cloned gene is then engineered to replace a substantial portion of the protein coding sequence with the neo gene. This construct is then introduced into mouse embryonic stem cells. After transfer to the uterus of a pseudopregnant mouse, the progeny are tested for the presence of the knockout allele. Progeny with the knockout allele are interbred. If the gene is essential for embryonic development, no homozygous knockout mice will be born. The arrested or spontaneously aborted fetuses can then be examined to determine how development has gone awry in fetuses that are homozygous for the knockout allele.

*41. A hypothetical disorder called G syndrome is an autosomal dominant disease characterized by visual, skeletal, and cardiovascular defects. The disorder appears in middle age. Because the symptoms of the disorder are variable, the disorder is difficult to diagnose. Early diagnosis is important, however, because the cardiovascular symptoms can be treated if the disorder is recognized early. The gene for G syndrome is known to reside on chromosome 7, and it is closely linked to two RFLPs on the same chromosome, one at the *A* locus and one at the *C* locus. The genes at the *G, A,* and *C* loci are very close together, and there is little crossing over between them. The following RFLP alleles are found at the *A* and *C* loci:

A locus: *A1, A2, A3, A4*
C locus: *C1, C2, C3*

Sally, shown in the following pedigree, is concerned that she might have G syndrome. Her deceased mother had G syndrome, and she has a brother with the disorder. A geneticist genotypes Sally and her immediate family for the *A* and *C* loci and obtains the genotypes shown on the pedigree.

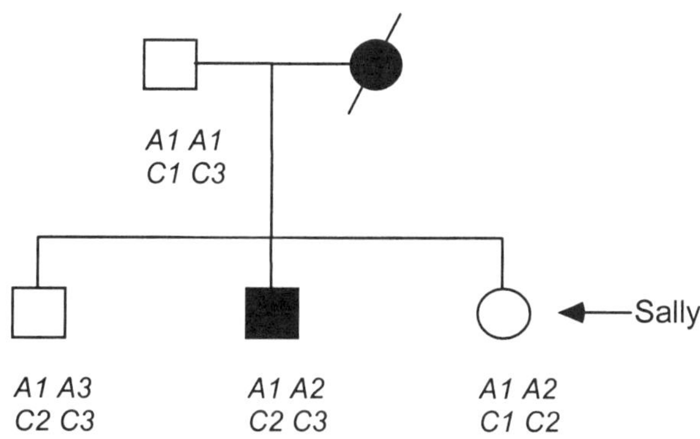

(a) Assume that there is no crossing over between the *A, C,* and *G* loci. Does Sally carry the gene that causes G syndrome? Explain why or why not.
First, we must determine the RFLP genotype of Sally's mother. Based on the RFLP genotypes of Sally and her siblings and their father, we deduce that Sally's mother must have had A2A3 and C2C2. The linkage relationships of these chromosomes are A1C1 and A1C3 from the father, and A2C2 and A3C2 from the mother. The mother passed on an A2C2 to Sally's brother with G syndrome, therefore the G syndrome allele must be linked with A2C2. Because Sally inherited the A2C2 chromosome from her mother, she must also have inherited the G syndrome allele, assuming that no crossover occurred between the A, C, and G loci.

(b) Draw the arrangement of the *A, C,* and *G* alleles on the chromosomes for all members of the family.

Father – A1 C1 g, A1 C3 g
Mother – A2 C2 G, A3 C2 g
Sally's unaffected brother – A1 C3 g, A3 C2 g
Sally's affected brother – A1 C3 g, A2 C2 G
Sally – A1 C1 g, A2 C2 G

CHALLENGE QUESTIONS

42. Suppose that a biotechnology firm hires you to produce a strain of giant fruit flies, by using recombinant DNA technology, so that genetics students will not be forced to strain their eyes when looking at tiny flies. You go to the library and learn that growth in fruit flies is normally inhibited by a hormone called shorty substance P (SSP). You decide that you can produce giant fruit flies if you can somehow turn off the production of SSP. SSP is synthesized from a compound called XSP in a single-step reaction catalyzed by the enzyme *runtase:*

    XSP————————>SSP
    runtase

    A researcher has already isolated cDNA for runtase and has sequenced it, but the location of the runtase gene in the *Drosophila* genome is unknown.

    In attempting to devise a strategy for turning off the production of SSP and producing giant flies by using standard recombinant DNA techniques, you discover that deleting, inactivating, or otherwise mutating this DNA sequence in *Drosophila* turns out to be extremely difficult. Therefore you must restrict your genetic engineering to gene augmentation (adding new genes to cells). Describe the methods that you will use to turn off SSP and produce giant flies by using recombinant DNA technology.

    One possible solution is to create a gene for expression of oligonucleotides in the cell that will act like oligonucleotide drugs. Given the DNA sequence of runtase, one could fuse a short portion of the cDNA for runtase in reverse orientation to a strong promoter, so that short antisense RNA molecules are produced in abundance. These antisense RNA molecules will hybridize with runtase mRNA and prevent its expression. Another route is to design a ribozyme whose targeting sequences are complementary to runtase mRNA. A DNA construct that would express a ribozyme can be transformed into the cell. The expressed ribozyme would then selectively target and cleave runtase mRNA.

43. A rare form of polydactyly (extra fingers and toes) in humans is due to an X-linked recessive gene, whose chromosomal location is unknown. Suppose a geneticist studies the family whose pedigree is shown here. She isolates DNA from each member of this family, cuts the DNA with a restriction enzyme, separates the resulting fragments by gel electrophoresis, and transfers the DNA to nitrocellulose by Southern blotting. She then hybridizes the nitrocellulose with a cloned DNA sequence that comes from the X chromosome. The pattern of bands that appears on the autoradiograph is shown below each person in the pedigree.

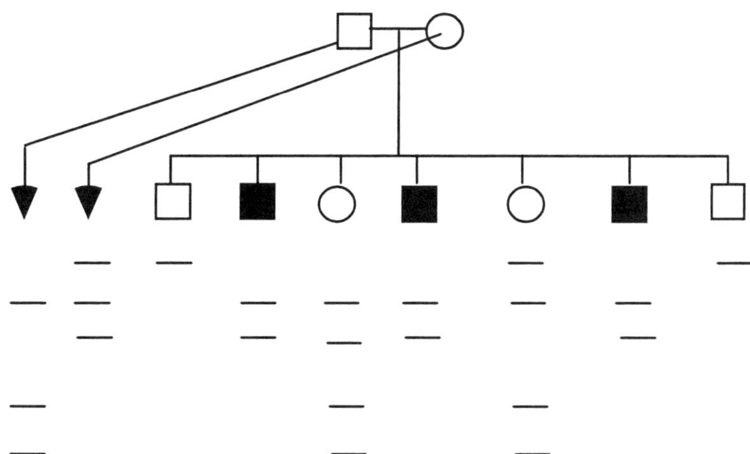

(a) For each person in the pedigree, give his or her genotype for RFLPs revealed by the probe. (Remember that males are hemizygous for X-linked genes, and females can be homozygous or heterozygous.)
The probe reveals three distinct RFLP alleles in this pedigree. I will designate these types as indicated on the pedigree diagram above: A = single high band, B = two intermediate bands, C = three bands (one intermediate plus two low bands). Father = CY; mother = AB; children from left to right: AY, BY, BC, BY, AC, BY, AY (Y indicates Y chromosome).

(b) Is there evidence for close linkage between the probe sequence and the X-linked gene for polydactyly? Explain your reasoning.
Yes, but with low confidence. All three sons who inherited the B allele of the RFLP locus also have polydactyly. Conversely, two sons who did not inherit the B allele do not have polydactyly. Therefore, the observed recombination frequency between the RFLP locus and the polydactyly locus is less than 1/5, or 20%. However, these data are far too limited to make a statistically reliable estimate of recombination frequency between these loci.

(c) How many of the daughters in the pedigree are likely to be carriers of X-linked polydactyly? Explain your reasoning.
Only one daughter inherited the RFLP B allele, so only one daughter is likely to be a carrier of X-linked polydactyly.

44. Much of the controversy over genetically engineered foods has centered on whether special labeling should be required on all products made from genetically modified crops. Some people have advocated labeling that identifies the product as made from genetically modified plants. Others have argued that only labeling should be required to identify the ingredients, not the process by which they were produced. Take one side in this issue and justify your stand.
Arguments for labeling products made from genetically modified plants should include risks of allergic reactions to the transgenic protein product, and desires of some consumers who wish to support traditional agricultural practices and oppose genetically modified foods on religious or philosophical grounds, or have concerns about impacts of genetically modified organisms on the diversity of the gene pool

and contamination of the gene pool of related species. A relevant precedent is the labeling of organically grown foods, free-range chickens, etc.

Arguments for labeling to identify only the ingredients center on the fact that many products from genetically modified organisms do not have genetic material and are indistinguishable from products made from non genetically modified organisms. For example, genes for pest resistance will not affect the composition or quality of soybean oil. Another argument is that genetic engineering simply accelerates traditional breeding methods. Modern crops, such as wheat and corn, result from centuries or millenia of breeding; genetic engineering achieves the same result far more quickly and with precision.

Chapter Nineteen: Genomics

COMPREHENSION QUESTIONS

1. (a) What is genomics and how does structural genomics differ from functional genomics?
   Genomics is the study of the content, organization, function, operation, and evolution of whole genomes. Structural genomics deals largely with the DNA sequence itself and its physical organization; functional genomics deals largely with the functions of the sequences in the genome, through RNA or protein molecules.

   (b) What is comparative genomics?
   Comparative genomics analyzes the similarities and differences among genomes, to achieve insights into genomic or organismal evolution, or relationships between groups of genes and physiological functions.

*2. What is the difference between a genetic map and a physical map? Generally, which has higher resolution and accuracy and why?
A genetic map locates genes or markers based on genetic recombination frequencies. A physical map locates genes or markers based on the physical length of DNA sequence. Because recombination frequencies vary from one region of the chromosome to another, genetic maps are approximate. Genetic maps also have lower resolution because recombination is difficult to observe between loci that are very close to each other. Physical maps based on DNA sequences or restriction maps have much greater accuracy and resolution, down to a single base pair of DNA sequence.

3. What is the purpose of the dideoxynucleoside triphosphate in the dideoxy sequencing reaction?
The dideoxynucleotide triphosphate causes chain termination wherever it is incorporated into a growing DNA strand. The result is a collection of newly synthesized DNA molecules whose lengths correspond to the positions of the dideoxynucleotide base (T, for example) in the DNA strand complementary to the template DNA strand.

*4. What is the difference between a map-based approach to sequencing a whole genome and a whole-genome shotgun approach?
The map-based approach first assembles large clones into contigs on the basis of genetic and physical maps, and then selects clones for sequencing based on their position in the contig map. The whole genome shotgun approach breaks the genome into short sequence reads, typically 600 – 700 bps, and then assembles them into contigs on the basis of sequence overlap using powerful computers to search for overlaps.

5. How are DNA fragments ordered into a contig by using restriction sites?
Restriction maps are compared among different clones to find overlaps. DNA fragments with overlapping restriction sites are ordered together to form a contig.

*6. Describe the different approaches to sequencing the human genome that were taken by the international collaboration and Celera Genomics.
The international collaboration took the ordered, map-based approach, beginning with construction of detailed genetic and physical maps. Celera took the whole-genome shotgun approach. Celera did make use of the physical map produced by the international collaboration to order their sequences in the assembly phase.

7. (a) What are expressed-sequence tags (EST)?
ESTs are single-pass sequence reads of randomly selected clones from a cDNA library.
(b) How are ESTs created?
First, mRNA is isolated from a whole organism, organ, tissue, or cell line. Reverse transcriptase is used to generate cDNAs. The cDNAs are cloned into plasmid or phage vectors. Sequencing primers based on the vector sequence flanking the cloning site are used to sequence the ends of the cDNA inserts.
(c) How are ESTs used in genomics studies?
ESTs are used to provide supporting evidence for gene predictions and annotations. ESTs can also show expression patterns for the corresponding gene.

8. What is a single-nucleotide polymorphism (SNP), and how are SNPs used in genomic studies?
SNPs are single base-pair differences in the sequence of a particular region of DNA from one individual compared to another of the same species or population. SNPs are useful as molecular markers for mapping and pedigree analysis and may themselves be associated with phenotypic differences.

9. What is a haplotype? How do different haplotypes arise?
A haplotype is a particular set of neighboring SNPs or other DNA polymorphisms observed on a single chromosome or chromosome region. They tend to be inherited together as a set because of linkage. Meiotic recombination within the chromosomal region can split the haplotype and create new recombinant haplotypes.

10. How are genes recognized within genomic sequences?
Genes can be recognized by analysis of the following:
(1) The presence of matching cDNA or EST sequences in the database.
(2) CpG islands
(3) Open reading frames (ORFs).
(4) Homology (sequence similarity) to previously characterized genes in the same or other species.
(5) Computer algorithms that use the preceding criteria to make gene predictions.

*11. What are homologous sequences? What is the difference between orthologs and paralogs?
Homologous sequences are derived from a common ancestor. Orthologs are sequences in different species that are descended from a sequence in a common ancestral species. Paralogs are sequences in the same species that originated by duplication of an ancestral sequence and subsequently diverged. Paralogs may have diverged in function.

12. Describe several different methods for inferring the function of a gene by examining its DNA sequence.
Homology: For protein-coding genes, the DNA sequence is translated conceptually into the amino acid sequence of the protein. The amino acid sequence of the protein then may yield clues to its function if it is similar to another protein of known function. For example, it is quite easy to recognize histones because their amino acid sequences are highly conserved among eukaryotes. Even if the whole protein is not similar, it may have regions, or domains, that are similar to other domains with known functions or properties. Finally, the amino acid sequence may contain small motifs or signatures that are characteristic of proteins with certain enzymatic activities or properties or subcellular localizations.

Phylogenetic profile: The coordinated absence or presence of clusters of genes in various species implies that the genes in the cluster have related functions. For example, genes required for nitrogen fixation would all be present in nitrogen-fixing species but absent in other species.

Gene fusions: In some species, genes of related function have undergone a fusion event to form a single multifunctional polyprotein. Then similar but separate component genes in other species can be presumed to have similar functions.

Gene clusters or operons: In bacteria, genes in metabolic or functional pathways are often clustered together into an operon. So, all the genes that are co-transcribed into a single polycistronic mRNA should have related functions.

13. What is a microarray and how can it be used to obtain information about gene function?
A microarray consists of thousands of DNA fragments spotted onto glass slides in an ordered grid (gene chips) or even proteins or peptides arrayed onto glass slides (protein chips). The identity of the DNA or peptide at each location is known. Gene chips are typically used in hybridization experiments with labeled mRNAs or cDNAs to survey the levels of transcript accumulation for thousands of genes, or even whole genomes, at one time. Peptide or protein chips can be used to identify protein-protein interactions or enzymatic activities or other properties of proteins.

14. Explain how a reporter sequence can be used to provide information about the expression pattern of a gene.
A reporter sequence is fused to a gene in such a way that the native gene regulatory sequences drive expression of the recombinant gene: reporter fusion product. Typically, genomic DNA sequence including the upstream promoter region and

other cis-acting regulatory sequences is ligated to the reporter gene sequence. This construct may then be used to create a transgenic organism expressing the recombinant reporter gene fusion. The reporter may have enzymatic activity (like beta-galactosidase) that is detectable with a substrate that forms a colored product, or with an antibody to the reporter protein, or the reporter may itself be fluorescent (like green fluorescent protein). The gene's own regulatory sequences specify the developmental pattern of expression of the reporter as they would the native gene. If the protein coding region of the gene is also included, in part or in full, the resulting translational fusion product can be used to study the subcellular localization of the protein.

*15. Briefly outline how a mutagenesis screen is carried out.
After random mutagenesis with chemicals or transposons, the mutant progeny population is screened for phenotypes of interest. The mutant gene can be identified by co-segregation with molecular markers or by sequencing the position of transposon insertion. To verify that the mutation identified is truly responsible for the phenotype, a mutation can be introduced into a wild-type copy of the gene and the phenotype observed in the offspring.

16. Eukaryotic genomes are typically much larger than prokaryotic genomes. What accounts for the increased amount of DNA seen in eukaryotic genomes?
The increased amount of DNA in eukaryotic genomes is due mostly to the increased amount of non-coding DNA, sometimes referred to as "junk" DNA, in introns and intergenic regions and to transposable elements. A relatively minor contribution to increased genome size is that eukaryotes, especially the complex multicellular species, generally encode more genes, and the average size of eukaryotic proteins may be larger than the average size of prokaryotic proteins.

17. What is one consequence of differences in the *G+C* content of different genomes?
Since higher G+C *content causes greater stability of the DNA duplex, one consequence is that high* G+C *content is characteristic of cells that are found at high temperatures, such as in hot springs. Increased* G+C *content also means that proteins will have a higher percentage of amino acids whose codons have G and C residues, such as glycine (GGN) and proline (CCN), alanine (GCN), and arginine (CGN).*

*18. What is horizontal gene exchange? How might it take place between different species of bacteria?
Horizontal gene exchange is transfer of genetic material across species boundaries. In bacteria, horizontal gene exchange may occur through uptake of environmental DNA through transformation, by conjugative plasmids with broad host range, or by transfection with bacteriophage with broad host range.

19. DNA content varies considerably among different multicellular organisms. Is this variation closely related to the number of genes and the complexity of the organism? If not, what accounts for the differences?

This question is almost a philosophical one because "complexity" of an organism is not well-defined, and thus difficult to quantify. However, we do know that the genomic DNA content can vary widely among related species, so there appears to be little relation between the "complexity" of an organism, the number of genes, and the DNA content. Large differences in DNA content may arise from differences in the frequency and size of introns, the abundance of DNA derived from transposable elements, and duplication of the whole or substantial parts of the genome in the evolutionary history of the species.

*20. More than half of the genome of *Arabidopsis thaliana* consists of duplicated sequences. What mechanisms are thought to have been responsible for these extensive duplications?
The Arabidopsis *genome appears to have undergone at least one round of duplication of the whole genome (tetraploidy) and numerous localized duplications via unequal crossing over.*

21. The human genome does not encode substantially more protein domains than do invertebrate genomes, and yet it encodes many more proteins. How are more proteins encoded when the number of domains does not differ substantially?
The human genome contains proteins with many more combinations of domains, often featuring multiple domains on a single protein.

22. What are some of the ethical concerns arising out of the information produced by the Human Genome Project?
The ethical questions concern privacy: Who will have access to a person's genetic profile, and what will be done with that information? Information concerning genetic susceptibility to disease may concern insurance companies and employers and could possibly be used to deny health insurance or employment. Other concerns include the concept of genetic determinacy: How does an individual or society interpret associations between genetic polymorphisms and various phenotypic traits, as susceptibility to various genetic diseases, and correlations with such complex traits such as intelligence and different aspects of behavior? Still other questions are raised about genetic engineering applied to the human germ-line as the ultimate form of eugenics. These are only some of the issues raised by the Human Genome Project and other technologies currently in development.

APPLICATION QUESTIONS AND PROBLEMS

*23. A 22-kb piece of DNA has the following restriction sites:

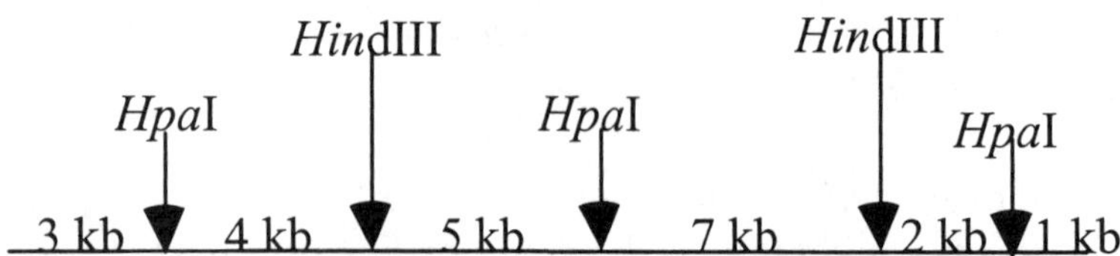

A batch of this DNA is first fully digested by *Hpa*I alone, then another batch is fully digested by *Hin*dIII alone, and finally a third batch is fully digested by *Hpa*I and *Hin*dIII together. The fragments resulting from each of the three digestions are placed in separate wells of an agarose gel, separated by gel electrophoresis, and stained by ethidium bromide. Draw the bands as they would appear on the gel.

12
9
7
5
4
3
2
1

Hpa*I* Hind*III* *Both*

*24. A piece of DNA that is 14 kb long is cut first by *Eco*RI alone, then by *Sma*I alone, and finally by *Eco*RI and *Sma*I together. The following results are obtained:

Digestion by *Eco*RI alone	Digestion by *Sma*I alone	Digestion by *Eco*RI and *Sma*I
3 kb fragment	7 kb fragment	2 kb fragment
5 kb fragment	7 kb fragment	3 kb fragment
6 kb fragment		4 kb fragment
		5 kb fragment

Draw a map of the *Eco*RI and *Sma*I restriction sites on this 14-kb piece of DNA, indicating the relative positions of the restriction sites and the distances between them.

We know that SmaI *cuts only once, in the middle of this piece of DNA, at 7 kb.* Eco*RI cuts twice. Comparing the* Eco*RI digest to the double digest, we see that neither the 3-kb nor the 5-kb fragments is cut by* Sma*I; only the 6-kb* Eco*RI fragment is cut by* Sma*I to 2-kb and 4-kb fragments. Therefore, the 6-kb* Eco*RI fragment is in the middle, and the 3-kb and 5-kb* Eco*RI fragments are at the sides.*

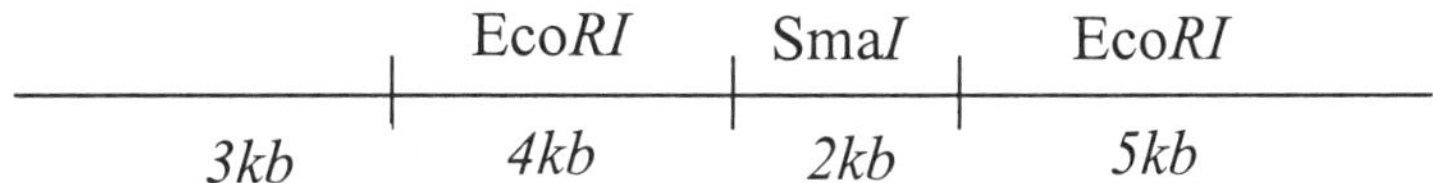

25. Suppose that you want to sequence the following DNA fragment:

Fragment to be sequenced: 5'—TCCCGGGAAA-primer site—3'

You first clone the fragment in bacterial cells to produce sufficient DNA for sequencing. You isolate the DNA from the bacterial cells and carry out the dideoxy sequencing method. You then separate the products of the polymerization reactions with gel electrophoresis. Draw the bands that should appear on the gel from the four sequencing reactions.

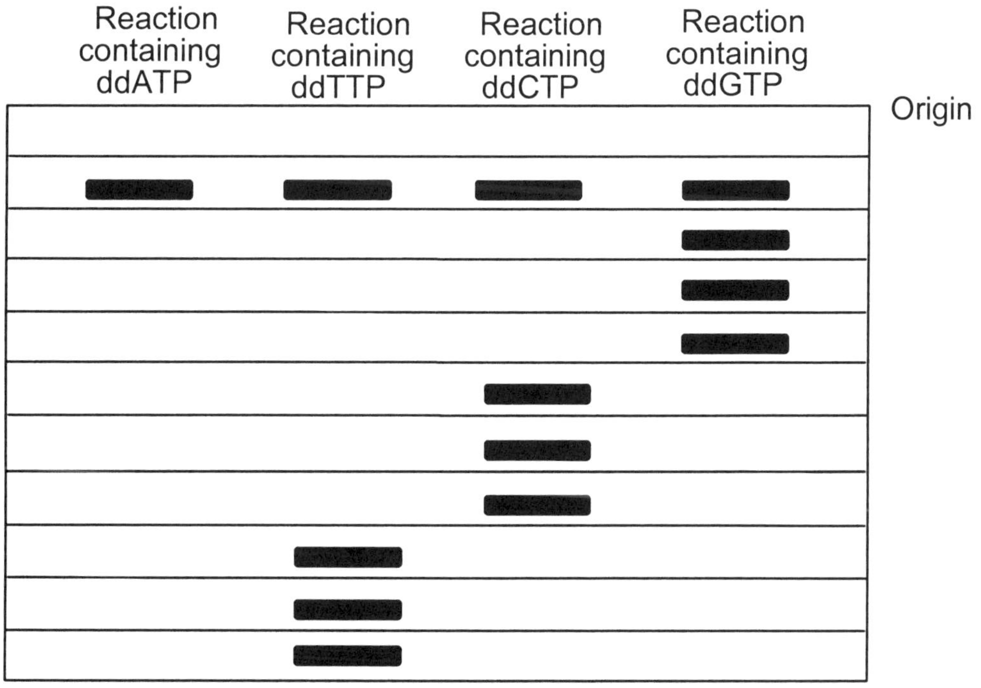

Reaction containing ddATP
Reaction containing ddTTP
Reaction containing ddCTP
Reaction containing ddGTP
Origin

Note that bands will appear on all four lanes at the 5′ terminus of the DNA template fragment; the chain will terminate in all four reactions at this position because this is the end of the template. Thus, the 5′ end nucleotide cannot be determined by looking at bands in the sequencing gel. If the dideoxynucleotides are labeled, then the labeled band will appear in only the ddA lane in the uppermost position.

*26. Suppose that you are given a short fragment of DNA to sequence. You clone the fragment, isolate the cloned DNA fragments, and set up a series of four dideoxy reactions. You then separate the products of the reactions by gel electrophoresis and obtain the following banding pattern:

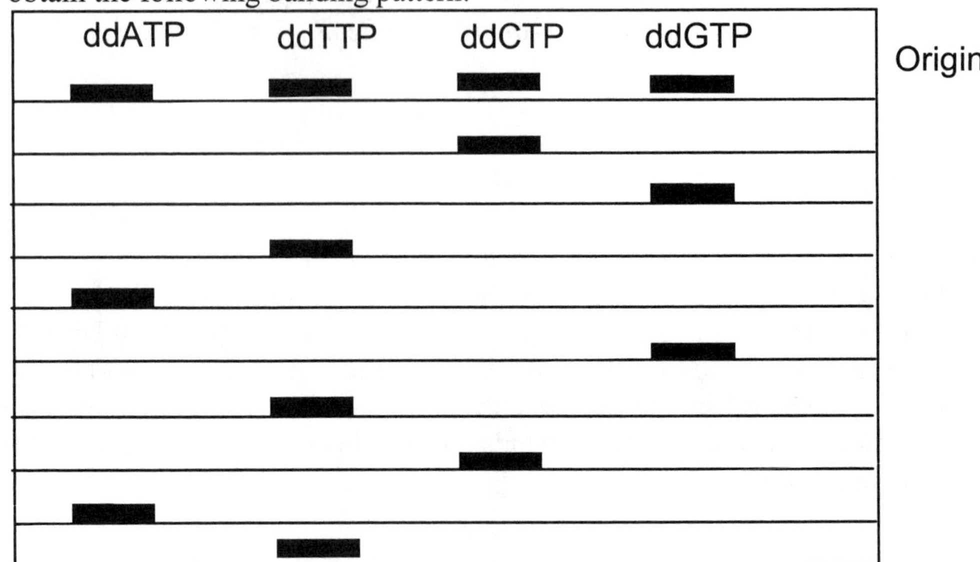

Write out the base sequence of the original fragment that you were given.

Original Sequence: 5'NGCATCAGTA3'

The base at the 5' end (N) cannot be determined because the chain stops in all four lanes.

27. Microarrays can be used to determine the levels of gene expression. In one type of microarray, hybridization of the red (experimental) and green (control) cDNAs is proportional to the relative amounts of mRNA in the samples. Red indicates the overexpression of a gene, green indicates the underexpression of a gene in the experimental cells relative to the control cells, yellow indicates equal expression in experimental and control cells, and no color indicates no expression in either experimental or control cells.

In one experiment, mRNA from a strain of antibiotic-resistant bacteria (experimental cells) is converted into cDNA and labeled with red fluorescent nucleotides; mRNA from a nonresistant strain of the same bacteria (control cells) is converted into cDNA and labeled with green fluorescent nucleotides. The cDNAs from the resistant and nonresistant cells are mixed and hybridized to a chip containing spots of DNA from genes 1 through 25. The results are shown in the illustration on the following page. What conclusions can you make about which genes might be implicated in antibiotic resistance in these bacteria? How might this information be used to design new antibiotics that are less vulnerable to resistance?

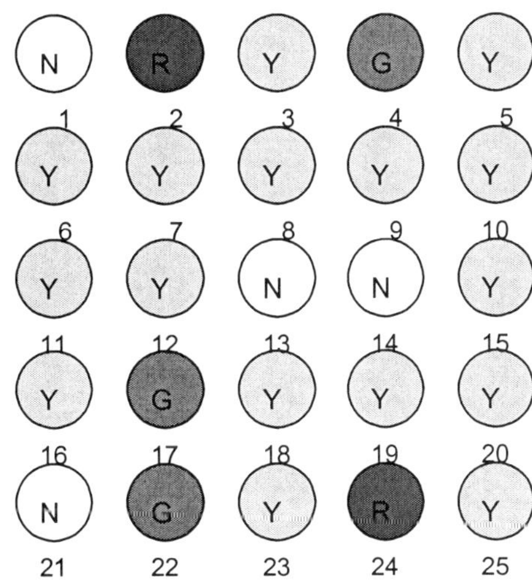

R = red
G = green
Y = yellow
N = no color

Genes 2 and 24 are expressed at far higher levels in the antibiotic-resistant bacteria than in the nonresistant cells. These genes may be involved in antibiotic resistance. Characterization of genes may lead to information regarding the mechanism of antibiotic resistance, and then to the design of new antibiotics that can circumvent this resistance mechanism.

*28. Genes for the following proteins are found in five different species whose genomes have been completely sequenced. On the basis of the presence-and-absence patterns of these proteins in the genomes of the five species, which proteins are most likely to be functionally related? (Hint: Create a table listing the presence or absence of each protein in the five species.)

	Species				
Proteins	*A*	*B*	*C*	*D*	*E*
P1	+	+	–	–	+
P2	+	+	+	–	–
P3	+	+	–	+	+
P4	+	–	+		+
P5	+	+	–	+	+

P3 and P5 are either both present or both absent in all species; therefore they are most likely to be functionally related.

29. The physical locations of several genes determined from genomic sequences are shown here for three bacterial species. On the basis of this information, which genes might be functionally related?

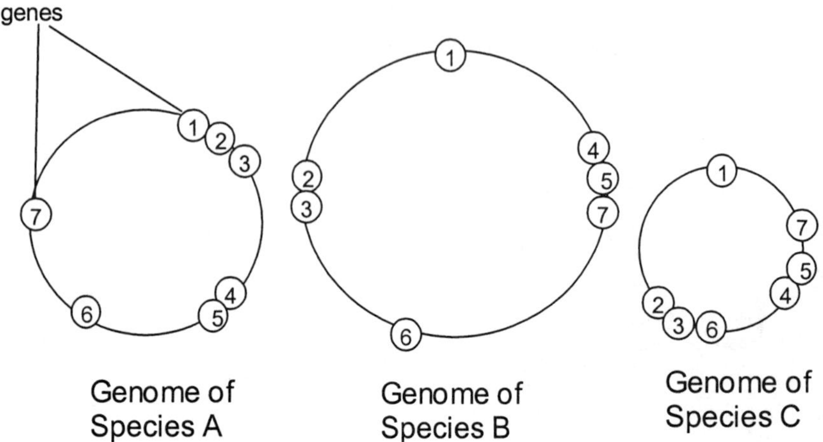

Two pairs of genes—2 and 3, and 4 and 5—are adjacent on all three genomes. Therefore, they may be pairs of functionally related genes.

30. The presence (+) or absence (–) of six sequence-tagged sites (STSs) in each of five bacterial artificial chromosome (BAC) clones (A–E) is indicated in the following table. Using these markers, put the BAC clones in their correct order and indicate the locations of the STS sites within them.

	STSs					
BAC clone	1	2	3	4	5	6
A	+	–	–	–	+	–
B	–	–	–	+	–	+
C	–	+	+	–	–	–
D	–	–	+	–	+	–
E	+	–	–	+	–	–

Solution:

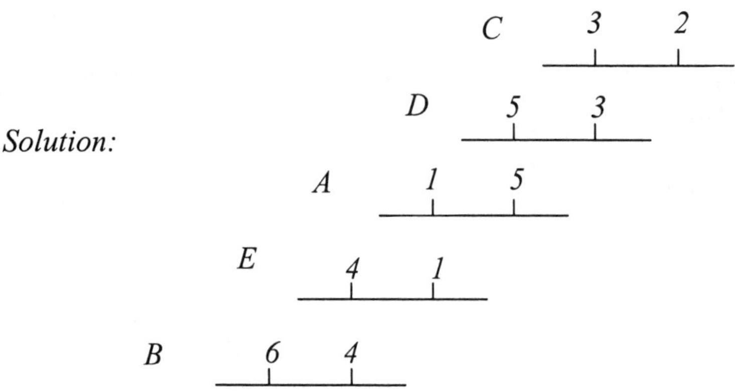

31. How does the density of genes found on chromosome 22 compare with the density of genes found on chromosome 21, two similar-sized chromosomes? How does the number of genes on chromosome 22 compare with the number found on the Y chromosome?
To answer these questions, go to the Ensembl Web site: http://www.ensembl.org. Under the heading "Ensemble Species," click "Human." On the left-hand side of the next page are pictures of the human chromosomes. Click on chromosome 22. You will be shown a picture of this chromosome and a histogram illustrating the density of total (uncolored bars) and known genes (colored bars). The number of known and novel (uncharacterized) genes is given in the upper right-hand side of the page.
Next, go to chromosome 21 by pulling down the "Change Chromosome" menu and selecting chromosome 21. Examine the density and total number of genes for chromosome 21. Do the same for the Y chromosome.
(a) Which chromosome has the highest density and greatest number of genes? Which has the fewest?
Chromosome 22 has the highest density and greatest number of genes, with over 450 known and more than 100 novel Ensembl genes, whereas the Y chromosome has the lowest density and fewest genes, with fewer than 100 known and fewer than 50 novel Ensembl genes.
(b) Examine in more detail the genes at the tip of the short arm of the Y chromosome by clicking on the top bar in the histogram of genes. A more detailed view will be shown. What known genes are found in this region? How many novel genes are there in this region?
The known genes SRY, RPS4Y, and ZFY are found in this region. No novel genes are found.

*32. Some researchers have proposed creating an entirely new, free-living organism with a minimal genome, the smallest set of genes that allows for replication of the organism in a particular environment. This organism could be used to design and create, from "scratch," novel organisms that might perform specific tasks, such as the breakdown of toxic materials in the environment.
(a) How might the minimal genome required for life be determined?
The minimal genome required might be determined by examining simple free-living organisms having small genomes to determine which genes they possess in common. Mutations can then be made systematically to determine which genes are essential for these organisms to survive. The apparently nonessential genes (those genes in which mutations do not affect the viability of the organism) can then be deleted one by one until only the essential genes are left: Elimination of any of these genes will result in loss of viability. Alternatively, essential genes could be assembled through genetic engineering, creating an entirely novel organism.

(b) What, if any, social and ethical concerns might be associated with the creation of novel organisms by constructing an entirely new organism with a minimal genome?
This synthetic organism would prove that humans have acquired the ability to create a new species or form of life. Humans would then be able to direct evolution as never before. Social and ethical concerns would revolve around whether human society has the wisdom to temper its power and whether such novel synthetic organisms can or will be used to develop pathogens for biological warfare or terrorism. After all, no person or animal would have been exposed previously or have acquired immunity to such a novel synthetic organism. Also, it would be uncertain how the new organism would affect the ecosystem if it was released or escaped. These are one person's concerns. Other people may have additional concerns.

33. What are some of the major differences between the ways in which genetic information is organized in the genomes of prokaryotes versus eukaryotes?
Most prokaryotic genomes consist of a single circular chromosome, whereas eukaryotic genomes have multiple linear chromosomes.

Prokaryotic genomes have none to few introns; complex eukaryotic genomes have numerous introns, with most genes having multiple introns.

Prokaryotic genomes have high gene density, approximately one gene per 1000 base pairs; most eukaryotic genomes have far lower gene density.

Prokaryotic genomes have functionally related genes organized into operons transcribed as polycistronic mRNAs from a single promoter; eukaryotic genomes generally have monocistronic mRNAs, each from its own promoter.

34. How do the following genomic features of prokaryotic organisms compare with those of eukaryotic organisms? How do they compare among eukaryotes?
(a) Genome size: *0.6 to 7 million bp in prokaryotes; 13 million (yeast) to 76 billion* (Amphiuma) *in eukaryotes (Table 11.3).*
(b) Number of genes: *From 500 to 7000 in prokaryotes; 6000 to 30,000 in eukaryotes.*
(c) Gene density (bp/gene): *Approximately 1000 bp/gene in prokaryotes; varies from 2000 bp/gene to greater than 100,000 bp/gene in eukaryotes.*
(d) *G+C* content: *From 26% to 69% in prokaryotes; less variable in eukaryotes (about 40% to 50%).*
(e) Number of exons: *With a few exceptions, prokaryotic genes have only one exon; multicellular eukaryotic genes typically have multiple exons.*

Chapter Twenty: Organelle DNA

COMPREHENSION QUESTIONS

*1. Briefly describe the general structures of mtDNA and cpDNA. How are they similar? How do they differ? How do their structures compare with the structures of eubacterial and eukaryotic (nuclear) DNA?

Mitochondrial and chloroplast DNAs have many characteristics in common with eubacterial DNAs. Most mitochondrial and chloroplast chromosomes are small, circular, and lack histone proteins—characteristics that are similar to eubacterial, but not eukaryotic, cells. Chloroplasts and some mitochondria produce polycistronic mRNA, another characteristic common to eubacteria. Chloroplasts, but not most mitochondria, typically possess Shine Dalgarno sequences. Antibiotics that inhibit eubacterial translation inhibit mitochondrial and chloroplast translation.

Eukaryotic nuclear genomes are typically composed of linear chromosomes and histone proteins. Eukaryotic nuclear DNA sequences also contain pre-mRNA introns, which are common in eukaryotes. Pre-mRNA introns are not found in chloroplast and mitochondrial genomes.

2. Explain why many traits encoded by mtDNA and cpDNA exhibit considerable variation in their expression, even among members of the same family.

Within a given eukaryotic cell, there can be thousands of copies of each type of organelle genome. Mutations that occur in one genome can give rise to populations of organelles that differ in DNA sequence within a cell, a phenomenon called heteroplasmy. During cytokinesis or cell division, the process of replicative segregation separates the organelles randomly into the daughter cells. In general, equal proportions of organelles containing mutant sequences and organelles containing wild-type sequences will be distributed to each progeny cell. However, because replicative segregation is random, occasionally one progeny cell may receive all wild-type organelles while the other progeny cell receives the mutant types. When only one type of genome is present within a cell, the condition is known as homoplasmy. Essentially, replicative segregation can lead to cells within a given individual and among family members that differ in organelle phenotype and genotype.

*3. What is the endosymbiotic theory? How does it help to explain some of the characteristics of mitochondria and chloroplasts?

The endosymbiotic theory proposes that mitochondria and chloroplast originated from formerly free-living bacteria that became endosymbiants within a larger eukaryotic cell.

Chloroplasts and mitochondria contain genomes that encode for proteins, tRNAs, and rRNAs. Size, circular chromosomes, and other aspects of genome structure in chloroplasts and mitochondria are similar to that of eubacterial cells. Moreover, the chloroplast and mitochondrial ribosomes are similar in size and function to eubacterial ribosomes. DNA sequences in mitochondrial and chloroplast genomes are most similar to those in eubacteria.

4. What evidence supports the endosymbiotic theory?
Several lines of evidence support the endosymbiotic theory. Perhaps the most convincing lines of evidence are sequence similarities of protein encoding and rRNA genes in mitochondria and chloroplasts to counterpart genes in specific groups of eubacteria. The mtDNA sequences are most similar to DNA from a group of bacteria called the α-proteobacteria, while the cpDNA sequences are most closely linked to the sequences from cyanobacteria. Another line of evidence that supports the endosymbiont theory includes sensitivity of protein synthesis in mitochondria and chloroplasts to antibiotics, such as streptomycin and tetracycline, inhibitors of protein synthesis in eubacteria but not eukaryotes. Conversely, some antibiotics, such as cycloheximide, inhibit eukaryotic protein synthesis but not that of the mitochondria and chloroplasts. A third line of evidence is found in mitochondrial genomes in which the AUG start codon encodes N-formyl methionine, which is typical of eubacteria translation initiation. Finally, the size and structure of ribosomes in both chloroplasts and mitochondria are similar to that of eubacterial ribosomes.

5. How are genes organized in the mitochondrial genome? How does this organization differ between ancestral and derived mitochondrial genomes?
The mitochondrial genome consists of circular DNA molecules that encode protein, rRNAs, and tRNAs. How the genes are organized varies among different eukaryotic species with ancestral mitochondrial genomes found in some plants and protists, whereas derived mitochondrial genomes are more common in animals and fungi.

Mitochondrial ancestral genomes typically contain more protein encoding genes and tRNA genes than derived genomes. The rRNA genes of ancestral genomes resemble those of eubacterial rRNA genes. Ancestral genomes include few introns, use universal codons, organize genes into clusters, and contain little non coding DNA sequences. Finally, DNA sequences of ancestral genomes more closely resemble eubacterial genomes than DNA sequences of derived genomes. Mitochondrial-derived genomes are typically smaller than ancestral genomes. Derived genomes may contain nonuniversal codons, and the rRNA genes differ from those of eubacterial rRNAs.

*6. What are nonuniversal codons? Where are they found?
Nonuniversal codons specify an amino acid in one organism that is not specified by that codon in most other organisms. Typically, they are found in mitochondrial genomes, and these exceptions vary in the mitochondria of different organisms.

7. How does replication of mtDNA differ from replication of nuclear DNA in eukaryotic cells?
DNA polymerase gamma is responsible for replication of mitochondrial DNA, whereas the DNA polymerases delta and alpha are necessary for nuclear DNA replication. In mitochondrial DNA, the replication origins for the different strands

of the double helix are found at different locations within the genome. Furthermore, replication of the two strands can be asynchronous with both strands exhibiting continuous replication. In nuclear DNA replication, the replication of both strands is synchronous and proceeds from a common origin with continuous replication of one strand and discontinuous replication on the other strand.

*8. The human mitochondrial genome encodes only 22 tRNAs, whereas at least 32 tRNAs are required for cytoplasmic translation. Why are fewer tRNAs needed in mitochondria?
During translation in mitochondria, the "wobble" at the third position of the codon appears more frequently. Most anticodons of the mitochondrial tRNAs can pair with more than one codon. Essentially, the first position of the anticodon can pair with any of the four nucleotides present at the third position of the codon.

9. What are some possible explanations for an accelerated rate of evolution in the sequences of vertebrate mtDNA?
Vertebrate mtDNA has a higher mutation rate than the vertebrate nuclear DNA and the mtDNA of plants, which could be responsible for the accelerated rate of evolution. Reasons for the increased level of mutations include a lack of DNA repair, increased errors in replication, and a higher rate of replication for mtDNA.

10. What are some of the advantages of using yeast for genetic studies?
Some advantages of using yeast for genetic studies:
- *Yeast are eukaryotic organisms so they share similarities in genetic and cellular systems with other more complex eukaryotic organisms*
- *Yeast are unicellular and like bacterial systems can be easily and inexpensively grown in the laboratory while requiring little space.*
- *Yeast can exist either in a haploid or diploid form. So, the phenotypes of recessive alleles can easily be identified in the haploid form. The interactions of alleles can be studied in the diploid form.*
- *Yeast meiotic products can be examined by tetrad analysis. The products of meiosis in yeast are contained with a structure called an ascus, which allows for direct examination of the products of meiosis.*
- *Yeast can be studied using many molecular biology techniques developed initially for studying bacteria.*
- *Yeast possess many genes and DNA sequences that are similar to genes in multicellular eukaryotes, including humans.*
- *The genome of Saccharomyces cerevisiae, a yeast, was completely sequenced.*
- *Yeast naturally contain a plasmid (2μ). This plasmid has been adapted as a vector for transferring genes or DNA sequences of interest into yeast.*
- *Yeast artificial chromosomes (YACs), which allow for the transfer of large DNA fragments into yeast, have been developed.*

*11. Briefly describe the organization of genes on the chloroplast genome.
The chloroplast genome is typically a double-stranded circular DNA whose organization and sequences resembles those of eubacterial genomes. Genes are located on both strands of cpDNA and may contain introns. The chloroplast genome usually encodes ribosomal proteins, 5 rRNAs, 30 to 35 tRNA genes, and proteins involved in photosynthesis, as well as proteins not involved in photosynthesis. A large inverted repeat is also found in the genomes of most chloroplasts.

12. What is meant by the term "promiscuous DNA"?
DNA sequences that have been exchanged among the nuclear, mitochondrial, or chloroplast genomes are referred to as promiscuous DNA.

APPLICATION QUESTIONS AND PROBLEMS

13. A wheat plant that is light green in color is found growing in a field. Biochemical analysis reveals that chloroplasts in this plant produce only 50% of the chlorophyll normally found in wheat chloroplasts. Propose a set of crosses to determine whether the light-green phenotype is caused by a mutation in a nuclear gene or a chloroplast gene.
Nuclear and chloroplast genes in wheat will exhibit different inheritance patterns. A nuclear gene is inherited biparentally, whereas a chloroplast gene is inherited uniparentally (or maternally). The differences in inheritance patterns will allow us to determine if the mutation has a nuclear gene or chloroplast gene origin. For the following crosses, the male plant refers to the pollen donor, while the female plant refers to the plant being pollinated and where fertilization will take place.

If the mutation is located within a chloroplast gene, then we would expect the following results, no matter which trait is dominant:
Wild-type wheat male × Light-green wheat female → offspring all light-green
Light-green wheat male × Wild-type wheat female → offspring all wild type

Matings between light-green offspring should always produce more light-green progeny, whereas matings between wild-type offspring should always produce wild-type progeny.

If the mutation is in a nuclear gene, then both parents can pass the light-green mutation to their offspring. If we assume that wild type is dominant, then we would expect the following results:
Wild-type wheat male × Light-green wheat female → offspring all wild type
Light-green wheat male × Wild-type Wheat → offspring all wild type

Separate matings between members of the F_1 progeny of each cross, should give progeny with a phenotypic ratio of 3:1 wild type to light-green.

If we assume that the light-green phenotype is dominant, then we would expect the following:
Wild-type wheat male × Light-green wheat female → offspring all light-green
Light-green wheat male × Wild-type wheat female → offspring all light-green

Separate matings between members of the F_1 progeny of each cross should give progeny with a phenotypic ratio of 3:1 light-green to wild type.

*14. A rare neurological disease is found in the family illustrated in the following pedigree. What is the most likely mode of inheritance for this disorder? Explain your reasoning.

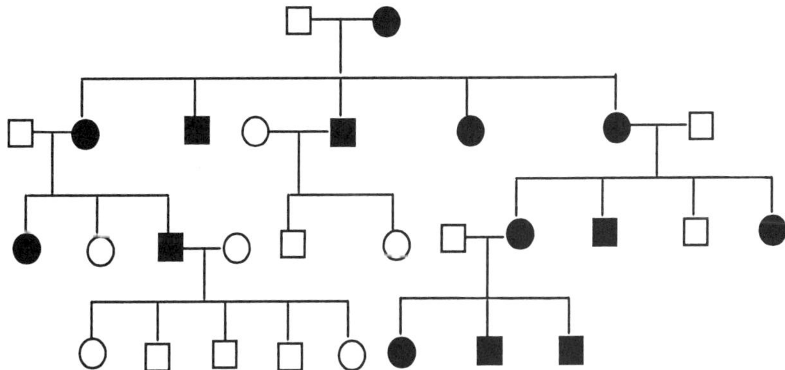

The pedigree indicates that the neurological disorder is a cytoplasmically inherited trait. Only females pass the trait to their offspring. The trait does not appear to be sex-specific in that males as well as females can have the disorder. These characteristics are consistent with cytoplasmic inheritance.

15. In a particular strain of *Neurospora,* a *poky* mutation exhibits biparental inheritance, whereas *poky* mutations in other strains are inherited only from the maternal parent. Explain these results.
In Neurospora, poky mutations are identified by the mutant organism's slow growth. Maternal parent inheritance indicates that the poky mutations have a mitochondrial origin because mitochondrial genomes are inherited maternally in Neurospora. Many poky mutations have been identified on the mitochondrial genome. For a Neurospora strain containing a poky mutation with biparental inheritance, the poky mutation probably is of nuclear origin. In Neurospora, nuclear genes exhibit biparental inheritance. This particular poky mutation most likely affects the energy producing pathways in the mitochondria as indicated by the poky phenotype, but the mutation is contained within a nuclear gene whose protein product targets mitochondria.

16. Antibiotics, such as chloramphenicol, tetracycline, and erythromycin inhibit protein synthesis in eubacteria but have no effect on protein synthesis encoded by nuclear genes. Cycloheximide inhibits protein synthesis encoded by nuclear genes but has no effect on eubacterial protein synthesis. How might these compounds be used to determine which proteins are encoded by the mitochondrial and chloroplast genomes?
Proteins that are synthesized by the mitochondrial and chloroplast genomes are typically sensitive to antibiotics that affect eubacterial proteins. The sensitivity is most likely due to the similarities of the translational machinery in eubacteria and in chloroplasts and mitochondria. To determine if a particular protein is coded for on the mitochondrial or chloroplast genomes, we can use antibiotics and determine

their effects on protein synthesis. For example, if the antibiotic cycloheximide does not inhibit the synthesis of the proteins in question, the lack of inhibition suggests the proteins are not encoded by the nuclear genome. Subsequently, if the antibiotics chloramphenicol, tetracycline, and erythromycin do inhibit the synthesis of the proteins, then this suggests again that the proteins are not of nuclear origin; they are synthesized either in the mitochondria or the chloroplast. This combination treatment of the different types of antibiotics—those that inhibit nuclear gene protein synthesis and those that inhibit eubacterial as well as organelle protein synthesis—should indicate if proteins are of nuclear or organelle origin.

*17. A scientist collects cells at various points in the cell cycle and isolates DNA from them. Using density gradient centrifugation, she separates the nuclear and mtDNA. She then measures the amount of mtDNA and nuclear DNA present at different points in the cell cycle. On the following graph, draw a line to represent the relative amounts of nuclear DNA that you expect her to find per cell throughout the cell cycle. Then draw a dotted line on the same graph to indicate the relative amount of mtDNA that you would expect to see at different points throughout the cell cycle.

Nuclear DNA levels should increase during only the S phase, before declining at cytokinesis. Mitochondrial DNA levels should increase throughout the cell cycle, before declining at cytokinesis.

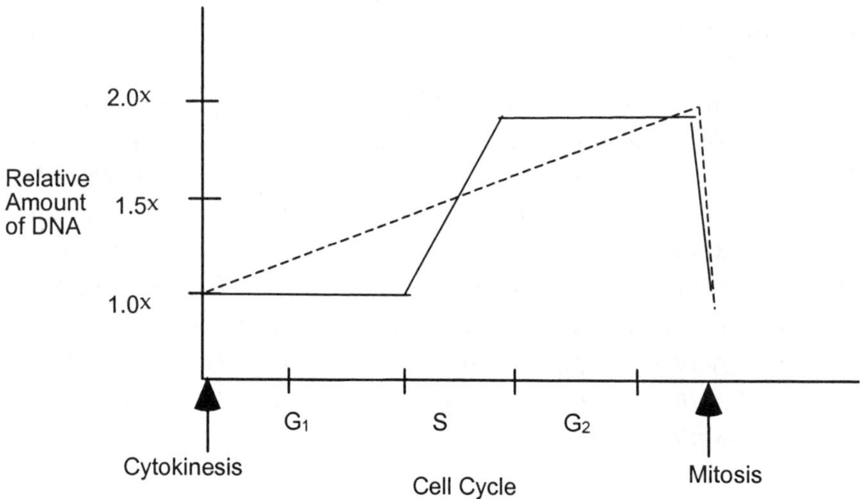

18. The introduction to Chapter 1 described how bones found in 1979 outside Ekaterinburg, Russia were shown to be those of Tsar Nicholas and his family, who were executed in 1918 by a Bolshevik firing squad in the Russian Revolution. To prove that the skeletons were those of the royal family, mtDNA was extracted from the bone samples, amplified by PCR, and compared with mtDNA from living relatives of the tsar's family. Why was DNA from the mitochondria analyzed instead of nuclear DNA? What are some of the advantages of using mtDNA for this type of study?

Because mitochondrial DNA in humans is inherited only maternally, the Tsar's living relatives of maternal descent should contain very similar mitochondrial DNA sequences. The mitochondrial sequences of the tsar's children and his wife can also be analyzed, but only by comparing their mitochondrial DNA with that of living maternal relatives to the Tsarina. Comparisons of nuclear DNA sequences from the tsar's family with living relatives would prove difficult because of the biparental transfer of nuclear genes and recombination between the nuclear chromosomes. Also, the possibility of similar sequences appearing by chance through matings of the tsar's contemporary relatives could not be ruled out. The nuclear DNA analysis could, however, indicate if the skeletons in the gravesite were related to each other.

An advantage of mitochondrial DNA typing for this type of study is the linear maternal inheritance. Essentially, any maternally related individual should possess very similar mitochondrial DNA sequences to the deceased individual. A second advantage is the number of mitochondrial genomes present. Multiple copies of each chromosome exist within each mitochondrion, and each cell contains multiple mitochondria, thus making it more likely that mitochondrial DNA survived.

19. From Figure 20.8, determine as best you can the percentage of human mtDNA that is coding (transcribed into RNA) and the percentage that is noncoding (not transcribed).
The largest region of noncoding DNA in human mitochondria is located within the D-loop. There are small sections of noncoding sequences (1 to 2 nucleotides in length) at other locations within the genome, but very few. Approximately 93% of the DNA in human mitochondria is coding, whereas only 7% is not.

CHALLENGE QUESTIONS

20. Mitochondrial DNA sequences have been detected in the nuclear genomes of many organisms, and cpDNA sequences are sometimes found in the mitochondrial genome. Propose a mechanism for how such "promiscuous DNA" might move between nuclear, mitochondrial, and chloroplast genomes.
The transfer of organelle DNA sequences to another location most likely involves multiple steps. Different models for transfer can be proposed. In one model, the nucleus or an organelle may uptake DNA directly from degraded organelles and incorporate the DNA into a genome by double-stranded recombination. A second model involves an RNA intermediate. The organelle gene is initially expressed, and a mature mRNA is produced. The mRNA molecule serves as a template for reverse transcriptase, which can produce a cDNA copy of the mRNA. The cDNA is then incorporated into either a nuclear chromosome or into the genome of a different type of organelle by recombination. In either model, once the DNA is incorporated, replication of this chromosome will propagate the "promiscuous DNA" within either the nuclear or organelle genomes of a cell and the subsequent progeny cells.

21. Steven A. Frank and Laurence D. Hurst argued that a cytoplasmically inherited mutation in humans that has severe effects in males but no effect in females will not be eliminated from a population by natural selection because only females pass on

mtDNA. Using this argument, explain why males with Leber hereditary optic neuropathy are more severely affected than females.
Leber hereditary optic neuropathy (LHON) is the result of mutations in mtDNA. Using the argument of Frank and Hurst, the most severe forms of LHON that appear in males will not be eliminated from the population if they do not also appear in females. If the particular mutation has little or no effect on females, then the females will be able to pass the mutation to their offspring. Male offspring who receive the mutation will be affected, but female offspring will not. The female offspring will then be able to pass the mutation along to their offspring. The selection is for healthy females who can pass the genes to their offspring. For mitochondrial genes, males have a limited role. So, it is possible that mutations with more severe effects in males than females will be present because there is little negative selection pressure against females with deleterious mutations in mitochondrial genes.

22. Several families have been described that exhibit vision problems, muscle weakness, and deafness. This disorder is inherited as an autosomal dominant trait and the disease-causing gene has been mapped to chromosome 10 in the nucleus. Analysis of the mtDNA from affected persons in these families reveals that large numbers of their mitochondrial genomes possess deletions of varying length. Different members in the same family and even different mitochondria from the same member possess deletions of different sizes, so the underlying defect appears to be a tendency for the mtDNA of affected persons to have deletions. Propose an explanation for how a mutation in a nuclear gene might lead to deletions in mtDNA.
Genes necessary for mitochondria DNA replication and repair are located on the nuclear genome. If a gene product encoded by chromosome 10 plays a role in either mitochondrial DNA maintenance or stability, then the lack of that functional gene product could result in damage such as deletions to the mitochondrial genome.

Chapter Twenty-One: Advanced Topics in Genetics: Developmental Genetics, Immunogenetics, and Cancer Genetics

COMPREHENSION QUESTIONS

*1. What experiments suggested that genes are not lost or permanently altered in development?
The ability to clone plants and animals from differentiated cells showed that the nuclei of these differentiated cells still retained all of the genetic information required for the development of a whole organism.

2. Briefly explain how the Dorsal protein is redistributed in the formation of the *Drosophila* embryo and how this redistribution helps to establish the dorsal–ventral axis of the early embryo.
Dorsal protein is distributed uniformly in the cytoplasm of the unfertilized egg. After fertilization, at the stage at which nuclei migrate to the periphery of the egg, Dorsal protein migrates into the nucleus along the future ventral side of the egg, but not along the future dorsal side of the egg. This gradient of nuclear-localized Dorsal protein activates transcription of mesodermal genes along the ventral surface of the embryo. Nuclei that lack Dorsal protein transcribe dorsal-specific genes such as decapentaplegic.

*3. Briefly describe how the *bicoid* and *nanos* genes help to determine the anterior–posterior axis of the fruit fly.
Maternally transcribed bicoid and nanos mRNAs are localized to the anterior and posterior ends of the egg, respectively. After fertilization, these mRNAs are translated, and the proteins diffuse to form opposing gradients: Bicoid protein concentrations are highest at the anterior, whereas Nanos protein concentrations are highest at the posterior. Bicoid protein at the anterior acts as a transcription factor to activate transcription of hunchback, a gene required for the formation of head and thoracic structures. Nanos protein at the posterior end inhibits translation of hunchback mRNA, thereby preventing the formation of anterior structures in the posterior regions.

*4. List the three major classes of segmentation genes and outline the function of each.
(1) Gap genes specify broad regions (multiple adjacent segments) along the anterior-posterior axis of the embryo. Interactions among the gap genes regulate transcription of the pair-rule genes.
(2) Pair-rule genes compartmentalize the embryo into segments and regulate expression of the segment polarity genes. Each pair rule gene is expressed in alternating segments.
(3) Segment polarity genes specify the anterior and posterior compartments within each segment.

5. What role do homeotic genes play in the development of fruit flies?
Homeotic genes specify segment identity—expression of the homeotic genes informs cells of their location or address along the anterior posterior axis.

6. How do class A, B, and C genes in plants work together to determine the structures of the flower?
In the first whorl, class A genes determine sepal identity. In the second whorl, expression of class A and class B genes causes petal development. In the third whorl, class B and class C genes together cause stamen development. In the fourth whorl, class C genes act alone to specify carpel development. Class A and class C genes are mutually inhibitory, so that class A and class C genes are not co-expressed in the same whorl.

*7. What is apoptosis and how is it regulated?
Apoptosis is programmed cell death, characterized by nuclear DNA fragmentation, shrinkage of the cytoplasm and nucleus, and phagocytosis of the remnants of the dead cell. Apoptosis is regulated by internal and external signals that regulate the activation of procaspases, cysteine proteases that are activated by proteolytic cleavage. Once activated, these caspases activate other caspases in a cascade and degrade key cellular proteins.

*8. Explain how each of the following processes contributes to antibody diversity.
(a) Somatic recombination. *Recombination produces many combinations of variable domains with junction segments and diversity segments.*
(b) Junctional diversity. *During recombination, the V, J, and D joining events are imprecise, resulting in small deletions or insertions and frameshifts.*
(c) Hypermutation. *The V gene segments are subject to somatic hypermutation—accelerated random mutation—that further diversifies antibodies.*

9. What is the function of the MHC antigens? Why are the genes that encode these antigens so variable?
MHC proteins present fragments of foreign or self antigens to T cells. Variability of the MHC proteins has the effect of making individuals virtually unique in terms of their MHC genotype. This variability, or high degree of polymorphism, may have evolved in response to selective pressure to present antigens from many different pathogenic organisms for recognition by T cells.

*10. Outline Knudson's multistage theory of cancer and describe how it helps to explain unilateral and bilateral cases of retinoblastoma.
The multistage theory of cancer states that more than one mutation is required for most cancers to develop. Most retinoblastomas are unilateral because the likelihood of any cell acquiring two rare mutations is very low, and thus retinoblastomas develop in only one eye. Bilateral cases of retinoblastoma occur in people born with a predisposing mutation, so that only one additional mutational event will result in cancer. Thus, the probability of retinoblastoma is higher and

likely to occur in both eyes. Because the predisposing mutation is inherited, people with bilateral retinoblastoma have relatives with retinoblastoma.

11. Briefly explain how cancer arises through clonal evolution.
A mutation that relaxes growth control in a cell will cause it to divide and form a clone of cells that are growing or dividing more rapidly than their neighbors. Successive mutations that cause even more rapid growth, or the ability to invade and spread, each produce progeny cells with more aggressive, malignant properties that take over the original clone.

*12. What is the difference between an oncogene and a tumor-suppressor gene? Give some examples of functions of proto-oncogenes and tumor suppressors in normal cells.
An oncogene stimulates cell division, whereas a tumor–suppressor gene puts the brakes on cell growth. Proto-oncogenes are normal cellular genes that function in cell growth and regulation of the cell cycle: from growth factors such as sis to receptors like ErbA and ErbB, protein kinases such as Src, and nuclear transcription factors like Myc. Tumor suppressors inhibit cell cycle progression: RB and P53 are transcription factors and NF1 is a GTPase activator.

13. What is haploinsufficiency? How might it affect cancer risk?
Haploinsufficiency is a condition where a normally recessive trait affects a heterozygous individual. Most cases of haploinsufficiency are thought to arise because the heterozygote expresses a reduced amount of the gene product. In most individuals, the reduced amount of gene product provides enough gene function to produce a normal phenotype. In rare cases, however, a combination of genetic and/or environmental factors creates a greater need for the gene function or otherwise reduces the effectiveness of the gene function, so that the reduced amount of gene product cannot meet the demand. Haploinsufficiency for a tumor suppressor gene would lead to an increased risk for cancer in the heterozygous individual compared to a homozygous wild-type individual.

14. Why do mutations in genes that encode DNA repair enzymes and chromosome segregation often produce a predisposition to cancer?
Mutations that affect DNA repair cause high rates of mutation that may convert proto-oncogenes into oncogenes or inactivate tumor-suppressor genes. Similarly, errors in chromosome segregation cause aneuploidy and chromosomal aberrations that cause loss of tumor-suppressor genes or add extra gene doses of proto-oncogenes.

*15. What role do telomeres and telomerase play in cancer progression?
DNA polymerases are unable to replicate the ends of linear DNA molecules. Therefore, the ends of eukaryotic chromosomes shorten with every round of DNA replication, unless telomerase adds back special non-templated telomeric DNA sequences. Normally, somatic cells do not express telomerase; their telomeres progressively shorten with each cell division until vital genes are lost and the cells

undergo apoptosis. Transformed cells (cancerous cells) induce the expression of the telomerase gene to keep proliferating.

16. How is DNA methylation related to cancer?
DNA methylation is associated with transcriptional repression. Methylation and silencing of tumor suppressor genes would increase the risk of cancer; demethylation and activation of proto-oncogenes would also increase the risk of cancer. Hypomethylation (loss of DNA methylation) may also increase the risk of cancer by increasing genomic instability, by mechanisms that are not yet clear.

APPLICATION QUESTIONS AND PROBLEMS

17. If telomeres are normally shortened after each round of replication in somatic cells, what prediction would you make about the length of telomeres in Dolly, the first cloned sheep?
The chromosomes in the mammary cell nucleus that provided the nuclear DNA for Dolly would have had shortened telomeres, compared to chromosomes in germ line cells (sperm and egg). If telomerase expression was not induced in the process of cloning and subsequent embryonic development, we would predict that Dolly's somatic cells will have undergone further loss of telomeric DNA, so that Dolly would have shorter telomeres than other sheep of the same age.

*18. Give examples of genes that affect development in fruit flies by regulating gene expression at the level of (a) transcription and (b) translation.
(a) The products of bicoid and dorsal affect embryonic polarity by regulating transcription of target genes.
(b) The nanos gene regulates translation of hunchback mRNA.

19. What would be the most likely effect on development of puncturing the posterior end of a *Drosophila* egg, allowing a small amount of cytoplasm to leak out, and then injecting that cytoplasm into the anterior end of another egg?
The posterior cytoplasm contains posterior determinants, such as maternal nanos mRNA. Injecting cytoplasm containing nanos mRNA into the anterior end of an egg would interfere with determination of anterior (head and thoracic) structures by hunchback, and result in an embryo lacking head structures.

*20. What would be the most likely result of injecting *bicoid* mRNA into the posterior end of a *Drosophila* embryo and inhibiting the translation of *nanos* mRNA?
Bicoid mRNA in the posterior end of the embryo would cause transcription of hunchback in the posterior regions. Without Nanos protein, the hunchback mRNA would be translated to create high levels of Hunchback protein in the posterior as well as in the anterior. The result would be an embryo with anterior structures on both ends.

21. What would be the most likely effect of inhibiting the translation of *hunchback* mRNA throughout the embryo?

If the translation of hunchback throughout the embryo were inhibited, no anterior structures would form. The embryo would be entirely posteriorized, perhaps forming all abdominal structures.

*22. Molecular geneticists have performed experiments in which they altered the number of copies of the *bicoid* gene in flies, affecting the amount of Bicoid protein produced.

(a) What would be the effect on development of an increased number of copies of the *bicoid* gene?
Females with an increased number of copies of the bicoid gene would have higher levels of bicoid maternal mRNA in the anterior cytoplasm of their eggs, and thus higher levels of Bicoid protein in the embryo after fertilization. The resulting Bicoid protein gradient would extend further to the posterior, resulting in the enlargement of anterior and thoracic structures.

(b) What would be the effect of a decreased number of copies of *bicoid*?
Conversely, a decreased number of copies of the bicoid gene would ultimately result in a reduced Bicoid protein gradient in the eggs. Thus, sufficient Bicoid protein concentrations for head structures would be found in a smaller, more anterior portion of the embryo, resulting in an embryo with smaller head structures.

23. What would be the most likely effect on fruit-fly development of a deletion in the *nanos* gene?
Female flies homozygous for deletions in the nanos gene would lay eggs lacking nanos mRNA. Upon fertilization, the resulting embryos would form all anterior and thoracic structures, with no posterior structures. If the egg does have maternal nanos mRNA (because the mother was a heterozygote), and even if the zygote was homozygous for a deletion in the nanos gene, the embryo would develop normally.

24. Give an example of a gene found in each of the categories of genes (egg-polarity, gap, pair-rule, and so forth) listed in Figure 21.12.
Egg-polarity (maternal effect): *bicoid, nanos*
Gap: *hunchback, Krüppel*
Pair-rule: *even-skipped, fushi tarazu*
Segment-polarity: *gooseberry*

25. Explain how: (a) the absence of class B gene expression produces the structures seen in class B mutants (Figure 21.14c); and (b) the absence of class C gene product produces the structures seen in class C mutants (Figure 21.14d).
(a) Class B genes are normally expressed in the second and third whorls. Class B genes work together with class A genes in the second whorl to specify petals, and class B genes work together with class C genes in the third whorl to specify stamens. In the absence of class B genes, the floral buds express only class A genes in the first and second whorls, which become sepals, and only class C genes in the third and fourth whorls, which become carpels.

(b) Class C genes are normally expressed in the third and fourth whorls. In the third whorl, class C genes work together with class B genes to specify stamens, and in the fourth whorl, class C genes alone specify carpels. Class C genes also inhibit expression of class A genes. With no class C gene expression, class A genes will be expressed in all four whorls, with the second and third whorls also expressing class B genes. The resulting pattern of floral development will produce abnormal floral structures that have sepal-petal-petal-sepal patterns.

26. What would you expect a flower to look like in a plant that lacked class A and class B genes? What about a plant that lacked class B and class C genes?
A plant that lacked class A and class B genes would express only class C genes in all four whorls, resulting in flowers with only carpels. A plant that lacked both class B and class C genes would express only class A genes in all four whorls, and result in flowers with only sepals.

27. What would be the flower structure of a plant within which expression of the following genes were inhibited:
(a) Expression of class B genes is inhibited in the 2nd whorl but not the 3rd whorl.
Sepal-sepal-stamen-carpel. The first and second whorls, expressing only class A genes, would both produce sepals. The third whorl with both class B and class C genes would produce stamens, and the fourth whorl with only class C genes would produce carpels.
(b) Expression of class C genes is inhibited in the 3rd whorl but not the 4th whorl.
Sepal-petal-petal-carpel. The first, second and fourth whorls are unaffected and produce the normal floral parts. The third whorl, without class C gene products, would express class A genes as well as class B genes to specify petals.
(c) Expression of class A genes is inhibited in the 1st whorl but not the 2nd whorl.
Carpel-petal-stamen-carpel. Without class A gene expression, the first whorl would express class C genes and produce carpels. The other whorls have normal gene expression and are unaffected.
(d) Expression of class A genes is inhibited in the 2nd whorl but not the 1st whorl.
Sepal-stamen-stamen-carpel. Lack of class A gene expression in the second whorl would allow class C gene expression, and result in stamens because of co-expression of class B and class C genes. The other whorls are unaffected and produce normal floral organs.

*28. In a particular species, the gene for the kappa light chain has 200 *V* gene segments and 4 *J* segments. In the gene for the lambda light chain, this species has 300 *V* segments and 6 *J* segments. Considering only the variability arising from somatic recombination, how many different types of light chains are possible?
Light chain genes undergo recombination to join one V gene segment to one J segment, in any combination. The number of different possible V and J combinations is given by the product of the number of V segments and the number of J segments for each light chain.
kappa light chain: $200 \times 4 = 800$
lambda light chain: $300 \times 6 = 1800$

Total light chains = kappa + lambda = 800 + 1800 = 2600

29. In the fictional book *Chromosome 6* by Robin Cook, a biotechnology company genetically engineers individual bonobos (a type of chimpanzee) to serve as future organ donors for clients. The genes of the bonobos are altered so that no tissue rejection takes place when their organs are transplanted into the client. What genes would need to be altered for this scenario to work? Explain your answer.
The MHC genes of the bonobos must be humanized. The T-cells of the immune system recognize antigen complexed with the body's own MHC molecules. Non-self MHC molecules trigger an immune response and tissue rejection. Therefore, the bonobo MHC molecules must be replaced with the human patient's own MHC molecules to prevent rejection.

*30. A couple has one child with bilateral retinoblastoma. The mother is free from cancer, but the father had unilateral retinoblastoma and he has a brother who has bilateral retinoblastoma.
(a) If the couple has another child, what is the probability that this next child will have retinoblastoma?
First, we summarize the information with a pedigree. The shaded boxes represent bilateral retinoblastoma; the striped box represents unilateral retinoblastoma.

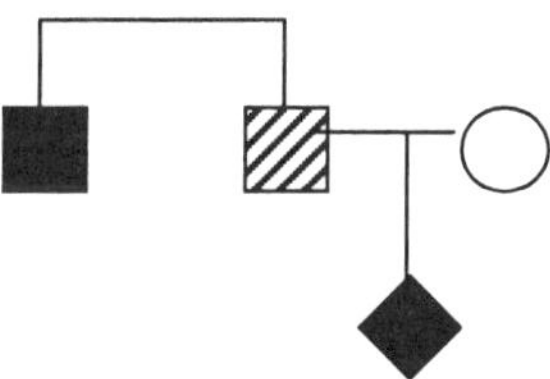

Familial retinoblastoma is caused by mutation of the RB tumor-suppressor gene (see Table 21.9). Because the loss of a functional RB allele means that only one additional mutation event will completely eliminate RB function and lead to retinoblastoma, loss-of-function RB mutations have dominant effects with regard to retinoblastoma. If the father with unilateral retinoblastoma is heterozygous for an RB mutation, then the chance of another child inheriting the mutant RB allele is ½. Of course, if the father is homozygous for the RB mutation, then the chance of the child having retinoblastoma is nearly 100%.
(b) If the next child has retinoblastoma, is it likely to be bilateral or unilateral?
Because retinoblastoma in this family is most likely an inherited disorder, a child with retinoblastoma will more likely have bilateral retinoblastoma. Unilateral retinoblastomas are usually spontaneous in origin, requiring two independent mutations in a single somatic retinal cell. Familial retinoblastomas occur in family members that inherited one of the two mutations required for retinoblastoma. As only one additional mutation is required in the somatic retinal cells, retinoblastoma occurs in both eyes and at earlier ages than spontaneous unilateral retinoblastomas.

(c) Propose an explanation for why the father's case of retinoblastoma was unilateral, while his sons and brother's cases were bilateral.
The father may have unilateral retinoblastoma because of incomplete penetrance of the mutation in the RB gene. Alleles at another locus or multiple other loci may have contributed to resistance to retinoblastoma in the father, so that he suffered retinoblastoma in only one eye. Alternatively, it may have been just good fortune (random chance) that one of his eyes was spared the second mutation event that led to retinoblastoma in his other eye.

31. Some cancers are consistently associated with the deletion of a particular part of a chromosome. Does the deleted region contain an oncogene or a tumor-suppressor gene? Explain why.
The deleted region contains a tumor-suppressor gene. Tumor suppressors act as brakes on cell proliferation. The deletion of tumor-suppressor genes will therefore permit the uncontrolled cell proliferation that is characteristic of cancer. Oncogenes, on the other hand, function as stimulators of cell division. Deletion of oncogenes will therefore prevent cell proliferation, and usually cannot cause cancer.

32. Cells in a tumor contain mutated copies of a particular gene that promotes tumor growth. Gene therapy can be used to introduce a normal copy of this gene into the tumor cells. Would you expect this therapy to be effective if the mutated gene were an oncogene? A tumor-suppressor gene? Explain your reasoning.
Gene therapy to introduce a normal copy of the gene into tumor cells will not work for oncogenes because oncogenes are dominant, activating mutations of proto-oncogenes. Gene therapy may work if the tumor arises from a mutation that inactivates a tumor-suppressor gene. Loss-of-function mutations are recessive; therefore, a normal copy of the gene will be dominant and restore regulation of cell proliferation in the tumor cells.

33. Some cancers have been treated with drugs that demethylate DNA. Propose an explanation for how these drugs might work. Do you think the genes causing cancers that respond to the demethylation are likely to be oncogenes or tumor-suppressor genes? Explain your reasoning.
Drugs that demethylate DNA would presumably activate expression of demethylated genes. Cancer growth and progression may be inhibited if these drugs are able to turn on expression of tumor suppressor genes that had been silenced by DNA methylation. If DNA demethylation turned on expression of oncogenes, cancer growth and progression would be accelerated.

CHALLENGE QUESTIONS

34. As we have learned in this chapter, the Nanos protein inhibits the translation of *hunchback* mRNA, thus lowering the concentration of Hunchback protein at the posterior end of a fruit-fly embryo and stimulating the differentiation of posterior characteristics. The results of experiments have demonstrated that the action of

Nanos on *hunchback* mRNA depends on the presence of an 11-base sequence that is located in the 3' untranslated region of *hunchback* mRNA. This sequence has been termed the Nanos response element (NRE). There are two copies of NREs in the trailer of *hunchback* mRNA. If a copy of NRE is added to the 3' untranslated region of another mRNA produced by a different gene, the mRNA now becomes repressed by Nanos. The repression is greater if several NREs are added. On the basis of these observations, propose a mechanism for how Nanos inhibits Hunchback translation.
Nanos may bind to the NREs, either directly or indirectly as a complex with other proteins. Multiple NRE binding elements may enhance the binding either by simply providing a higher concentration of binding sites, or through cooperativity (the binding of protein to one NRE enhances the binding of protein to other NREs). The complex of Nanos (and other proteins) bound to the NREs at the 3' untranslated region may target the mRNA for rapid degradation. Alternatively, the NRE-bound protein complexes may interfere with ribosome binding or translational initiation at the 5' end of the mRNA, although this postulates that mRNAs assume a circular topology where the 3' and 5' ends of the mRNA are in close proximity.

35. Offer a possible explanation for the widespread distribution of *Hox* genes among animals.
One explanation is that the Hox genes arose in the common evolutionary ancestor of the animals that have Hox genes. This ancestor used the ancestral Hox gene for body patterning. The complex of Hox genes and the overall function was then retained in the descendant lineages.

36. Many cancer cells are immortal (will divide indefinitely) because they have mutations that allow telomerase to be expressed. How might this knowledge be used to design anticancer drugs?
Because cancer cells depend on telomerase activity to preserve their telomeres, drugs that target telomerase enzymatic activity may limit the ability of cancer cells to divide indefinitely.

37. Bloom syndrome is an autosomal recessive disease that exhibits haploinsufficiency. A recent survey showed that people heterozygous for mutations at the *BLM* locus are at increased risk of colon cancer. Suppose you are a genetic counselor. A young woman whose mother has Bloom syndrome is referred to you; the young woman's father has no family history of Bloom syndrome. The young woman asks whether she is likely to experience any other health problems associated with her family history of Bloom syndrome. What advice would you give her?
The young woman must be heterozygous for the mutation at the BLM locus because her mother was homozygous for the mutation. Although the young woman does not have Bloom syndrome, haploinsufficiency at this locus will result in some increased risk of colon cancer. Her cells will have a reduced amount of the BLM helicase involved in DNA double-strand break repair, and will be more susceptible to mutations that may lead to cancer.

Chapter Twenty-Two: Quantitative Genetics

COMPREHENSION QUESTIONS

*1. How does a quantitative characteristic differ from a discontinuous characteristic?
Discontinuous characteristics have only a few distinct phenotypes. In contrast, a quantitative characteristic shows a continuous variation in phenotype.

2. Briefly explain why the relation between genotype and phenotype is frequently complex for quantitative characteristics.
Quantitative characteristics are polygenic, so that many genotypes are possible. Moreover, most quantitative characteristics are also influenced by environmental factors. Therefore, the phenotype is determined by complex interactions of many possible genotypes and environmental factors.

*3. Why do polygenic characteristics have many phenotypes?
Many genotypes are possible with multiple genes. Even for the simplest two-allele loci, the number of possible genotypes is equal to 3^n, *where* n *is the number of loci or genes. Thus, for 3 genes, we have 27 genotypes, 4 genes yields 108, and so forth. If each genotype corresponds to a unique phenotype, then we have the same numbers of phenotypes: 27 possible phenotypes for 3 genes and 108 possible phenotypes for 4 genes. Finally, the phenotype for a given genotype may be influenced by environmental factors, leading to an even greater array of phenotypes.*

*4. Explain the relation between a population and a sample. What characteristics should a sample have to be representative of the population?
A sample is a subset of the population. To be representative of the population, a sample should be randomly selected and sufficiently large to minimize random differences between members of the sample and the population.

5. What information do the mean and variance provide about a distribution?
The means is the center of the distribution. The variance is how broad the distribution is around the mean.

6. How is the standard deviation related to the variance?
The standard deviation is the square root of the variance.

*7. What information does the correlation coefficient provide about the association between two variables?
The magnitude or absolute value of the correlation coefficient reports how strongly the two variables are associated. A value close to +1 or –1 indicates a strong association; values close to zero indicate weak association.

8. What is regression? How is it used?
Regression is a mathematical relationship between correlated variables. Regression is used to predict the value of a variable from the value of a correlated variable.

*9. List all the components that contribute to the phenotypic variance and define each component.
V_G – *Component of variance due to variation in genotype.*
V_A – *Component of variance due to additive genetic variance.*
V_D – *Component of variance due to dominance genetic variance.*
V_I – *Component of variance due to genic interaction variance.*
V_E – *Component of variance due to environmental differences.*
V_{GE} *Component of variance due to interaction between genes and environment.*

*10. How do the broad-sense and narrow-sense heritabilities differ?
The broad-sense heritability is the portion of phenotypic variance that is due to all types of genetic variance, including additive, dominance, and genic interaction variances. The narrow-sense heritability is just that portion of the phenotypic variance due to additive genetic variance.

11. Briefly outline some of the ways that heritability can be calculated.
Elimination of variance components from the equation: $V_P = V_G + V_E + V_{GE}$. *By either eliminating genetic variance* ($V_G = 0$) *with genetically identical individuals, or by eliminating* V_E *with individuals raised in identical environments, we can determine the values of* V_E *or* V_G, *respectively. If* V_P *can be determined under conditions of genotypic variance, then the missing term* V_G *or* V_E *can be calculated by simple subtraction.*

Parent-offspring regression: The mean phenotypic values of the parents are plotted against the mean phenotypic values of the offspring for a series of families. The narrow-sense heritability equals the regression coefficient.

Comparison of phenotypes for different degrees of relatedness: Compare phenotypes of monozygotic and dizygotic twins. Twice the difference in correlation coefficients of monozygotic and dizygotic twins yields an estimate of the broad-sense heritability.

Response to selection: The response to selection is equal to the product of the narrow-sense heritability and the selection differential.

12. Briefly discuss common misunderstandings or misapplications of the concept of heritability.
(1) *Heritability is the portion of phenotypic* variance *due to genetic variance; it does not indicate to what extent the phenotype itself is determined by genotype.*
(2) *Heritability applies to populations; it does not apply to individuals.*
(3) *Heritability is determined for a particular population in a particular environment at a particular time. Heritability determined for one population does not apply to other populations, or even the same population facing different environmental conditions at a different period.*

(4) *A trait with high heritability may still be strongly influenced by environmental factors.*
(5) *High heritability does not mean that differences between populations are due to differences in genotype.*

13. Briefly explain how genes affecting a polygenic characteristic are located with the use of QTL mapping.
Two homozygous, highly inbred strains that differ at many loci are crossed and the F_1 *are interbred. Quantitative traits are measured and correlated with the inheritance of molecular markers throughout the genome. The correlations are used to infer the presence of a linked QTL.*

*14. How is the response to selection related to the narrow-sense heritability and the selection differential? What information does the response to selection provide?
The response to selection (R) = *narrow-sense heritability* (h^2) × *selection differential* (S). *The value of* R *predicts how much the mean quantitative phenotype will change with different selection in a single generation.*

15. Why does the response to selection often level off after many generations of selection?
After many generations, the response to selection plateaus because of two factors. First, the genetic variation may be depleted—all the individuals in the population now have the alleles that maximize the quantitative trait; with no genetic variation, there can be no selection or response to selection. Second, even if genetic variation persists, artificial selection may be limited by an opposing natural selection.

APPLICATION QUESTIONS AND PROBLEMS

*16. For each of the following characteristics, indicate whether it would be considered a discontinuous characteristic or a quantitative characteristic. Briefly justify your answer.

(a) Kernel color in a strain of wheat, in which two codominant alleles segregating at a single locus determine the color. Thus, there are three phenotypes present in this strain: white, light red, and medium red.
This is a discontinuous characteristic because only a few distinct phenotypes are present and it is determined by alleles at a single locus.

(b) Body weight in a family of Labrador retrievers. An autosomal recessive allele that causes dwarfism is present in this family. Two phenotypes are recognized: dwarf (less than 13 kg) and normal (greater than 13 kg).
This is a discontinuous characteristic because there are only two phenotypes (dwarf and normal) and a single locus determines characteristic.

(c) Presence or absence of leprosy; susceptibility to leprosy is determined by multiple genes and numerous environmental factors.
This is a quantitative characteristic because susceptibility is a continuous trait that is determined by multiple genes and environmental factors. It is an example of a quantitative phenotype with a threshold effect.

(d) Number of toes in guinea pigs, which is influenced by genes at many loci.
A quantitative characteristic because it is determined by many loci. The number of toes is an example of a meristic characteristic.

(e) Number of fingers in humans; extra (more than five) fingers are caused by the presence of an autosomal dominant allele.
A discontinuous characteristic because there are only a few distinct phenotypes determined by genes at a single locus.

*17. Listed below are the numbers of digits per foot in 25 guinea pigs. Construct a frequency distribution for these data.
4, 4, 4, 5, 3, 4, 3, 4, 4, 5, 4, 4, 3, 2, 4, 4, 5, 6, 4, 4, 3, 4, 4, 4, 5

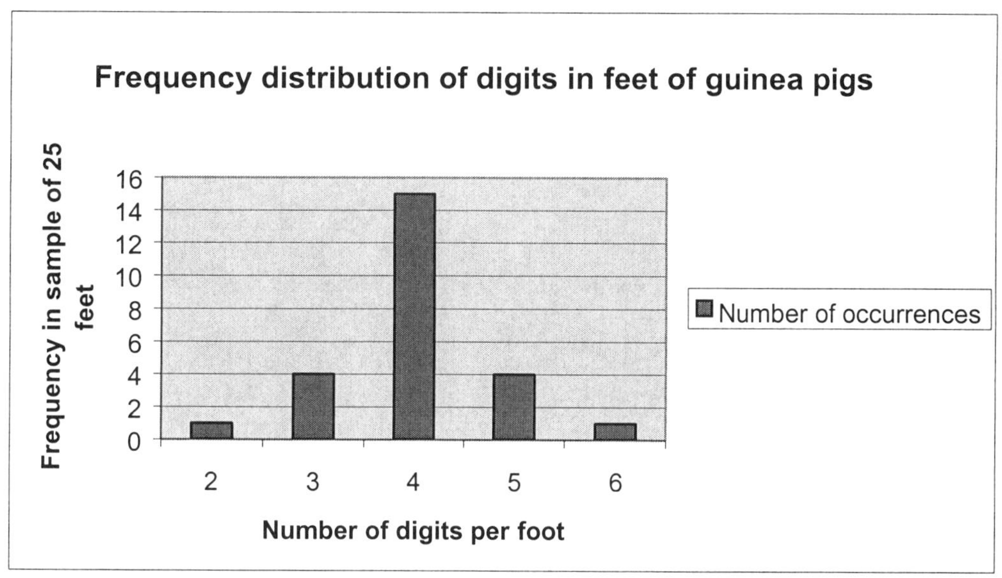

18. Ten male Harvard students were weighed in 1916. Their weights are given in the following table. Calculate the mean, variance, and standard deviation for these weights.

Weight (kg) of Harvard students (class of 1920)
51
69
69
57
61
57
75
105
69
63

x_i	$\lvert x_i - mean \rvert$	$(x_i - mean)^2$
51	16.6	275.56
69	1.4	1.96
69	1.4	1.96
57	10.6	112.36
61	6.6	43.56
57	10.6	112.36
75	7.4	54.76
105	37.4	1398.76
69	1.4	1.96
63	4.6	21.16
sum = 676		$\Sigma = 2024.4$
mean = 67.6		

The sum of the weights is 676, divided by 10 students, yields a mean of 67.6 kg.

The variance is $s^2 = \dfrac{\sum (x_i - \bar{x})^2}{n-1}$ *= 2024.4/9 = 224.9*

The standard deviation = $s = \sqrt{s^2}$ = *15*

19. Among a population of tadpoles, the correlation coefficient for size at metamorphosis and time required for metamorphosis is –.74. On the basis of this correlation, what conclusions can you make about the relative sizes of tadpoles that metamorphose quickly and those that metamorphose more slowly?
Size at metamorphosis and time required for metamorphosis are inversely correlated (negative correlation coefficient). The greater the time required for metamorphosis, the smaller the size, and vice versa. Therefore, tadpoles that metamorphose quickly are larger than tadpoles that metamorphose slowly.

*20. A researcher studying alcohol consumption in North American cities finds a significant, positive correlation between the number of Baptist preachers and alcohol consumption. Is it reasonable for the researcher to conclude that the Baptist preachers are consuming most of the alcohol? Why or why not?
No; correlation does not mean causation. A number of other factors may be responsible for the observed correlation. For example, Baptist preachers may feel their services are more needed in areas with high alcohol consumption. The reader should think of other plausible scenarios.

21. Body weight and length were measured on six mosquito fish; these measurements are given in the following table. Calculate the correlation coefficient for weight and length in these fish.

Wet weight (g)	Length (mm)
115	18
130	19
210	22
110	17
140	20
185	21

The correlation coefficient r *is calculated from the formula* $r = \frac{\mathrm{cov}_{xy}}{s_x s_y}$.

First, we calculate the covariance of the two traits and their standard deviations.

x_i	y_i	$x_i - \bar{x}$	$y_i - \bar{y}$	$(x_i - \bar{x})(y_i - \bar{y})$	$(x_i - \bar{x})^2$	$(y_i - \bar{y})^2$
115	18	−33.33	−1.5	50	1111.11	2.25
130	19	−18.33	−0.5	9.17	336.11	0.25
210	22	61.67	2.5	154.17	3802.78	6.25
110	17	−38.33	−2.5	95.83	1469.44	6.25
140	20	−8.33	0.5	−4.17	69.44	0.25
185	21	36.67	1.5	55	1344.44	2.25
$\bar{x} = 148.3$		$\bar{y} = 19.5$		$\Sigma = 360$	$\Sigma = 8133.33$	$\Sigma = 17.5$
				$cov_{xy} = 72$	$s^2_x = 1626.67$	$s^2_y = 3.5$
					$s_x = 40.33$	$s_y = 1.87$

$$r = \frac{72}{(40.33)(1.87)} = 0.95$$

*22. The heights of mothers and daughters are given in the following table.

Height of mother (inches)	Height of daughter (inches)
64	66
65	66
66	68
64	65
63	65
63	62
59	62
62	64
61	63
60	62

(a) Calculate the correlation coefficient for the heights of the mothers and daughters.

The correlation coefficient r *is calculated from the formula* $r = \frac{cov_{xy}}{s_x s_y}$.

x	y	$x_i - \bar{x}$	$y_i - \bar{y}$	$(x_i - \bar{x})(y_i - \bar{y})$	$(x_i - \bar{x})^2$	$(y_i - \bar{y})^2$
64	66	1.3	1.7	2.21	1.69	2.89
65	66	2.3	1.7	3.91	5.29	2.89
66	68	3.3	3.7	12.21	10.89	13.69
64	65	1.3	0.7	0.91	1.69	0.49
63	65	0.3	0.7	0.21	0.09	0.49
63	62	0.3	−2.3	−0.69	0.09	5.29
59	62	−3.7	−2.3	8.51	13.69	5.29
62	64	−0.7	−0.3	0.21	0.49	0.09
61	63	−1.7	−1.3	2.21	2.89	1.69
60	62	−2.7	−2.3	6.21	7.29	5.29
$\bar{x} = 62.7$	$\bar{y} = 64.3$			$\Sigma = 35.9$	$\Sigma = 44.1$	$\Sigma = 38.1$

$cov(x,y) = 3.99$ $\quad s^2_x = 4.9$ $\quad s^2_y = 4.23$

$s_x = 2.2$ $\quad s_y = 2.1$

$b = 0.81$

$b(\bar{x}) = 51$

$a = 13$

$$r = \frac{3.99}{(4.2)(4.1)} = .80$$

(b) Using regression, predict the expected height of a daughter whose mother is 67 inches tall.

In the regression equation, y = a + bx, b *is given by the formula* $b = \frac{cov_{xy}}{s_x^2}$ *and the value of* a *by the equation* $a = \bar{y} - b\bar{x}$. *If the mother is 67 inches tall, the regression equation* y = a + bx *becomes* y = *13.26 + 0.81(67) = 67.8 inches.*

*23. Assume that plant weight is determined by a pair of alleles at each of two independently assorting loci (*A* and *a*, *B* and *b*) that are additive in their effects. In addition, assume that each allele represented by an uppercase letter contributes 4 g to weight and each allele represented by a lowercase letter contributes 1 g to weight.

(a) If a plant with genotype *AABB* is crossed with a plant with genotype *aabb*, what weights are expected in the F_1 progeny?

All the F_1 progreny will have genotype AaBb, *so they should all have 4 + 4 + 1 + 1 = 10 grams of weight.*

(b) What is the distribution of weight expected in the F_2 progeny?

We can group the 16 expected genotypes by the number of uppercase and lowercase alleles:
4 uppercase: AABB = *1/16 with 16 grams*
3 uppercase: 2 AaBB, *2* AABb = *4/16 with 13 grams*
2 uppercase: 4 AaBb, aaBB, AAbb = *6/16 with 10 grams*
1 uppercase: 2 Aabb, 2 aaBb = *4/16 with 7 grams*
0 uppercase: aabb = *1/16 with 4 grams*

*24. Assume that three loci, each with two alleles (*A* and *a*, *B* and *b*, *C* and *c*), determine the differences in height between two homozygous strains of a plant. These genes are additive and equal in their effects on plant height. One strain (*aabbcc*) is 10 cm in height. The other strain (*AABBCC*) is 22 cm in height. The two strains are crossed, and the resulting F_1 are interbred to produce F_2 progeny. Give the phenotypes and the expected proportions of the F_2 progeny.
The AABBCC *strain is 12 cm taller than the* aabbcc *strain. We therefore calculate that each dominant allele adds 2 cm of height above the baseline 10 cm of the all-recessive strain. The F_1, with genotype* AaBbCc, *therefore will be 10 + 6 = 16 cm tall. The seven different possible phenotypes with respect to plant height and the expected frequencies in the F_2 are listed in the following table:*

# dominant alleles	*Height (cm)*	*Proportion of F_2 progeny*
6	*22*	*1/64*
5	*20*	*6/64*
4	*18*	*15/64*
3	*16*	*20/64*
2	*14*	*15/64*
1	*12*	*6/64*
0	*10*	*1/64*
		total = 64/64

The proportions can be determined by counting the numbers of boxes with one dominant allele, two dominant alleles, and so on from an 8 × 8 Punnett square.

*25. A farmer has two homozygous varieties of tomatoes. One variety, called *Little Pete*, has fruits that average only 2 cm in diameter. The other variety, *Big Boy*, has fruits that average a whopping 14 cm in diameter. The farmer crosses *Little Pete* and *Big Boy*; he then intercrosses the F_1 to produce F_2 progeny. He grows 2000 F_2 tomato plants and doesn't find any F_2 offspring that produce fruits as small as *Little Pete* or as large as *Big Boy*. If we assume that the differences in fruit size of these varieties are produced by genes with equal and additive effects, what conclusion can we make about the minimum number of loci with pairs of alleles determining the differences in fruit size of the two varieties?
That six or more loci are involved. Generally, $(1/4)^n$ of the F_2 progeny should resemble one of the homozygous parents, where n *is the number of loci with pairs of alleles that determine the differences in the trait. If 5 genes were involved, the farmer should have found approximately 1/1000 of the F_2 that resembled either*

Little Pete *or* Big Boy. *Because he did not, we can conclude that at least 6 loci are involved in the difference in fruit size between the two varieties:* $(1/4)^6 = 1/4096$ *would be expected to resemble one of the parents if six loci were involved.*

26. Seed size in a plant is a polygenic characteristic. A grower crosses two pure-breeding varieties of the plant and measures seed size in the F_1 progeny. He then backcrosses the F_1 plants to one of the parental varieties and measures seed size in the backcross progeny. The grower finds that seed size in the backcross progeny has a higher variance than seed size in the F_1 progeny. Explain why the backcross progeny are more variable.
The F_1 progeny all have the same genetic makeup: they are all heterozygotes for the loci that differ between the two pure-breeding strains. The backcross progeny will have much greater genetic diversity as a result of the genetic diversity from meiosis of the F_1 heterozygotes.

27. Suppose that you just learned that the heritability of blood pressure measured among a group of African-Americans from Detroit, Michigan is 0.40. In your own words, explain what this heritability value means. What does it tell us about genetic and environmental contributions to blood pressure?
This heritability value indicates that approximately 40 percent of the variance in blood pressure among this group of individuals is due to genetic differences. Therefore, about 60 percent of the variance is due to differences in environment. This heritability value applies only to this specific group and cannot be extrapolated to other populations. It tells us nothing about relative contributions of genetics and environment to absolute blood pressure, but rather tells us about their relative contributions to differences in blood pressure for this group in this environment.

*28. Phenotypic variation in tail length of mice has the following components:
Additive genetic variance (V_A) = .5
Dominance genetic variance (V_D) = .3
Genic interaction variance (V_I) = .1
Environmental variance (V_E) = .4
Genetic-environmental interaction variance (V_{GE}) = .0
(a) What is the narrow-sense heritability of tail length?
Narrow-sense heritability is $V_A/V_P = .5/1.3 = .38$.
(b) What is the broad-sense heritability of tail length?
Broad-sense heritability is $V_G/V_P = (V_A + V_D + V_I)/V_P = .9/1.3 = .69$.

29. The narrow-sense heritability of ear length in Reno rabbits is .4. The phenotypic variance (V_P) is .8 and the environmental variance (V_E) is .2. What is the additive genetic variance (V_A) for ear length in these rabbits?
Narrow-sense heritability = $V_A/V_P = .4$
Given that $V_P = .8$, $V_A = .4(.8) = .32$

*30. Assume that human ear length is influenced by multiple genetic and environmental factors. Suppose you measured ear length on three groups of people, in which group A consists of five unrelated individuals, group B consists of five siblings, and group C consists of five first cousins.

(a) Assuming that the environment for each group is similar, which group should have the highest phenotypic variance? Explain why.

Group A, because unrelated individuals have the greatest genetic variance.

(b) Is it realistic to assume that the environmental variance for each group is similar? Explain your answer.

No. Siblings from the same family and who are raised in the same house should have smaller environmental variance than group A of unrelated individuals.

31. A characteristic has a narrow-sense heritability of .6.

(a) If the dominance variance (V_D) increases and all other variance components remain the same, what will happen to the narrow-sense heritability? Will it increase, decrease, or remain the same? Explain.

The narrow-sense heritability will decrease. Narrow-sense heritability is V_A/V_P. *Increasing the* V_D *will increase the total phenotypic variance* V_P. *If* V_A *remains unchanged, then the proportion* V_A/V_P *will become smaller.*

(b) What will happen to the broad-sense heritability? Explain.

The broad-sense heritability V_G/V_P *will increase.* V_G *is the sum of* $V_A + V_D + V_I$. V_P *is the sum of* $V_A + V_D + V_I + V_E$. *Increasing the numerator and denominator of the fraction by the same arithmetic increment will result in a larger fraction, if the fraction is smaller than 1.*

(c) If the environmental variance (V_E) increases and all other variance components remain the same, what will happen to the narrow-sense heritability? Explain.

The narrow sense heritability V_A/V_P *will decrease because the total phenotypic variance* V_P *will increase if* V_E *increases.*

(d) What will happen to the broad-sense heritability? Explain.

The broad-sense heritability V_G/V_P *will decrease because* $V_P = V_G + V_E$ *will increase.*

32. Flower color in the pea plants that Mendel studied is controlled by alleles at a single locus. A group of peas homozygous for purple flowers is grown in a garden. Careful study of the plants reveals that all their flowers are purple, but there is some variability in the intensity of the purple color. If heritability were estimated for this variation in flower color, what would it be? Explain your answer.

The plants are homozygous for the single color locus; therefore, there is no genetic variance: $V_G = 0$. *Because heritability is* V_G/V_P, *if* V_G *is zero, then heritability is zero.*

*33. A graduate student is studying a population of bluebonnets along a roadside. The plants in this population are genetically variable. She counts the seeds produced by 100 plants and measures the mean and variance of seed number. The variance is 20. Selecting one plant, the graduate student takes cuttings from it, and cultivates these cuttings in the greenhouse, eventually producing many genetically identical clones of the same plant. She then transplants these clones into the roadside population, allows them to grow for one year, and then counts the number of seeds produced by each of the cloned plants. The graduate student finds that the variance of seed number among these cloned plants is 5. From the phenotypic variance of the genetically variable and genetically identical plants, she calculates the broad-sense heritability.

(a) What is the broad-sense heritability of seed number for the roadside population of bluebonnets?

In the genetically identical population, $V_G = 0$, *and* $V_P = V_E = 5$. *In the original population,* $V_G = V_P - V_E = 20 - 5 = 15$. *The broad-sense heritability is then* $V_G/V_P = 15/20 = 0.75$.

(b) What might cause this estimate of heritability to be inaccurate?

This estimate may be inaccurate if the environmental variance of the genetically identical population is different from the environmental variance of the genetically diverse population.

34. Many researchers have estimated heritability of human traits by comparing the correlation coefficients of monozygotic and dizygotic twins. One of the assumptions of using this method is that two monozygotic twins experience environments that are no more similar to each other than those experienced by two dizygotic twins. How might this assumption be violated? Give some specific examples of ways that the environments of two monozygotic twins might be more similar than the environments of two dizygotic twins.

One obvious way a monozygotic twins may have a more similar environment is if the dizygotic twins differ in sex. Dizygotic twins also differ more in physical traits than monozygotic twins. Such differences, in hair color, eye color, height, weight, and others, lead to different preferences in clothing, whether or when eye glasses or braces are required, and differences in preferred activities such as different aptitudes for sports.

35. A genetics researcher determines that the broad-sense heritability of height among Baylor University undergraduate students is 0.90. Which of the following conclusions would be reasonable? Explain your answer.

(a) Because Sally is a Baylor University undergraduate student, 10% of her height is determined by non genetic factors.

(b) Ninety percent of variation in height among all undergraduate students in the United States is due to genetic differences.

(c) Ninety percent of the height of Baylor University undergraduate students is determined by genes.

(d) Ten percent of the variation in height of Baylor University undergraduate students is determined by variation in non genetic factors.

(e) Because the heritability of height among Baylor University students is so high, any change in the students environment will have minimal impact on their height.

Heritability is the proportion of total phenotypic variance that is due to genetic variance, and applies only to the particular population. Thus, the only reasonable conclusion is (d). Statement (a) is not justified because the heritability value does not apply to absolute height, but to the variance in height among Baylor undergraduates. Statement (b) is not justified because the heritability has been determined only for Baylor University students; students at other universities, with different ethnic backgrounds and from different regions of the country may have different heritability for height. Statement (c) is again not justified because the heritability refers to the variance in height rather than absolute height. Statement (e) is not justified because the heritability has been determined for the range of variation in non genetic factors experienced by the population under study; environmental variation outside this range (such as severe malnutrition) may have profound effects on height.

*36. The length of the middle joint of the right index finger was measured on 10 sets of parents and their adult offspring. The mean parental lengths and the mean offspring lengths for each family are listed in the following table. Calculate the regression coefficient for regression of mean offspring length against mean parental length and estimate the narrow sense heritability for this characteristic.

Mean parental length (mm)	Mean offspring length (mm)
30	31
35	36
28	31
33	35
26	27
32	30
31	34
29	28
40	38
33	34

The narrow-sense heritability is equal to the regression coefficient b *of a regression of the means of the parents and the means of the offspring.*

x	y	$x_i-\bar{x}$	$y_i-\bar{y}$	$(x_i-\bar{x})(y_i-\bar{y})$	$(x_i-\bar{x})^2$	$(y_i-\bar{y})^2$
30	31	−1.7	−1.4	2.38	2.89	1.96
35	36	3.3	3.6	11.88	10.89	12.96
28	31	−3.7	−1.4	5.18	13.69	1.96
33	35	1.3	2.6	3.38	1.69	6.76
26	27	−5.7	−5.4	30.78	32.49	29.16
32	30	0.3	−2.4	−0.72	0.09	5.76
31	34	-0.7	1.6	−1.12	0.49	2.56
29	28	−2.7	−4.4	11.88	7.29	19.36
40	38	8.3	5.6	46.48	68.89	31.36
33	34	1.3	1.6	2.08	1.69	2.56

$\bar{x} = 31.7$ $\bar{y} = 32.4$ $\Sigma = 112.2$ $\Sigma = 140.1$ $\Sigma = 114.4$

$cov(x,y) = 12.47$ $s^2_x = 15.6$ 12.7

$$b = 0.80$$
$$b(\bar{x}) = 25.4$$
$$a = 7.0$$

From the above table, the narrow-sense heritability = b = *0.8.*

*37. Mr. Jones is a pig farmer. For many years, he has fed his pigs the food left over from the local university cafeteria, which is known to be low in protein, deficient in vitamins, and downright untasty. However, the food is free and his pigs do not complain. One day a salesman from a feed company visits Mr. Jones. The salesman claims that his company sells a new, high-protein, vitamin-enriched feed that enhances weight gain in pigs. Although the food is expensive, the salesman claims that the increased weight gain of the pigs will more than pay for the cost of the feed, increasing Mr. Jones' profit. Mr. Jones responds that he took a genetics class when he went to the university, and that he has conducted some genetic experiments on his pigs; specifically, he has calculated the narrow-sense heritability of weight gain for his pigs and found it to be .98. Mr. Jones says that this heritability value indicates that 98% of the variance in weight gain among his pigs is determined by genetic differences, and therefore the new pig feed can have little effect on the growth of his pigs. He concludes that the feed would be a waste of his money. The salesman does not dispute Mr. Jones' heritability estimate, but he still claims that the new feed can significantly increase weight gain in Mr. Jones' pigs. Who is correct and why?
The salesman is correct because Mr. Jones' determination of heritability was conducted for a population of pigs under one environmental condition: low nutrition. His findings do not apply to any other population or even to the same population under different environmental conditions. High heritability for a trait does not mean that environmental changes will have little effect.

38. Joe is breeding cockroaches in his dorm room. He finds that the average wing length in his population of cockroaches is 4 cm. He picks six cockroaches that have the largest wings; the average wing length among these selected cockroaches is 10 cm. Joe interbreeds these selected cockroaches. From previous studies, he knows that the narrow-sense heritability for wing length in his population of cockroaches is 0.6.
(a) Calculate the selection differential and expected response to selection for wing length in these cockroaches.
From equation 22.21: $R = h^2 \times S$, *where* S *is the selection differential. In this case,* S = *10 cm – 4 cm = 6 cm, and we are given that the narrow-sense heritability* h^2 *is 0.6. Therefore, the response to selection* R = *0.6(6 cm) = 3.6 cm.*
(b) What should be the average wing length of the progeny of the selected cockroaches?

The average wing length of the progeny should be the mean wing length of the population plus R: *4 cm + 3.6 cm = 7.6 cm.*

39. Three characteristics in beef cattle—body weight, fat content, and tenderness—are measured and the following variance components are estimated:

	Body weight	Fat content	Tenderness
V_A	22	45	12
V_D	10	25	5
V_I	3	8	2
V_E	42	64	8
V_{GE}0	0	1	

In this population, which characteristic would respond best to selection? Explain your reasoning.
Tenderness would respond best because it has the highest narrow-sense heritability. The response to selection is given by the equation $R = h^2 \times S$, *where the narrow-sense heritability* h^2 *is equal to* V_A/V_P.

*40. A rancher determines that the average amount of wool produced by a sheep in his flock is 22 kg per year. In an attempt to increase the wool production of his flock, the rancher picks five male and five female sheep with the greatest wool production; the average amount of wool produced per sheep by those selected is 30 kg. He interbreeds these selected sheep and finds that the average wool production among the progeny of the selected sheep is 28 kg. What is the narrow-sense heritability for wool production among the sheep in the rancher's flock?
We use the equation $R = h^2 \times S$. *The value of* R *is given by the difference in the average wool production of the progeny of the selected sheep compared to the rest of the flock: 28 kg – 22 kg = 6 kg. The value of* S *is the difference between the selected sheep and the flock: 30 kg – 22 kg = 8 kg. Then* $h^2 = R/S = 6/8 = 0.75$.

41. A strawberry farmer determines that the average weight of individual strawberries produced by plants in his garden is 2 g. He selects the 10 plants that produce the largest strawberries; the average weight of strawberries among these selected plants is 6 g. He interbreeds these selected strawberry plants. The progeny of these selected plants produce strawberries that weigh 5 grams. If the farmer were to select plants that produce an average strawberry weight of 4 grams, what would be the predicted weight of strawberries produced by the progeny of these selected plants?
Here we can use the equation $R = h^2 \times S$. *R, the response to selection, is the difference between the mean of the starting population and the mean of the progeny of the selected parents. In this case,* $R = 5\text{ g} - 2\text{ g} = 3\text{ g}$. *S, the selection differential, is the difference between the mean of the starting population and the mean of the selected parents; in this case* $S = 6\text{ g} - 2\text{ g} = 4\text{ g}$. *Substituting in the equation, we get* $3\text{ g} = h^2(4\text{ g})$; $h^2 = 0.75$. *If the selected plants averaged 4 g, then S*

would be 2 g and R = 0.75(2 g) = 1.5 g. Therefore, the predicted average weight of strawberries from the progeny plants would be 2 g + 1.5 g = 3.5 g.

42. The narrow-sense heritability of wing length in a population of *Drosophila melanogaster* is .8. The narrow-sense heritability of head width in the same population is .9. The genetic correlation between wing length and head width is –.86. If a geneticist selects for increased wing length in these flies, what will happen to head width?
The head width will decrease. These two traits have high negative genetic correlation. Therefore, selection for one trait will affect the other trait inversely.

43. Pigs have been domesticated from wild boars. Would you expect to find higher heritability for weight among domestic pigs or wild boars? Explain your answer.
Wild boars will probably have higher heritability than domestic pigs. Domestic pigs, because of many generations of breeding and selection, are likely to have less variance, and more homozygosity, for genes that affect commercial traits such as weight.

CHALLENGE QUESTIONS

44. We have explored some of the difficulties in separating genetic and environmental components of human behavioral characteristics. Considering these difficulties and what you know about calculating heritability, propose an experimental design for accurately measuring the heritability of musical ability.
For the purpose of this discussion, let us assume that we have a reliable and accurate method of quantifying musical ability. I propose a study comparing musical abilities in individuals with different degrees of relatedness. I would compare two groups: one group would consist of monozygotic (identical) twins raised apart; the second group would consist of dizygotic (fraternal, or nonidentical) twins raised apart. Both groups should have comparable environmental variance, but the monozygotic twins share 100% of their genes, whereas the dizygotic twins share only 50% of their genes. By correlating the musical abilities of the two groups, we can estimate the broad-sense heritability with equation 22.20:

$$H^2 = 2(r_{MZ} - r_{DZ})$$

where r_{MZ} is the correlation coefficient of musical ability in the monozygotic group and r_{DZ} is the correlation coefficient in the dizygotic group.

45. A student who has just learned about quantitative genetics says, "Heritability estimates are worthless! They do not tell you anything about the genes that affect a characteristic. They do not provide any information about the types of offspring to expect from a cross. Heritability estimates measured in one population cannot be used for other populations, so they do not even give you any general information about how much of a characteristic is genetically determined. I cannot see that heritabilities do anything other than make undergraduate students sweat during

tests." How would you respond to this statement? Is the student correct? What good are heritabilities, and why do geneticists bother to calculate them?
One of the most valuable aspects of heritability is that it allows geneticists to predict the response to selection, either natural or artificial. In breeding plants and animals for desired QTL traits, knowing the heritability of the trait in the breeding population allows the geneticist (breeder) to make better predictions about the effectiveness of any artificial selection program.

Another important application concerns susceptibility to disease in humans (or livestock or cultivated plants). Knowing to what extent susceptibility to a particular disease is influenced by genes or by environment is essential in making public health policy decisions. If a large part of the variance is due to environmental factors, then the overall health of the population may be improved by addressing environmental improvements (e.g., reducing cigarette smoke to combat the incidence of lung cancer).

46. A geneticist selects for increased size in a population of fruit flies that she is raising in her laboratory. She starts with the two largest males and the two largest females and uses them as the parents for the next generation. From the progeny produced by these selected parents, she selects the two largest males and the two largest females and mates them. She repeats this procedure each generation. The average weight of flies in the initial population was 1.1 mg. The flies respond to selection, and their body size steadily increases. After 20 generations of selection, the average weight is 2.3 mg. However, after about 20 generations, the response to selection in subsequent generations levels off, and the average size of the flies no longer increases. At this point, the geneticist takes a long vacation; while she is gone, the fruit flies in her population interbreed randomly. When she returns from vacation, she finds that the average size of the flies in the population has decreased to 2.0 mg.
(a) Provide an explanation for why the response to selection leveled off after 20 generations.
The response most likely leveled off because of an opposing natural selection. Too large a body mass may have led to reduced viability or fertility in these flies.
(b) Why did the average size of the fruit flies decrease when selection was no longer applied during the geneticist's vacation?
The genetic variance of the population with regard to body mass had not been depleted; these flies were not all homozygous for the QTL loci. Therefore, in the absence of any selection, the average body mass of the flies would show random drift. Natural selection against the highest body mass would cause a reduction in average body mass in the absence of artificial selection for higher body mass.

47. Manic-depressive illness is a psychiatric disorder that has a strong hereditary basis, but the exact mode of inheritance is not known. Previous research has shown that siblings of patients with manic-depressive illness are more likely also to develop the disorder than are siblings of unaffected individuals. A recent study demonstrated that the ratio of manic-depressive brothers to manic-depressive sisters is higher when the patient is male than when the patient is female. In other words, relatively

more brothers of manic-depressive patients also have the disease when the patient is male than when the patient is female. What does this new observation suggest about the inheritance of manic-depressive illness?

These observations suggest that an X-linked locus or loci may influence manic-depressive illness. Males inherit their X-chromosome genes only from their mother. Females inherit X-chromosome genes from both parents. Therefore, the brothers of an affected male inherited their X chromosome alleles from the same parent, the mother. On the other hand, an affected female may have inherited a contributory X-linked QTL locus allele from either parent; if this allele came from her father, there is no chance that her brothers inherited the same X-linked allele.

Chapter Twenty-Three: Population and Evolutionary Genetics

COMPREHENSION QUESTIONS

1. What is a Mendelian population? How is the gene pool of a Mendelian population usually described? What are the predictions given by the Hardy-Weinberg law?
A Mendelian population is a group of sexually reproducing individuals mating with each other and sharing a common gene pool. The gene pool is usually described by genotype frequencies and allele frequencies. The Hardy-Weinberg law states that a large population mating randomly with no effects from selection, migration, or mutation will have the following relationship between the genotype frequencies and allele frequencies:
$f(\text{AA}) = p^2$; $f(\text{Aa}) = 2\text{pq}$; $f(\text{aa}) = \text{q}^2$, *where* p *and* q *equal the allelic frequencies*
Moreover, the allele frequencies do not change from generation to generation, as long as the above conditions hold.

*2. What assumptions must be met for a population to be in Hardy-Weinberg equilibrium?
Large population, random mating, and not affected by migration, selection, or mutation.

3. What is random mating?
Random mating takes place when each genotype mates according to its frequency, so that the frequency of an AA *individual mating with another* AA *individual will be* $f(\text{AA})^2$, *the frequency of* AA *mating with* Aa *will be* $2f(\text{AA})f(\text{Aa})$, *the mating of* Aa *with* Aa *will be* $f(Aa)^2$, *and so forth.*

*4. Give the Hardy-Weinberg expected genotypic frequencies for (a) an autosomal locus with three alleles and (b) an X-linked locus with two alleles.
(a) If the frequencies of alleles A1, A2, *and* A3 *are defined as* p, q, *and* r, *respectively:*
$f(\text{A1A1}) = \text{p}^2$
$f(\text{A1A2}) = 2\text{pq}$
$f(\text{A2A2}) = \text{q}^2$
$f(\text{A1A3}) = 2\text{pr}$
$f(\text{A2A3}) = 2\text{qr}$
$f(\text{A3A3}) = \text{r}^2$

(b) For an X-linked locus with two alleles:
$f(\text{X}^1\text{X}^1) = \text{p}^2$ *among females;* $\text{p}^2/2$ *for the whole population*
$f(\text{X}^1\text{X}^2) = 2\text{pq}$ *among females;* pq *for the whole population*
$f(\text{X}^2\text{X}^2) = \text{q}^2$ *among females;* $\text{q}^2/2$ *for the whole population*
$f(\text{X}^1\text{Y}) = \text{p}$ *among males;* p/2 *for the whole population*
$f(\text{X}^2\text{Y}) = \text{q}$ *among males;* q/2 *for the whole population*

5. Define inbreeding and briefly describe its effects on a population.
Inbreeding is preferential mating between genetically related individuals. Inbreeding increases homozygosity and reduces heterozygosity in the population.

6. What determines the allelic frequencies at mutational equilibrium?
At mutational equilibrium, the allelic frequencies are determined by the forward and reverse mutation rates.

*7. What factors affect the magnitude of change in allelic frequencies due to migration?
The proportion of the population due to migrants (m) and the difference in allelic frequencies between the migrant population and the original resident population.

8. Define genetic drift and give three ways that it can arise. What effect does genetic drift have on a population?
Genetic drift is change in allelic frequencies resulting from sampling error. It may arise through a long-term limitation on population size, founder effect that occurs when the population is founded by a small number of individuals, or a bottleneck effect when the population undergoes a drastic reduction in population size. Genetic drift causes changes in allelic frequencies and loss of genetic variation because some alleles are lost as other alleles become fixed. It also causes genetic divergence between populations because the different populations undergo different changes in allelic frequencies and become fixed for different alleles.

*9. What is effective population size? How does it affect the amount of genetic drift?
The effective population size N_E *differs from the actual population size if the sex ratio is disproportionate, or if a few dominant individuals of either sex contribute disproportionately to the gene pool of the next generation. When the sex ratio differs from 1:1, the effective population size can be calculated as follows:* $N_E = 4 \times$ *number of males × number of females/(number of males + number of females).*

The smaller the effective population size, the greater the magnitude of the genetic drift.

10. Define natural selection and fitness.
Natural selection is the differential reproduction of genotypes. Fitness is the relative reproductive success of a genotype.

11. Briefly discuss the differences between directional selection, overdominance, and underdominance. Describe the effect of each type of selection on the allelic frequencies of a population.
Directional selection is when one allele has greater fitness than another. Overdominance is when the heterozygote has greater fitness than either of the homozygotes. Underdominance is when the heterozygote has less fitness than either of the homozygotes. Directional selection will cause the allele with greater fitness to increase in frequency and eventually reach fixation. Overdominance establishes a balanced equilibrium that maintains both alleles. Underdominance results in an

unstable equilibrium that will degenerate once disturbed and move away from equilibrium until one allele is fixed.

12. What factors affect the rate of change in allelic frequency due to natural selection?
The intensity of selection (the difference in fitness between the different genotypes) and the dominance relationships affect the rate of change. The rate of change also depends on the allelic frequency: the rate of decline of a deleterious allele will be proportional to the allelic frequency.

*13. Compare and contrast the effects of mutation, migration, genetic drift, and natural selection on genetic variation within populations and on genetic divergence between populations.
Mutation increases genetic variation within populations and increases divergence between populations because different mutations arise in each population.
Migration increases genetic variation within a population by introducing new alleles, but decreases divergence between populations.
Genetic drift decreases genetic variation within populations because it causes alleles eventually to become fixed, but it increases divergence between populations because drift takes place differently in each population.
Natural selection may either increase or decrease genetic variation, depending on whether the selection is directional or balanced. It may increase or decrease divergence between populations, depending on whether different populations have similar or different selection pressures.

14. Give some of the advantages of using molecular data in evolutionary studies.
DNA and protein sequences are genetic; they can be studied and compared in all organisms; genomes provide a vast and growing amount of data in public databases; the molecular differences are quantifiable; comparisons of DNA and protein sequences provide information about evolutionary history.

*15. What is the key difference between the neutral-mutation hypothesis and the balance hypothesis?
The neutral-mutation hypothesis proposes that most molecular variation is adaptively neutral. The balance hypothesis proposes that most genetic variation is maintained by balanced selection, favoring heterozygosity at most loci.

16. Discuss some of the methods that have been used to study variation in DNA.
Three different types of DNA variation have been studied. (1) Restriction site variation has been analyzed by restriction endonuclease digestion of PCR-amplified fragments of DNA, and analysis of resulting DNA fragments by gel electrophoresis. (2) Microsatellite length variation has also been analyzed, most frequently by PCR amplification of the microsatellite locus and gel electrophoresis to separate the PCR products of different lengths. (3) Variation in DNA sequences, such as base substitutions that do not result in changes in either restriction fragment length or PCR product length, may be characterized by sequencing of PCR-amplified DNA from the locus in question.

17. Draw a simple phylogenetic tree and label the following: node, branch, and outgroup.

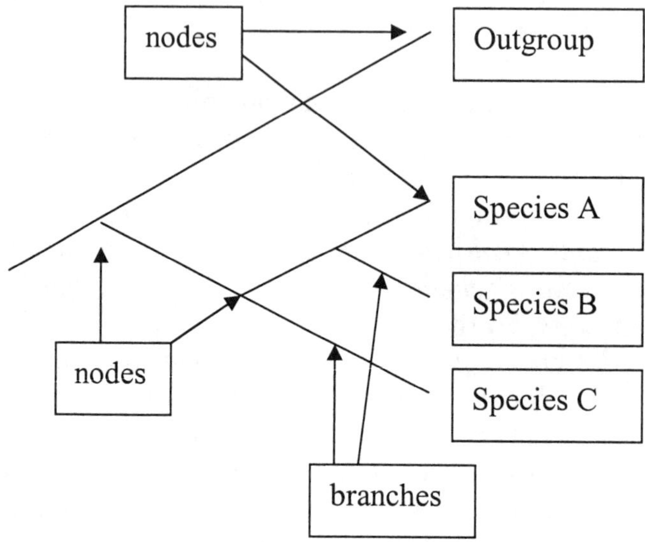

18. Briefly describe the difference between the distance approach and the parsimony approach to reconstruction of phylogenetic trees.
Distance approaches rely on overall similarity between species or gene sequences; most similar species are grouped together. Parsimony approaches try to reconstruct an evolutionary pathway that requires the minimum number of changes to arrive at the modern species or genes from a common ancestor.

19. Outline the different rates of evolution that are typically seen in different parts of a protein-encoding gene. What might account for these differences?
Nucleotide substitutions that do not change the amino acid sequence occur far more frequently than those that do change the amino acid sequence. Changes also occur more rapidly in those parts of the protein sequence that are not essential for the function of the protein. Nucleotide changes occur most rapidly in intron sequences, except for those positions that affect splicing. Selective pressure to maintain the function of proteins essential or advantageous for the survival and reproduction of the organism accounts for the differential rates of mutation and evolution.

*20. What is the molecular clock?
The molecular clock is idea that the rate at which nucleotide changes take place in a DNA sequence is relatively constant and therefore the number of nucleotide substitutions that have taken place between two organisms can be used to estimate the time since they shared a common ancestor.

21. What is exon shuffling? How can it lead to evolution of new genes?
Exon shuffling is a hypothesis based on the observation that exons often correspond to functional domains of proteins. New genes with novel functions may arise through recombination events that bring together different exons encoding different functional modules of proteins.

22. What is a multigene family? What processes produce multigene families?
Multigene families are a group of genes in the same genome that are related by descent from a common ancestral gene. Multigene families arise through gene duplication events with subsequent diversification. Some members acquire mutations that render them nonfunctional and become pseudogenes.

23. Define horizontal gene transfer and list some of the processes that may cause it.
Horizontal gene transfer, also called lateral gene transfer, is transmission of genetic information across species boundaries. Horizontal gene transfer occurs frequently in bacteria, through transformation and phage-mediated transduction. In eukaryotes, horizontal gene transfer may occur through endocytosis events (e.g., mitochondrial and chloroplast genes), viruses, parasites, and by human intervention (genetic engineering).

APPLICATION QUESTIONS AND PROBLEMS

24. How would you respond to someone who said that models are useless in studying population genetics because they represent oversimplifications of the real world?
While it is important to keep in mind that models do represent simplifications, they nevertheless provide important and valuable information and predictions about the effects of size, mutation, migration, inbreeding, and selection on the gene pool of the population.

*25. Voles (*Microtus ochrogaster*) were trapped in old fields in southern Indiana and were genotyped for a transferrin locus. The following numbers of genotypes were recorded.

T^ET^E	T^ET^F	T^FT^F
407	170	17

Calculate the genotypic and allelic frequencies of the transferrin locus for this population.
The total number of voles is 594.
$f(T^ET^E) = 407/594 = .685$
$f(T^ET^F) = 170/594 = .286$
$f(T^FT^F) = 17/594 = .029$
$f(T^E) = f(T^ET^E) + f(T^ET^F)/2 = .685 + .143 = .828$
$f(T^F) = f(T^FT^F) + f(T^ET^F)/2 = .029 + .143 = .172$

26. Orange coat color in cats is due to an X-linked allele (X^o) that is codominant to the allele for black (X^+). Genotypes of the orange locus of cats in Minneapolis and St. Paul, Minnesota, were determined and the following data were obtained.

X^oX^o females	11
X^oX^+ females	70
X^+X^+ females	94
X^oY males	36
X^+Y males	112

Calculate the frequencies of the X^o and X^+ alleles for this population.

We add up all the X^o or X^+ alleles and divide by the total number of X^o and X^+ alleles.

The number of X^o alleles = $2(X^oX^o) + (X^oX^+) + (X^oY) = 22 + 70 + 36 = 128$

The number of X^+ alleles = $2(X^+X^+) + (X^oX^+) + (X^+Y) = 188 + 70 + 112 = 370$

$f(X^o) = 128/(128 + 370) = 128/498 = .26$

$f(X^+) = 370/498 = .74$

27. A total of 6129 North American Caucasians were blood typed for the MN locus, which is determined by two codominant alleles, L^M and L^N. The following data were obtained:

Blood type	Number
M	1787
MN	3039
N	1303

Carry out a chi-square test to determine whether this population is in Hardy-Weinberg equilibrium at the MN locus.

The total number of individuals is 6129.

$f(L^M) = p = (1787 + 3039/2)/6129 = .54$

$f(L^N) = q = (1303 + 3039/2)/6129 = .46$

A population in Hardy-Weinberg equilibrium should have the following genotype frequencies:

$f(L^ML^M) = p^2 = (.54)^2 = 0.29 = 1777/6129$

$f(L^ML^N) = 2pq = 2(.54)(.46) = .50 = 3045/6129$

$f(L^NL^N) = q^2 = (.46)^2 = 0.21 = 1287/6129$

Now we set up a chi-square test:

Blood type	*Observed*	*Expected*	O–E	$(O–E)^2$	$(O–E)^2/E$
M	*1787*	*1777*	*10*	*100*	.056
MN	*3039*	*3045*	*6*	*36*	.012
N	*1303*	*1287*	*16*	*256*	.20

Chi-squared = $\Sigma(O–E)^2/E = .268$

The number of degrees of freedom is the number of genotypes minus the number of alleles = 3 − 2 = 1.

Looking at a chi-square table, we see the p *value is easily greater than .05, so we do not reject the hypothesis that this population is in Hardy-Weinberg equilibrium with respect to the MN locus.*

28. Genotypes of leopard frogs from a population in central Kansas were determined for a locus that encodes the enzyme malate dehydrogenase. The following numbers of genotypes were observed:

Genotype	Number
$M^1 M^1$	20
$M^1 M^2$	45
$M^2 M^2$	42
$M^1 M^3$	4
$M^2 M^3$	8
$M^3 M^3$	6
Total	125

(a) Calculate the genotypic and allelic frequencies for this population.

$f(M^1M^1) = 20/125 = .16$
$f(M^1M^2) = 45/125 = .36$
$f(M^2M^2) = 42/125 = .34$
$f(M^1M^3) = 4/125 = .032$
$f(M^2M^3) = 8/125 = .064$
$f(M^3M^3) = 6/125 = .048$
$f(M^1) = p = .16 + .36/2 + .032/2 = .16 + .18 + .016 = .356$
$f(M^2) = q = .34 + .36/2 + .064/2 = .34 + .18 + .032 = .552$
$f(M^3) = r = .048 + .032/2 + .064/2 = .048 + .016 + .032 = .096$

(b) What would be the expected numbers of genotypes if the population were in Hardy-Weinberg equilibrium?

For population in Hardy-Weinberg equilibrium:
$f(M^1M^1) = p^2 = (.356)^2 = .127; .127(125) = 16$
$f(M^1M^2) = 2pq = 2(.356)(.552) = .393; .393(125) = 49$
$f(M^2M^2) = q^2 = (.552)^2 = .305; .305(125) = 38$
$f(M^1M^3) = 2pr = 2(.356)(.096) = .068; .068(125) = 8$
$f(M^2M^3) = 2qr = 2(.552)(.096) = .106; .106(125) = 13$
$f(M^3M^3) = r^2 = (.096)^2 = .009; .009(125) = 1$

Genotype	*Observed*	*Expected*	O–E	$(O–E)^2$	$(O–E)^2/E$
M^1M^1	*20*	*16*	*4*	*16*	*1*
M^1M^2	*45*	*49*	*–4*	*16*	*.33*
M^2M^2	*42*	*38*	*4*	*16*	*.42*
M^1M^3	*4*	*8*	*–4*	*16*	*2*
M^2M^3	*8*	*13*	*–5*	*25*	*1.9*
M^3M^3	*6*	*1*	*5*	*25*	*25*

Chi-squared = 30.65
Degrees of freedom = # genotypes – # alleles = 6 – 3 = 3

The p *value is much lower than 0.05; this population is not in Hardy-Weinberg equilibrium for this locus.*

29. Full color (*D*) in domestic cats is dominant over dilute color (*d*). Of 325 cats observed, 194 have full color and 131 have dilute color.
    (a) If these cats are in Hardy-Weinberg equilibrium for the dilution locus, what is the frequency of the dilute allele?
    f(dilute) = *f*(dd) = q^2 = *131/325* = *.403;* q = *.635*
    (b) How many of the 194 cats with full color are likely to be heterozygous?
    If q = *f*(d) = *.635, then* p = *1* – q = *.365*
    f(Dd) = 2pq = *2(.365)(.635)* = *.464; .464(325)* = *151 heterozygous cats*

30. Tay-Sachs disease is an autosomal recessive disorder. Among Ashkenazi Jews, the frequency of Tay-Sachs disease is 1 in 3600. If the Ashkenazi population is mating randomly for the Tay-Sachs gene, what proportion of the population consists of heterozygous carriers for the Tay-Sachs allele?
    If q = *the frequency of the Tay-Sachs allele, then* q^2 = *1/3600;* q = *1/60* = *.017*
    The frequency of the normal allele = p = *1* – q = *.983*
    The frequency of heterozygous carriers = 2pq = *2(.983)(.017)* = *.033; approximately 1 in 30 are carriers.*

31. In the plant *Lotus corniculatus,* cyanogenic glycoside protects the plants against insect pests and even grazing by cattle. This glycoside is due to a simple dominant allele. A population of *L. corniculatus* consists of 77 plants that possess cyanogenic glycoside and 56 that lack the compound. What is the frequency of the dominant allele that results in the presence of cyanogenic glycoside in this population?
    The frequency of the recessive allele = q; *if the population is in Hardy-Weinberg equilibrium (or if the population went through a round of random mating), then the frequency of homozygous recessives* = q^2 = *56/(77+56)* = *56/133* = *.42;* q = *.65. Then* p = *.35* = *the frequency of the dominant allele.*

*32. Colorblindness in humans is an X-linked recessive trait. Approximately 10% of the men in a particular population are colorblind.
    (a) If mating is random for the color-blind locus, what is the frequency of the color-blind allele in this population?
    For males, f(color blind) = p = *.1.*
    (b) What proportion of the women in this population is expected to be colorblind?
    For females, f(homozygotes) = p^2 = *.01, or 1%.*
    (c) What proportion of the women in the population is expected to be heterozygous carriers of the color-blind allele?
    For females, f(heterozygotes) = 2pq = *2(.1)(.9)* = *.18, or 18%.*

*33. The human MN blood type is determined by two codominant alleles, L^M and L^N. The frequency of L^M in Eskimos on a small Arctic island is .80. If the inbreeding coefficient for this population is .05, what are the expected frequencies of the M, MN, and N blood types on the island?

If $f(L^M) = p = .80$*; then* $f(L^N) = q = 1 - p = .20$.
$f(L^ML^M) = p^2 + Fpq = (.80)^2 + .05(.80)(.20) = .64 + .008 = .648$
$f(L^ML^N) = 2pq - 2Fpq = 2(.80)(.20) - 2(.05)(.80)(.20) = .32 - .016 = 0.304$
$f(L^NL^N) = q^2 + Fpq = (.20)^2 + .05(.80)(.20) = .04 + .008 = .048$

34. Demonstrate mathematically that full sib mating ($F = ¼$) reduces the heterozygosity by ¼ with each generation.
For a randomly mating population, the heterozygote frequency = 2pq;
for inbreeding populations, the heterozygote frequency = 2pq – 2Fpq.
If F = ¼, *then the heterozygote frequency* = 2pq – 2(¼)pq = 1.5pq.
The reduction in heterozygote frequency is ½pq, *or* ¼ *of the randomly mating heterozygote frequency of* 2pq.

35. The forward mutation rate for piebald spotting in guinea pigs is 8×10^{-5}; the reverse mutation rate is 2×10^{-6}. Assuming that no other evolutionary forces are present, what is the expected frequency of the allele for piebald spotting in a population that is in mutational equilibrium?
Here we use equation 23.15: $\hat{q} = \frac{\mu}{\mu + \upsilon}$; *where* $\mu = 8 \times 10^{-5}$ *and* $\upsilon = 2 \times 10^{-6}$*; the frequency at equilibrium is then* $8 \times 10^{-5}/(8 \times 10^{-5} + 2 \times 10^{-6}) = 8/8.2 = .98$.

*36. In German cockroaches, curved wing (*cv*) is recessive to normal wing (cv^+). Bill, who is raising cockroaches in his dorm room, finds that the frequency of the gene for curved wings in his cockroach population is .6. In the apartment of his friend Joe, the frequency of the gene for curved wings is .2. One day Joe visits Bill in his dorm room, and several cockroaches jump out of Joe's hair and join the population in Bill's room. Bill estimates that 10% of the cockroaches in his dorm room now consists of individual roaches that jumped out of Joe's hair. What will be the new frequency of curved wings among cockroaches in Bill's room?
The proportion of migrants in the new population = m = .1. *The frequency of the allele for curved wings in the old population* = q_{old} = .6; *the allele frequency in the migrant population is* $q_{migrants}$ = .2:
$q_{new} = mq_{migrants} + (1 - m)q_{old} = .1(.2) + .9(.6) = .56$
Now that we have calculated the allelic frequencies, we can calculate the new genotype frequency assuming random mating:
$f_{new}(cv,cv) = q_{new}^2 = (.56)^2 = .31$

37. A population of water snakes is found on an island in Lake Erie. Some of the snakes are banded and some are unbanded; banding is caused by an autosomal allele that is recessive to an allele for no bands. The frequency of banded snakes on the island is .4, whereas the frequency of banded snakes on the mainland is .81. One summer, a large number of snakes migrate from the mainland to the island. After this migration, 20% of the island population consists of snakes that came from the mainland.

(a) Assuming that the mainland population and the island population are in Hardy-Weinberg equilibrium for the alleles that affect banding, what is the frequency of the allele for bands on the island and on the mainland before migration?
Because banding is recessive, the frequency of banded snakes = q^2.
On the island before the migration, $q^2 = .04$; $q = .2$.
On the mainland, $q^2 = .81$; $q = .9$.

(b) After migration has taken place, what will be the frequency of the banded allele on the island?
After the migration, $q_{new} = mq_{migrants} + (1 - m)q_{old} = .2(.9) + .8(.2) = .34$

*38. Calculate the effective size of a population with the following numbers of reproductive adults:

(a) 20 males and 20 females
$N_e = 4(20)(20)/(20+20) = 40$

(b) 30 males and 10 females
$N_e = 4(30)(10)/(30+10) = 30$

(c) 10 males and 30 females
$N_e = 4(10)(30)/(10+30) = 30$

(d) 2 males and 38 females
$N_e = 4(2)(38)/(2+38) = 7.6$

39. Pikas are small mammals that live at high elevation in the talus slopes of mountains. Populations located on mountain tops in Colorado and Montana in North America are relatively isolated from one another because the pikas do not occupy the low-elevation habitats that separate the mountain tops and do not venture far from the talus slopes. Thus, there is little gene flow between populations. Furthermore, each population is small in size and was founded by a small number of pikas.
A group of population geneticists propose to study the amount of genetic variation in a series of pika populations and to compare the allelic frequencies in different populations. On the basis of biology and the distribution of pikas, what do you predict the population geneticists will find concerning the within- and between-population genetic variation?
The small population sizes and the founder effects would cause strong effects from genetic drift. The geneticists will find large variation between populations in allele frequencies. Within populations, the same factors coupled with inbreeding will cause loss of genetic variation and a high degree of homozygosity.

40. In a large, randomly mating population, the frequency of the allele (*s*) for sickle-cell hemoglobin is .028. The results of studies have shown that people with the following genotypes at the beta-chain locus produce the average numbers of offspring given:

Genotypes	Average number of offspring produced
SS	5
Ss	6
ss	0

(a) What will be the frequency of the sickle-cell allele (*s*) in the next generation?

f(s) = q = *.028;* f(S) = p = *1* – q = *.972*

The current genotype frequencies are:

$f(\text{SS}) = p^2 = .945$

$f(\text{Ss}) = 2pq = .054$

$f(\text{ss}) = q^2 = .001$

$W_{SS} = 5/6 = .83;\ s_{SS} = .17$

$W_{Ss} = 6/6 = 1.0;\ s_{Ss} = 0$

$W_{ss} = 0/6 = 0;\ s_{ss} = 1.0$

$\overline{W} = p^2W_{SS} + 2pqW_{Ss} + q^2W_{ss} = .79 + .054 + 0 = .84$

For the next generation, each genotype will have the following reproductive contribution:

$$\frac{p^2W_{SS}}{\overline{W}},\ \frac{2pqW_{Ss}}{\overline{W}},\ \textit{and}\ \frac{q^2W_{ss}}{\overline{W}}\ \textit{for}\ \text{SS, Ss,}\ \textit{and}\ \text{ss,}\ \textit{respectively.}$$

f(SS) = *.94*

f(Ss) = *.064*

f(ss) = *0*

The new allele frequency is then: q' = *.064/2* = *.032.*

(b) What will be the frequency of the sickle-cell allele at equilibrium?

We use the equilibrium equation for overdominance:

$$\hat{q} = f(s) = \frac{s_{SS}}{s_{SS} + s_{ss}} = .17/(.17 + 1.0) = .17/1.17 = .145$$

41. Two chromosomal inversions are commonly found in populations of *Drosophila pseudoobscura:* Standard (*ST*) and Arrowhead (*AR*). When treated with the insecticide DDT, the genotypes for these inversions exhibit overdominance, with the following fitnesses:

Genotype	Fitness
ST/*ST*	.47
ST/*AR*	1.0
AR/*AR*	.62

What will be the frequency of *ST* and *AR* after equilibrium has been reached?

$s_{ST/ST} = .53$

$s_{ST/AR} = 0$

$s_{AR/AR} = .38$

$\hat{q} = f(\text{AR}) = .53/(.53 + .38) = .53/.91 = .58$

*42. In a large, randomly mating population, the frequency of an autosomal recessive lethal allele is .20. What will be the frequency of this allele in the next generation?

If q = *.2, then* p = *.8. After one round of random mating, the genotype frequencies in the next generation will be:*

$f(\text{AA}) = \text{p}^2 = .64$
$f(\text{Aa}) = 2\text{pq} = .32$
$f(\text{aa}) = \text{q}^2 = .04$, *but these die.*
Adjusting for the death of all the homozygotes for the recessive lethal allele, the new genotype frequencies f′ in the survivors are:
$f'(\text{AA}) = f(\text{AA})/[f(\text{AA}) + f(\text{Aa})] = .64/(.64 + .32) = .64/.96 = 2/3 = .67$
$f'(\text{Aa}) = f(\text{Aa})/[f(\text{AA}) + f(\text{Aa})] = .32/.96 = 1/3 = .33$
Here we are simply taking the frequencies of the AA *and* Aa *genotypes and dividing each by the total survivors, those with* AA *and* Aa *genotypes.*
Then the new allelic frequency $q' = f'(\text{aa}) + f'(\text{Aa})/2 = 0 + 1/6 = .17$; *note that because the genotype* aa *is lethal,* $f'(\text{aa}) = 0$.

43. A certain form of congenital glaucoma results from an autosomal recessive allele. Assume that the mutation rate is 10^{-5} and that persons having this condition produce, on average, about 80% of the offspring produced by persons who do not have glaucoma.
    (a) At equilibrium between mutation and selection, what will be the frequency of the gene for congenital glaucoma?
    The fitness of individuals with glaucoma is .8; the selection coefficient is .2.
    For a recessive allele, $\hat{q} = \sqrt{\frac{u}{s}} = \sqrt{(10^{-5}/.2)} = .0071$
    (b) What will be the frequency of the disease in a randomly mating population that is at equilibrium?
    Frequency of homozygotes $= \text{q}^2 = 5 \times 10^{-5}$

CHALLENGE QUESTION

44. The Barton Springs salamander is an endangered species found only in a single spring in the city of Austin, Texas. There is growing concern that a chemical spill on a nearby freeway could pollute the spring and wipe out the species. To provide a source of salamanders to repopulate the spring in the event of such a catastrophe, a proposal has been made to establish a captive breeding population of the salamander in a local zoo. You are asked to provide a plan for the establishment of this captive breeding population, with the goal of maintaining as much of the genetic variation of the species as possible in the captive population. What factors might cause loss of genetic variation in the establishment of the captive population? How could loss of such variation be prevented? Assuming that it is feasible to maintain only a limited number of salamanders in captivity, what procedures should be instituted to ensure the long-term maintenance of as much of the variation as possible?
    Genetic variation in the zoo salamander colony could be reduced because of a founder effect from the limited number of individuals used to establish a breeding colony. Genetic variation would be reduced further by inbreeding and genetic drift. Given that only a limited number of salamanders can be maintained in the zoo colony, regular introduction of wild salamanders from the spring into the colony will keep mixing in fresh genotypes. A continual influx of migrants from the spring

will over time effectively increase the number of individuals sampled, keep the gene pool of the zoo colony close to the gene pool of the spring, and mitigate inbreeding. It is also important to maintain an 50:50 sex ratio, as deviations from 50:50 causes a reduction in the effective number of adults. Matings between the zoo colony should be carefully planned to avoid inbreeding.

CASE STUDY QUESTIONS

1. The frequency of PKU in Iceland is 0.0001. (a) Assuming that this population is in Hardy-Weinberg equilibrium at the *PAH* locus, calculate the frequency of mutant and normal alleles in this population. (b) What is the expected frequency of heterozygote carriers of PKU alleles in this population?
   (a) For populations in equilibrium, the frequency of homozygous individuals is the square of the allele frequency: $0.0001 = q^2$. Therefore, the allele frequency is 0.01.
   (b) The equilibrium frequency of heterozygous individuals is 2pq. The carrier frequency is therefore 2(0.99)(0.01) = 0.02, or about 2% of the population.

2. Three copies of the *A300S* mutant PKU allele are found in a particular population. All three copies are found on different haplotypes. Are these copies likely to be identical by descent or due to different mutational events? Explain your answer.
   These copies more probably arose from different mutational events. If the copies were identical by descent from the same mutational event, closely linked markers would be co-inherited with the mutation, because the recombination between closely linked markers is rare. Thus copies of mutations that are identical by descent should have the same haplotypes. Since these mutations occur on different haplotypes, they likely arose from different mutational events.

3. The mutation rate for a particular PKU allele is 0.000004. If no other evolutionary forces are acting other than mutation and selection, what is the expected equilibrium frequency of this allele?
   The expected frequency of a recessive allele at equilibrium between mutation and selection is given by the equation: $\hat{q} = \sqrt{\frac{\mu}{s}}$

   Assuming that the fitness of PKU homozygous individuals is zero, then s = 1. Thus the equilibrium frequency is the square root of the mutation rate, or 0.002.

4. What is overdominance? How might overdominance explain the relatively high frequency of PKU alleles found in certain human populations?
   Overdominance is the situation where heterozygous individuals have a higher fitness than homozygous individuals for either the mutant or normal alleles. At low allele frequencies, most of the copies of the PKU alleles are present in heterozygous individuals. A selective advantage for heterozygotes would more than offset the selection against the homozygous recessive PKU genotype, and therefore cause an increase in the PKU allele frequency. The PKU alleles will be maintained in the population in a stable equilibrium, where the equilibrium allele frequencies are determined by the relative fitness of the two homozygotes: $q = s_{11}/(s_{11} + s_{22})$.

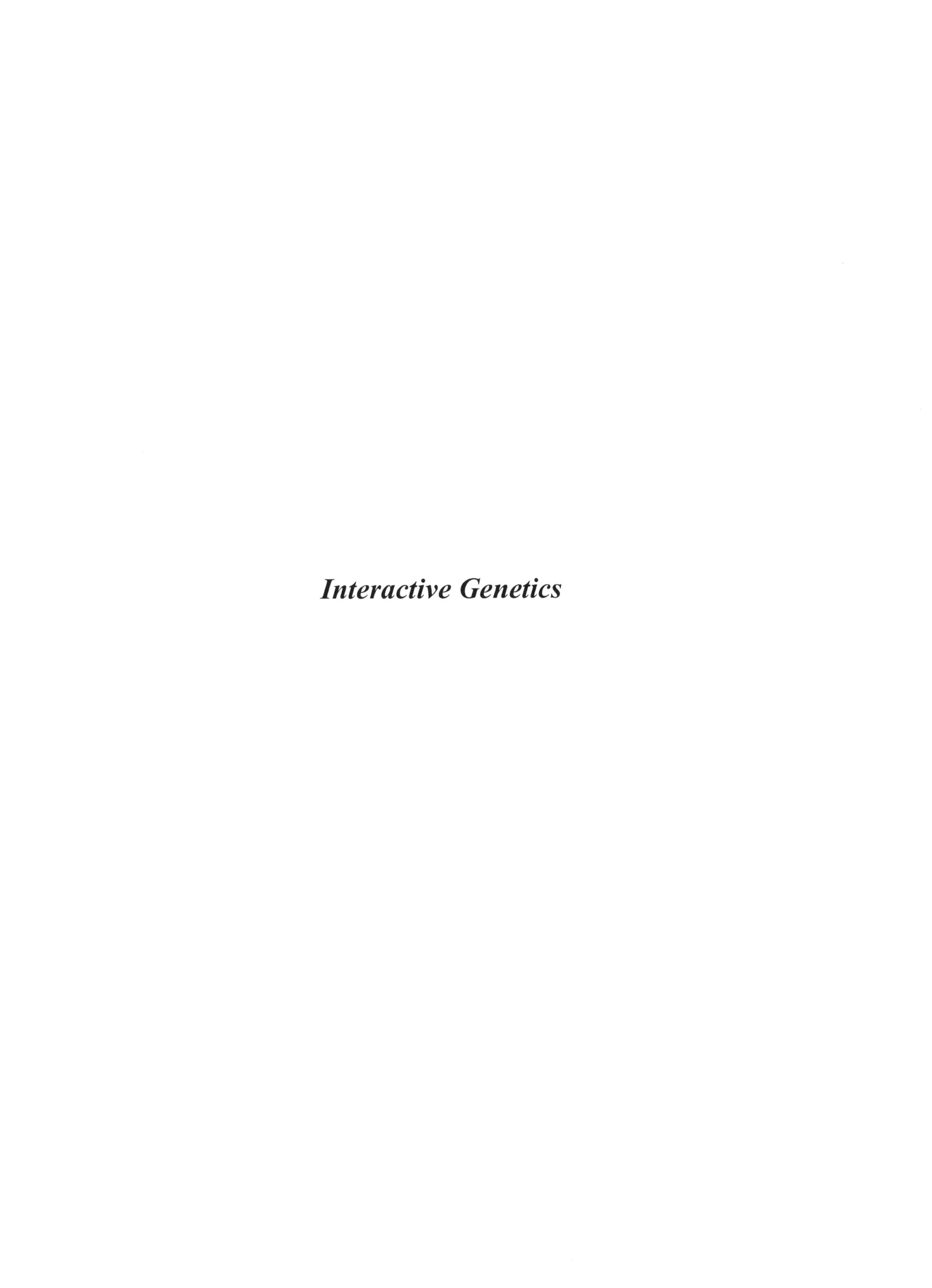
Interactive Genetics

Mendelian Analysis

Goals for Mendelian Analysis:

1. Describe the mode of inheritance of a phenotypic difference between two strains:
   - Distinguish dominance and recessiveness in traits
   - Determine whether the phenotype difference is due to a single gene with two alleles
   - Write genotypes and predict phenotypes
   - Predict the number of possible gamete types, their kinds and ratios
2. Distinguish self-crosses from test crosses.
3. Apply these concepts to simple human pedigrees for select traits:
   - Calculate probabilities using product and sum rules
   - Use the binomial expansion to calculate the combinations of possible outcomes

Mendelian Problems

Problem 1

Let's begin with a simple cross between two pure breeding peas, a purple flowered pea and a white flowered pea. How many traits or characters are different between the purple and white parents?

In this cross the pollen from the purple parent is used to fertilize the ovules of the white parent. How many genetically distinct gametes are produced from the purple parent?

How many genetically distinct gametes does the white flowered parent produce?

All the progeny of the parental cross are purple flowered pea plants. However, when these peas are self-crossed, both purple and white peas appear in the next or F_2 generation. Which phenotype is the dominant phenotype?

Using *A* and *a* to designate the different forms of the inherited trait (alleles), assign genotypes to the three generations of the cross described above. How many genetically distinct gametes are produced from the F_1? In the simple Punnett square shown below, fill in the genotype of the progeny using *A/a*.

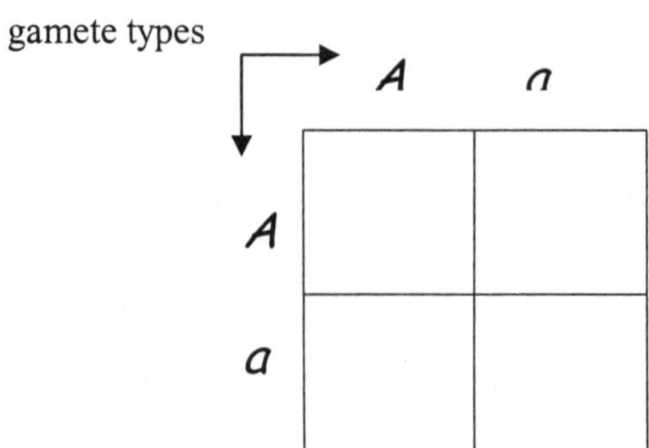

Indicate the phenotype expected for each genotype in the Punnett square above. What is the ratio of purple flowered progeny to white flowered progeny in the F_2?

What genotypic ratios would you predict?

Problem 2

The grass tree is the common name for the Australian genus *Xanthorrhoea* in the lily family, *Liliaceae*. Grass trees grow to a height of about 4.5m (15 feet) and bear long, narrow leaves in a tuft at the top of the trunk. White flowers or yellow flowers can be produced in a dense spike above the leaves in a given tree. A tree producing yellow flowers was self-crossed. The seeds were collected and planted to determine flower color of each progeny tree. Twenty-eight trees were found to produce yellow flowers and eight trees were found to produce white flowers.

What is the expected genetic (phenotypic) ratio of yellow flowering grass trees to white flowering grass trees among the progeny?

Which phenotype is dominant, white or yellow flowering trees?

Using the symbols of *Y/y*, write the genotypes of the original yellow flowering tree and the progeny.

Yellow × Yellow

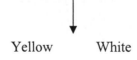

↓

Yellow White

Predict the phenotypic ratios for the crosses shown below:

Cross	Phenotypic Ratios	
	Yellow	White
White tree × white tree		
Original yellow tree × white tree		
Pure breeding yellow tree × white tree		
Original yellow tree self-crossed		

Problem 3

Two pure breeding strains of peas, one giving wrinkled, yellow seeds and the other round, green seeds, were crossed and all the resulting peas were round and yellow. These round seeds were then planted and the flowers self-fertilized. The peas produced from this selfing contained four phenotypic classes.

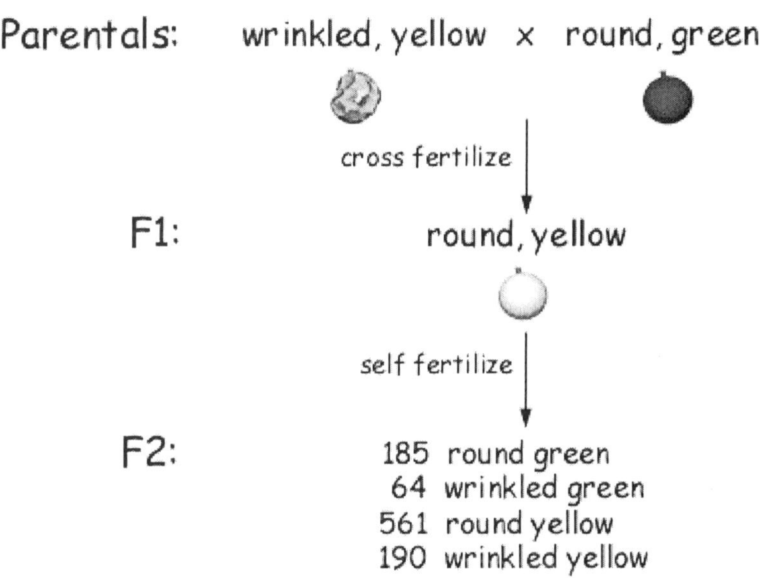

How many traits or characters are different between the pure breeding parents?

Using *A* and *a* to designate pea shape and *B* and *b* to designate pea color, assign genotypes to the three generations in the diagram of the cross above.

How many genetically distinct gametes are produced from the F_1?
Fill out the Punnett square shown below.

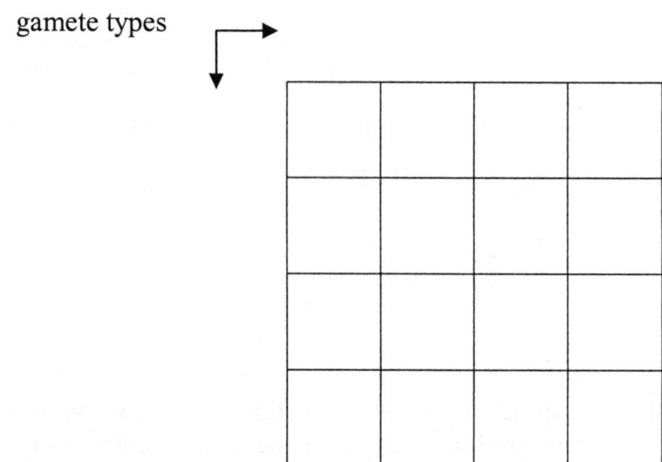

How many genotypes above correspond to the round and yellow phenotype?

How many genotypes above correspond to the wrinkled yellow phenotype?

How many genotypes above correspond to the round green phenotype?

How many genotypes above give rise to a green wrinkled pea?

Using the Punnett square above, what is the ratio of round peas to wrinkled peas (ignore the color)?

What is the ratio for yellow versus green?

Determine the ratio of round yellow peas to round green peas to wrinkled yellow peas to wrinkled green peas expected from this dihybrid cross.

Problem 4

You have been sent to Planet TATACCA to conduct basic genetic experiments on a new species: **X**. The two traits that you are observing are the shape and attachment of the ear. Pointy ears are dominant to round, and attached ear lobes are dominant to detached.

Assuming that genes responsible for these phenotypes assort independently, write out the genotype of the parents in each of the crosses. Assume homozygosity unless there is evidence otherwise. ***P*** and ***R*** stand for the pointy and rounded phenotypes, respectively, and ***A*** and ***D*** stand for the attached and detached phenotypes. Use *P* and *p* for the pointy and round ear alleles and *A* and *a* for the attached and detached.

Parental Genotype	Parental Phenotype	Number of Progeny			
		PA	*PD*	*RA*	*RD*
	PA × *PA*	447	152	147	50
	PA × *PD*	107	98	0	0
	PA × *RA*	95	0	91	0
	RA × *RA*	0	0	153	48
	PD × *PD*	0	149	0	51
	PA × *PA*	140	46	0	0
	PA × *PD*	151	147	48	50

Problem 5

Geneticists working with the common fruit fly, *Drosophila melanogaster*, are studying the mode of inheritance of vestigial wings. Vestigial wings are a mutant phenotype in which wing development is curtailed. Wild-type or normal flies display normal wing development. Geneticists working with flies label genes and the alleles, and therefore the flies, after the mutant phenotype. Pure breeding vestigial flies have the genotype of *vgvg* (two copies of the mutant allele) and the phenotype is denoted as *vg* for vestigial. Pure breeding normal flies are denoted vg^+vg^+ for their genotype and their phenotype is described as vg^+.

After sorting flies for several hours, a freshman volunteering in the laboratory accidentally sneezes, mixing up all three piles of flies. Originally, one pile of flies displayed vestigial wings, a second pile of flies had normal wings but was heterozygous, and the third pile had normal wings and was homozygous. The principal investigator of the laboratory calls on you, as an expert geneticist, to resort the flies. By performing several crosses, you are able to determine from which pile each fly originated.

Which phenotype is recessive, vestigial wings or wild-type wings? (See the table below).

Using the symbols of *vg+* and *vg*, write the genotypes of the flies used in each cross in the table below.

Genotype	Phenotype	Number of Progeny	
		vg+	*vg*
	normal × normal	628	209
	vestigial × normal	333	340
	normal × vestigial	454	0
	vestigial × vestigial	0	121
	normal × normal	92	0

Problem 6

One day as you were traveling along a semi-deserted highway, your car has an electrical surge and suddenly stops. You can't get your car to run again so you decide to get out and walk to the nearest phone to call for a tow truck. While you are walking you see a dairy farm in the distance. As you get closer you notice that these are not ordinary cattle. Then, you see the signs indicating a government research facility. Because you are extremely interested in the genetics of these odd cows, you take a few with you as you run off. Amazingly, you and your new bovines have made it to your family's farm safely. You decide to breed the cattle to obtain a better understanding of the genetic basis for their coat color. The results from your crosses are found on the table below.

Parental Genotypes	Parental Phenotypes	Green Progeny	Purple Progeny
	purple × purple	0	98
	green × green	94	0
	purple × green	51	49
	purple × green	97	0
	green × green	71	27

Which color results from possessing a dominant allele?

Using *G* and *g* as allele designations in the form of *Gg* × *Gg*, assign the most probable genotypes for the parents in each cross above. Assume homozygosity unless there is evidence otherwise.

Besides having either a purple or a green coat, your cattle have two other odd traits that you have observed to be inherited in a Mendelian manner. Your new bovines have either yellow eyes (*Y*) or blue eyes (*B*) and produce either regular white milk (*W*) or chocolate milk (*C*). Consider the table below. Which is the recessive eye color?

What type of milk is produced when a cow possesses a dominant allele?

Parental Phenotypes	Blue Chocolate	Yellow Chocolate	Blue White	Yellow White
BC* × *BC	10	4	3	1
BW* × *YC	7	8	0	0
BC* × *BW	4	0	3	0
BW* × *BW	0	0	15	5
YC* × *YC	0	22	0	7
BC* × *BC	12	4	0	0
BC* × *YC	10	11	3	4

Using *C* to indicate the allele for chocolate milk production, *c* for white milk production, *B* for blue eyes, and *b* to indicate the allele for yellow eyes, write the parental genotypes for each of the crosses shown above. Assume homozygosity unless there is evidence otherwise.
What proportion of the offspring of two parental cattle each of the genotype *Gg Bb Cc* (green with blue eyes and chocolate milk producing) will be *Gg bb cc*?

Consider the following cross: *Gg Bb Cc* × *gg Bb cc*

What fraction of the progeny do you expect to phenotypically resemble the first parent?

If 100 progeny resulted from crosses such as this, how many would you expect to exhibit new genotypes (i.e., do not genotypically resemble either parent)?

Pedigree and Probabilities

Problem 1

Part a. For the pedigree shown below, state whether the condition depicted by darkened symbols is dominant or recessive. Assume the trait is rare. Assign genotypes for all individuals using *A*/*a* designations.

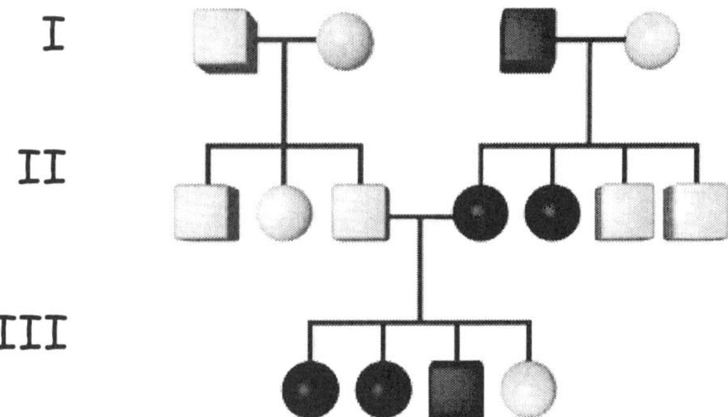

Part b. Assume the inherited trait depicted in the pedigree below is rare and state whether the condition is dominant or recessive. Assign genotypes for all individuals using *A*/*a* designations.

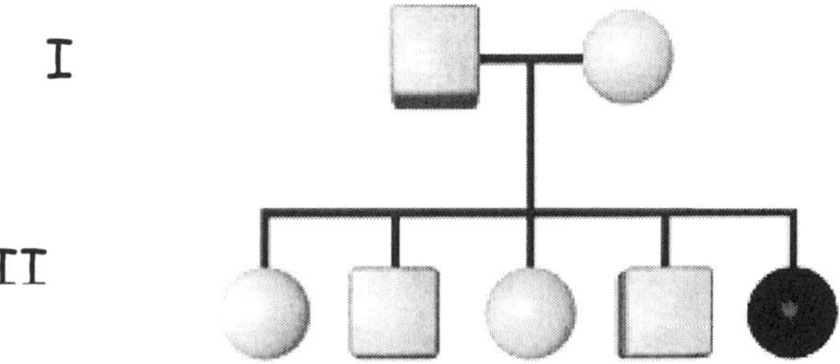

Part c. Is the pedigree shown below consistent with a dominant or recessive trait?

In analyzing this pedigree, would you conclude that this trait is caused by a rare or common allele?

Assign allele designations using *A*/*a*.

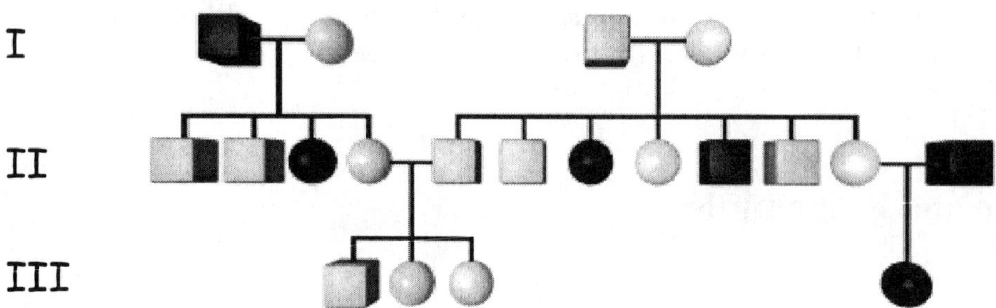

Problem 2

You have a single six-sided die. What is the chance of rolling a six?

What is the chance you will roll a five or a six? (HINT: Should you use the product rule or the sum rule to determine the probability of rolling one or the other?)

You are now given two additional dice. If you roll all three dice simultaneously, what is the chance of obtaining a five on all three dice?

Rolling your three dice again, what is the chance of obtaining no fives at all?

What is the chance of obtaining two fours and one three on any of the dice in a single roll?

What is the chance of obtaining the same number on all three dice?

What is the chance of rolling a different number on all three dice?

Problem 3

Preparing for a long night studying genetics, you open a big bag of candy-coated peanuts. Two friends are supposed to join you so, so you divide the bag into thirds. As you wait, you get hungry. But since you want to keep the bowls equal, you eat one from each bowl.

Bowl 1	Bowl 2	Bowl 3
30 blue candies	40 blue candies	30 blue candies
30 yellow candies	50 yellow candies	20 yellow candies
40 brown candies	10 green candies	50 red candies

What is the chance that you will select three blue candies?

What is the probability of selecting a brown, a green, and a red candy?

If you pick one candy from each bowl, what is the chance you will pick two blue candies and a yellow candy?

If you pick one candy from each bowl, what is the probability of obtaining no yellow candies?

If you pick one candy from each bowl, what is the chance of picking at least one yellow?

The probability of having two peanuts in a single candy is 2%. If you eat 300 candies, how many double peanuts do you expect to find?

Problem 4

The ability to taste the chemical phenylthiocarbamide (PTC) is an autosomal dominant phenotype. Lori, a taster woman, marries Russell, a taster man. Lori's father and her first child, Nicholas, are both nontasters.

What is the probability their second child will be a nontaster girl?

What is the probability their second child will be a taster boy?

What is the probability that their next two children will be nontaster girls?

Problem 5

A rare recessive allele inherited in a Mendelian manner causes phenylketonuria (PKU), which can lead to mental retardation if untreated. Fortunately, with a diet low in phenylalanine and tyrosine supplementation, normal development and lifespan are possible. Mike, a phenotypically normal man whose father had PKU marries Carol, a phenotypically normal woman whose brother had the disorder. The couple wants to have a large family and come to you for advice on the probability that their children will have PKU.

What is the probability that the couple's first child will have PKU?

Their first child, Greg, has PKU and the couple wants to have five additional children. What is the probability that out of their next five children only two will have PKU?

What is the possibility that they will have no more than one affected child in the five additional children they are planning?

Problem 6

John and Maggie are expecting a child. John's great grandmother (mother's lineage) and Maggie's brother have a rare autosomal recessive condition. What is the chance that their child will be affected?

John and Maggie have just discovered they are going to have twins. What is the chance that both twins will be affected if they are identical twins?

If the twins are dizygotic twins (non-identical or two-egg twins), what is the chance they will both be affected?

If they are dizygotic twins, what is the chance that at least one of them will be affected?

Chromosomal Theory of Inheritance

Goals for Chromosomal Inheritance:

1. Understand the classical evidence that genes are located on chromosomes:
   - Similarity of behavior of chromosomes in divisions with the behavior of genes (alleles) in inheritance
   - Identification of genes with sex-linked inheritance and chromosomes with sex-linked inheritance
2. Distinguish mitosis from meiosis figures, as well as the species diploid chromosome number, by inspection of simple diagrams with chromosome size, shape, and copy number as the cues.
3. Connect genetic inheritance with chromosome behavior during divisions:
   - Assign alleles to chromosomes: sister chromatids have identical alleles, homologous chromosomes (in a heterozygote) have different alleles
   - Homologous chromosomes pair and disjoin from each other in the first meiotic division, illustrating segregation of the alternate alleles
   - Sister chromatids separate at mitosis, illustrating the constancy of the genotype in both daughter cells
   - Genes that are on different chromosome pairs, i.e., non-homologous pairs, always assort independently
4. Be able to identify sex-linked inheritance in a pedigree.

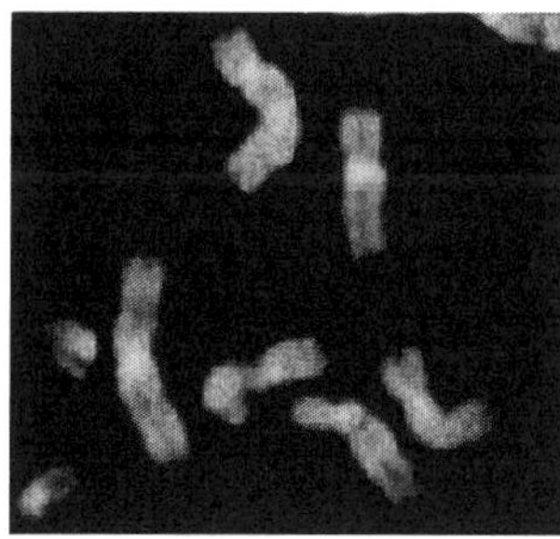

Mitosis and Meiosis

Problem 1

Many cells undergo a continuous alternation between division and nondivision. The interval between each mitotic division is called interphase. Which stages (G1, S, G2 or M) make up interphase?

It was once thought that the biochemical activity during interphase was devoted solely to cell growth and specific functions the cell normally performs. However, it is now known that another biochemical step critical to the next mitosis occurs during interphase: the replication of the DNA of each chromosome. During which stage of the cell cycle is DNA replicated?

The concentration of DNA contained in a single set of chromosomes is frequently referred to as "c." Diploid somatic cells have two sets of chromosomes and alternate between 2c and 4c. A cell has 2c before DNA replication. Then, it has twice that amount, 4c, until it undergoes mitotic division. After mitosis, the two resulting cells will both have 2c.

The job of the mitotic division phase is to segregate the replicated chromosomes into two cells, each with the same chromosome and genetic complement as the parent cell. The four phases of mitosis are: prophase, metaphase, anaphase, and telophase.

DNA replication (in S phase) produces a second double helix. During prophase, the sister double helices condense, forming the X shaped chromosomes typically seen under the microscope.

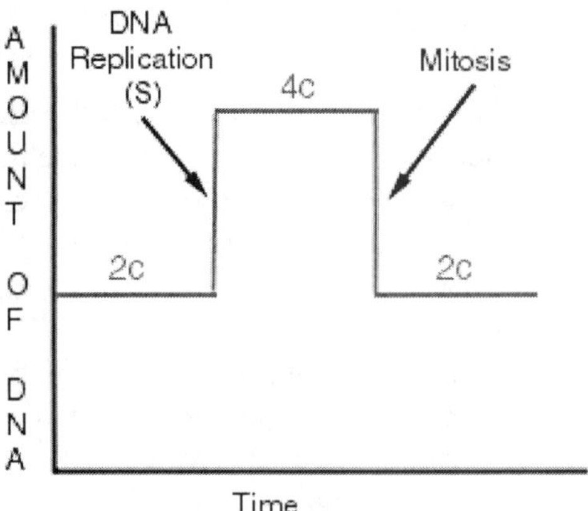

Draw a cell containing one pair of homologous chromosomes after it undergoes DNA replication (i.e., S phase) as seen in metaphase.

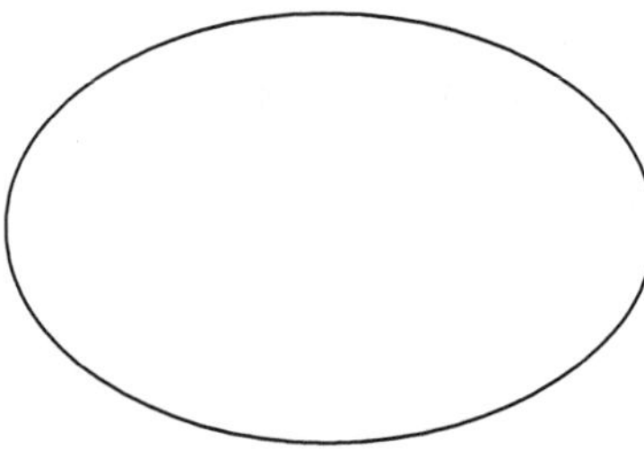

Suppose one homologue carries the *a* allele and the other homologue carries the *A* allele. Label each chromatid in your drawing.

Put each of the pictures below (containing two pairs of homologous chromosomes) in correct order and label with the appropriate phase. Note: Homologous chromosomes do not have to be next to each other during mitosis.

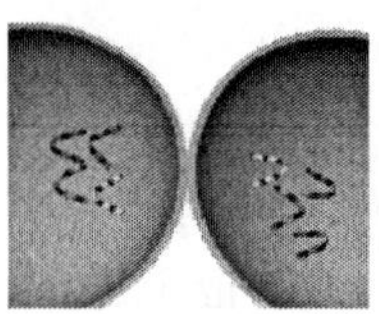 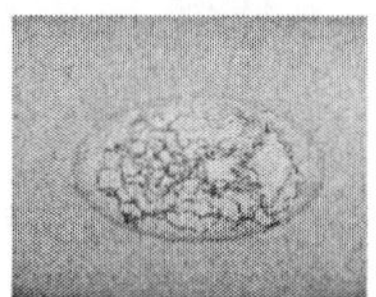 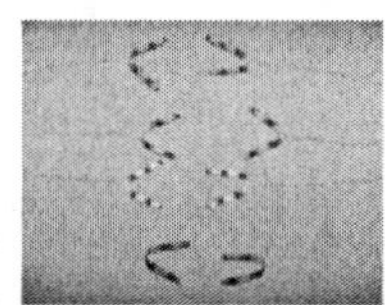 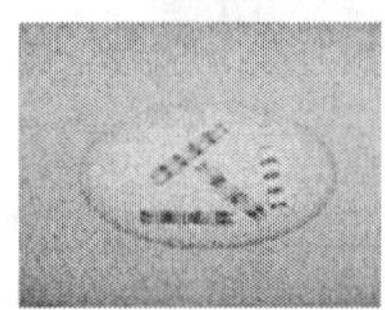 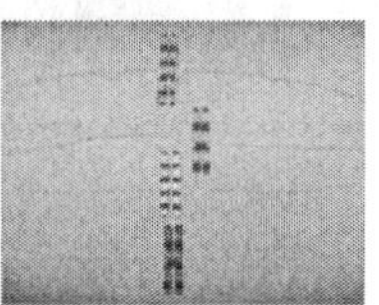

__________ __________ __________ __________ __________

Draw two daughter cells generated by a mitotic division with the appropriate homologues (labeled with *A* and *a*).

Problem 2

The results Mendel observed from his early crosses established the idea that alleles segregate. Unfortunately, Mendel couldn't figure out the biological mechanism behind this principle.

About 100 years ago, scientists realized that gametes (sex cells) were the result of a specialized cell division. This division process began with a cell that had two sets of chromosomes (a diploid cell) and ended with four cells with only one set of chromosomes (a haploid gamete). Today we know that the segregation of homologous chromosomes within this division process, meiosis, provides the mechanism for the segregation of alleles.

During sexual reproduction, two gametes (one from the mother, one from the father) combine in fertilization to form a diploid cell. A diploid fly has eight chromosomes. Without meiosis, how many chromosomes would the first generation progeny contain?

The purpose of meiosis is to convert one diploid cell into four haploid cells, each containing a single complete set of chromosomes. Meiosis can be divided into two cycles: Division I and Division II (or meiosis I and meiosis II). Two cells result from meiosis I. For a cell with a single pair of homologous chromosomes (labeled with *A/a*), draw the chromosomes found in the two cells formed by the first meiotic division. Then draw the four products of meiosis, with the appropriately labeled chromosome.

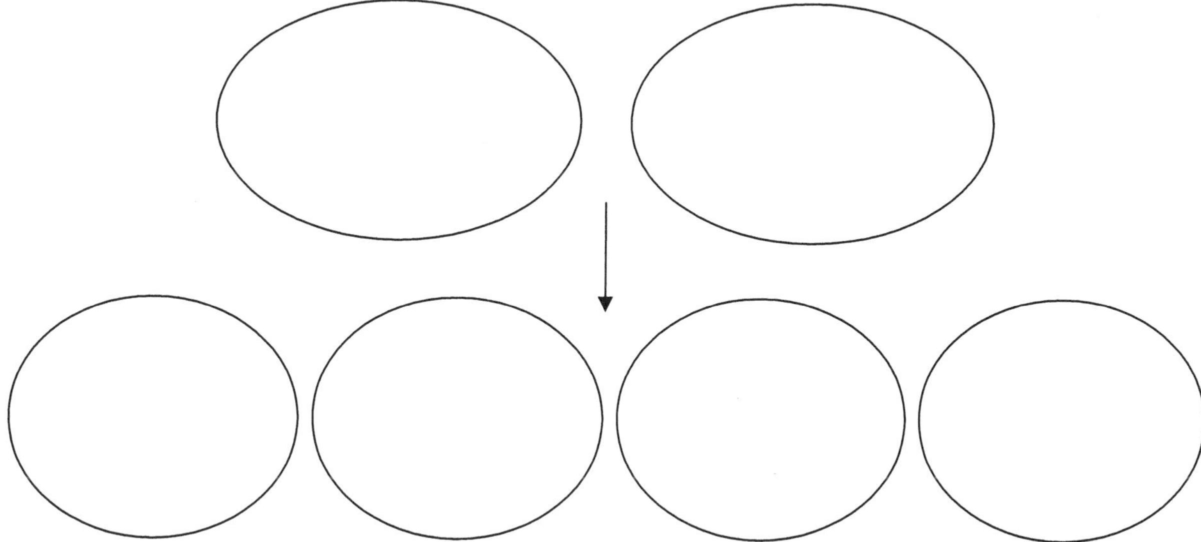

How many types of gametes with different alleles are formed in this example?

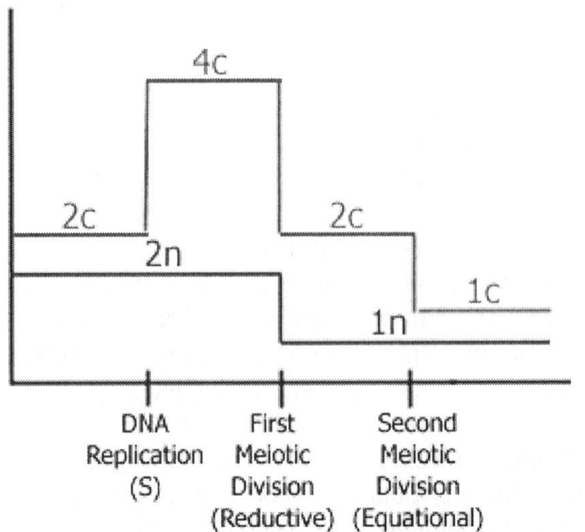

Regardless of whether there are one or two chromatids per chromosome, n indicates the number of chromosomes in a cell. Recall that c refers to the concentration of DNA in a cell.

vv

a

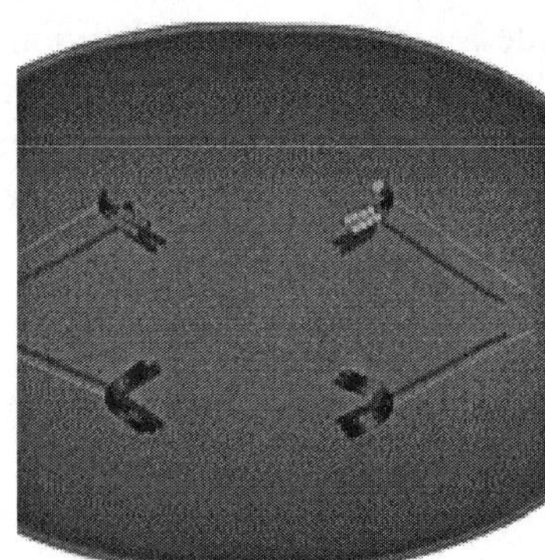

What phase of meiosis best describes the cell in Figure a?

How many pairs of homologous chromosomes are there in this cell?

How many chromatids are present?

What is the value of c at this stage of meiosis?

b

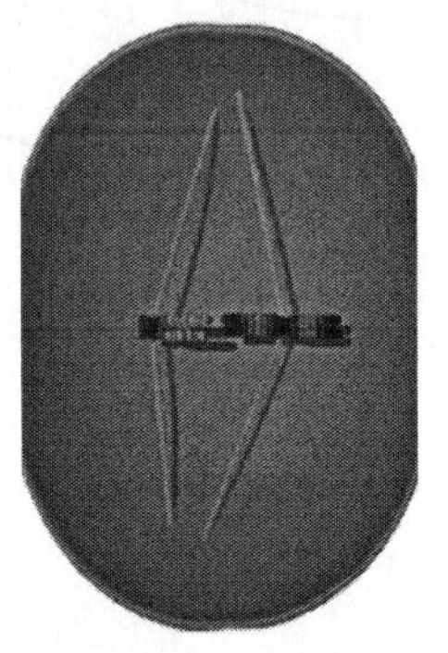

What phase of meiosis best describes the cell in Figure b?

Does the cell shown here have 1n or 2n chromosomes?

Is the amount of DNA best described as c, 2c, or 4c?

c

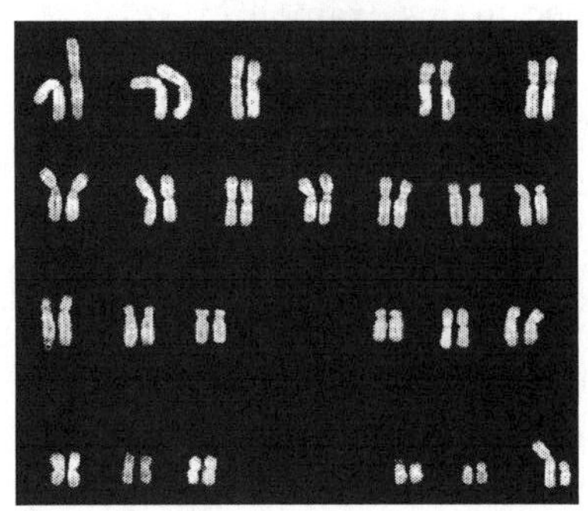

A karyotype (Figure c) is the ordered visualization of a complete set of chromosomes from metaphase. Karyotypes can be used to determine the number of chromosomes a species has, as well as any abnormalities an individual chromosome might have. This is the karotype of a male gorilla. How many chromosomes does a gorilla have? How many chromosomes are found in a gorilla gamete?

Autosomes are the same in both males and females. Sex chromosomes (bottom right corner) differ according to sex (XX in females, XY in males). X and Y chromosomes behave as homologues in males (the X and Y chromosomes pair and segregate at meiosis I). How many chromosomally different gamete types does a male produce?

Problem 3

Through the analysis of dihybrid crosses, Mendel was able to deduce that the genes he was studying were assorting independently, giving rise to gametic ratios of 1:1:1:1. However, in 1865, the physical nature of this independent assortment was not understood. After the discovery of meiosis in the 1880s, scientists recognized that genes located on different chromosomes should assort independently. A physical basis for Mendel's hypothesis was now possible!

The genetic material must be duplicated (or replicated) prior to its assortment. Draw two pairs of homologous chromosomes (labeled with *A/a* and *B/b*) in the cell below as they would appear after DNA replication (in prophase I).

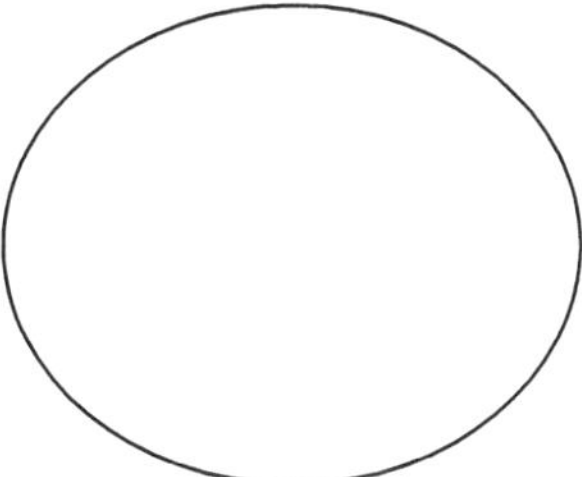

Two cells result from the first division of meiosis. Draw the chromosomes as they would appear in the cells after meiosis I. Remember that there are two different, equally probable ways for the chromosomes to assort (just draw one).

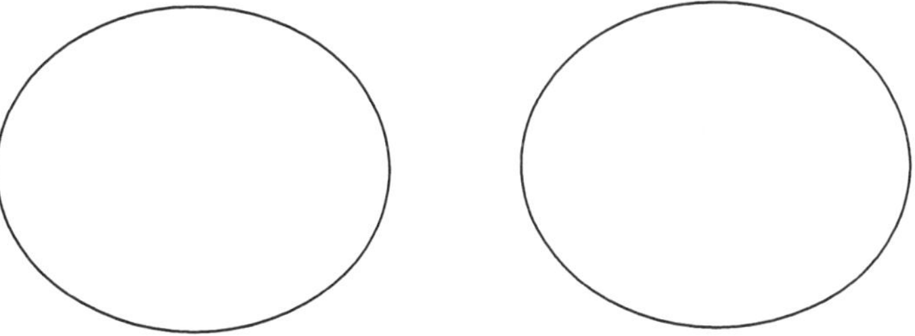

Write the possible gametes produced by a female with the genotype *AaBb*.

What are the possible gametes produced by a male with the genotype *AaBb?*

Notice that the gametes produced are identical to the meiotic products for an *AaBb* × *AaBb* cross. Recall that the Punnett square accurately predicts the genotypic frequencies when the genes involved assort independently of one another. In further testing Mendel's law of independent assortment, you cross an *Aa Bb Dd* female mouse with an *aa bb dd* male. How many possible gamete types can the female produce?

Suppose instead of seeing the number of classes of mice you expected, you find only four. You find 30 *abd*, 33 *aBd*, 29 *AbD*, and 28 *ABD* mice. In order to explain this enigma, we will look at one possible arrangement of genes on the mother's chromosomes.

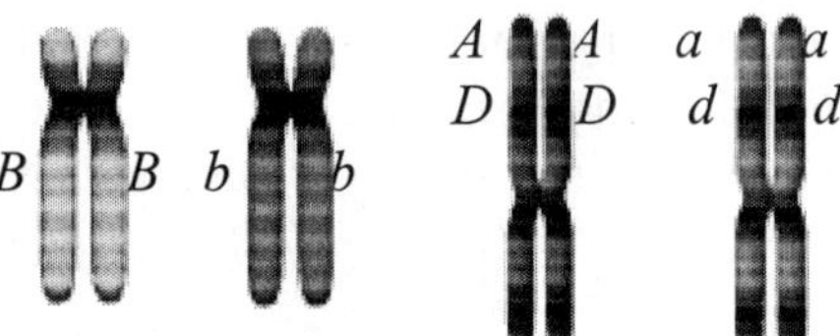

Two cells result from the first division of meiosis. Draw the chromosomes shown above to create the cells after meiosis I. Note that even though the two chromosomes are assorting randomly, the *A* and *D* alleles travel together since they are on the same chromosome.

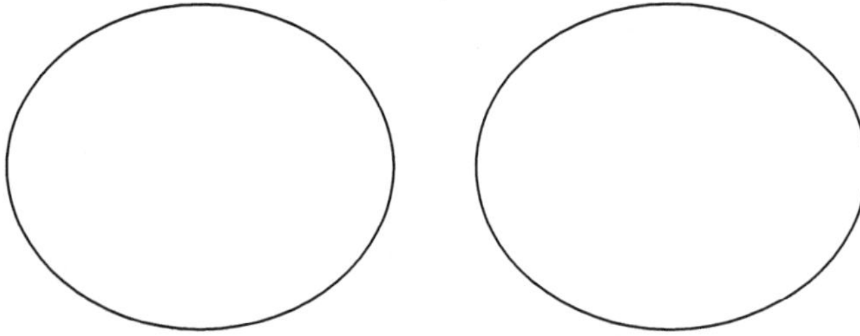

Completion of meiosis will produce four gamete types. What are the four possible genotypes (consider both possible assortments)?

This is an example of tightly linked genes on a chromosome where no crossing over can occur (see topic Linkage).

Problem 4

Problem 4 is a series of cartoons of cells at different stages of meiosis and mitosis and should be done on the CD-ROM.

X-Linked Inheritance
Problem 1

Is the pedigree below consistent with a dominant or a recessive trait? Assume the trait is rare.

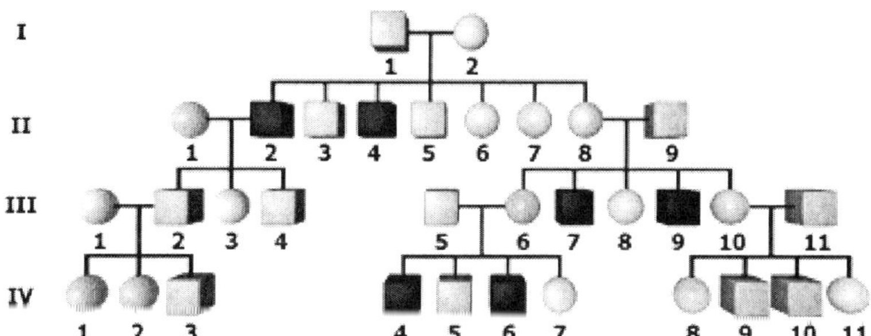

Is this pedigree consistent with X-linked or autosomal inheritance? Assume the trait is rare.

In the pedigree above, is there any evidence of father to son transmission of the trait?

What is the genotype of individual II-2? Write *A* or *a* for the X-linked alleles and Y for the Y chromosome.

What is the genotype of individual III-6? Write *A* or *a* to designate the X-linked dominant or recessive alleles.

Suppose in the pedigree above, individual II-8 was a male instead of a female and individual II-9 a female instead of a male. All other individuals are unchanged. Would this change the mode of inheritance deduced from this pedigree?

Is this pedigree still consistent with a recessive trait?

Can you tell whether the trait, in this hypothetical pedigree, is common or rare?

Now that the mode of inheritance is different, what is the genotype of individual II-2? Use *A* or *a* for the dominant and recessive alleles, respectively.

Problem 2

Hemophilia is a rare, recessive X-linked disease that usually affects only males. Consider the pedigree below of a family of normal parents with a son who has hemophilia.

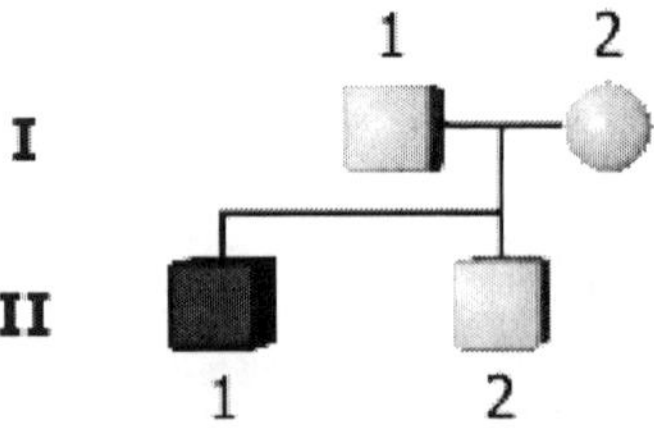

What is the father's genotype? Write *H* for X^H, *h* for X^h, or Y.

What is the mother's genotype? Write *H* for X^H and *h* for X^h.

They are going to have another child, whom they know is male (individual II-3). What is the chance that he will be affected?

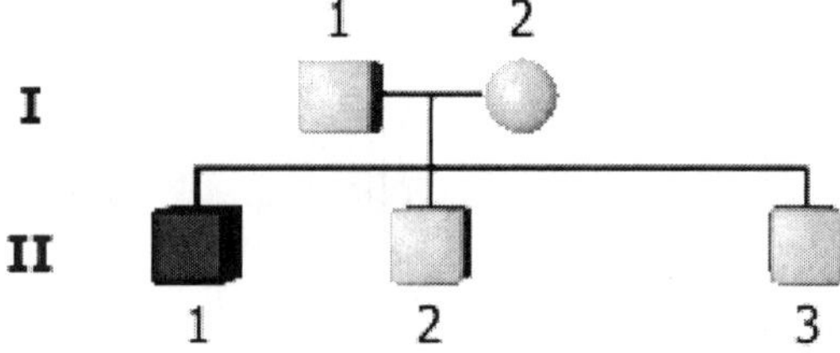

What is the chance that a daughter (individual II-5) will be a carrier?

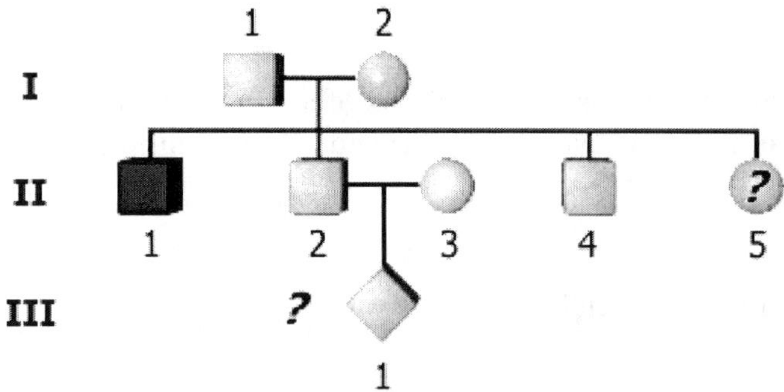

What is the probability that the Child III-1 will be affected?

What is the probability that the Child III-1 will be a carrier girl?

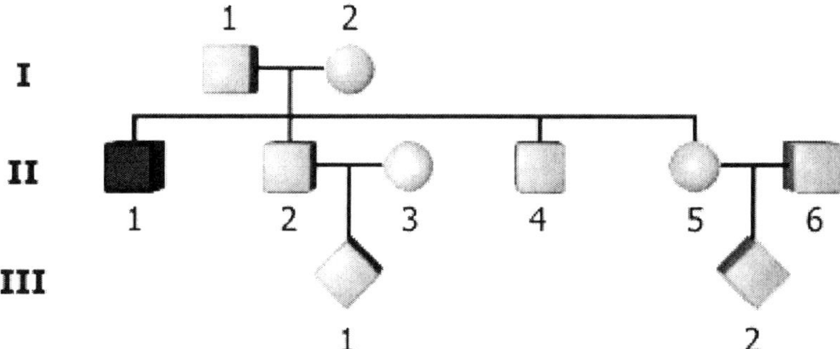

What is the probability that the Child III-2 will be an affected boy?

Problem 3

The pedigree below belongs to a family with a rare trait known as vitamin D resistant rickets. Is this pedigree consistent with a dominant or recessive mode of inheritance?

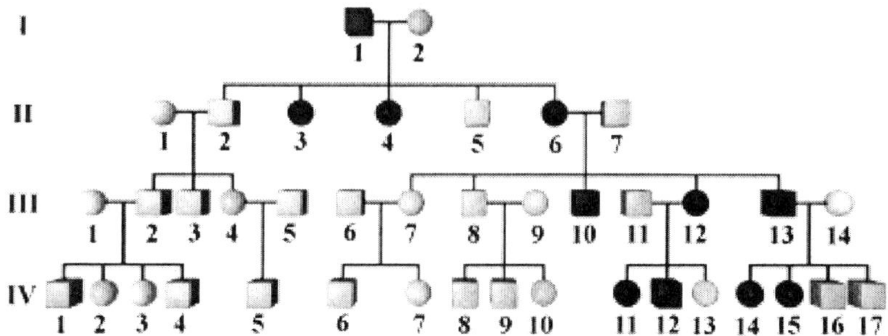

Is this pedigree consistent with an autosomal or an X-linked trait?

Assuming the trait is X-linked, what is II-6's genotype? Use *A* or *a* to designate the dominant or recessive X-linked alleles.

What is the genotype of individual II-5? Use *A* or *a* to designate the dominant or recessive X-linked alleles.

Suppose that III-14 has just found out she is going to have another girl. What is the probability that the child will be affected?

Individuals III-11 and III-12 are going to have another child but they don't know its sex. What is the probability they will have an affected son?

Problem 4

Red green colorblindness is a common X-linked recessive trait in humans that affects both males and females. Affected individuals are unable to distinguish red from green. Its prevalence accounts for the fact that you see some affected females as well as affected males.

Sickle cell anemia is an autosomal recessive disorder with a high incidence in the African American population (1/400). It is a structural hemoglobin abnormality that causes sickling of red blood cells with resulting complications.

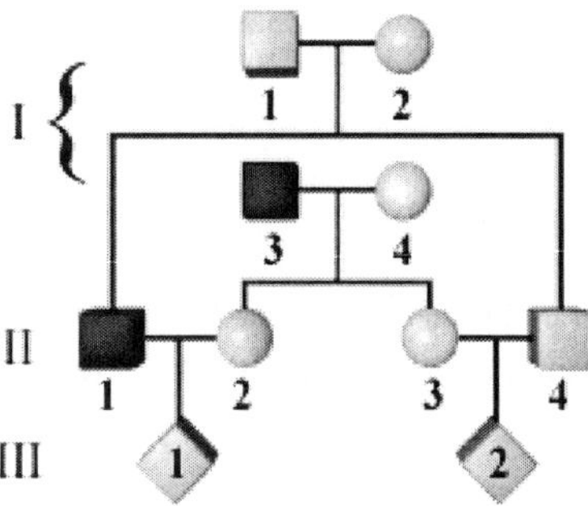

Consider this pedigree of two brothers marrying two sisters. All the individuals in generation II, namely II-1, II-2, II-3, and II-4 are carriers of the sickle cell allele. Individuals with filled blue squares are colorblind. Is there a greater chance that Child III-1 will have sickle cell anemia than Child III-2?

What is the chance that Child III-1 will be colorblind and have sickle cell anemia?

What is the probability that Child III-2 will be colorblind and affected with sickle cell anemia?

If the child (III-2) from the second marriage is colorblind, what is the child's gender?

Problem 5

Nondisjunction is a rare event that occurs in meiosis when paired chromosomes or sister chromatids do not disjoin properly. When this occurs in meiosis I, both homologues end up in one daughter cell. When nondisjunction occurs in meiosis II, one of the four meiotic cells ends up with both sister chromatids from one chromosome, and the other meiotic cell involved doesn't have any copies of that chromosome.

In humans, nondisjunction is generally lethal except when it involves chromosome 21 (leading to Down syndrome) or the sex chromosomes. In the following examples, nondisjunction of the sex chromosomes will be examined in conjunction with the X-linked gene that causes red green colorblindness. Affected individuals are unable to distinguish red color from green. In severely affected individuals, everything appears gray.

A man who is colorblind has a daughter with Turner syndrome who is also colorblind. You want to explain the origin of this daughter through nondisjunction in one parent.

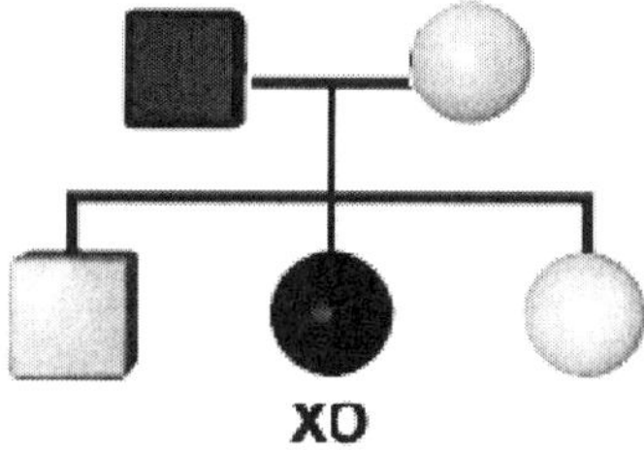

Assuming the mother is not a carrier of the colorblindness allele, use the genetic marker of colorblindness in this family to determine which parent had the nondisjunction event.

Can you tell at what stage of meiosis (meiosis I or meiosis II) nondisjunction occurred?

A woman who is colorblind has a son with Klinefelter (XXY) syndrome who is not colorblind. You want to explain the origin of this son through nondisjunction in one parent. Start by assigning a genotype to each individual.

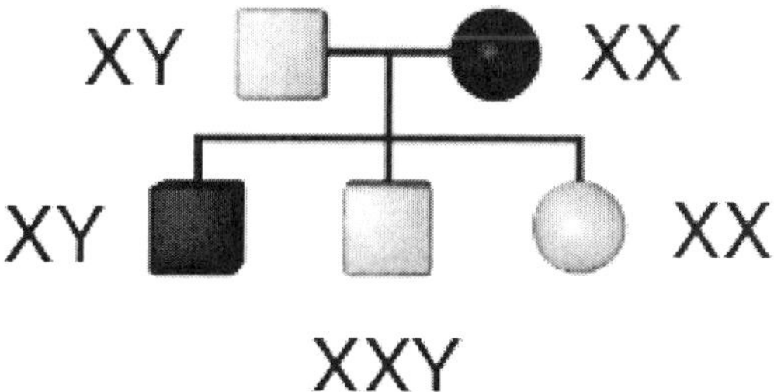

Can you use the genetic marker of colorblindness in this family to determine which parent had the nondisjunction event? Indicate that parent in the pedigree above.

Can you determine whether the nondisjunction event occurred in meiosis I or meiosis II of the parent's gametic line?

In the *Drosophila* fruit fly, the gene for white eyes is located on the X-chromosome. The white-eyed allele is recessive to the wild type. In crosses between white-eyed females and wild-type males almost all (regular) daughters have red eyes, except for about 1/2000 (0.05%) exceptional white-eyed daughters. Almost all (regular) sons have white eyes except for about 1/2000 exceptional red-eyed sons that are sterile.

Upon inspection of the flies' karyotypes, the white-eyed exceptional daughters were found to carry XXY. (Regular females are XX and males are XY.) Does this finding indicate that the Y chromosome determines sex in fruit flies?

Indicate the correct alleles of the exceptional females, one allele at a time. Follow the *Drosophila* nomenclature rules (see Fly Lab below) and choose between *w* and *W* for the white-eyed allele, and between *w*+ and *W*+ for the wild-type allele.

When the exceptional red-eyed sons are inspected, they are found to be XO, having a single X chromosome. Does this fact agree with the conclusion that the presence of Y does not determine sex in the *Drosophila* fruit flies?

Do these unusual findings of white-eyed exceptional daughters and red-eyed exceptional sons represent nondisjunction in the *Drosophila* father or *Drosophila* mother?

Can we tell whether the nondisjunction occurred in the first meiotic division or in the second meiotic division?

Fly Lab

Welcome to the fly lab! The following problems allow you to play the role of a geneticist working with the common fruit fly, *Drosophila melanogaster*. Since *Drosophila* nomenclature may be confusing at first, here is a brief review:

Genes are named after the mutant phenotype in *Drosophila*. Therefore, if the mutant gene results in the lack of eye formation, the gene could be named "eyeless." (Note that this differs from the Mendelian designations used for peas, for example, where the gene is typically named after the dominant trait.)

When allele symbols are assigned, lower case is used if the mutant allele is recessive (for example, *ro* for rough eyes). Capitalizing the first letter indicates that the mutant allele is inherited as a dominant trait (for example, *N* for notched wings).

Wild-type alleles are designated with a superscript + (eg., ro^+ or N^+). In the following problems simply indicate the + after the allele symbol (i.e., *ro*+ or *N*+).

The problems below are set up as simulations on the CD-ROM. Here, the results are provided.

Problem 1

In *Drosophila*, the brown mutant is characterized by a brown eye color compared with the brick red, wild-type color. A wild-type fly is crossed with a brown-eyed fly to produce the F_1 generation and then the F_1 flies are crossed to produce the F_2 generation.

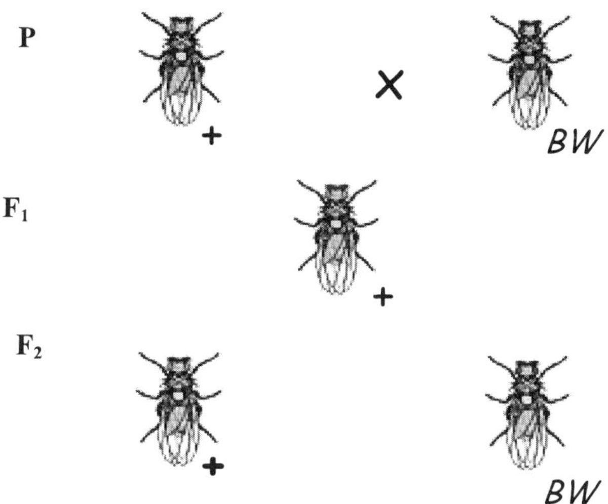

Is the brown mutation dominant or recessive?

Using the letters *BW* and proper Drosophila nomenclature (*bw* for recessive and *Bw* for dominant), indicate the proper mutant allele designation below.

What is the genotype of each of the flies above?

Problem 2

In *Drosophila*, the lobed mutation is characterized by smaller eyes compared to the wild type. A wild-type fly is crossed with a lobed fly to produce the F_1 generation (all lobed). Then the F_1 flies are crossed with each other to produce the F_2 generation (768 lobed and 259 wild type).

Is the lobed mutation dominant or recessive?

How would you indicate the mutant allele using the correct *Drosophila* nomenclature?

What is the genotype of each of the parental, F_1, and F_2 flies?

What fraction of the mutant F_2 flies is homozygous?

Problem 3

In *Drosophila*, the white mutant is characterized by white eyes compared to the brick red, wild-type color. A wild-type female fly is crossed with a white-eyed male to produce 974 wild-type F_1 flies. At this point what can you conclude about the mutation (is it dominant or recessive)?

The reciprocal cross is then performed: A wild-type male is mated with a white-eyed female to produce 436 white-eyed males and 421 wild-type females.

What can you now conclude about the mode of inheritance (autosomal or X-linked)?

Now go back to each of the crosses above and write the genotype of all the flies.

Problem 4

In *Drosophila*, the bar mutant is characterized by eyes that are restricted to a narrow, vertical bar. When a bar female is mated to a wild-type male, all the F_1 flies are bar. However, when a bar male is mated to a wild-type female, 857 bar females and 905 wild-type males are observed.

What is the mode of inheritance of the bar mutant?

What is the genotype of each of the flies in the two crosses above?

Problem 5

In *Drosophila*, the ebony mutant is characterized by an ebony body color and purple is characterized by purple eyes. Mating an ebony, purple female with a wild-type male yields all wild-type progeny. The reciprocal cross gives the same results.

What is the mode of inheritance for the ebony mutation?

What is the mode of inheritance for the purple mutation?

What is the genotype of the F_1 flies?

Mating the F_1 flies together yields 226 wild type, 74 ebony, 78 purple, and 25 ebony, purple flies. What is the ratio of progeny for each of the phenotypic classes?

Which F_2 fly should you use for a testcross of the F_1 flies?

How many different phenotypic classes, and in what ratios, do you expect from this cross?

Problem 6

In *Drosophila*, the sable mutant is characterized by a sable body color and dumpy is characterized by shorter, oblique wings. In a cross between a sable, dumpy female and a wild-type male, all the female progeny are wild type and the male progeny are sable. When the F_1 siblings are mated, the F_2 consists of 338 wild type, 336 sable, 114 dumpy, and 110 sable, dumpy (both male and female).

What is the mode of inheritance of the sable mutant?

What is the mode of inheritance of the dumpy mutation?

In order to understand the unusual ratios and the lack of apparent linkage to sex of the sable phenotype in the F_2, assign genotypes to the parental and F_1 generations. Write the genotypes above, using *s* or *s*+ or Y to indicate sable alleles and *dp* or *dp*+ to indicate dumpy alleles. Predict the proportion of each phenotype you would expect in the F_2.

Genotype/Phenotype

Goals for Genotype/Phenotype:

1. Understand the following phenomena that lead to variations on Mendelian phenotypic ratios:
   - Incomplete dominance
   - Codominance
   - Epistasis
   - Homozygous lethality
2. Recognize combinations of 9:3:3:1 phenotypic ratios, where the phenotype comes from two genes and involves epistasis.
3. Distinguish from phenotypic ratios whether multiple genes or a single gene with multiple alleles are involved in determining the phenotypes.

Problem 1

Gardeners at the Japanese Botanical Garden discovered that after planting only red and ivory snapdragons some plants with pink flowers appeared among the progeny. The gardeners decided to experiment with the snapdragon flowers and carried out a number of additional crosses. In this diagram, the parents are shown along the margins with progeny types inside the boxes.

Parents	**Red**	**Ivory**	**Pink**
Red	red	pink	red, pink
Ivory	pink	Ivory	ivory, pink
Pink	red, pink	pink, ivory	red, ivory, pink

Which of the parents are homozygous?

Which of the parents are heterozygous?

By crossing the pink plants we should be able to distinguish whether their color is caused by one or more gene differences between the red and ivory parents. Take a moment to predict what offspring you expect for one gene versus two genes.

From the original results, the pink × pink cross yielded three phenotypes among the progeny. The numbers of each of the phenotypes are listed below.

F_2: 261 red
489 pink
243 ivory

What is the ratio suggested by these results?

How many <u>genes</u> differ between the red and ivory parents in determining flower color?

Assume the gene differing between the red and ivory parents has two alleles, *P1*, associated with the presence of red pigment, and *P2*, associated with the absence of red pigment. What are the genotypes of the different progeny types in the F_2 above?

Which of the following concepts best explains the observed results of the F_1 and F_2 progeny: incomplete dominance, codominance, *P1* dominant, *P2* dominant, multiple alleles, or multiple genes?

Problem 2

The tools we use to infer genotypes are phenotypes and the results of crosses. A powerful advance is the ability to look at phenotypes that are closer to gene activity. Here we describe an example using gel electrophoresis of proteins.

Proteins can be separated based on size and/or charge using electrophoresis through a gel matrix. Migration distance through the gel serves as a phenotypic marker that could help determine an individual's genotype. Such is the case with the normal and the sickle cell hemoglobin molecules. With sickle cell, heterozygous individuals produce both normal and sickle cell hemoglobin molecules. When the normal and the abnormal hemoglobin molecules are run through a gel, they migrate at different rates and can be visually separated.

Part a. The family below has a history of sickle cell anemia. The hemoglobin electrophoresis pattern for each child is shown in the lane below that child.

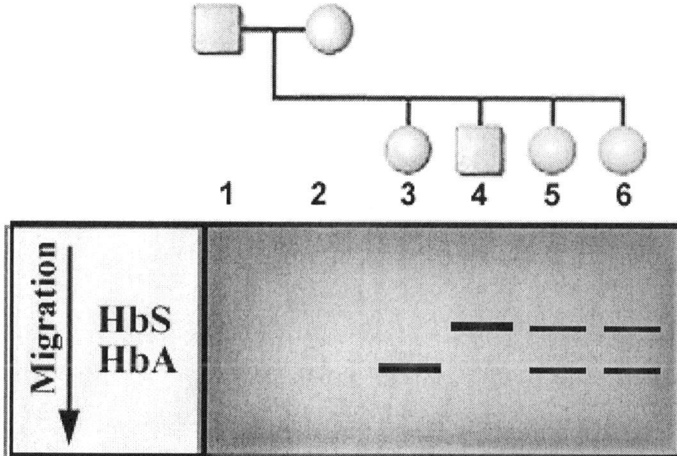

Which individuals in this family are homozygous for the sickle cell allele?

Which individuals from this family are heterozygous?

Imagine that this family is one out of a large number of families with heterozygous parents tested for their hemoglobin structure. Predict the ratios of children expected to be homozygous for *HbA*, homozygous for *HbS*, and heterozygous.

Which of the following concepts can best explain the expected phenotypic ratio: *HbA* dominant, *HbS* dominant, incomplete dominance, or codominance?

Part b. Gel electrophoresis can be used to screen individuals for the *HbS* allele.

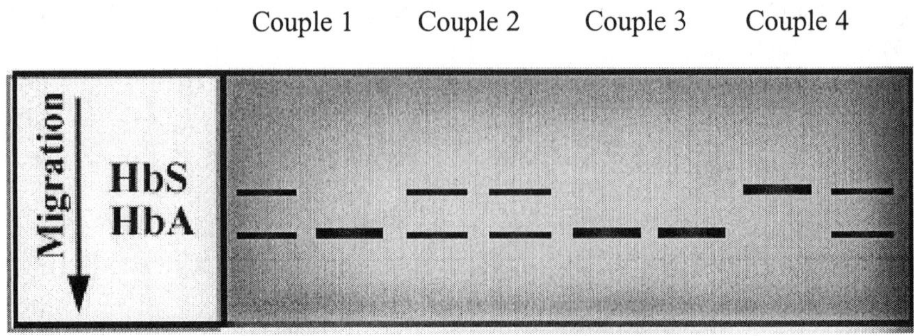

Which of the couples above are at risk of having an affected child?

What is the probability that Couple 2 will have a child with sickle cell anemia?

Part c. Use the Southern blot below to determine which of the three males (4, 5, or 6) could be the children's father.

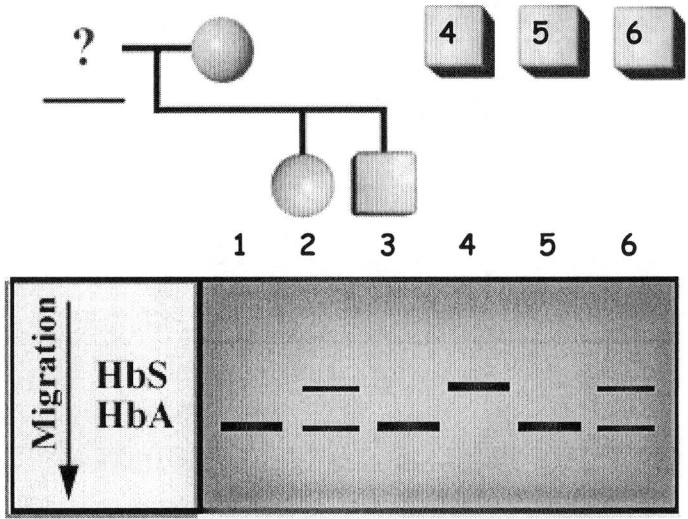

If this couple has another child, what is the chance the child will be anemic?

Problem 3

Karl Landsteiner was an Austrian-American physician who discovered that human blood differed in the capacity of serum to agglutinate red blood cells. By 1902, he and his group divided human blood into the groups A, B, AB, and O. Landsteiner concluded that two genes, A and B, control the ABO blood system he discovered. He proposed that each gene had two alleles, the presence and the absence of that allele.

Genotype	Phenotype
A– bb	A
aa B–	B
A– B–	AB
aa bb	O

Since you have been volunteering in Labor and Delivery for quite some time, you decide to compare the blood types you would expect using Landsteiner's two-gene hypothesis to the blood types you have observed. What genotype do you expect the children of two type O parents will have?

What is the expected phenotype of their children?

From your observations, 500 O × O parents have 763 type O children, whereas no type A, type B, or type AB children are observed. Is this consistent with the two-gene hypothesis?

One day you realize that in all 503 O × AB couples you have never seen any AB or O children. You have observed 600 type A and 650 type B children. Is this what you would predict based upon Landsteiner's hypothesis?

*Additional related questions are on the CD-ROM.

Problem 4

Suppose you are studying a novel bird species that displays a variation in feathers (blue, green, teal, and purple). Starting with pure breeding males and females from each phenotypic class, you perform the crosses diagrammed below.

Parents	F_1	F_2
teal × green	teal	¾ teal, ¼ green
green × blue	green	¾ green, ¼ blue
teal × purple	teal	¾ teal, ¼ purple
green × purple	green	¾ green, ¼ purple
blue × purple	purple	¾ purple, ¼ blue

Based on the data shown above, is feather color in this novel species segregating as if it were associated with multiple genes or multiple alleles?

Using the data shown above, which allele is dominant, blue or green?

Blue or purple? Green or purple? Green or teal?

What is the order of alleles corresponding to increasing dominance?

Cross	Parental Phenotypes	Phenotypes of Progeny			
		Blue	Green	Purple	Teal
1	blue × green	0	4	4	0
2	green × purple	0	3	3	0
3	teal × blue	0	4	0	4
4	teal × purple	0	4	0	3
5	green × green	0	6	2	0

For each of the crosses shown above, deduce the parental genotypes where *f* represents the feather gene and the following superscript designations represent the different alleles:

b - blue; *g* - green; *p* - purple; *t* - teal

*Assume homozygosity unless otherwise indicated.

Problem 5

A geneticist discovered two pure breeding lines of ducks. One line had white eyes and quacked with a "quack-quack." The other line had orange eyes and had a deeper quack, "rock-rock." In order to determine the mode of inheritance of these characteristics, she mated the two types of ducks and found that all F_1 ducks had yellow eyes, and uttered "quack-quack." When the F_1 ducks were interbred, the F_2 ducks were found in the following ratios:

24	yellow-eyed, quack-quack
12	white-eyed, quack-quack
12	orange-eyed, quack-quack
6	yellow-eyed, rock-rock
3	white-eyed, rock-rock
3	orange-eyed, rock-rock
2	yellow-eyed, squawk
1	white-eyed, squawk
1	orange-eyed, squawk

How many genes are involved in the inheritance of eye color?

P		orange-eyed × white-eyed
F_1		yellow-eyed
F_2	2	yellow-eyed
	1	orange-eyed
	1	white-eyed

Use *A1/A2*, 1/ *B B2*, etc. to designate eye color alleles, assign genotypes to the individuals in the cross described above.

How many genes are involved in the inheritance of quacking?

P		quack-quack × rock-rock
F_1		quack-quack
F_2	12	quack-quack
	3	rock-rock
	1	squawk

Use *B/b*, *C/c*, etc. to designate quacking phenotype, assign genotypes to the individuals in the cross described above.

Problem 6

Recall from Problem 4 that blood types are classified by their surface antigens. Type A blood has antigen A on its surface, type B has antigen B, type AB has both antigen A and antigen B, and type O has neither antigen on its surface.

Ellen and Carl have just had a baby boy. Ellen's blood type is B, Carl's is AB and the baby's is O. Having a good knowledge of blood typing, you realize that something is not quite right with this story. You happen to know that Ellen's parents' blood types are B and O. From this, what is Ellen's genotype?

What is Carl's genotype?

Does it appear that Ellen and Carl can have a child with O blood?

Ellen's ex-boyfriend, Mark, has type O blood. Could Mark be the baby's father?

Ellen insists that Carl is the father, so you look into their family history (you may notice that Ellen and Carl are cousins). The results are seen in the pedigree below. Which parents have phenotypes that are incompatible with the blood types of their children?

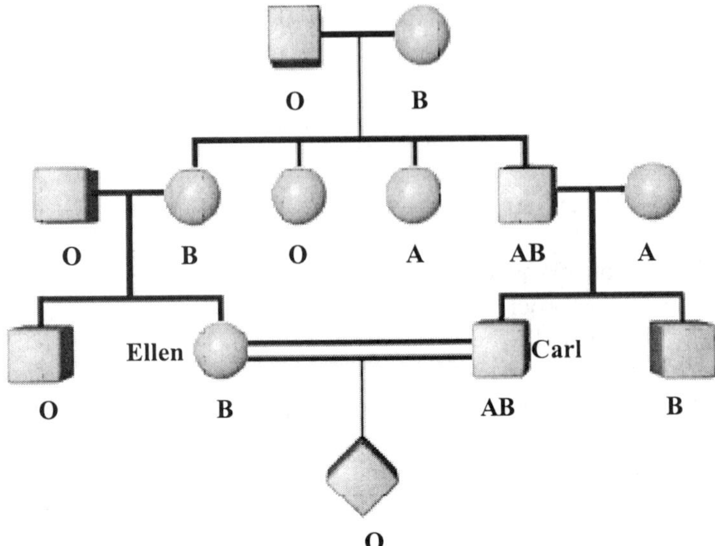

One of your friends suggests that this may be an example of lack of penetrance, i.e., the phenotype does not reflect the genotype. However, since Ellen and Carl share a grandfather who also has a suspicious O phenotype, you wonder if it's possible that they both inherited a recessive allele that is epistatic to the ABO blood antigens.

Upon further research you find there is such a rare recessive mutation, *h*, that is epistatic to the ABO system gene. Individuals who are homozygous for *h* cannot synthesize A or B antigens, so they have an O phenotype, referred to as the Bombay phenotype.

Assign genotypes for the *H* gene, assuming this is the reason for the unusual phenotypes (use *H/h*). With this information, determine the genotype for the ABO locus for individual I-1.

Does it now appear that Carl could be the father?

Does this information conclusively prove that Carl is the father?

In the hope of solving the paternity issue, you decide to run a Southern blot, probing for the alleles of the *H* gene.

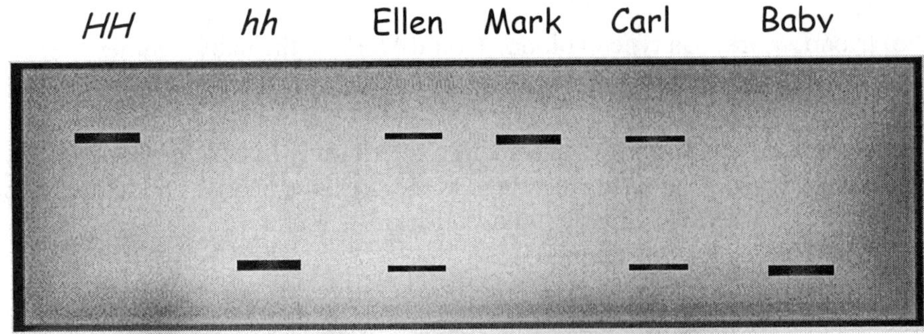

With this information, can you definitively say who is the baby's father?

Problem 7

Carolyn and Jeff are both cat lovers and neighbors in a local apartment complex. Recently, Carolyn's cat, Grace, escaped from her apartment and was seen mating with Jeff's ferocious feline, Chuck. Although Carolyn and Jeff were initially upset with the scandalous behavior of their pets, they felt reassured that Grace and Chuck would produce a beautiful and profitable litter of kittens. Weeks later, Carolyn and Jeff discovered the litter of kittens and were surprised to find one cat with curled ears.

As their close friend and local genetics expert, they turn to you to provide a genetic explanation for this strange occurrence. At this point, can you determine whether the curled-ear cat occurred as a result of a spontaneous dominant mutation or whether both parents were heterozygous for a recessive mutant allele?

In order to determine whether the curled-ear mutation is dominant or recessive to the normal condition, you mate the curled-ear cat with an unrelated normal cat. From this mating, *two curled-ear cats and two normal cats* are produced. Based on these results, is the curled-ear mutant allele dominant or recessive to the wild-type allele?

The allele responsible for curled ears in cats is located at the Ear locus. Using the allele symbols *c* and *c+*, assign genotypes to the cats in the cross described above. Be sure to follow the format shown below when assigning genotypes:

Homozygous individuals: *cc* or *c+c+*
Heterozygous individuals: *cc+* or *c+c*

What ratio of curled-ear cats to normal cats do you expect from the mating of two heterozygous curled-ear cats?

After performing the mating between two heterozygous curled-ear cats, you wait several weeks and examine the litter of kittens. The mating produced four curled-ear cats and two normal cats. From these results, can you be certain about the mode of inheritance of curled ears?

What results do you expect for a mating between a homozygous curled-ear cat and a normal cat?

What results do you expect for a mating between a heterozygous curled-ear cat and a normal cat?

You decide to perform 170 matings between curled-eared cats to determine whether curled ears are inherited in a simple Mendelian manner. You obtain 1020 kittens from your 170 matings between heterozygous curled-ear cats. If curled ears are inherited according to our predicted 3 to 1 ratio, how many curled-ear cats do you expect to see out of the 1020 kittens?

You only observe 677 curled-ear cats and 343 normal cats from the 170 matings. After re-counting the kittens several times, you are certain that the number of curled-ear cats is accurate. What is our observed ratio of curled-ear cats to normal-ear cats?

What is the most likely genetic explanation for the results we have gathered?

Linkage

Goals for Linkage Analysis:

1. Recognize the difference between linkage and independent assortment in dihybrid crosses.
2. Relate parental chromosome input to linkage phase to calculate recombination frequencies.
3. Be able to make maps from recombination frequencies.
4. Use three-factor crosses to distinguish order and distances between genes.

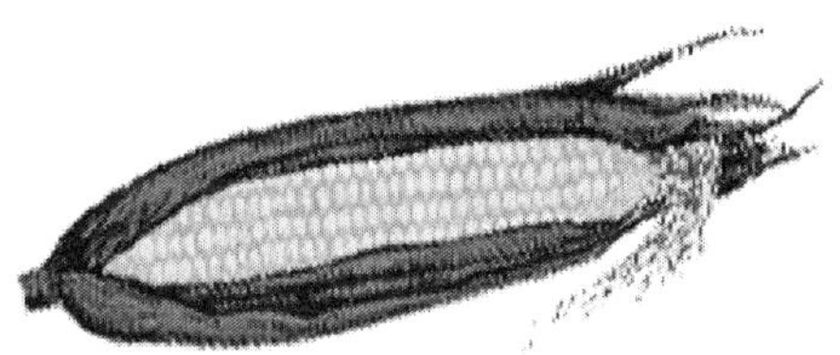

Problem 1

Consider two corn mutants: dwarf and glossy. Dwarf (gene symbol *d*) is a recessive trait characterized by short, compact plants. Glossy (gene symbol *gl*) is also recessive and is characterized by a bright leaf surface. F_1 progeny from a cross of pure breeding parents were back-crossed to the recessive parent. The results of this cross are shown below.

F_2:		
	286	wild type
	89	glossy
	97	dwarf
	277	glossy dwarf

Since the ratio of phenotypic classes is not the expected 1:1:1:1 for a test cross, what concept best explains the observed phenotypic ratio?

What phenomenon explains the four phenotypic classes of unequal ratios instead of two phenotypic classes as predicted by the tight linkage?

Complete the genotype of each F_2 progeny class. Which phenotypic classes carry the recombinant genotypes?

What approximate percentage of progeny is descended from distinguishable recombinant gametes?

What is the map distance between dwarf and glossy determined from these numbers?

Using the known map distance between glossy and dwarf determined above, what is the expected number of each F_2 phenotypic class out of a total of 1000 F_2 progeny?

Problem 2

In *Drosophila*, the forked phenotype is characterized by short bristles with split ends. The scalloped phenotype is characterized by scalloped wings at the margins and thicker wing veins. Both genes are X-linked and are marked with recessive mutant alleles. In the cross below, an F_1 is generated by mating a wild-type female with a scalloped, forked male to produce all wild-type progeny. The females are then mated to their fathers (a test cross) to generate an F_2.

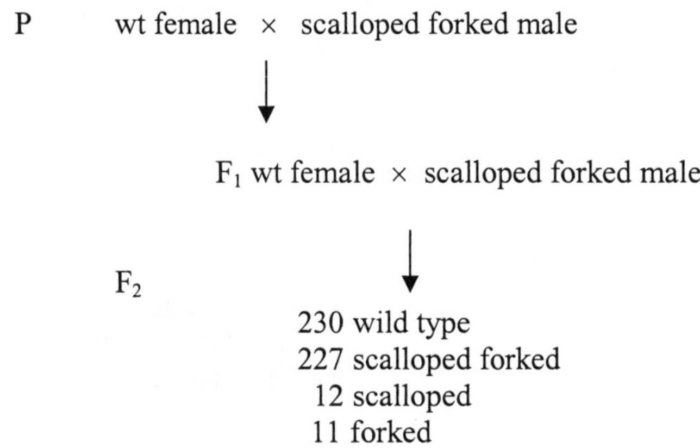

What concept best explains the observed phenotypic ratio of the F_2 progeny?

Which of the phenotypes belongs to F_2 progeny that that are descended from distinguishable crossover bearing (recombinant) gametes?

What approximate fraction of progeny is descended from distinguishable crossover bearing gametes?

What is the map distance between forked and scalloped?

Problem 3

The *Drosophila* forked phenotype is X-linked recessive (as we saw in Problem 2) and is characterized by short bristles with split ends. The miniature phenotype, also X-linked recessive, is characterized by reduced wing size.

A miniature female is crossed with a forked male. The F_1 progeny consists of wild-type females and miniature males. These are then mated to yield the following F_2:

510 miniature females	418 miniature males
490 wild-type females	412 forked males
	87 miniature forked males
	83 wild-type males

What is the approximate distance in map units (mu) between forked and miniature that you calculate from this cross?

When a wild-type female is crossed with a miniature forked male, and the wild-type F_1 females are testcrossed, the F_2 are as shown below:

84 miniature
86 forked
417 miniature forked
413 wild type

Is the distance you find in mu the same in this cross as the distance you found for the markers in trans?

In Problem 2 you found that the distance between the scalloped gene and the forked gene is 5 mu. In this problem you found that the distance between miniature and forked is 17 mu. Is this information sufficient to find the gene order of the three genes on the X chromosome?

When a wild-type female is crossed with a miniature scalloped male, all the progeny are wild type. A testcross of the heterozygous females yields the following F_2:

51 miniature
45 scalloped
354 miniature scalloped
350 wild type

What is the approximate map unit (mu) distance between miniature and scalloped that you calculated from this cross?

Now that you know the distances between each pair of genes, construct a map of the three genes.

Problem 4

In *Drosophila*, the spineless mutant is characterized by shortened bristles compared to the longer bristles of the wild type. The radius incomplete mutant is characterized by an incomplete wing vein pattern. Both genes are autosomal and are marked by recessive mutant alleles.

A wild-type female is mated with a spineless, incomplete male. Then, an F_1 wild-type female is crossed to the parental male to generate the following F_2:

449 wild type
452 spineless incomplete
52 spineless
48 incomplete

In this cross you did not obtain the expected 1:1:1:1 ratio expected for a test cross of unlinked genes. This implies that the genes are linked. Use these data to calculate the distance between spineless and incomplete.

A wild-type male is mated with a spineless, incomplete female. Then, the F_1 wild-type male is mated with the parental female. The following F_2 progeny are observed:

361 wild-type flies
357 spineless incomplete flies

Note that very different results are observed in this F_2 compared to the first cross. To help explain these unusual results, let's first assign genotypes to the F_1 male and tester female.

Write the genotypes of the F_1 flies in the cross above. Use the symbols *sp, sp+* for spineless and *ri* or *ri+* for radius incomplete. Indicate linkage by writing the two linked alleles on one side of a / (e.g., *a+b+/ab*).

After examination of these genotypes, does it make sense that these two genes are on the same chromosome?

Geneticists made a novel discovery when working with *Drosophila melanogaster.* Scientists found that male flies do <u>not</u> undergo recombination. Therefore, test cross of the F_1 male yields only two gamete types. The distance between genes cannot be measured using F_1 males. What gametes are produced in the F_1 male fly?

The consequence of no recombination in males is that crosses to map linked genes must use F_1 females as the dihybrid parent. This is true only in *Drosophila melanogaster*. In other species there is frequently a difference in recombination between the sexes, but it is not usually 0 in males.

Problem 5

In *Drosophila*, the X-linked genes singed (*sn*), characterized by bent bristles, miniature (*m*), reduced wing size, and tan (*t*), a tan body color, are marked with recessive alleles. In this problem you will use three-factor crosses to determine the gene order and distance between these markers. (Mating the F_1 flies is essentially a test cross, allowing one to analyze all of the F_2 flies directly).

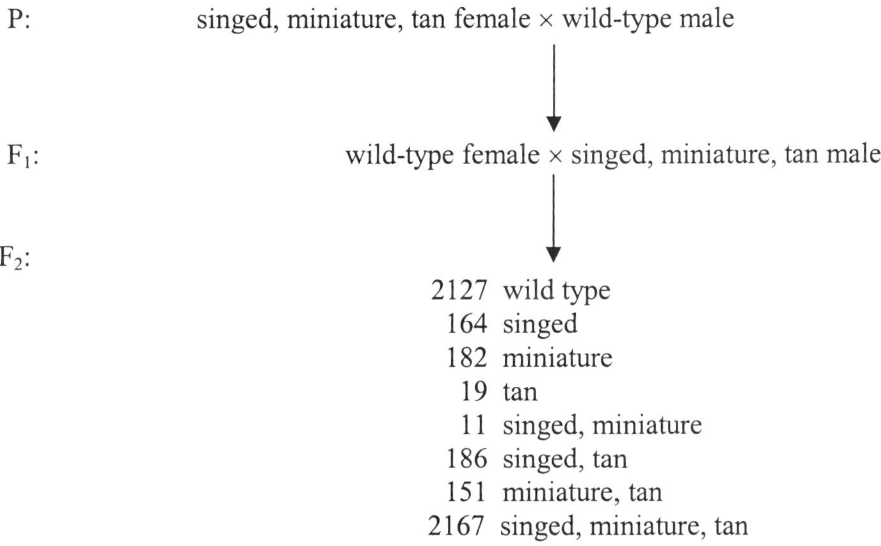

How many gamete types do you expect the F_1 female to produce?

Which two classes of gamete types are the produced most often by the F_1 female? (This is reflected in the phenotypic classes of the progeny.) These are the parental types.

Which of the reciprocal gamete types are the produced least often in the F_1 female? (This is reflected in the phenotypic classes of the progeny.) These are the double crossover types.

Compare the double crossover gametes with the parental gametes to determine which gene is in the middle. Write the correct gene order below.

Using your data, determine the distance between the singed gene and the tan gene, and the distance between the tan gene and the miniature gene.

Problem 6

Chromosome 13 of the mouse carries the locus for flexed tail with the alleles *f* and *f*+ and the locus for extra toes with the alleles *Et* and *Et*+. The flexed tail mutant is a recessive trait while the extra toes mutant is dominant. The next exercises will use 24 map units as the known distance between *Et* and *f* to predict the expected number of F_2 progeny from crosses.

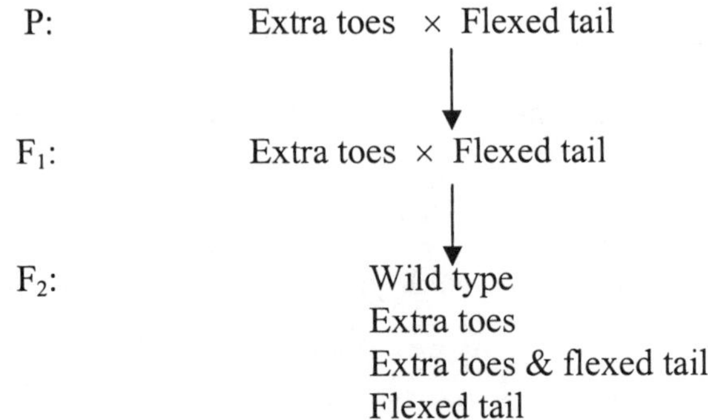

Find the genotype of each mouse above. Both parental mice are from pure breeding stocks. Use *Et* and *Et*+ for the extra toes gene, and *f* or *f*+ for the flexed tail gene.

Which F_2 progeny represent recombinants?

The distance between *Et* and *f* is 24 mu. From a total of a 1000 F_2 mice obtained from crosses such as this, how many are expected for each phenotypic class? (Start with the number of wild-type progeny.)

Wild type
Extra toes
Extra toes & flexed tail
Flexed tail

Let's now try to find the expected progeny for a three point cross. Chromosome 13 of mouse also carries the locus for satin fur texture with the alleles *sa* and *sa*+. The satin fur mutant is a recessive trait. Use the following map unit distances to map *sa*: *sa* – *Et*: 4 mu, *sa* – *f*: 20 mu. Draw a map of the three markers.

Use this map to determine the number of predicted progeny of each phenotypic class shown below with parents of genotypes *Et f*/*Et f* and *sa*/*sa*. The F_2 progeny classes from these crosses are as follows:

Wild type
Flexed tail, satin with extra toes
Flexed tail with extra toes
Satin with extra toes
Flexed tail, satin
Extra toes
Flexed tail
Satin

Bacterial Genetics

Goals for Bacterial Genetics:

1. Be familiar with the mechanism of conjugation and transduction as methods of gene transfer.
2. Determine genotype from ability to grow on different media.
3. Determine type of media needed for selection of exconjugants.
4. Be familiar with replica plating and interpretation of resulting data.
5. Map genes relative to other genes by interrupted mating, natural gradient of transfer, and variations of these techniques.

Conjugation
Problem 1

Various strains of bacteria were incubated on plates containing minimal media plus several amino acids. From the pattern of growth, answer the following questions regarding the genotype of each strain.

What are the genotypes of colonies 1–4 with respect to *arg*, *met*, *asp*, and *ile*?

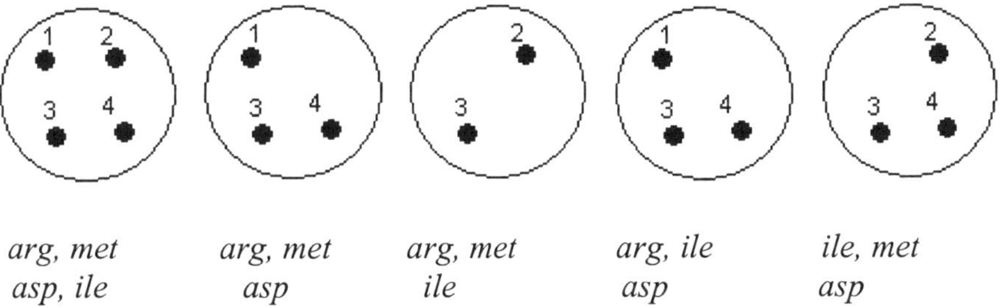

The following plates show bacterial strains that have been replica plated on minimal media plus the indicated mixture of amino acids and antibiotics.

Determine the genotypes of colonies 1–4 with respect to *arg*, *met*, *pen*, and *str*.

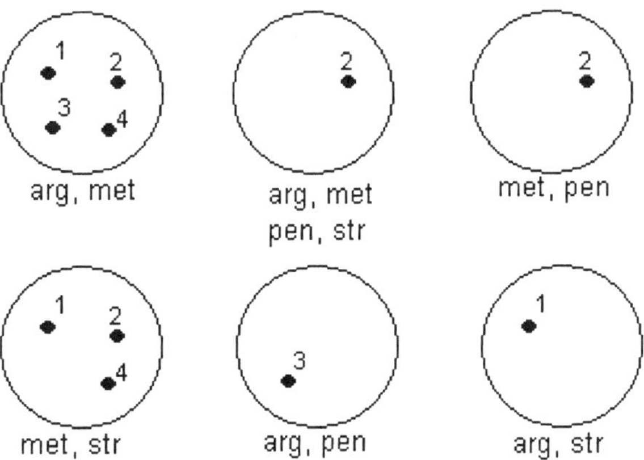

The plates shown below contain different sugars as carbon sources. Cells were first plated

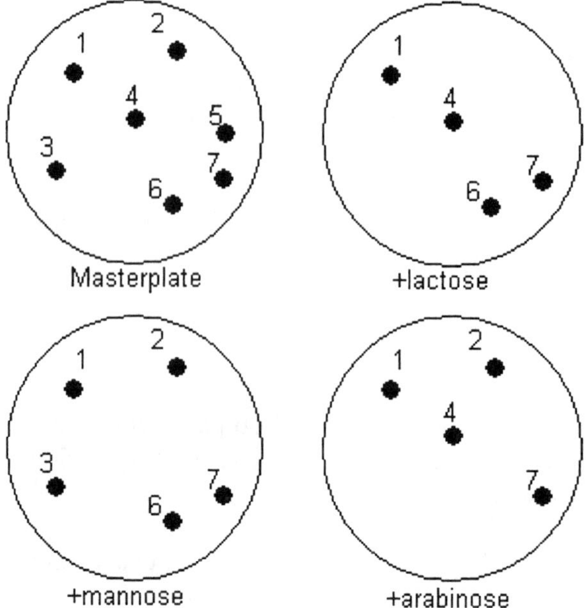

on minimal media with glucose as the carbon source (Masterplate) and then replica plated onto plates with lactose, mannose, or arabinose (no glucose).

Which colony has the genotype *lac− man+ ara+*?

Which colony has the genotype *lac+ man− ara+*?

Which colony has the genotype *lac+ man+ ara−*?

Which colony has the genotype *lac− man+ ara−*?

Which colony has the genotype *lac− man− ara−*?

Problem 2

Lederberg and Tatum made crosses by mixing pairs of strains. They used a C strain which was *arg−met+* and an E strain which was *arg+met−*. However, they did not know whether each of these strains was F−, F+ or Hfr. An example of the type of crosses they performed is shown below.

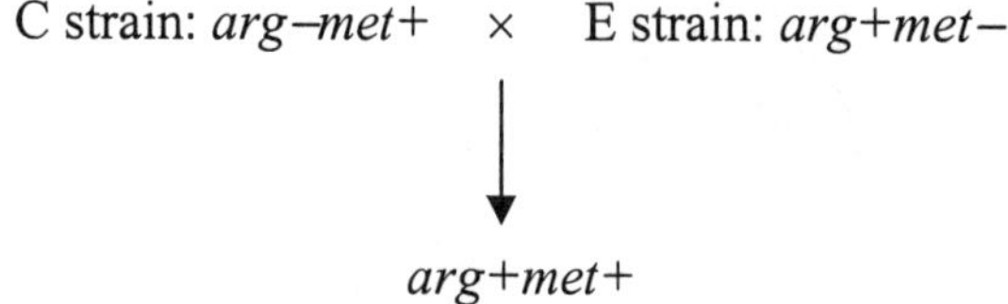

What type of cross would give you many exconjugants?

Which of the following crosses will give you only a few exconjugants?

What kind of crosses will give no *arg+ met+* exconjugants? (select from the choices below)

F+ × F+ **F− × Hfr** **F+ × Hfr**

F− × F− **F+ × F−** **Hfr × Hfr**

The table shown here contains the results of all possible crosses (0 = none; F = few; M = many). From the questions you just answered, you should now be able to identify which strains are F+, F−, and Hfr.

	E1	E2	E3	E4	E5
C1	F	0	M	F	0
C2	0	M	0	0	M
C3	0	F	0	0	F
C4	F	0	M	F	0
C5	0	F	0	0	F

Bonus Question

What we are looking for in our exconjugants are the recombinants of the E strains, *arg+met−*, with the C strains, *arg−met+*. What type of media should be used to select for the prototrophic exconjugants?

Problem 3

From the time that Hfr and F− cells are combined, the number of markers transferred for each mating pair depends on how long the two cells stay conjoined. The mating can be interrupted at specified times after the start of mating by vigorous shaking, frequently accomplished by blending. Without such mixing the disruption of the mating pairs occurs naturally generating a gradient of transfer.

In the following experiment, a prototrophic Hfr strain (streptomycin sensitive) is mated to a streptomycin resistant F− strain auxotrophic for methionine, leucine, and cysteine. A pipette is used to distribute cells to each of the plates shown below, each containing minimal media with streptomycin, methionine, and cysteine added in order to select only for leucine prototrophs.

Aliquots of culture A (Hfr cells only), culture B (Hfr cells + F− cells), and culture C (F− cells only) are added to the three plates containing streptomycin, methionine, and cysteine. The plates are then incubated. The results are shown in the following illustration. Genotypes: (Hfr: *str-s, met+, leu+, cys+*) (F−: *str-r, met−, leu−, cys−*).

A: ***str, met, cys***
No colonies

B: ***str, met cys***
375 colonies

C: ***str, met, cys***
No colonies

Why were no colonies observed after plating the Hfr strain on plate A above?

Why were no colonies observed after plating the F- strain on plate C above?

Why were cells able to grow on plate B?

The mating mixture produced 375 exconjugant colonies that replaced the *leu–* allele of the F– strain with the *leu+* allele of the Hfr strain. Additional markers may have also been transferred and could be checked by replica plating.

To measure the time of entry of each of the markers, matings between the Hfr and F– strains can be interrupted at various time points and plated on selective media. Below is a graph of colonies versus time.

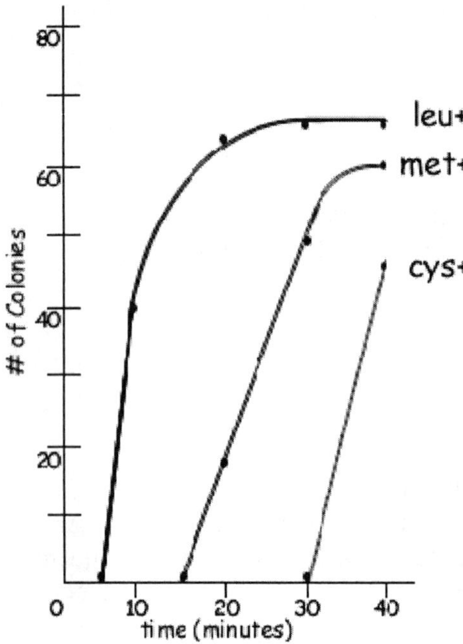

Draw a map below to indicate the order of the markers in the Hfr strain with respect to the origin.

Problem 4

An Hfr strain with genotype *met– leu+ his+ trp+* that transfers the *met* gene very late was mated with an F– *met+ leu– his– trp–* strain. After mating for 30 minutes, cells were plated on minimal media (MM) with the added nutrients listed below each plate. The number of colonies that grew on each plate is indicated.

Plate	**1**	**2**	**3**
Supplements	his, trp	his, leu	leu, trp
Colonies	250	50	500

What is the purpose of the methionine marker in this cross?

What markers are selected for on Plate 1?

What markers were selected on Plate 2?

What markers were selected on Plate 3?

Those markers closest to the origin are transferred first and yield the highest number of recombinant colonies. Based on the number of colonies on each plate above, determine the order of the markers in the Hfr strain with respect to the origin.

Problem 5

A small bacterial genome was mapped using three different Hfr strains and an F– str^r *ala– ade– bio– his– ile– val–*. Each of the Hfr strains was str^s and contained the wild-type alleles for all the markers. The matings were interrupted after 30 minutes and the exconjugants were selected on the following plates (the indicated nutrient is left out of the otherwise complete media + streptomycin).

Hfr: str^s *ala+ ade+ bio+ his+ ile+ val+* × F–: str^r *ala– ade– bio– his– ile– val–*

Mating	**Missing Nutrient**					
	Ala	**Ade**	**Bio**	**His**	**Ile**	**Val**
HfrA × F-	300	0	0	0	900	750
HfrB × F-	0	400	0	0	700	875
HfrC × F-	200	0	956	724	0	0

In the table above, some of the plates have no colonies. How can this be explained?

Consider the data from the Hfr A × F– cross. Which of the markers can be ordered using this cross?

What is the first marker to be transferred by Hfr A?

Order the three markers starting with *ile*.

What is the order of the markers indicated by the results of the Hfr B × F– cross?

What is the order of the markers indicated by the results of the third cross?

In every problem so far we have represented the bacterial chromosome as a circle. The original recognition of this fact comes from comparing the maps of different Hfr strains. As you can see from your results, the maps can only be reconciled as permutations from a common circular map, with the different Hfr strains having different origin points and directions of transfer. Use this information to construct a circular map.

Transduction
Problem 1

A generalized transducing phage is grown on a prototrophic strain and then used to transduce a recipient that is *arg– leu– gln– gua– thr–*. In this experiment *arg+* transductants will be selected by plating on media lacking arginine but containing all other supplements required by the recipient strain. This will be the master plate.

DONOR: *arg+ leu+ gln+ gua+ thr+*
RECIPIENT: *arg– leu– gln– gua– thr–*

MASTERPLATE: leu, gln, gua, thr (supplements)

Fifty-nine colonies grew on this master plate. To test for co-transduction of the unselected markers, the colonies were replica plated to different media, omitting one supplement at a time. The results are shown in the following table.

Supplements to Minimal Media					
Plate	Leucine	Guanine	Glutamine	Threonine	Colonies
1	–	+	+	+	0
2	+	–	+	+	0
3	+	+	–	+	24
4	+	+	+	–	0

What does the absence of growth indicate on Plate 1?

What does the presence of growth indicate on Plate 3?

What is the co-transduction frequency (in percent) of *arg* and *gln*?

To develop a genetic map, let us now repeat the transduction experiment and make a new master by plating on media lacking leucine but containing all other supplements required by the recipient strain. This is our second master plate. Seventy-three colonies grew on this plate and were replica plated onto various plates containing the media indicated in the table below.

Supplements to Minimal Media					
Plate	Arginine	Guanine	Glutamine	Threonine	Colonies
1	–	+	+	+	0
2	+	–	+	+	22
3	+	+	–	+	0
4	+	+	+	–	45

What is the co-transduction frequency (in percent) of *leu* and *gua*?

What is the co-transduction frequency (in percent) of *leu* and *gln*?

To finish our experiment, we will do one more transduction. In this third master plate our media lacks guanine but contains all other supplements required by the recipient strain. Sixty-five colonies grow on this plate which is replica plated onto various plates containing the media indicated in the table below.

Supplements to Minimal Media					
Plate	Arginine	Leucine	Glutamine	Threonine	Colonies
1	–	+	+	+	0
2	+	–	+	+	20
3	+	+	–	+	13
4	+	+	+	–	0

What is the co-transduction frequency of *gua* and *gln*?

Using these co-transduction frequencies you have determined, determine the order of the three markers: *arg*, *gua,* and *gln*.

Problem 2

In a transduction experiment, the donor strain is *kan*^{r} *lys*+ *arg*+ and the recipient strain is *kan*^{s} *lys*– *arg*–. Transductants are plated on MM (minimal media) supplemented with kan (kanamycin), lys (lysine), and arg (arginine) and then replica plated on the plates shown below.

Supplements to Minimal Media				
Plate	**Kan**	**Lys**	**Arg**	**# of colonies**
Master	+	+	+	500
Replica 1	+	–	–	20
Replica 2	+	+	–	21
Replica 3	+	–	+	200

In the table below, fill in the number of colonies for each of the four genotypes.

Genotype	**# of colonies**
kanr arg+ lys+	
kanr arg+ lys–	
kanr arg– lys+	
kanr arg– lys–	

What is the co-transduction frequency of *kan* and *arg*?

What is the co-transduction frequency of *kan* and *lys*?

Problem 3

This problem is a simulation and can only be done on the *Interactive Genetics* CD-ROM.

Problem 4

Frequently, mutations are so close together that it's impossible to determine their order with respect to a nearby marker by simply using conjugation or transduction.

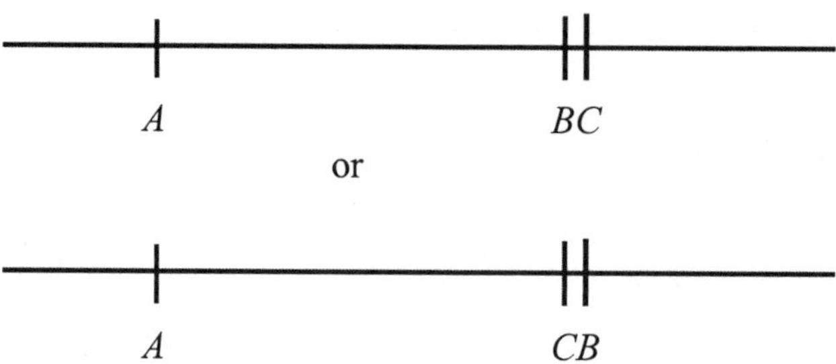

In the illustration above it is difficult to determine if *B* is closer to *A* or *C* is closer to *A*. The co-transduction frequency between *A* and *B* would be about the same as the co-transduction frequency between *A* and C. However, reciprocal three-factor transductions can be used to determine the order of these markers.

Unlike in plants and animals, reciprocal transductions in bacteria do not refer to switching markers between sexes. Instead, in bacteria we refer to switching the markers between the donor and recipient. That is, in Transduction I the donor may be *B+C–* and the recipient *B–C+*, while in Transduction II (the reciprocal) the donor may be *B–C+* and the recipient *B+C–*. Having the markers in trans is essential for this experiment since it forces a crossover between the two tightly linked markers (*B* and *C*) to generate a prototroph. The outside marker will be *A+* in the donor and *A–* in the recipient in all transductions.

In the two reciprocal transductions shown below the order is assumed to be *ABC*. Draw an X in the appropriate places to generate prototrophic (*A+B+C+*) transductants for each experiment.

Transduction I: Donor: *A+B+C–* Recipient: *A–B–C+*

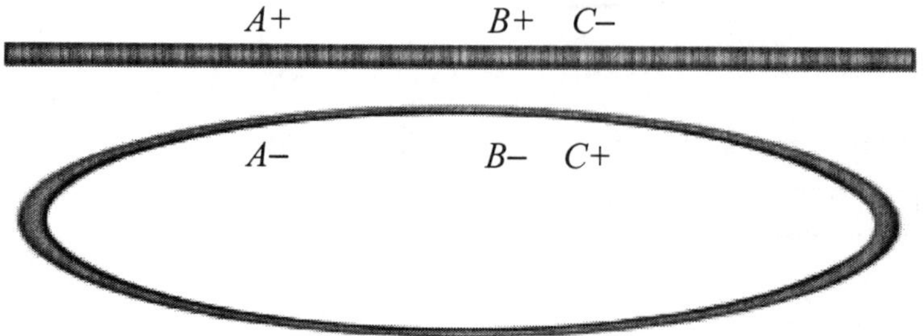

Transduction II: Donor: *A+B–C+* Recipient: *A–B+C–*

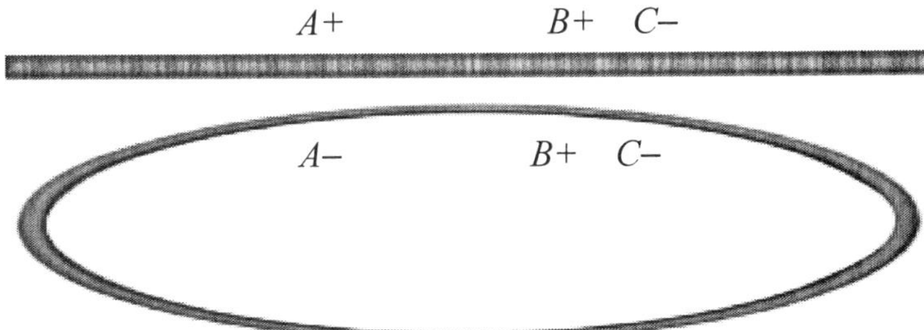

Notice that in Transduction I, prototrophs were generated with only two crossovers, whereas in Transduction II, four crossovers were required to generate *A+B+C+*. This is true only when the order is *ABC*.

Now we will see what happens if the order is *ACB*. Again, draw X's in the appropriate places to generate prototrophic (*A+B+C+*) transductants.

Transduction I: Donor: *A+B+C-* Recipient: *A–B–C+*

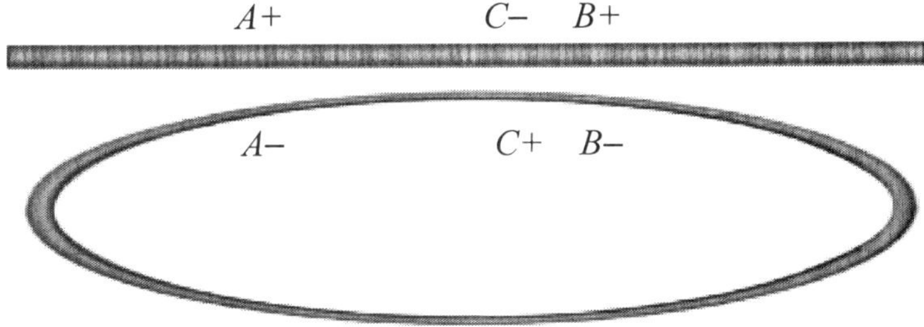

Transduction II: Donor: *A+B-C+* Recipient: *A–B+C–*

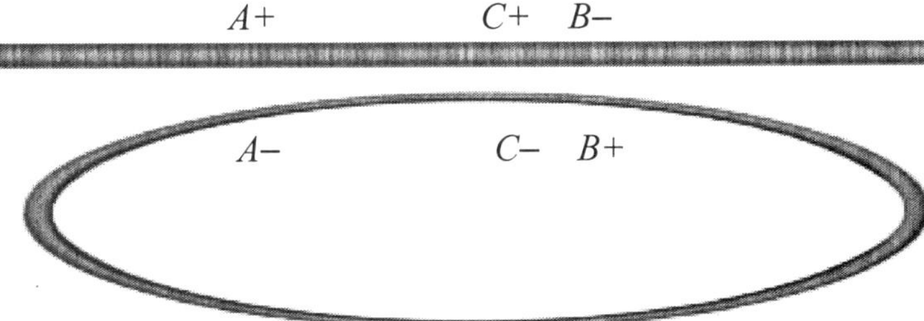

Notice that Transduction I prototrophs were generated with four crossovers, whereas in Transduction II only two crossovers were required to generate *A+B+C+*.

A minimum of two crossovers is required to incorporate genes from the donor fragment into the circular recipient chromosome. This results in the frequencies of transduction we usually observe. However, a situation that requires four crossovers will yield many fewer transductants or none at all.

You need to compare reciprocal transductions in order to determine which experiment gives you normal transduction frequencies and which gives you reduced frequencies. This example asks you to determine the frequencies using two different orders, although there is only one order.

Reciprocal transductions were used to order three mutations (*trp1, trp2,* and *trp3*) required for metabolism of tryptophan with respect to a nearby marker tyrosine (*tyr+*). For each pair of mutants (*trp1* and *trp2; trp2* and *trp3; trp1* and *trp3*) a pair of reciprocal transductions will be made to order them with respect to *tyr*.

	Donor	Recipient
Transduction I	*tyr+ trpx– trpy+*	*tyr– trpx+ trpy–*
Transduction II	*tyr+ trpx+ trpy–*	*tyr– trpx– trpy+*

Experiment	x y	I	II
1	*trp1 trp2*	800	5
2	*trp2 trp3*	3	750
3	*trp1 trp3*	659	8

Which mutation, *trp1* or *trp2*, is closer to *tyr*?

Which mutation, *trp1* or *trp3*, is closer to *tyr*?

Which mutation, *trp2* or *trp3*, is closer to *tyr*?

Draw a map below to indicate the order of the three mutations with respect to *tyr*.

Biochemical Genetics

Goals for Biochemical Genetics:

1. Determine the number of genes coding for steps in a biosynthetic pathway. Use auxotrophic mutant strains to determine the number of complementation groups for a pathway.
2. Determine a biosynthetic pathway from nutritional studies that supplement with compounds that may be intermediates in the pathway.
3. Determine whether any of the genes are linked to each other.
4. Distinguish the metabolism of amino acids between anabolic (biosynthetic) and catabolic (degradative) reactions. Most human biochemical disorders show up from the loss of a catabolic enzyme. Auxotrophic mutant strains of microorganisms result from the loss of an anabolic enzyme.

Problem 1

Six *Neurospora* mutants were isolated that require vitamin B1 to grow. To determine whether the six mutants have mutations in different genes, complementation studies were performed. Heterokaryons were formed and tested for growth on minimal media.

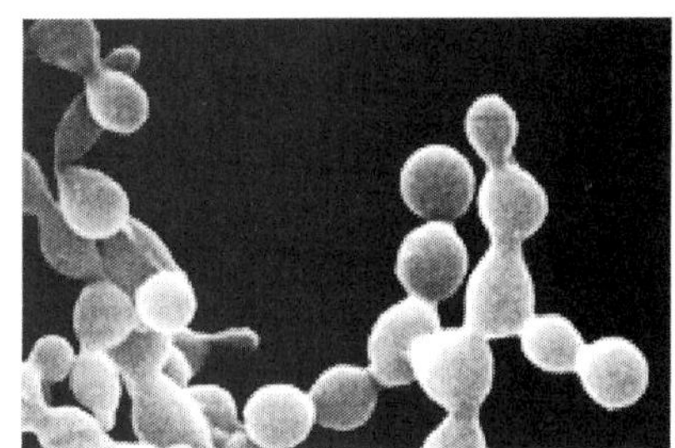

The following table shows the experimental results; + indicates heterokaryon growth and – indicates no heterokaryon growth in the absence of vitamin B1.

	A	B	C	D	E	F
A	–	+	+	–	+	+
B	+	–	+	+	–	–
C	+	+	–	+	+	+
D	–	+	+	–	+	+
E	+	–	+	+	–	–
F	+	–	+	+	–	–

Which mutants belong to the same complementation group as mutant A?

Which mutants belong to the same complementation group as mutant B?

Which mutants belong to the same complementation group as mutant C?

Which mutants belong to the same complementation group as mutant D?

How many complementation groups are there?

Would a heterokaryon formed from mutant A and the double mutant C,E grow on minimal media? The double mutant strain contains two gene mutations: the gene mutation in strain C and the gene mutation in strain E.

Would a heterokaryon formed from the double mutants A,B and C,F grow on minimal media?

Would a heterokaryon formed between the three double mutants A,E and B,C and C,D grow on minimal media?

Problem 2

You are a graduate student working in a *Neurospora* lab and have recently isolated five *Neurospora* methionine auxotrophs that each contain a single gene mutation. You set out to determine if any of the mutated genes in the *Neurospora* strains are linked. You decide to do this by crossing mutant strains and examining the phenotypes of the progeny.

You form a transient diploid by mating mutant strains (represented by the cross $ab+ \times a+b$). The diploids then undergo meiosis immediately by sporulation. A diploid cell with a crossover in the interval between the linked *a* and *b* genes (in this example) will yield four genetically distinct spores. Three of them are still auxotrophs and only the $a+b+$ spore is phenotypically distinct as a prototroph. The frequency of prototrophs in a large collection of spores can be used to determine the map distance between two mutants.

The results of your crosses with the five *Neurospora* auxotrophic strains (A–E) in all combinations are shown below. For each cross, 1000 ascospores were plated on minimal media. The table shows the number of methionine prototrophs that you recovered from each cross.

	A	B	C	D	E
A	0	150	250	215	40
B		0	250	65	110
C			0	250	250
D				0	175
E					0

Since strains A–E are methionine auxotrophs, the prototrophic ascospores must have been produced from independent assortment or recombination between two mutated genes during meiosis. How many prototrophic ascospores would you expect from a cross if the recombinant progeny were produced by independent assortment?

For each of the linked genes in the table above, determine the map distance between them. Then, combine the information from each of the two factor crosses to assemble a map of the entire linkage group.

Problem 3

Wild-type *Neurospora* is orange when exposed to light during growth. You have isolated three albino mutants that are completely white even when grown in the presence of light. Using heterokaryon complementation studies, you find that these three strains have mutations in different genes, which you name *al-1*, *al-2*, and *al-3*.

You suspect that the albino mutants are white because they are unable to make the carotenoid pigment (known to cause the orange color). To test your idea, you seed the albino mutants on media supplemented with carotenoid pigment.

You are thrilled to find that supplementation with carotenoid pigment results in orange colored hyphae for all three albino mutants! Assuming carotenoid synthesis is affected in these mutant strains, you set out to determine which step in the biosynthetic pathway is blocked by each of the mutations.

Luckily, three precursors in the carotenoid pigment biosynthetic pathway had previously been discovered, although their order was unknown. To determine the order in which the precursors are converted into the carotenoid pigment, you grow the albino mutants on media supplemented with each of the three precursors.

You find that supplementing media with these different precursors restores wild-type color in some of the albino mutant. You compile your results into a simple table, where + indicates wild-type color and – indicates white color.

	Carotenoid pigment	**GGPP**	**Phytoene**	**PPP**
al-1	+	–	–	–
al-2	+	–	+	+
al-3	+	–	+	–

Use the data from your experiments to determine the order of the precursors as they appear in the biosynthetic pathway and which step in the biosynthetic pathway is blocked by each mutant.

Problem 4

Saccharomyces cerevisiae has played a fundamental role in human history and culture. For centuries, yeast has been used by humans for the rising of bread and for the fermentation of wines and beers. Today yeast is used as a model organism for both genetic and cellular biology studies. Yeast are eukaryotes that grow simply as either single cells or colonies.

Saccharomyces cerevisiae can exist as either a haploid or a diploid. Both haploids and diploids are able to grow and divide by mitosis, through a process called budding. This differs from *Neurospora* where the diploid zygote is transient and quickly undergoes meiosis to form ascospores. In the haploid state there are two mating types, a and α. When an a cell and an α cell are brought together, they fuse to form a diploid cell. First the cellular membranes fuse, then the nuclei (yeast do not normally form stable heterokaryons). The diploid cell can grow and divide indefinitely. However, when nutrients are depleted or when environmental conditions become unfavorable, the diploid cells will undergo meiosis to form four haploid spores (a process called sporulation). The spores can be separated manually and tested directly for phenotype. Since they are haploid, their phenotype directly reflects their genotype.

Ten mutant yeast strains have been isolated that cannot grow on medium with galactose as the sole carbon source (galactose medium). The ten strains are named A-J. Each mutant was separately crossed to the others to form a set of diploid strains. The ability of the diploids to grow on galactose medium was used as a measure of complementation. The results of this experiment are given in the chart below.

Gal Mutants

	A	B	C	D	E	F	G	H	I	J
A	–	+	+	+	+	–	–	+	–	+
B		–	+	+	+	+	+	–	+	+
C			–	+	+	+	+	+	+	+
D				–	–	+	+	+	+	–
E					–	+	+	+	+	–
F						–	–	+	–	+
G							–	+	–	+
H								–	+	+
I									–	+
J										–

Are mutants A and B in the same complementation group?

Are mutants A and F in the same complementation group?

How many complementation groups are present?

To determine whether the four genes identified were linked to one another, each diploid strain was sporulated and 1000 random spores were analyzed for growth on galactose medium. The number of spores that were able to grow on galactose medium is indicated in the table below.

	A	B	C	D	E	F	G	H	I	J
A	0	250	0	0	0	0	0	250	0	0
B		0	250	250	250	250	250	0	250	250
C			0	0	0	0	0	250	0	0
D				0	0	0	0	250	0	0
E					0	0	0	250	0	0
F						0	0	250	0	0
G							0	250	0	0
H								0	250	250
I									0	0
J										0

Since we know certain strains contain mutations in the same gene, we can simplify the chart above to reflect the same data by grouping together mutants from the same complementation group. We will arbitrarily give each complementation group a *gal* gene designation as follows: Mutants A, F, G, I—*gal1*; Mutants B, H—*gal4*; Mutant C—*gal7*; Mutants D, E, J—*gal10* (these designations are given based on actual *gal* genes in yeast).

	gal1	***gal4***	***gal7***	***gal10***
gal1	0	250	0	0
gal4		0	250	250
gal7			0	0
gal10				0

Is *gal1* linked to *gal4*?

gal1 shows no wild-type recombinants with either *gal7* or *gal10*. How can you explain this observation?

Analysis of 100,000 haploid spores produced from each of the crosses above were plated on galactose medium. Use this data to determine the map distance between *gal1* and *gal7*, *gal1* and *gal10*, and *gal7* and *gal10*.

	gal1	***gal4***	***gal7***	***gal10***
gal1	0	25000	35	16
gal4		0	25000	25000
gal7			0	18
gal10				0

Problem 5

This problem is a simulation and can only be done on the *Interactive Genetics* CD-ROM.

Population Genetics

Goals for Population Genetics:

1. Be able to relate allele frequencies at a gene to population homozygote and heterozygote frequencies.
2. Be able to use Chi-square test to determine if a population is in Hardy-Weinberg equilibrium.
3. Combine family studies with population frequencies to predict the chance a child will be homozygous for a disease allele.

Problem 1

Do you know your genotype with respect to your ABO or MN blood types? Will you consider these genotypes when you decide to marry? Most people are unaware of the alleles they carry for the majority of their genes. How do you study the genetics of animals in a natural environment, where family units are usually impossible to discern (e.g., fish). Because of the lack of pedigree data, individuals are regarded only as samples from the larger population.

A population, consisting of interbreeding individuals in a prescribed geographical area, contains a reservoir of all the gene copies (alleles) that will give rise to the individuals in the next generation. In these and similar examples, the population's allele frequencies are used to predict the genotype frequencies of the individuals.

The reservoir of alleles for a single gene is referred to as the gene pool for that gene. The genotypes of individuals can be considered a random (unbiased) sampling from the gene pool, with a gamete representing a single sample from the gene pool. In the following sets of questions we are going to use a bowl of ping-pong balls (whose different colors represent different alleles) to simulate random sampling.

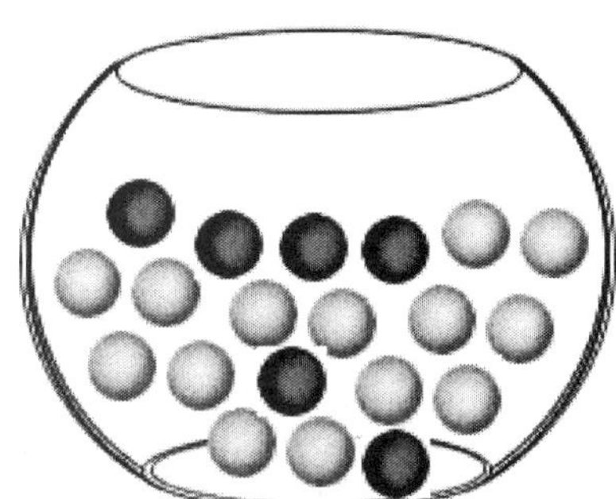

To represent diploid individuals subsequent questions will use pairs of samplings. What is the frequency of black ping-pong balls in the adjacent bowl?

Note that the probability of randomly obtaining a black ball, represented by p, is the frequency of black balls in the bowl.

What is the frequency of the white ping-pong balls in this bowl?

The frequency of white balls, represented by q, is the probability of randomly obtaining a white ball. Here black and white are the only colors of Ping-Pong balls in this bowl, and $p + q = 1$. To represent diploid individuals subsequent questions will use pairs of samplings.

What is the probability of picking two black balls (homozygous black)?

What is the probability of picking two white balls (homozygous)?

What is the probability of picking one white ball and one black ball (heterozygous) in either order?

As you have probably noticed, the sum of the probabilities found equals 1, that is $p^2 + 2pq + q^2 = 1$. This equation, which is derived from $(p + q)^2 = 1$ (when there are two alleles for a certain trait), describes the expected genotype frequencies of a population in Hardy-Weinberg equilibrium.

Problem 2

Consider this hypothetical population of 25 individuals of the FISH species. The sum total of all alleles present in the population is referred to as the population's gene pool. We will examine only one gene locus with two possible alleles: *F* for green color, and *f* for white color. In this population, the solid green fish are homozygous for *F*, the solid white fish are homozygous for *f*, and the spotted fish are heterozygous. Answer the following questions using the observed phenotypes.

Using the fish above, calculate p, the frequency of the green allele.

What is q, the frequency of the white allele? Recall, in the population above, ten fish are *FF*, five fish are *ff*, and ten fish are *Ff*.

To determine if this population is in equilibrium you need to determine the expected numbers using the Hardy-Weinberg equilibrium equation. To begin with, what is the expected frequency of *FF* individuals?

What is the expected number of FF individuals in this population?

What is the expected number of *ff* fish? What is the expected number of *Ff* individuals?

To find whether or not the population depicted above is in HWE (Hardy-Weinberg equilibrium), we will perform a chi-square (χ^2) test, using expected and observed values of each genotype. The χ^2 test is used to determine whether deviation from expected values are due to chance alone. If the deviations are too large we conclude that something besides chance is involved and the population is not in HWE. What is the value of χ^2 for this example? This χ^2 value can be converted into a probability value. To do this, we need the number of degrees of freedom (df) for the particular χ^2 test. The df equals the number of variables - 1. Although there are three genotypic classes, their numbers are determined by only two variables (p and q). Therefore, the degrees of freedom is 1. Using a χ^2 distribution chart, determine the probability (the p value) of obtaining these deviations due to chance alone. From your χ^2 test, is the population depicted above in Hardy-Weinberg equilibrium (HWE)?

Problem 3

Men are from Mars. Women are from Venus. When Mars and Venus collide, a new population arises. The allele frequencies on Mars and Venus differ. In the following problem, two alleles of a gene are represented by gray and blue ping-pong balls.

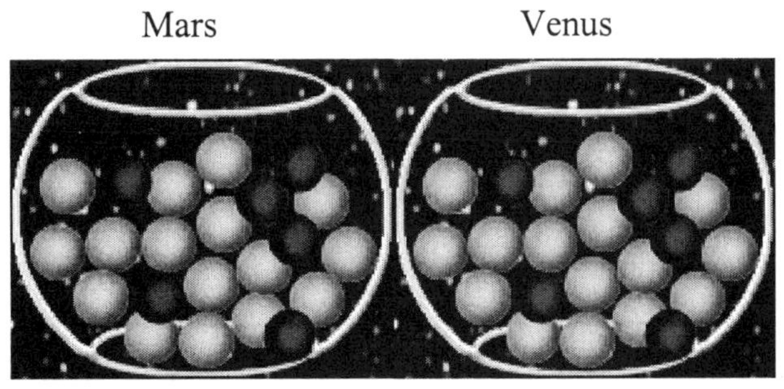

Predict the genotype frequency for homozygous gray in the new population.

What proportion of homozygous blue individuals do you expect in the new population?

What is the proportion of heterozygotes you expect the new population to have?

If the new population breeds only among themselves, will the next generation be in Hardy-Weinberg equilibrium?

Problem 4

The ability to taste the chemical phenylthiocarbamide (PTC) is an autosomal dominant phenotype. Tasters can detect it as extremely bitter, while nontasters cannot detect it at all. The genotype frequency of the homozygous recessive individuals (nontasters) in the U.S. population is 0.35, or 35%. Note that this population is in Hardy-Weinberg equilibrium (HWE) for this trait. What is the allele frequency for the nontaster allele?

q is commonly used to represent the allele frequency of the recessive (nontaster) allele. q^2, therefore, represents the genotype frequency of the homozygous recessive, nontaster individuals with a value of 0.35. What is the allele frequency for the dominant taster allele?

p is commonly used to represent the allele frequency of the dominant (taster) allele. After finding allele frequencies, we will use these values to find genotype frequencies. Answer the following questions using p and q. What is the frequency of heterozygotes in the general population?

What is the frequency of heterozygotes among the PTC taster population only?

A PTC taster man whose mother is a taster while his father is not, marries a taster woman. Recall that PTC tasting is an autosomal dominant trait. What is the probability that their child will be a nontaster?

Problem 5

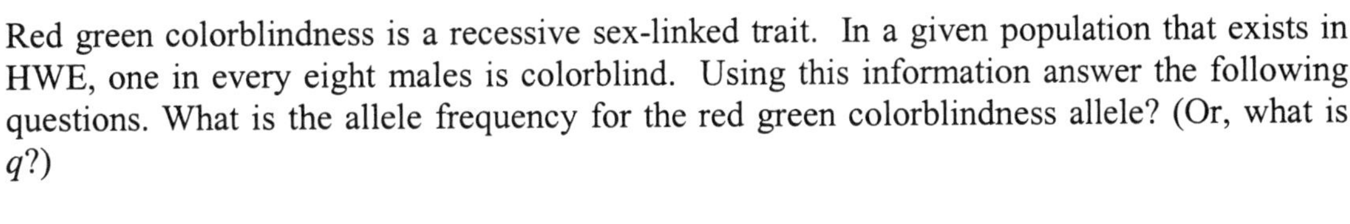

Red green colorblindness is a recessive sex-linked trait. In a given population that exists in HWE, one in every eight males is colorblind. Using this information answer the following questions. What is the allele frequency for the red green colorblindness allele? (Or, what is q?)

What proportion of all women are colorblind?

In what proportion of marriages will all the males be colorblind but none of the females?

In what proportion of the marriages will all the males be normal and the females be heterozygous?

Molecular Markers

Goals for Molecular Markers:

1. Recognize the kinds of variation in DNA sequences between homologous chromosomes that can be used as codominant alleles, including restriction fragment site polymorphisms and variations in the number of repeat sequences.
2. Understand the techniques (Southern Analysis and PCR) used to detect these variations within a defined chromosome region.
3. Understand how molecular markers are used for linkage studies to locate a disease gene and to start the positional cloning of the gene.
4. Use molecular markers to determine the genotype of an individual for disease prediction, for relationship studies between individuals, and for forensic identification of unknown individuals.

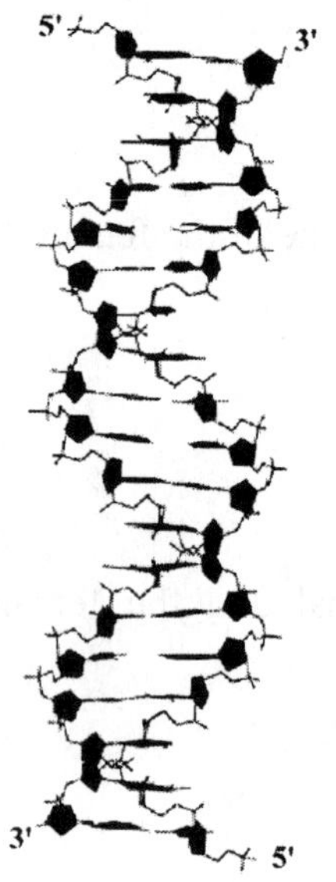

Problem 1

Restriction fragment length polymorphisms are identified by screening DNA isolated from members of families, using an array of different restriction enzymes. Random human genomic clones are then used as probes in a Southern blot. In the following problems you must identify the probe and restriction enzyme that gives rise to a polymorphism and then determine if the RFLP is linked to the inherited trait.

The DNA sample from each person is split into six tubes and digested with the following enzymes:

Apa I, Bam HI, Eco RI, Hind III, Sal I, Xma I

Gel electrophoresis separates the DNA fragments according to size. Denaturation of the DNA and transfer to nitrocellulose allows specific bands to be detected by hybridization with a radioactive probe (Southern blot). Choose the probe and restriction enzyme below that has an RFLP linked to the disease gene indicated by the pedigree.

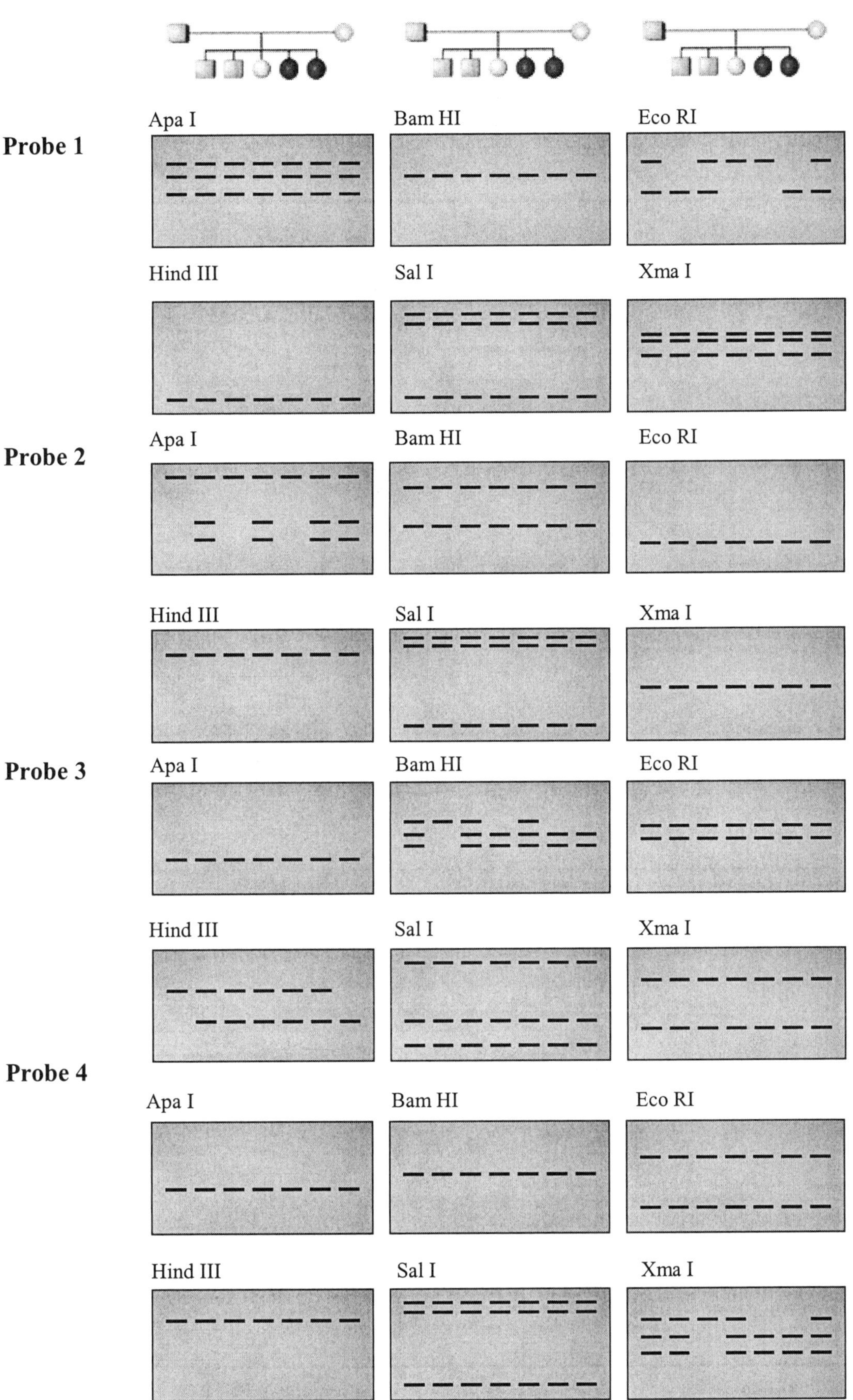
Probe 1
Apa I
Bam HI
Eco RI
Hind III
Sal I
Xma I
Probe 2
Apa I
Bam HI
Eco RI
Hind III
Sal I
Xma I
Probe 3
Apa I
Bam HI
Eco RI
Hind III
Sal I
Xma I
Probe 4
Apa I
Bam HI
Eco RI
Hind III
Sal I
Xma I

Problem 2

Part a

In the pedigree below, the presence of the rare recessive disease phenylketonuria (PKU) is indicated by filled symbols. Using the *A/a* above, assign genotypes based on the pedigree analysis.

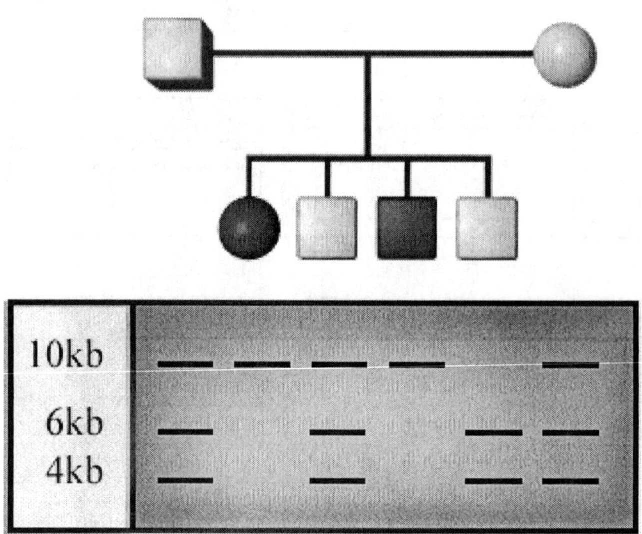

DNA was obtained from each of the individuals in the pedigree above, digested with a restriction enzyme and probed with a fragment known to hybridize to a linked RFLP. Analysis of the Southern blot above in conjunction with the pedigree allows determination of the linkage between the *A/a* gene and the RFLP shown. Focus first on the affected children. Use this information to determine the coupling (linkage) in each of the parents and then in the unaffected children. On the parental chromosomes shown below, fill in the galactosemia alleles (*A/a*) and the RFLPs (10 or 6-4) on the chromosomes to indicate linkage.

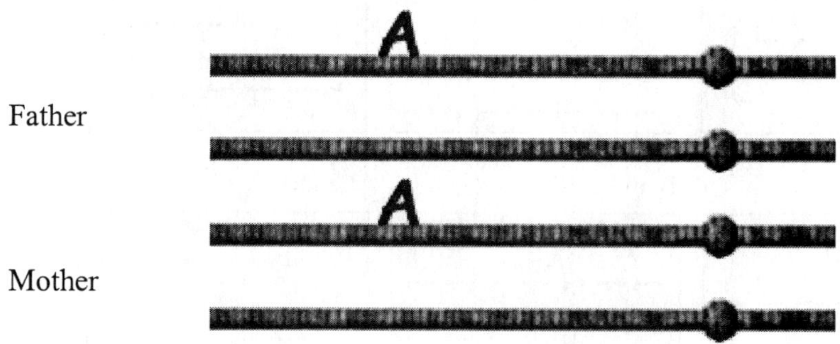

Part b

As a second example, let's look at linkage between the gene causing galactosemia and a nearby RFLP. In the pedigree above, the presence of the rare disease galactosemia is indicated by filled symbols. Using *A* and *a*, assign genotypes to each individual.

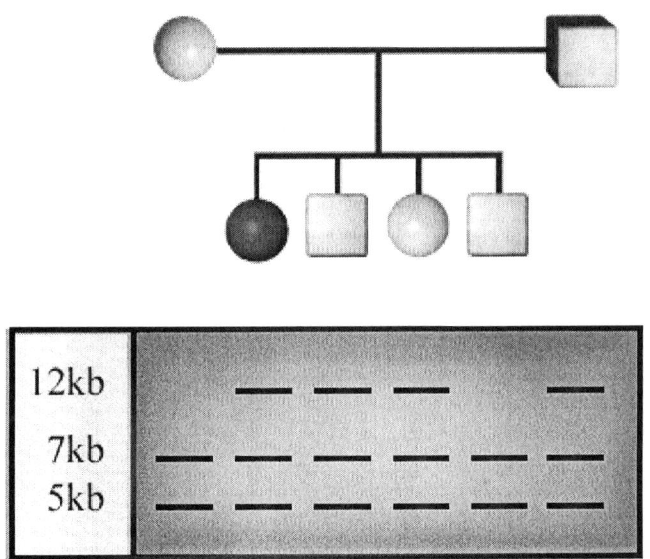

DNA was isolated from each of the individuals in the pedigree above and a probe from a linked RFLP was used in a Southern analysis to determine the genotype of each of the members of this family. On the parental chromosomes shown below indicate the linkage between the galactosemia alleles (*A*/*a*) and the RFLPs (12 or 7-5).

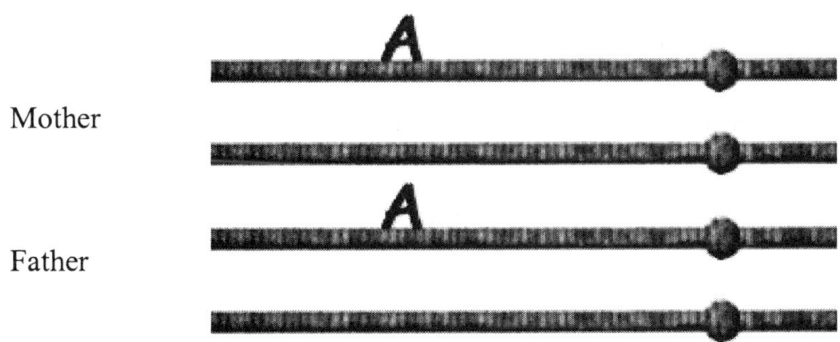

What is the probability that Child 3 is a carrier of galactosemia? (Assume the RFLP is tightly linked to the galactosemia gene and no crossing over occurs between them).

What is the probability that Child 4 is a carrier of galactosemia? (Again assume tight linkage and no crossing over.)

Problem 3

Part a

The pedigree shown below traces the inheritance of polydactyly, associated with a dominant, autosomal allele (*D*). Southern analysis reveals a closely linked RFLP. On the chromosomes shown below, indicate this linkage by typing in the appropriate *D* or *d* alleles and the linked 8 kb (8) or 6-2 kb (6-2) alleles.

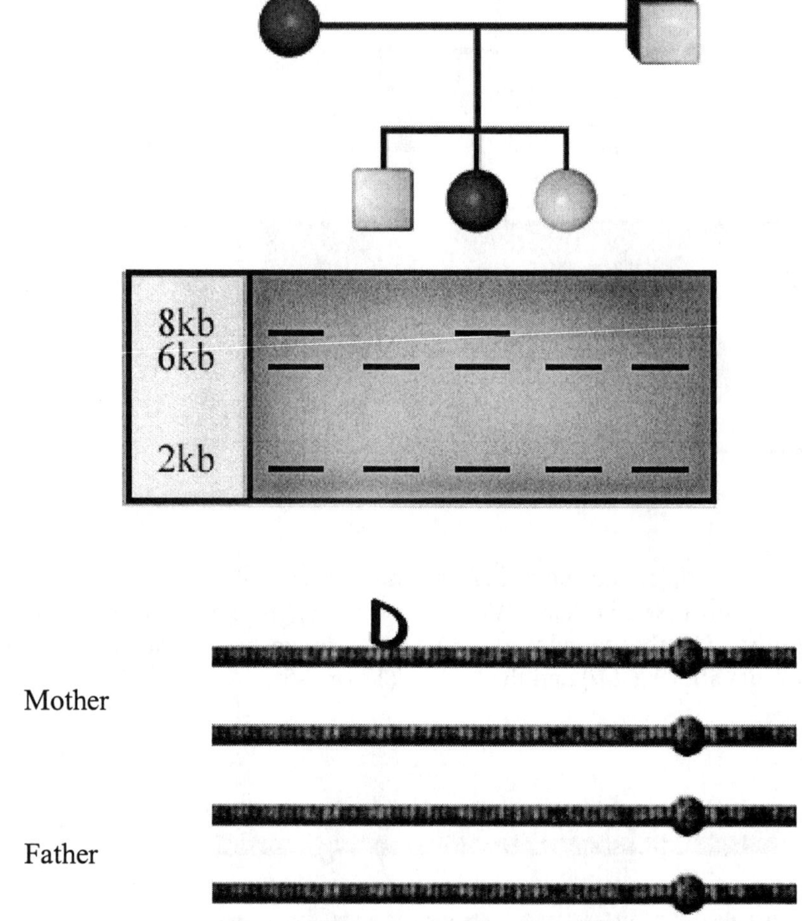

Mother

Father

Use the Southern blot above to determine which children are heterozygous for polydactyly.

Child 1 marries a normal female and they are expecting a child. What is the probability the child will be affected?

Child 2 marries a normal male and they are expecting a child. What is the probability the child will be affected?

Would analysis of this RFLP in the fetus help them determine whether their child will have polydactyly?

Part b

Red-green colorblindness in humans is caused by an X-linked recessive allele (*c*). An STRP (small tandem repeat polymorphism) was found that is closely linked to the colorblindness gene. Assume no recombination in this problem. PCR was used to amplify this region of the genome for the individuals in the pedigree. Assign the appropriate alleles (*C*, *c* or *Y*) and linked STRPs (6 or 7) on the parental chromosomes below.

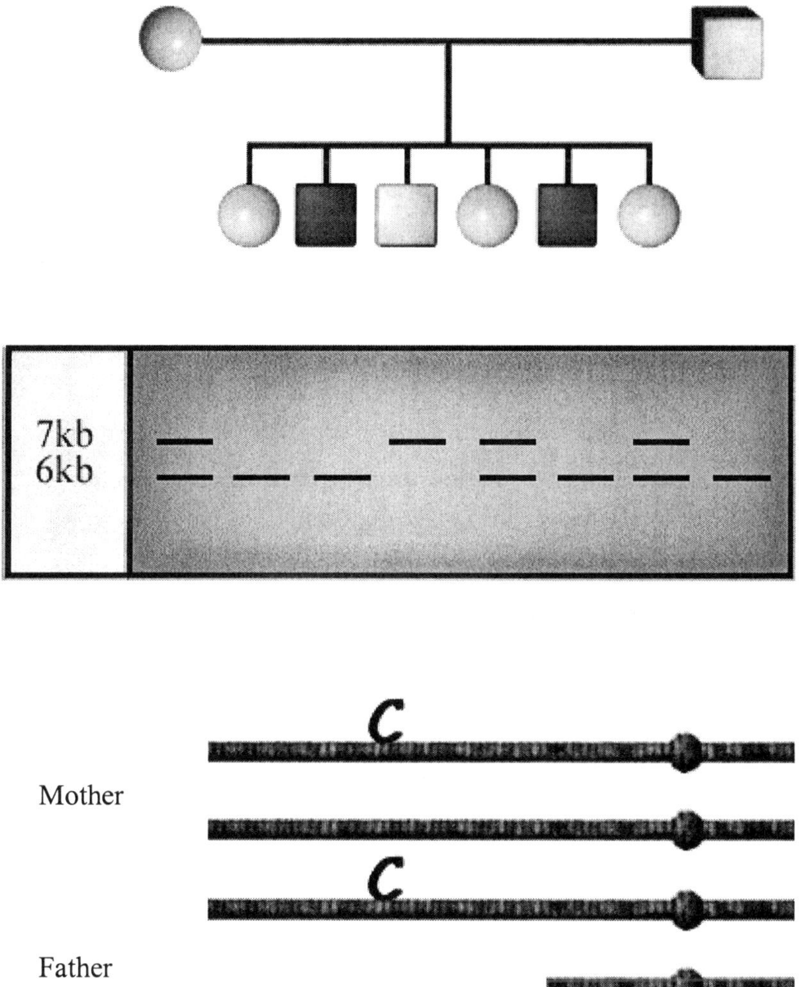

Use the gel above to determine which children are heterozygous.

Child 1 marries a normal male and they are expecting a child. What is the probability their child will be colorblind?

They find out that their baby will be a boy. Would analysis of this STRP in the baby help them determine whether their child will be colorblind?

Problem 4

You have recently identified two molecular probes (A and B) that hybridize to chromosome 12 in yeast. Although the loci are linked, you suspect they may be far enough apart to measure recombination. To test this, you mate two haploid strains to produce a diploid, which is then induced to undergo meiosis. You examine 100 meiotic haploid spores by Southern blotting and find the results shown on the next page.

Pattern	# spores	Probe A	Probe B
1	44	8 kb, 1kb	4 kb
2	7	9 kb	4 kb
3	42	9 kb	4.4 kb
4	7	8.1 kb	4.4 kb

You notice immediately that four patterns are discernable and that these are found at different frequencies. Using the information in the table, determine the genotype of the diploid cell formed from the two haploid parents. Write the linked alleles on the chromosomes below.

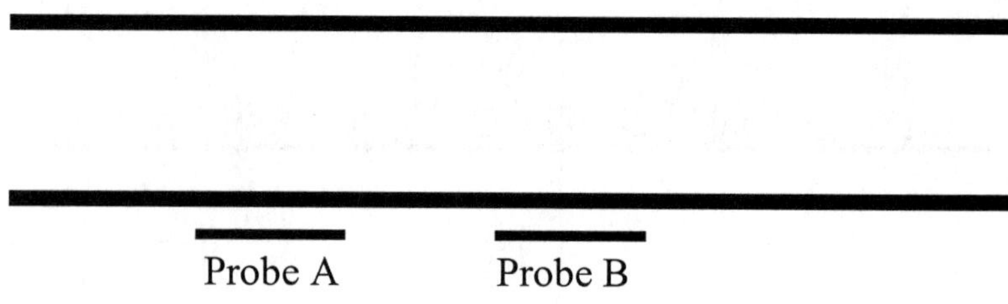

Using the data in the table, determine the distance between locus A and locus B.

Problem 5

Individuals with Duchenne's muscular dystrophy (DMD) are missing an important structural "glue" of skeletal muscle. This makes them susceptible to muscle tears and leads to the progressive death of muscle tissue. Many patients show pseudo-hypertrophy (false muscle enlargement), especially in the calves, as muscles die and are replaced by fat and connective tissue. Almost all are confined to a wheelchair by the age of 10, and most die in their twenties, as their respiratory muscles fail.

Although the gene leading to DMD has been cloned and sequenced (dystrophin located on the X chromosome) and many mutations can be detected directly, not all alleles have been identified. In these cases, a linked polymorphism can be used to determine probabilities. Suppose in the following family an STRP located 6 map units away has been identified.

Answer the following questions taking into account any recombination that may occur. Note that all known disease alleles are recessive, appearing primarily in males.

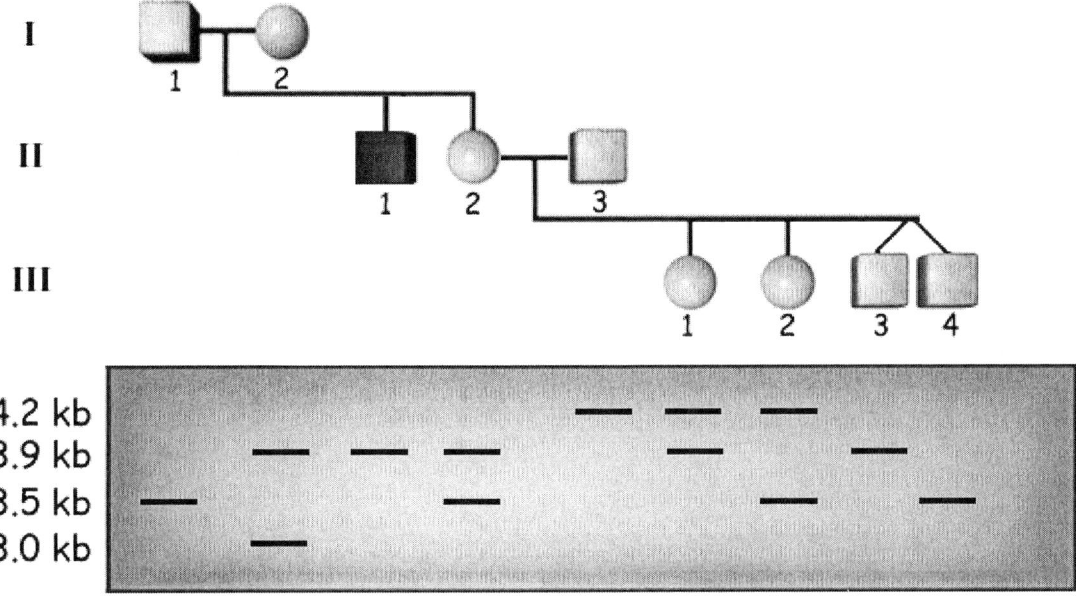

Roya (II-2) and Fardad (II-3) are in genetic counseling, after finding out she is pregnant with twin boys. Roya's brother has DMD, and it is believed to be inherited from their mother. What is the probability that Roya is a carrier? Remember, the polymorphism shown is known to be 6 map units from the DMD gene. Assume there is no crossing over in the mother I-2 leading to the child II-1.

Roya has two unaffected girls. What is the probability that her daughter Sonya (III-1) is a carrier for DMD?

What is the probability that her second daughter Kate (III-2) is a carrier for DMD?

Roya is concerned about her twin boys. The first question she asked the genetic counselor was if the twins were identical or fraternal. Based on the gel above, how would you answer this question?

What is the chance the first twin (III-3) has DMD?

What is the chance the second twin (III-4) has DMD?

Problem 6

Megan, a nineteen year old, was found severely beaten off the side of a small rural road by a patrol officer. She was taken immediately to the nearest hospital, where she was treated for her injuries and it was discovered that she had also been raped. A semen sample was recovered and stored for DNA analysis.

Two men seen in the area at the time of the crime were brought in for questioning. They claimed to have been together all through the night and knew nothing about the young woman. Blood samples were taken from both men and four STRPs were examined and compared with the semen sample.

Southern analysis of both the suspects' blood samples (1 and 2) and the semen sample (S) was performed using well characterized STRPs. Allele frequencies determined using the FBI databases for the U.S. population were then used for probability determinations (only alleles *A*, *B*, and *C* shown). It is assumed these alleles are present in Hardy-Weinberg equilibrium. From the data below, can either suspect be excluded?

STRP 1

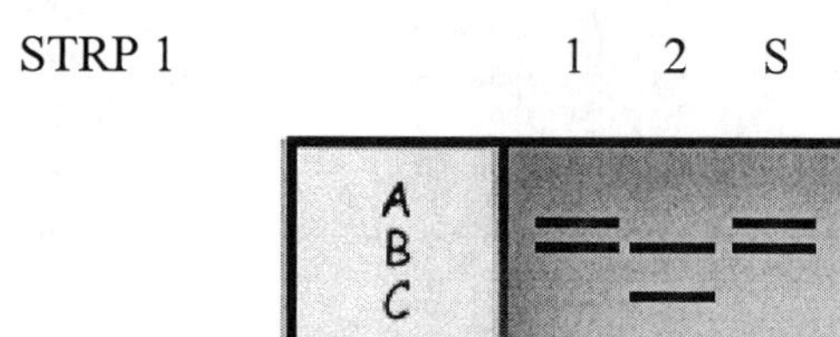

Frequency in US Population

Allele	Frequency
A	1/3
B	1/9
C	1/15

What is the probability that an individual would have the genotype consisting of STRP alleles *A* and *B*? Does this prove that Suspect 1 committed the rape?

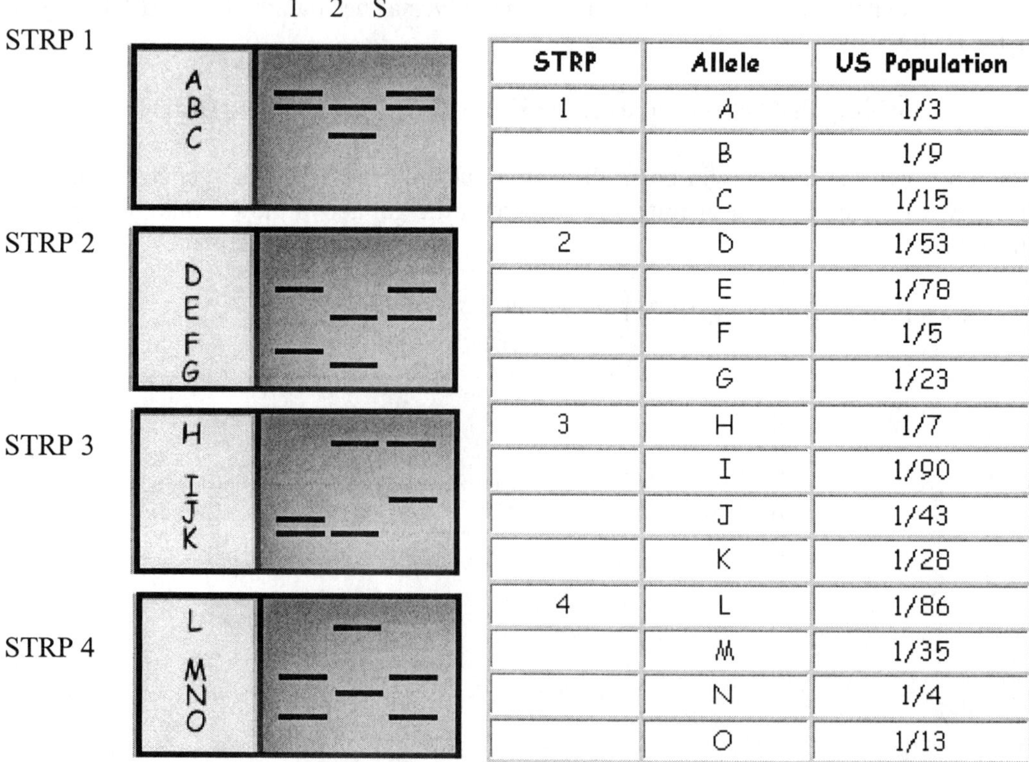

STRP	Allele	US Population
1	A	1/3
	B	1/9
	C	1/15
2	D	1/53
	E	1/78
	F	1/5
	G	1/23
3	H	1/7
	I	1/90
	J	1/43
	K	1/28
4	L	1/86
	M	1/35
	N	1/4
	O	1/13

An additional three independent STRP loci were typed (as shown above). Looking at this additional data, did Suspect 1 rape the victim?

What is the probability of an individual in the United States having the set of alleles found in the semen sample for all four STRPs?

In a population of six billion (the current population of the earth), how many individuals will have this pattern?

Medical Genetics

Goals for Medical Genetics:

1. Understand differences in classification of genetic diseases as chromosomal, single gene, or multifactorial in origin.
2. Understand how to use pedigree information to determine the mode of inheritance.
3. Interpret karyotypes to identify genetic diseases associated with chromosomal disorders.
4. Be able to understand and use in diagnosis molecular genetic techniques (ASO, FISH, and PCR) that recognize mutant DNA causing genetic diseases.

In this section, you will use case studies to see how the genetic techniques you have studied are used in medicine to help doctors form a diagnosis. To aid you in your diagnosis, the patient's clinical description and family history will be available to you. You will be able to perform various chromosomal and molecular genetic tests. At any point in your investigation, you can use the genetic reference database, and whenever you are ready, you may submit a diagnosis. You may want to take notes along the way to keep track of the information you have gathered.

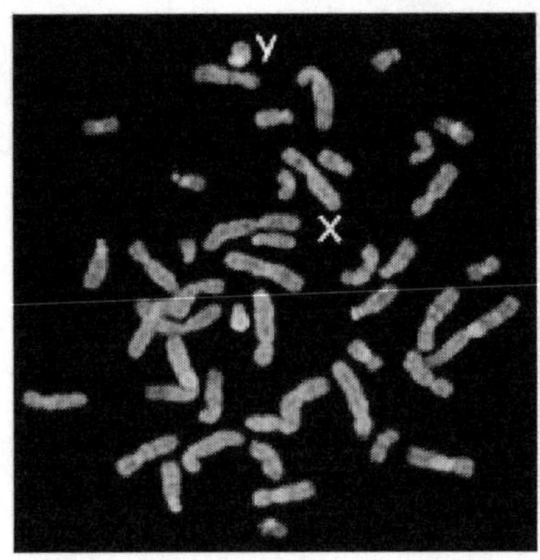

Case I

Frank and Mary Smith asked that their six-year-old son George be seen by a pediatrician. His parents report that that he was slower than their other children to make his developmental milestones. They had attributed his delays to a heart defect that he had when he was born that had required surgery in infancy. He has now started school and is having trouble because he is acting out and having some behavioral problems. They are concerned that he is not ready for school.

The pediatrician is concerned about the developmental delays but notes that George's language skills appear normal or even advanced for a six year old. He notes facial dysmorphisms (unusual facial features) that seem different from his family, including full lips and puffy eyes. The pediatrician suggests that genetic tests be performed to rule out the following conditions for which further information is given:

Angelman syndrome
Di George syndrome
Down syndrome
Fragile X
PKU
Prader-Willi syndrome
William syndrome

Case II

Han Chen and Yuh Nung Lee are referred to the Westside Fertility Clinic after a series of miscarriages. Han Chen is 43 and her husband is 38. Neither reports any major health problems. Han Chen was born in Taiwan. Four brothers and sisters survive in her immediate family. Two siblings died shortly after birth and one was stillborn. Yuh Nung is an only child with no family history of infant mortality.

The clinic first evaluates Yuh Nung's sperm count and finds it is normal. During this evaluation Han Chen became pregnant, but loses the fetus after three months. Fetal tissue was obtained and examined for chromosomal abnormalities in the cytogenetics laboratory. Refer to the chromosomal section under disease reference for the list of chromosomal disorders that are considered.

Case III

Sara and David Goldenstein have brought their daughter Rebecca in for her six-month checkup. They are concerned because she has stopped gaining weight, though she seems to have a healthy appetite. She has had frequent and numerous colds and suffers from diarrhea. Sara and David are first cousins, but there is no other family data that is useful.

The physician finds no evidence of mental retardation. Based on the relatedness of the parents, a recessive single Mendelian gene is suspected of causing the symptoms. Refer to the single gene section under disease reference for the list of disorders that may be considered in this case.

Case IV

Jan de Broek (age 60) was referred for medical examination by social services after his arrest for vagrancy and disorderly behavior. Social services reports that he has lost his job and apartment and is currently homeless. Jan's family was contacted and states that his behavior seems similar to both his father's and uncle's at this age (they are both deceased).

The physician records several of the symptoms associated with senile dementia. These include episodes of severe irritability, forgetfulness, anxieties, ataxia, and alcohol abuse.

Behavioral changes are among the hardest to diagnose; age-related causes may include alcoholism, Alzheimer disease, and Huntington disease. Schizophrenia is usually an early onset disorder, but should be considered if the patient history is uncertain. Further information is given in the disease reference list.

Molecular Biology

Gene Expression

Goals for Gene Expression:

1. Understand the molecular mechanisms by which the genetic information in a DNA sequence is converted to an amino acid sequence, or protein.
2. Be able to compare the similarities and differences between prokaryotic and eukaryotic transcription and the proteins involved in this process.
3. Understand the mechanism of splicing and how introns and exons are mapped in eukaryotic genes.
4. Understand how RNA is translated into a protein and how changes in a gene's sequence can give rise to defective proteins.

Problem 1

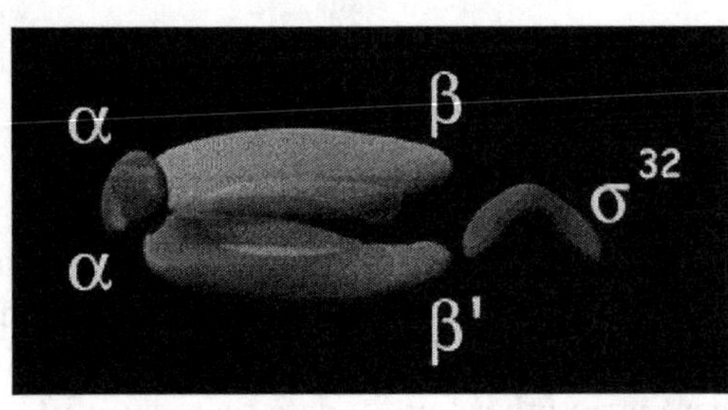

RNA Polymerase holoenzyme in the cell contains a sigma factor that is 70 kD (σ^{70}). However, there are several other sigma factors that recognize different promoter sequences and respond to specific cellular signals. One such sigma factor (σ^{32}) is activated in response to heat shock and recognizes promoters encoding chaperone proteins and proteases. Expression of these genes helps the cell deal with the heat denatured proteins to prevent further damage.

Comparison of promoter sequences found near genes that are transcribed under heat shock conditions revealed the recognition sequence of sigma-32. Shown below are ten such promoter sequences. Determine the best consensus sequence for this promoter.

GCCTATATA
GCCCAACTT
CCCCATGTA
CCGCATTGA
CGCCACGTA
CCCGCTATT
CGCCATCTA
ACTCTTTTT
CCCTAGATA
CGCCATGTA

Promoters that have sequences that are a good match with the consensus bind RNA polymerase more often and lead to an increased amount of transcription. These are considered strong promoters. Which promoter sequence above has the best match to the consensus sequence? Which promoter would be considered the weakest promoter?

5' AGCCTAGCTCCATATAGAACGATCATCTAAG 3'
3' TCGGATCGAGGTATATCTTGCTAGTAGATTC 5'

You have recently found a new heat shock gene above and decide to use the consensus sequence to give a first approximation of the transcription start site (+1). Which nucleotide in the sequence above represents +1?

Write the appropriate substrates in a chemical equation below to describe formation of the first phosphodiester bond from this promoter.

Which end of the mRNA has the triphosphate group?

Only one of the two DNA strands is used as a template for RNA synthesis. Which strand is used in this example?

Problem 2

Exon-intron structure of genes can be determined in a number of ways. One method involves the comparison of <u>cDNA</u> with genomic DNA. This can be done either by DNA sequencing or Southern analysis. In this problem, the gene structure of calcitonin will be examined by Southern analysis.

Calcitonin is a peptide hormone synthesized by the thyroid gland and serves to decrease circulating levels of calcium and phosphate. This is achieved primarily by inhibiting bone decalcification (or resorption).

mRNA was isolated from human thyroid cells and converted to cDNA using reverse transcriptase. Isolation of the calcitonin cDNA was done by hybridization with the calcitonin genomic DNA (previously cloned). Ten subclones from the regions of the genomic DNA shown below were isolated and labeled for use as probes (1–10).

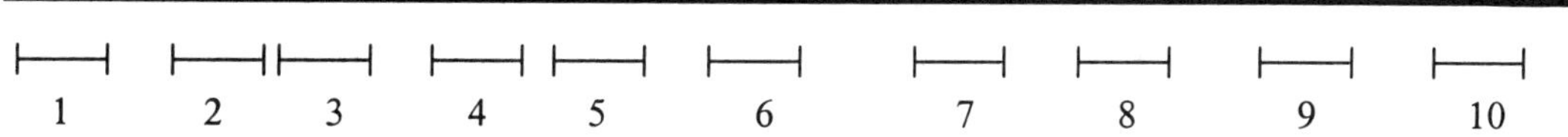

In order to determine where the calcitonin gene is located in the original genomic clone, Southern blots of the thyroid cDNA clone were hybridized with each of the ten probes (lane t, vector plus cDNA insert). A control of total genomic DNA digested with Not1 is included in lane g. See CD-ROM for the hybridization results.

The Poly-A site was found to be on the right side of the genomic DNA above, orienting the gene in a left to right direction. Use this information and the results from the Southern analysis to create an exon-intron map for the calcitonin gene.

Surprising results were observed when the same probes were used in conjunction with a cDNA clone isolated from neurons. Go to the CD-ROM to observe the results of Southern analysis on the neuronal cDNA clone (n) versus genomic DNA (g). Additional questions are available on the CD-ROM.

Problem 3

Hemoglobin is a tetrameric protein ($\alpha_2\beta_2$) that functions to carry oxygen in all vertebrates and some invertebrates. Each subunit of hemoglobin is associated with a prosthetic group, iron containing heme (in white above), which provides the ability to bind oxygen. Disorders associated with alterations to, or the disrupted synthesis of, hemoglobin are the most common genetic diseases in the world.

The β-globin gene is a relatively small gene, composed of three exons and two introns. One major class of hereditary disorders associated with hemoglobin is characterized by alterations in the amino acid sequence of β-globin but not the amount (structural variants). A second major class is characterized by a reduced level of hemoglobin (thalassemias), due to either mutations affecting transcription or splicing. In the following problem, β-globins were obtained from mutant hemoglobins. Use your knowledge of gene expression to help determine the cause of each of the mutations described.

Blood samples were collected from individuals homozygous for several distinct hemoglobin disorders. Protein from each sample was electrophoresed through either a native polyacrylamide gel or an SDS gel (shown below) and the separated proteins transferred to nitrocellulose. β-hemoglobin was then detected using antibodies (Western analysis).

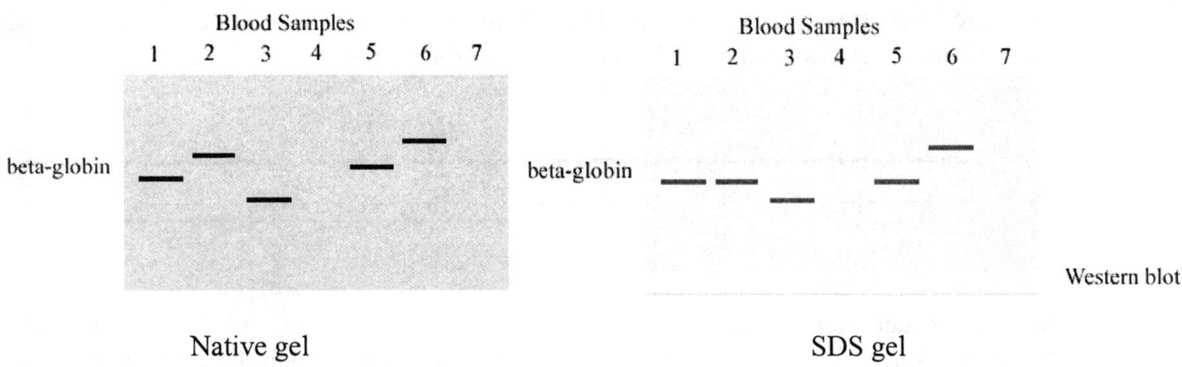

The first lane (blood sample 1) contains wild-type β-globin. Which lanes from 2–7 above represent structural changes in β-globin?

Compare the sample in lane 2 on a native versus SDS gel. Do these results suggest a change in the length of the polypeptide or the charge of the polypeptide?

Compare the sample in lane 3 on a native versus SDS gel. Do these results suggest a change in the length of the polypeptide or the charge of the polypeptide?

DNA and RNA samples were then examined by Southern and Northern blotting to detect changes in the DNA or RNA.

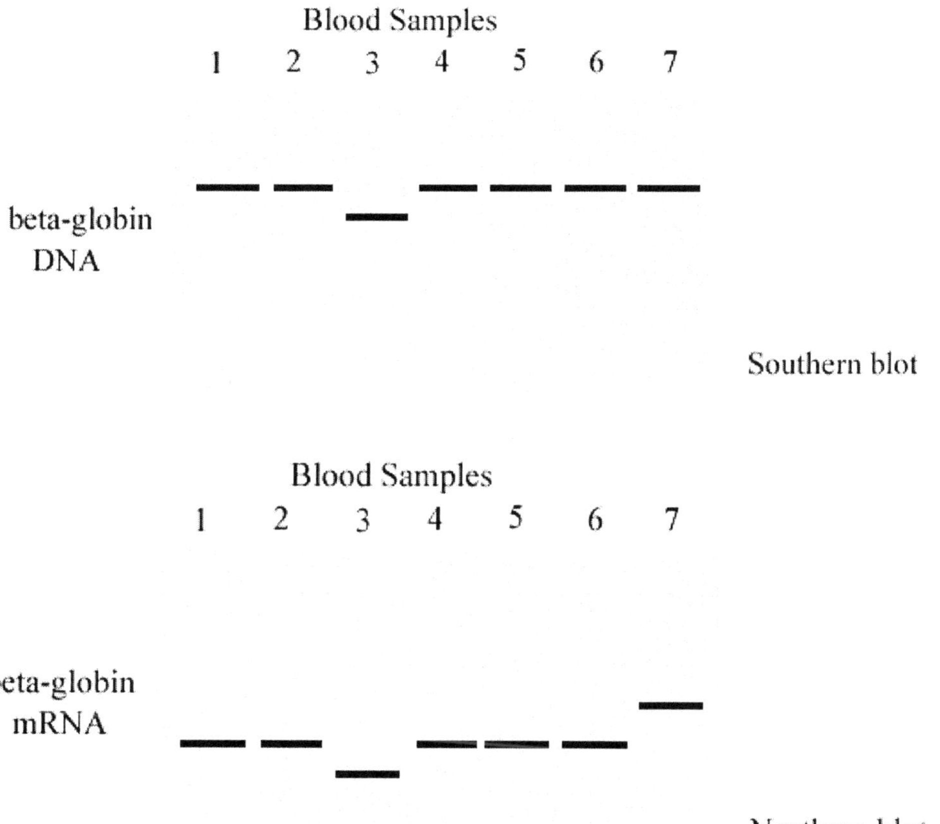

Again, the first lane contains a wild-type sample. To begin, let us examine sample 2 in detail. You have already determined the β-globin polypeptide differs from the wild type by a charge change. Does the Southern blot detect any difference in sample 2 from wild type?

Does the Northern blot detect any changes in mRNA length in sample 2?

DNA from sample 2 was analyzed by dideoxy sequencing and compared to the wild-type gene. A single mutation was found in exon 1, as shown below (only the nontemplate strand is shown). Use the genetic code to determine whether this mutation represents a nonsense mutation (stop codon), missense mutation (amino acid substitution) or a frameshift mutation (insertion or deletion).

Wild type ATG GTG CAC CTG ACT CCT GAG GAG AAG TCT GCC

Sample 2 ATG GTG CAC CTG ACT CCT AAG GAG AAG TCT GCC

Now look at samples 3–7 and determine which fit the following descriptions. Which of the samples above could be explained by a nonsense mutation?

Which of the samples can be explained by a deletion of more than a few nucleotides?

Sickle-cell anemia is known to be caused by a mutation that changes a glutamate codon at amino acid 6 to a valine. Which sample above is consistent with the sickle cell allele (*HbS*)?

Many thalassemias are caused by mutations that effect splice sites. Which sample above would be consistent with a splicing mutation?

The only sample that has not yet been identified is sample 6. What explanation is consistent with the results obtained for sample 6 above?

Molecular Biology

Gene Cloning

Goals for Gene Cloning:

1. Review the use of restriction enzymes and be able to use restriction enzymes to make a physical map of a clone.
2. Review basic cloning techniques, including vectors, library construction, and screening techniques.
3. Review the DNA sequencing, Southern analysis, and PCR amplification.
4. Be able to design a cloning strategy, beginning with understanding what is to be cloned, why it is to be cloned, and what will be done with the clone. The cloning strategy consists of the origin of DNA (genomic or cDNA library), choice of cloning vector, and choice of screening technique to recognize clone.

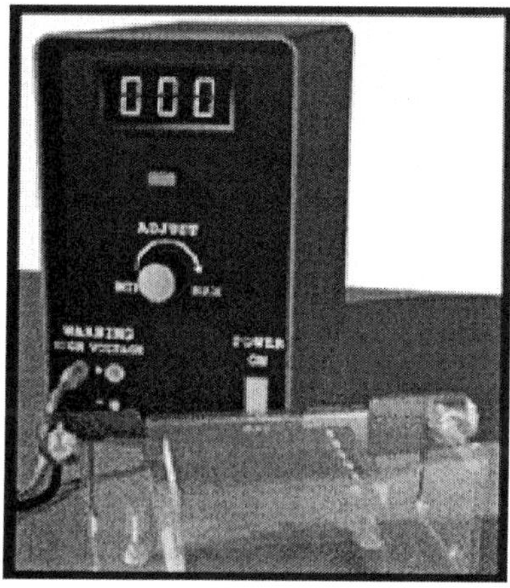

Problem 1

A restriction map provides information as to the presence and position of one or more restriction sites. In many ways it is like the road map to a DNA molecule, and it is used in many recombinant DNA techniques.

Digestion with a single enzyme can provide information about the number of restriction sites and the distance between them, but the order of the sites cannot be discerned. However, if a second enzyme is used both alone and in combination with the first enzyme, a map can be constructed.

In this problem, you will determine the restriction map for a linear DNA sample using the enzymes Eco RI and Hind III. Three samples of DNA are first digested with either Eco RI, Hind III, or both Eco RI and Hind III. The resulting fragments (restriction fragments) are separated by size using gel electrophoresis.

The results are shown here.

(-) electrode

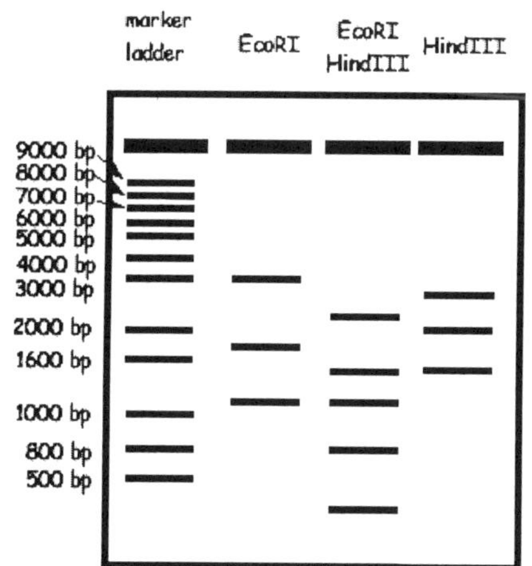

(+) electrode

Since the DNA fragments in each lane are derived from the same linear DNA fragment, the sum of the size of fragments in each lane should be the same and reflect the size of the original DNA fragment. How many base pairs are in the original DNA molecule?

Focus first on the Eco RI single digest. Since we know our sample DNA is linear, how many EcoRI sites are present in the DNA?

When the same DNA sample is digested with Eco RI and Hind III, five fragments are generated. Let's look at this one band at a time. Since the 3000 bp Eco RI fragment is not present in the double digest, it must contain at least one Hind III site. Which fragments in the double digest add up to 3000? Which bands in the double digest add up to 1800? Now there is only one fragment left in the double digest. This is the same size as one of the Eco RI fragments in the single digest. It must correspond to the band in the Eco RI lane.

Now, let's focus on the Hind III digest. The largest band, 2500 bp, disappears in the double digest. Which two bands in the double digest add up to 2500 bp? Which two bands in the double digest add up to 2000 bp? That leaves just the 1500 bp Hind III fragment, which is also present in the double digest. This means that there are no Eco RI sites inside this fragment and may indicate the end fragment on a linear DNA.

Since this is a linear DNA fragment, let's start with one of the fragments that is the same in both the single and double digest. This could be an end fragment. Although we could start with either the 1200 or 1500 bp fragment, let's begin with placing the 1200 bp fragment on the left side and build the map from there. Create a map using the information gained from the single and double digests above.

Problem 2

You have just identified a patient who appears to have an altered hemoglobin protein (β-subunit). Analysis of the patient's blood by Southern, Northern, and Western blotting reveals no detectable change of the DNA, a larger RNA, and smaller protein. You surmise that the mutation affects hemoglobin splicing. With a clone of the wild-type hemoglobin in hand, and cultured skin cells from the patient (as a source of DNA or RNA), select a type of insert, a type of vector, and a method in which to identify and study the mutant hemoglobin gene.

Problem 3

After many months in the laboratory, you have finally obtained a pure preparation of DNA polymerase from a novel fungus. However, you have a very limited amount and cannot do all of the biochemical experiments you would like. Your advisor suggests that you use your purified polymerase to clone the gene encoding the polymerase. After consulting several other graduate students, you come up with a plan. Select a type of insert, a type of vector, and a method to identify the DNA polymerase gene.

Problem 4

Polydactyly is known to be caused by a dominant allele and is characterized by extra fingers and/or toes. Recently, an RFLP has been identified which is linked to polydactyly and you decide to use this information to clone the gene. Determine what type of insert you would use in making your library, what type of cloning vector would be optimal, and how you would go about identifying the correct clone.

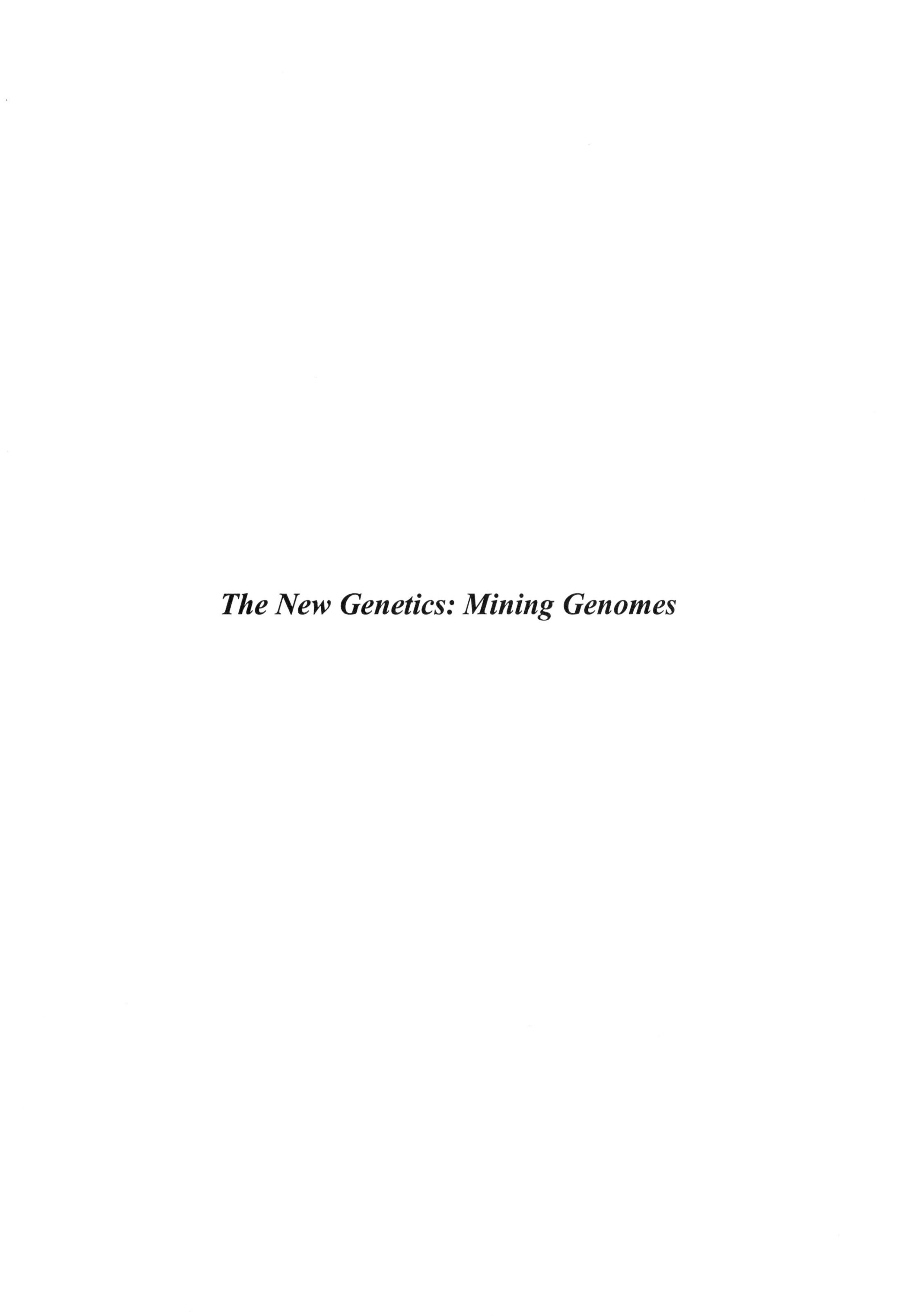

The New Genetics: Mining Genomes

6. Mining Genomes: Introduction to Bioinformatics and the National Center for Biotechnology Information

Starting link: **http://www.ncbi.nlm.nih.gov/.**

Introduction

In this tutorial, you will be introduced to the diverse suite of powerful interactive bioinformatics tools at the National Center for Biotechnology Information (NCBI). NCBI is part of the National Library of Medicine, which is managed by the National Institutes of Health. It was founded in 1988 and has rapidly expanded since then.

Today it is an extensive collection of interconnected and interactive tools and databases. According to their Web site, "NCBI has been charged with creating automated systems for storing and analyzing knowledge about molecular biology, biochemistry, and genetics; facilitating the use of such databases and software by the research and medical community; coordinating efforts to gather biotechnology information both nationally and internationally; and performing research into advanced methods of computer-based information processing for analyzing the structure and function of biologically important molecules."

What kinds of people use the databases and tools? What kind of questions are they asking? Almost everyone doing biological research uses at least some of these tools in the course of their work. Scientific writers, medical doctors, genome scientists, bio-prospectors, population ecologists, and others access NCBI to search the published literature, identify new genes, understand genetic variations, and carry out many other different types of analyses.

There is so much information and so many programs available at NCBI that it can be daunting for a student to become oriented on the site. Luckily, though, the curators and managers of the site are aware of how difficult it might be for newcomers, and incorporate many features to make it easier to use the site. These features include introductory-level descriptions of the programs and their use, diagrams of how the tools are interconnected, and an intuitive interface layout. With this introductory tutorial and a little practice, you'll be amazed at how rapidly you can use the NCBI programs!

Some of NCBI sites are educational, others are for data retrieval and analysis, and others are tools to find information such as peer-reviewed journal articles on a specific topic. In this tutorial, we'll look at these three different types of sites in turn.

Objectives

- to introduce the types of studies that encompass bioinformatics.
- to familiarize you with the NCBI Web site.

Educational Resources at NCBI

Open the NCBI homepage on your browser at **http://www.ncbi.nlm.nih.gov/.**

The NCBI site has many different educational resources, including descriptions of NCBI programs and tools, extending to primers on topics in genetics, and even online textbooks. As you can see, there are links to many different resources available on this page. Across the top of the page are links to the most frequently used data analysis tools on the site, including *PubMed* (literature searches), *Entrez* (database searches), *BLAST* (sequence comparisons), and *OMIM* (human genetics information). In a column along the right side of the page is a list of links to other programs and tools for analysis, such as *Clusters of orthologous groups* (conserved protein database), *Human map viewer* (graphical displays of the human genome), and *Unigene* (a collection of entries on specific genes). Along the left side of the page is yet another column of links, including many informational sites. From the list on the left side of the page, select **About NCBI.**

Some of the educational resources at NCBI can be accessed from this page, and details about the various entries can be seen by **mousing over** the individual headings. For our purposes, select **A Science Primer.** Again, **mousing over** the headings will display a short description of the content of each of the entries. Select **Bioinformatics.**

As you can see, this site describes what bioinformatics is, what biological databases are, and some of the types of questions that scientists address using NCBI resources. It explains some studies concerning evolution, three-dimensional protein structure, and genome mapping, and it introduces the use of one of the tools available at NCBI, the *Map Viewer.* These informational sites will undoubtedly be of interest and help as you explore the NCBI site.

There are also resources available here that will help you with your studies in genetics during the course of the semester. Return to the *Science Primer* page by clicking **A Science Primer** at the top of the page.

If you were to select the *Molecular Genetics* link from this page, you would be taken to a comprehensive site that describes and illustrates many of the basic concepts in molecular genetics. It also has descriptions of many different kinds of laboratory techniques that are used in molecular genetics studies.

Even more valuable, though, are the online textbooks that NCBI makes available to the public free of charge. At the top of this page, select **Databases and Tools.** This takes you to a site where you can access many of the different analysis programs available at NCBI. Details about the various entries can be seen by **mousing over** the individual headings. To see the online textbooks that are available, select **Literature Databases,** and from the resulting page select **Bookshelf**.

As you can see, numerous scientific books can be accessed from this page, including textbooks in genetics and cell biology. These books can be used to explore topics that you are interested in or to supplement the textbook that you are using. Entering a search term into the text box on this page will search all of the displayed textbooks for relevant information; alternatively, you can click on a textbook's cover to browse through it.

Data Retrieval and Analysis at NCBI

So far we have looked at some of the educational resources that are available at NCBI. However, this site is designed to allow investigators to carry out research as well as to retrieve information from other studies. Let's look now at what some of these tools are and what they can do. Instead of simply going to some of these sites and discussing how they function, let's see them in action!

Let's say that you have just read an article about an intriguing set of experiments using a strain of mice called Doogie mice. You can go to the site of a *Scientific American* article about these experiments by going to
http://www.sciam.com/article.cfm?articleID=00075ED9-33D5-1C75-9B81809EC588EF21

Reading through this article, you will learn that Doogie mice were created by Dr. Joseph Z. Tsien at Princeton University. These mice have extra copies of a gene called Grin2b, which is referred to in the article by an early name, NR2B. Grin2b encodes an NMDA receptor, and these extra copies are modified so that they are expressed more strongly when the mouse is mature. These modifications allow the mice to learn more effectively than the parent strain, and to remember things longer. This result is remarkable— complex phenomena such as learning and memory can be altered (especially, improved!) by increasing the amount of a single protein. The last line of the abstract for their publication reads 'Our results suggest that genetic enhancement of mental and cognitive attributes such as intelligence and memory in mammals is feasible.'

These experiments bring up many interesting questions. Do humans have a corresponding gene? Could manipulation of this gene help to treat human disorders such as Alzheimer's disease? Have mutations in this gene been shown to be involved with any human disorders? Do other organisms have a similar gene? Is it desirable to genetically enhance the intelligence and memory of humans or other mammals? Are there other consequences of altering this protein? Other issues are also raised by these experiments-- some of the possible social, legal, and ethical ramifications of raising the intelligence level of other species have been explored in an interesting series of science fiction books by the astrophysicist and author David Brin.

Entrez Searches at NCBI

Let's return now to the NCBI homepage **http://www.ncbi.nlm.nih.gov/** and introduce some of the NCBI programs by addressing a few of the questions raised above. As noted before, there are a number of links on this page. From the bar across the top of the page, select **Entrez** to use the primary search engine for NCBI resources. At the top of the page is a text box to enter search terms. Enter **Grin2b** in this box.

To the left of the search box is a pull-down menu. The default setting on the menu is PubMed. By selecting the menu bar you will be able to see many other databases to search. Descriptions of these databases are listed on this page. Look through these to familiarize yourself with the types of options that you have. Set the menu to **OMIM** (Online Mendelian Inheritance in Man, a database that summarizes information available for human genes), and then select the **Go** button.

Online Mendelian Inheritance in Man (OMIM)

This takes you to a list of OMIM entries that contain the term Grin2b. The first result, #138252, is for Grin2b itself. Select ***138252** to go to the OMIM Grin2b entry. As you can see, the OMIM site describes what is known about this gene, and summarizes some of the experiments that led to this understanding. It discusses attempts to use animal models to understand the functioning of this gene in humans (this section is pretty interesting), and it lists a number of peer-reviewed manuscripts that have been published concerning Grin2b and its protein product.

Now, let' see if we can find the actual protein sequence of Grin2b. At the top of the OMIM page is a box to enter search terms. This is an Entrez search box; the default setting is for OMIM because you are on the OMIM site, but you can use the pull-down menu to search other databases. Enter **Grin2b** in this box, change the pull-down menu to **Protein**, and select **Go.** This takes you to a list of protein entries, including those from the mouse and from the human. Select the entry labeled **Q01097**, for the mouse gene.

This entry gives details on this protein: who determined its sequence; where and when the sequence was published, and what the actual sequence is. If you scroll down to the bottom of the entry, the protein sequence itself is displayed, using the single letter amino acid code, in which M = methionine, K = lysine, etc. At 1483 amino acids, this is a relatively large protein.

There are a variety of ways that a researcher might use this sequence, for example, in studies comparing the normal sequence to related sequences in Alzheimer's, Parkinson's, or schizophrenic patients. The sequence might also be used in studies looking at the evolution of this protein, or in trying to assign a function to new, highly similar genes uncovered in genome sequencing projects.

Additionally, you could search the databases using this protein sequence as a query in order to find out how many similar proteins have been studied, or whether there are related but distinct proteins in humans. Let's carry out this type of analysis. Use your mouse to **highlight** the amino acid sequence of the mouse Grin2b protein. Then **copy** the amino acid sequence (don't worry about the numbers, it's ok if you copy them, too, as the program will ignore them).

BLAST Searches at NCBI

Now let's return to the homepage of NCBI by clicking the **NCBI icon** at the very top left of the page.

Select **BLAST** from the list of links at the top of the main page. BLAST stands for Basic Local Alignment Search Tool. This program will compare a query amino acid sequence

(BLASTp) or a query nucleotide sequence (BLASTn) against sequences that are in the databases. From this page, select **Standard protein-protein BLAST [blastp]**. **Paste** the Grin2b sequence into the Search text box and select the **BLAST!** button.

When the next page appears, select the **FORMAT!** button. A new browser window will open with the results of your search. It may take a minute or two for your search results to be returned, depending on how many other people are carrying out BLAST searches at the same time.

When the results are returned, you should see a box containing a series of long red bars (these are database protein entries that matched your query sequence across its entire length) and then a series of smaller red bars (these are entries that matched well across a large portion of your query).

As you **scroll down** the page you will see a listing of the actual sequences that your query matched, and then below that, a set of alignments in which your query and the database match are compared. Under the heading 'Sequences producing significant alignments,' look at the first sequence. This sequence is the best match for your query because it is actually the same entry that you copied your query sequence from.

Selecting the link labeled **gi|14549168|sp|Q01097|NME2_MOUSE** will take you to the original entry that you copied the sequence from. Selecting one of the links from the other matches would take you to their corresponding entries in the protein sequence database. Now, **close** the window that the BLAST results appeared in.

BLAST searches are used for a number of different purposes, such as identifying the function of a newly isolated gene. By finding what characterized genes a new gene is closely related to, you can make inferences about the new gene's possible function. Alternatively, if you are interested in performing experiments to determine the function of an unknown human gene, you might want to find out whether there is a corresponding gene in mice. There are a variety of BLAST searches, and a number of ways of using the output of BLAST searches. We will explore BLAST extensively in a later tutorial.

LocusLink

Now let's return to the homepage of NCBI by clicking the **NCBI icon** at the top of the page. Again, enter **Grin2b** in the search box, but this time set the pull-down menu to **LocusLink**. Then select **Go.** Several hits for your search should be returned. **Select the box** for the human gene (signified by '*Hs*', for *Homo sapiens*). This entry is numbered 2904. Then select **View Loci**. This will take you to the LocusLink entry for Grin2b.

As you can see, the entry contains a variety of types of information concerning this gene and protein product (most of it with hyperlinks to other relevant NCBI sites). There is a description of Grin2b, and a section detailing studies that have led to the proposed functional role of the protein. There is also information on the map location of the gene (which chromosome, and where on that chromosome) as well as links to sequence database entries containing that or related sequences.

PubMed Searches at NCBI

Across the top of the LocusLink page is a bar of colored segments. The individual segments take you to other regions of NCBI that deal with Grin2b. For example, one of the most-used features of NCBI is the PubMed site. A PubMed search allows you to instantly search through the Medline bibliographic database at the National Library of Medicine. This database contains citations and abstracts from more than 4,600 biomedical journals, with more than 11 million entries dating back to the 1960s. Select the red **PUB** bar to carry out a PubMed search using the query Grin2b.

This takes you to a list of peer-reviewed articles about Grin2b. There are not a lot of articles here, however, because most studies of this protein have used the alternate name, NR2B. Enter **Grin2b OR NR2B** in the search box, and select **Go**. Note that the 'OR' is a Boolean search term and needs to be in the upper case.

This returns several hundred journal articles. **Select the authors of a journal article**. This takes you to an abstract for that article. Depending on the journal that the article was published in, selecting a link above the abstract may take you to a full-text version of the original article (some journals only provide access to paying subscribers). Remember that you can also carry out PubMed searches using *Entrez.*

MapViewer

The last stop that we'll make on our NCBI tour is at the MapViewer. Return to the homepage of NCBI by clicking the **NCBI icon**.

From the list of links along the right side of the page, select **Human map viewer.** This will take you to a page that displays the human genome as a set of individual chromosomes labeled from 1 to 22 (longest to shortest), plus the sex chromosomes. Let's find where the Grin2b protein is. In the *Search for* box at the top of the page, enter **Grin2b** and then select **Find.**

This should return a depiction of the chromosomes in which a single red bar is adjacent to chromosome 12, indicating that there is only one copy of this gene in the human genome, and that it is located in this region of chromosome 12.

Underneath the depiction of chromosome 12 is a hyperlink to chromosome 12. **Select** this link to go to a much more detailed depiction of chromosome 12. This will take a minute to load, but eventually a complex graphic will load on your screen. There are many different kinds of information depicted in this graphic, most of which are too detailed to describe in this introductory tutorial. One of the things that is shown is the location of the Grin2b gene (highlighted in pink). The genes that are adjacent to Grin2b are also shown.

To return to the LocusLink site for this gene, select the pink-highlighted **GRIN2B** link in the *symbol* column. This returns you to the LocusLink site that you visited earlier. From LocusLink, selecting the yellow MAP bar would take you back to the Map Viewer site for Grin2b. There are also many other ways of moving from one NCBI program to another. As you get more accustomed to the site, you will be able to find information and ask questions much more rapidly. Future tutorials will explore many of these sites in greater detail.

Finally, let's return to the homepage of NCBI, by selecting the **NCBI icon** on the top of the page. Once on the homepage, select the link entitled **Site Map** at the top of the list on the left. This takes you to a long list of resources at NCBI, with descriptions of each of the tools and programs. This would be a good place to start playing around and exploring on your own.

That concludes your introduction to the National Center for Biotechnology Information—hopefully, your NCBI visit sparked your imagination and gave you a glimpse into the fascinating world of bioinformatics.

Review Questions

The following questions will help to review and extend your tutorial.

1. What kinds of information are found on the OMIM site? The Bookshelf site?
2. What kinds of questions can be asked using BLAST searches?

Thought and Application Questions

1. Use the PubMed site to search for recently published articles by Dr. Joseph Z. Tsien. Note that the format for author queries is **Tsien JZ**. Has Dr. Tsien published any papers in the last two years? If so, list the title and authors for one of these publications.
2. When you searched PubMed using the search term **Grin2b OR NR2B**, several hundred-literature articles were returned. Most of these articles are not written with an undergraduate audience in mind. Sometimes, it is useful to narrow your search by looking for review articles (articles that summarize and interpret a number of primary sources). How many hits do you get for the search **Grin2b OR NR2B AND review**?

9. Mining Genomes: The Human Genome Project

Starting link: **http://www.ncbi.nlm.nih.gov/.**

Introduction

The sequencing of the Human Genome was a major effort, requiring the cooperation of numerous government agencies and years of commitment by hundreds of capable, hardworking scientists. In the 1980s, when the project was taken on within the United States Department of Energy, there were numerous critics who felt that the utility of the project was unclear and the prospects for completing the project were dim. Others were concerned that funding would be diverted from more important projects. Today there is widespread agreement that completion of the Genome Project is of immense value to biomedical research, and many believe that it may represent one of the greatest achievements of humanity. The new technologies that were developed along the way to allow rapid sequencing and analysis have found a variety of new and promising uses, and the sequence itself is being used in numerous unforeseen applications.

Our genome is roughly 3 billion bases long, and only about 5% of the genome encodes proteins and functional RNAs (the remainder is noncoding). That 5% has approximately 30,000 genes, and these genes are thought to encode as many as 80,000 different proteins. One of the reasons that the number of proteins is greater than the number of genes is that the primary transcript from a gene can undergo different kinds of processing at different times.

An important benefit of the human genome sequence data is in the identification and localization of genes that are involved with a variety of disorders. This research will play a central role in the development of new means for the testing and diagnosis of genetic disorders. It will also be critical for developing new treatments and therapies for numerous debilitating conditions, including cancer and heart disease. In this tutorial, we will tour the Human Genome sequence using some of the bioinformatics tools available at the National Center for Biotechnology Information.

Objectives

- to explore the Human Genome sequence and some of the uses of knowing that sequence.
- to introduce some of the genomics tools available at the National Center for Biotechnology Information, including Map Viewer, LocusLink, and Genes and Disease.

Who Sequenced the Human Genome?

The Human Genome was sequenced simultaneously by different public and private entities. The public entities included groups from many countries in an effort spearheaded by the U.S. Department of Energy (DOE). Although the DOE seems like an odd choice (as opposed to, for example, the National Institutes of Health), it had experience leading and directing large groups of engineers and multidisciplinary scientists in massive undertakings. This type of project fit the DOE's capabilities, even though it clearly has nothing whatsoever to do with the energy policy, supply, or security of the U.S. (Note: it can be entertaining to read how the DOE justifies its involvement in the project, mumbling something about how radiation can cause mutations, and because they study nuclear energy......) To learn about the history of the Human Genome Project and some of its offshoots such as ethical, legal, and social implications, you might want to explore the DOE's Web site at **http://www.ornl.gov/hgmis/.**

Relatively late in the game (1998), a private corporation was also formed to sequence the human genome. This corporation was named *Celera* to signify swiftness, and in fact it was quite speedy at producing annotated sequence data. Celera was formed as a partnership between Perkin-Elmer, the company that made the fastest automated sequencers of the time, and J. Craig Venter, a visionary scientist and entrepreneur. Celera's success lay in its approach as well as its timing—whereas the DOE program spent an enormous amount of energy mapping and cloning the genome, Celera used a 'shotgun-sequencing' approach in which it generated massive amounts of sequence data and then used new computer programs to assemble the data into contiguous sequences.

To illustrate this approach, I've chopped up a sentence into random fragments below. Reconstruct the following linear sentence by connecting the fragments in overlapping regions.

ithout learning ha
agination withou
ho has imaginatio
One who has im
ning has wings b
hout learning has win
ings but no feet.
ho has imagination withou
earning has wing
thout learning ha
ut no feet. One who has ima

How Much of it is Finished?

Most people, upon hearing that a 'working draft' of the human genome has been released to the public, wonder what that phrase actually means. How much of the genome has been sequenced? The answer to this question depends on how accurate you require the sequence data to be before it is considered to be 'finished'. For example, the accuracy of the most common sequencing methods is currently ~99%, that is, perhaps 1/100 of the bases in a given reaction will be called incorrectly. If a particular base is identical in two different reactions, the probability that it was called incorrectly is $1/100 \times 1/100 = 1/10,000$. The more redundant the sequencing of a particular region, the more accurate that region is. Also, do you want to say that the sequencing is 'done' simply when you have the order of the nucleotides? Most people agree that the sequence isn't finished until it has been annotated, and the genes and other features have been described.

More information on how much of the human genome has been sequenced can be found at **http://www.ncbi.nlm.nih.gov/genome/seq/**. Up-to-date information can also be found at **http://www.strategicgenomics.com/Genome/.**

How Much did it Cost?

The answer to this question is unclear, because many different groups, public and private, and in many different countries, participated in the project. It also is debatable which expenses should be included as a part of the project. Ballpark estimates from the late 1980s were that it would cost about $3 billion, and it appears that those estimates were probably close to accurate. It may have cost slightly less than this.

How Many Diseases have a Known Genetic Basis?

The answer to this question is also somewhat uncertain. Clearly, sequencing the human genome will help researchers understand genetic diseases and develop new treatments for them, including common disorders such as some cancers. The actual number of inherited diseases that have a known genetic basis depends on many factors, such as what you define as a disease, which is sometimes a matter for considerable debate. For example, is deafness a disease to be cured? Mental retardation? Baldness? Given this limitation, as of mid-2002, there are approximately 1300 disorders that have a known genetic basis.

The Human Genome Project at NCBI

Now go to the NCBI Web site at **http://www.ncbi.nlm.nih.gov/.**

This is the gateway to a diverse and rapidly-expanding collection of bioinformatics tools and data that are available for public use. Most of these programs are fully cross-linked, so that complex and comprehensive analyses can be performed in a rapid and intuitive manner. Across the top of the page some of the most commonly used programs available from NCBI are listed, and links to many other programs are listed down the right side of the page. We are going to begin our tour of the human genome at one of the links on the right side. Select the link to **Genes and disease**.

Genes and Disease

The photograph on this page shows an example of chromosome painting, a technique in which each of the different chromosomes is colored with a different labeled probe. This approach is useful for analyzing chromosomal aberrations such as deletions, duplications, and translocations of parts of specific chromosomes. If you look at some of the chromosomes in the middle region, you will see that they are labeled with different colors, indicating that translocations have occurred. This approach can also be used to look at conditions such as anueploidy, which involve alterations in the numbers of chromosomes. This technique is useful for studying evolutionary questions as well, for example, how gene order and location on chromosomes evolve as mammals diverge, and how polyploidization has affected plant genomes.

This site provides background information relating to a number of specific disorders, and can be used in a variety of ways. Down the left side of the page are links to several different groups of disorders, such as cancers, disorders of the immune system, and disorders of the nervous system. Selecting one of those links will take you to a page with details on specific conditions and genes. Across the top of the screen is a bar with the numbers 1 through 22, representing the autosomal chromosomes, and the letters X and Y, representing the sex chromosomes. Selecting one of the numbers takes you to an ideogram of that chromosome and links to some of the genes that have been associated with particular disorders. For example, select chromosome **15** from the list.

This takes you a depiction of chromosomes 13 through 16. The banding pattern on the chromosomes reflects the reproducible pattern that is seen when a particular chromosome is dyed with Giemsa stain; this banding pattern can easily be observed in a light microscope and allows scientists to identify specific regions of a chromosome. The constricted region of each chromosome (near the top of chromosomes 13-15 and closer to the middle of chromosome 16) represents the centromere. The short arm is called the *p* arm (for *petite)* and the long arm is called the *q* arm (because that comes after *p*, of course). The different arms are then divided into regions and the bands within each region are numbered consecutively starting closest to the centromere. Thus, on chromosome 14, the gene that is associated with Alzheimer's disease is said to be located at 14q24 (chromosome 14, the 4th band of the second region on the q arm).

Now, select the link to **UBE3A Angelman syndrome** on chromosome 15.

Angelman Syndrome

Angelman syndrome (AS) is a disorder caused by the deletion of a section of chromosome 15. This page describes the disorder and discusses what is known about its genetic basis. If you **mouse-over** the image of chromosome painting from an AS person, you will see a karyotype of the chromosomes. Using this technique, it is relatively easy to see that three entities, instead of two, are dyed with the chromosome-15 specific probe. Of the three that are marked with the chromosome-15 probe, on the left side is a normal chromosome 15, and to the right is the partially deleted chromosome followed by the deleted region.

Did the chromosomes in this figure come from a male or female?

Researchers have found that the deleted region is 15q11-q13, which contains the gene for ubiquitin ligase (UBE3A). It is thought that Angelman syndrome results from problems with protein degradation during brain development that take place when UBE3A is disrupted. Now we will look at this region more closely using the Map Viewer function.

One way to view this region would be to select the link to UBE3A on chromosome 15 from the list of links on the left. Let's take another route, though, so that you can learn how to find the locations of genes that aren't documented in the Genes and Disease site. At the top right of the page, select the Map View link labeled **Map**.

Map Viewer

This takes you to a self-explanatory depiction of the human genome, separated into individual chromosomes (and 'MT' for our mitochondrial genome). There are various ways to get to the region of chromosome 15 that we are interested in. For example, we could search for UBE3A, or we could select the link called 'Blast search the human genome' and search using the nucleotide sequence of ubiquitin ligase. For now, let's select the link to chromosome **15** under the drawing of that chromosome.

Map View of Chromosome 15

You should see a blue column on the left side of the page, with an ideogram depicting the Giemsa bands of chromosome 15. Most of the page has a white background, and in this area there are a series of columns displaying different views of chromosome 15. At the top of the page is a description of the map view you are looking at, including the number of genes on this chromosome (1,175) and the region that is displayed. On the left side is a more detailed ideogram, with the bands and regions labeled. Find the centromere, where the numbering for the p and q arms begins, in order to better understand how this labeling process works.

Directly to the right of the ideogram are a variety of other displays, all relating to different aspects of chromosome 15 (labeled in dark blue, at the top of the column). The first is a *Contig* map. A *contig* is a contiguous region of DNA for which the sequence of nucleotides has been determined. Part of the Human Genome Project's goal was to find overlapping regions between different contigs and assemble the overlapping contigs into complete chromosomal sequences, like the exercise we did at the beginning of this tutorial. The Contig view gives details on the contigs that were used to assemble the sequence of chromosome 15.

There are dozens of different possible ways that you can view this chromosome, depending on what type of information you are interested in. There are maps depicting different types of sequence information, as well as maps depicting cytogenetic analysis and genetic linkage data. To explore what some of your options are, select the yellow **Maps & Options** link near the top of the page.

This opens a new window with menus that allow you to customize the maps you are looking at. It will show the default settings at *Ideogram, Contig, Hs_Unigene*, and *Gene.* Click on **Contig** in the right menu, and then click **REMOVE**. Do the same for **Hs_Unigene**. From the menu on the left, select **Mitelman Breakpoint,** and then click **ADD**. Do the same for **Morbid/Disease.** Then select **Apply**, and, after the new map view has loaded, select **Close.**

These views allow you to look at chromosome 15 in other ways. For example, the Mitelman Breakpoint Map portrays chromosomal breakpoints that have been observed in different cancers (cancerous cells frequently exhibit chromosome breaks and other chromosomal abnormalities). The Morbidity Map notes the map location of disease genes that are described in the *Online Mendelian Inheritance in Man*, an extensive catalog of human genes and genetic disorders, with interactive links to other tools at the NCBI. The Genes_Sequence

Map notes the locations of genes that have been localized to the contigs using a variety of approaches.

Zoom in to Focus on Region 15q11 - 15q13

In the blue region on the left side of the page is a tool that allows you to zoom in to get a more detailed view of part of the chromosome. Select the **15q11-q13** region on the ideogram of the chromosome, and then select **Zoom in x8** on the pop-up menu that will appear. When the new view of the map loads, there is a red indicator to the right of the zoom ideogram that shows you what region of chromosome 15 is being displayed.

On the Mitelman breakpoint map, moving your **mouse over** the indicated breakpoints will display information on the types of cancer related to chromosome abnormalities involving that site, as well as the number of cases in which that particular break has been observed. Clicking on the numbered links in the Morbidity Map will link you to the corresponding OMIM entry.

Spend some time exploring the structure of Map Viewer, and then return to the Map Viewer *Homo sapiens* genome view home page by going to **http://www.ncbi.nlm.nih.gov/cgi-bin/Entrez/hum_srch?chr=hum_chr.inf&query.**

The first time that you used this page, you knew that Angelman syndrome involved a break in chromosome 15 in the 15q11-q13 region, and you used the map to go to this region. You can also use this site to determine the genomic location of a gene that you are interested in, as well as how many different copies of that sequence are present.

For example, the model for the development of cancer is that a single cell accumulates multiple mutations that allow the cell to undergo uncontrolled proliferation. These mutations interfere with normal progression through the cell cycle and inhibit apoptosis functions. Normal progression through the cell cycle is controlled at some levels by cyclins. Cyclins are proteins whose concentration varies at different parts of the cell cycle. Cyclins usually exert their effect by activating or inhibiting kinases that regulate activities such as transcription. Let's determine how many cyclins are known to be coded for in the human genome, and the chromosomal location of the cyclin genes.

In the text box next to the phrase Search For, enter **cyclin**. Then select **Find**.

This takes you to a genome view in which the chromosomal locations of various cyclin genes are depicted. For example, there are five cyclins on chromosome 4, none on chromosome 9, and one on chromosome 19.

LocusLink

Let's use this site to introduce the last feature that you will be shown in this tutorial, LocusLink. Select the underlined **19** link that will take you to a map view page of chromosome 19.

In this view, you once again have the ideogram of the chromosome on the left, and other portrayals of the chromosome to the right. In the heading, it shows that 1696 genes have been identified on chromosome 19. Note that this is considerably more than on the longer chromosome 15. This is an example of the phenomenon that genes are not equally disbursed around the genome—some regions are gene-rich and others are gene-poor.

As you scroll down the page, you will see a pink-highlighted entry. This entry is the cyclin gene that you found in your search. Under the Genes_seq map view, select the entry **CCNE1** to go to the LocusLink entry for this gene, which encodes the protein *cyclin E1*. You can also access LocusLink from the NCBI home page, under the *Hot Spots* list on the right side of the page.

LocusLink is an important tool at NCBI that ties together many different kinds of information about specific genetic loci. From this site you can find a description of this gene, its chromosomal location, and the function of the gene product, as well as whether or not there are homologous genes in genomes from other organisms. Each of these features is linked to other NCBI data related to this cyclin. There are also links to several representative GenBank entries for cyclin E1. Scroll through the entry to get a feel for the types of information that are available. Then scroll back to the top of the page. Across the top in the colored boxes are a variety of ways of accessing other information about this cyclin. For example, selecting **PUB** carries out a PubMed search to find peer-reviewed articles that relate to this cyclin, and selecting **OMIM** takes you to the Online Mendelian Inheritance in Man site on cyclin E1. Selecting **MAP** would return you to the Map Viewer, and selecting **HOMOL** would take you to a site detailing the homologous genes that have been identified in other organisms.

That concludes your introduction to the Human Genome Project at NCBI!

Review Questions

1. How many bases are there in the human genome? How many have been sequenced?
2. How many genes are in the human genome? How similar are different humans at the genetic level?
3. What are some of the difficulties with determining how many genes have been associated with specific genetic diseases?

Thought and Application Questions

1. Many functional proteins, such as hemoglobin, are composed of different polypeptide subunits. Are the genes for hemoglobin alpha subunits, beta subunits, and delta subunits located near one another? On the same chromosome? Where are the genes for these subunits located?
2. Scientists recently altered the expression of a single gene in mice, and the resulting mice were shown to have dramatically improved memories. The altered mice were called *Doogie* mice, after a television sitcom named "Doogie Hauser, M.D.", about a particularly intelligent young boy. *Scientific American* has a site describing Doogie

Mice, which can be found at **http://www.sciam.com/article.cfm?articleID=00075ED9-33D5-1C75-9B81809EC588EF21.**
The protein whose expression is altered in the Doogie mice is called Grin2b. Do humans have a gene that is homologous to the mouse Grin2b gene? If so, on what chromosome and where on that chromosome is this homologous gene located? What is the percent identity between the two gene sequences?
In your view, would it be ethical to attempt to manipulate the expression of this gene in a human that had Alzheimer's disease? In your view, would it be ethical to attempt to manipulate the expression of this gene in a human with memory abilities in the normal range, in an attempt to increase those abilities? Explain.

3. Approximately one in ten American women will contract breast cancer at some point in her lifetime. However, 80 to 90 percent of women who have a mutation in the gene BRCA1 or BRCA2, and a strong family history of this disease, will contract breast cancer. These mutations may account for as many as half of early onset cases (before age 45), and some mutations in these genes are also linked to an increased risk of ovarian cancer. The BRCA genes are thought to be involved in the repair of DNA damage, such as that caused by UV radiation. Where are BRCA1 and BRCA2 located? Are there homologous genes in mice?
4. How many genes are located on the X chromosome? How many are located on the Y chromosome?

10. Mining Genomes: Introduction to GenBank

Starting link: **http://www.ncbi.nlm.nih.gov/.**

Introduction

This tutorial will introduce you to some of the genetics databases that are most frequently used by contemporary researchers. These databases contain enormous numbers of nucleotide and amino acid sequences that have been discovered by researchers around the world. Because of the large amount of sequence information and the complexity of finding and understanding entries relevant to a particular project, a number of specialized tools have been developed to 'mine' the databases so that specific information can be retrieved. While the databases are most commonly used as a component of a larger research program, a growing number of researchers are simply analyzing the information that is available in the databases. In this project, we will explore some of the tools available at the National Center for Biotechnology Information, which is managed by the National Library of Medicine of the United States.

Objectives

- to obtain sequence data from GenBank, the main repository for nucleotide and amino acid sequence data.
- to find specific information within a GenBank entry.
- to find primary literature using PubMed.

Project

Let's begin by looking at the Web site for the National Center for Biotechnology Information (NCBI), which is operated by the National Library of Medicine. Open the Web site at **http://www.ncbi.nlm.nih.gov/** in your browser window.

There are many tools available through NCBI. Access to several of these programs (PubMed, Entrez, BLAST, OMIM) is available from this page. There are also guided tours and tutorials for several of the programs, as well as archives of interesting review articles. Several of the sequence databases are of interest to us, including GenBank.

GenBank is the NIH genetic sequence database, an annotated collection of all publicly available DNA sequences. GenBank is part of the International Nucleotide Sequence Database Collaboration, which is comprised of the DNA DataBank of Japan (DDBJ), the European Molecular Biology Laboratory (EMBL), and GenBank at NCBI. These three organizations share information freely and completely, and exchange data on a daily basis.

Searching GenBank

We are going to begin our look at GenBank by searching for a specific entry. You are probably aware of Michael Crichton's book *Jurassic Park*, in which scientists bring dinosaurs back to life using ancient DNA retrieved from fossils. While it is unlikely that dinosaur DNA has survived to the present, many researchers do study ancient DNA. We will begin by searching for Neandertal sequences in the database.

Neandertals existed as recently as 30,000 years ago, and are closely related to modern humans. Various competing theories disagree as to whether Neandertals and humans interbred, or whether Neandertals represent a separate evolutionary branch. From the NCBI homepage, you can search a number of different databases using their search engine, called Entrez. Search GenBank by entering the term **Neandertal** into the search field, and clicking the **Go** button. Keep the Entrez search drop-down menu on **Nucleotide**.

You should now be looking at a list of GenBank entries that have something to do with Neandertal. (There are four as of the time that this is being written, but the databases change rapidly!). Open the entry labeled **AF282972** (the first one) by clicking on it. 'AF282972' is the *accession number* for this entry. Each entry has its own unique accession number, which is assigned by GenBank at the time of submission.

As you see, there is a lot more information here than just a collection of A's, C's, G's, and T's. Each entry has a *definition* that tells you what is contained within the entry, as well as *source information* that details the taxonomy of the organism whose DNA was sequenced. Here, the sequence is from the mitochondrion of an ancient species related to modern humans—*Homo sapiens neanderthalensis.* Mitochondria are organelles that carry out respiratory functions in cells, and a particular individual inherits its mitochondria only from its mother. Because mitochondria exhibit strict maternal inheritance, the sequences are not altered over time by meiotic recombination. Mitochondrial DNA sequences are therefore more useful than nuclear DNA sequences to researchers studying evolution.

There is also information about the researchers that submitted the sequence data, and frequently there is a reference to a research article in which the data is discussed and interpreted. That reference here is 'A view of Neandertal genetic diversity' in the journal *Nature Genetics* 26 (2), 144-146 (2000), and one of the authors is 'Paabo, S'.

As you **scroll** down the entry, you come to a section labeled *features*, with a subheading of 'Location/Qualifiers'. This important section serves as a roadmap to the sequence entry. Here it is a very brief roadmap, because there are only 288 nucleotides in this entry, and they aren't part of a protein-coding sequence. In larger entries, for example those containing thousands of nucleotides, the 'Features' section tells you which nucleotides are thought to encode proteins, what the deduced amino acid sequences are, and what the encoded proteins are thought to be. Following all of that descriptive information is the actual sequence entry. There it is—genetic data from a Neandertal!

Retrieving GenBank Entries Using Accession Numbers

It's also useful to be able to retrieve GenBank entries using accession numbers. So much sequence data is being generated that most scientific reports don't actually include the sequence data itself. Instead, they refer the reader to the accession number of the relevant

GenBank entry. Accession numbers are an easy way to access just the information that you are interested in. For example, let's say that you come across an article by Dr. Paabo that discusses the use of human mitochondrial DNA sequences to understand the origin of modern humans. What happens if you search 'mitochondrion' in GenBank?

Click on the **NCBI icon** at the top left of the window to return to the GenBank homepage. Then, search GenBank using the term **mitochondrion**.

You probably obtained around 100,000 hits for your search. This is because many people are interested in using mitochondrial DNA sequences for different kinds of studies, including evolutionary studies of diverse organisms (there are entries detailing mitochondrial sequences from the wallaroo, the northern brown bandicoot, the white-fronted capuchin, and the slow loris). This should also give you an idea of how incredibly extensive these databases actually are. It would take a long time to narrow down your search to find a particular entry.

Instead, let's retrieve a sequence whose accession number is referenced in a research article by Dr. Paabo. The accession number of the entry that we are interested in is AF347006. Use Entrez to search the GenBank nucleotide sequences using the term **AF347006**. Then, **click** on the displayed number in order to access the entry.

This entry contains a lot more information than the previous one that we examined. In fact, this entry is the complete genome of a human mitochondrion! There are more than 16,000 bases in the human mitochondrial genome, so of course this entry is more complex than the one we examined initially. However, the organization of this entry is just the same as the first one that we examined.

The *definition* section tells us that this entry contains a complete human mitochondrial genome. Another research article is referenced as well – 'Mitochondrial genome variation and the origin of modern humans', in the journal *Nature* 408 (6813), 708-713 (2000). When we **scroll** to the roadmap part of the entry (the section labeled *features*), we see that this is a much longer section than in the first entry, reflecting the increased complexity of the entry. The beginning of the features section is similar—there is information about the organism that the sequence came from. Here it tells us that the mitochondrion that was sequenced was taken from a Saami person. The Saami are northern Scandinavia's indigenous people, and this sequence was obtained as part of a project investigating the origin of modern humans. In that research project, scientists determined the complete mitochondrial genome sequence of 53 people of diverse ethnic backgrounds.

As we **scroll** further down the features section, there are notes detailing what is encoded by specific sections of the genome. This is organized starting with the first nucleotide in the entry and ending with the last. You can readily see that much of the genome encodes transfer RNAs. These tRNAs are used for synthesis of proteins in the mitochondria. Some of the sections are labeled 'CDS', which stands for codons; these sections encode proteins.

Take a look at the first protein listed. As you can see, the numbers 3306-4262 are next to the CDS link. These are the nucleotides that encode the protein. "NADH dehydrogenase subunit 1" is the name of the protein. The letters on the translation line are the linear sequence of amino acids that are encoded by the gene. The single letter code for the amino acids is used, i.e, M = methionine, C = cysteine, S = serine.

If you now **scroll** down to the bottom of the entry, you will see the nucleotide sequence of the entire Saami mitochondrial genome. The protein we just looked at is encoded by nucleotides 3306-4262.

Now you understand the basics of searching GenBank and how GenBank entries are organized. Note that you can use more complex search approaches, such as the Boolean search terms 'AND', 'OR', and 'NOT'. For example, you could have found the entry above by using the search string 'Saami AND Paabo'.

Perhaps along the way you have become interested in the idea of using mitochondrial DNA sequences for evolutionary studies, or for understanding human origins. In the next part of this project, we'll continue our focus on human origins as we explore another database at NCBI—PubMed.

The PubMed Database

The PubMed database allows you to retrieve information about and from published scientific articles. The left side of your browser should still be open to the GenBank entry with accession number AF347006. As you recall, one of the components of this entry is a reference to a scientific article that interpreted the sequence information in the entry. If you **scroll** to that section of the entry, you will see a referenced article entitled 'Mitochondrial genome variation and the origin of modern humans' that was published in the journal *Nature*. The volume and pages are also listed. Using this information you could go to your library and look up the article and see if it is of interest. However, there is a much easier way to access scientific literature.

Below the article information there is a line that reads PubMed 11130070. Click on the highlighted reference number **11130070**. This should take you to a PubMed entry for that article. The authors and their affiliations are listed, and more importantly, the abstract of the article is available for you to read. The abstract should give you a much better idea of whether you are genuinely interested in the article or not.

But wait! There's more. Above the abstract is a rectangular link with the name of the journal that the article was published in, ***Nature***. If you click on that box, a new window will open, taking you to the journal's Web site, where at the click of another button, you can get the entire text of the article. The article can be read online, or you can download a pdf (portable document file) of the article free of charge.

At the risk of sounding like a television sales pitch—wait! There's more. Click on the **full text** version (as opposed to the pdf) so that you can look at the article. If you **scroll** to the end of the article you will see that the reference section is hyperlinked—so that you can read any of the abstracts and many of the full text articles of the referenced reports. Although you can't yet access *every* article in *every* journal, more and more journals are making their contents available online.

Many university libraries have subscriptions for electronic access of journals. If a particular journal can't be accessed through PubMed, you should see if your library subscribes to the journal. Frequently, you can gain access to the article via the Web site of your institution's library.

Clearly this is a powerful and rapid means of accessing scientific literature. PubMed is usually the best way to begin your efforts, so let's return to the PubMed site. **Close** the browser window with the *Nature* article in it. Then, click the **PubMed** link on the thin black bar at the top part of the left side of this browser.

Searching PubMed

Earlier we mentioned Michael Crichton's *Jurassic Park.* Let's say that we wanted to learn more about efforts to study dinosaur DNA, or even whether such a thing as studying dinosaur DNA is possible. We can enter the search term 'dinosaur' at the PubMed browser – but this will return hundreds of references, most of which we aren't interested in. A better approach would be the use of Boolean connectors, for example, 'dinosaur AND DNA'. This returns only a few articles, most of which are specifically focused on the topic.

Now you give it a turn. Let's say that you want to know about ancient DNA that has been extracted from insects that were encased in amber. Amber is fossilized tree sap, and some amber samples are hundreds of millions of years old. Amber offers interesting insights into the past –the insects are very well preserved, and there have even been efforts to culture bacteria from the guts of ancient amber-encased insects. Dr. Raul Cano has published on culturing bacteria from a 120 million year old amber sample.

Which of the following searches would you try first?

Compare the results of the following searches.
'insect DNA'
'ancient AND DNA'
'ancient AND DNA AND amber'
'Cano RJ' (*note the format used by PubMed to search authors)*
'ancient AND DNA AND amber AND Cano RJ'

Finding Articles Targeted at a General Audience

During your efforts to learn about science, you will be required to write term papers on technical topics. Many of the articles that you find will be difficult for you to understand, because they were not targeted at an undergraduate audience. One way to find articles that were written for a general audience is to include the word 'review' in your search. For example, if you were going to write a paper about ancient DNA, you might search 'ancient AND DNA AND review'.

During the course of this tutorial, you have been introduced to some of the widely used databases at the National Center for Biotechnology Information. Hopefully, you've realized how powerful and easy to use these programs are. Along the way, you probably also noticed that there were many other programs interlinked with these databases. Other *Mining Genomes* tutorials will introduce you to some of these other programs, and you are encouraged to explore on your own. The designers of the NCBI site have made it intuitive and fun to use. Additionally, the following review questions will help to reinforce and extend what you've learned in this tutorial.

Review Questions

1. What is an accession number? How are accession numbers used?
2. Why are mitochondrial sequences useful for evolutionary studies? What advantage does this type of sequence have over nuclear sequences for these studies?

Thought and Application Questions

1. Discussions of ancient DNA frequently return to the topic of *Jurassic Park*, in which various undesirable scary things occur as the result of retrieving ancient DNA and using it to bring dinosaurs back to life. Although that particular scenario is unlikely, there are important ethical and social issues relating to ancient DNA. For example, there are efforts to extract DNA from frozen woolly mammoth cells and add it to enucleated elephant eggs in an effort to clone woolly mammoths and bring them back to life. Similar efforts have been proposed with the dodo, a large bird that was hunted to extinction in the late 1600s. Is it unethical (or unwise) to attempt to return extinct species to the world, given that it is difficult to predict the consequences? Is it unethical to *not* try to return extinct species to the world, if human activities were responsible for the extinction? Explain your answer.
2. Are there DNA sequences that have been recovered from the dodo or the woolly mammoth? Use the GenBank site to find relevant entries. For one of the entries, answer the following questions.
   What is the accession number?
   What kind of DNA sequence does the entry list?
   Who submitted the sequence data?
   Does the DNA sequence encode a protein? If so, what proteins are encoded? If not, what kind of function does the DNA have?
3. Assume that you have been given the assignment of writing an article on efforts to clone extinct species. Using the PubMed database, find three relevant articles that you might use for your report. Can you download pdf versions of any of these papers from your library's Web site?
4. Most of us have favorite animals—animal species that we are particularly drawn to. List three of your favorite organisms. Are mitochondrial sequences available from these organisms? If so, list the accession numbers.

11. Mining Genomes: Introduction to BLAST

Starting link: **http://www.ncbi.nlm.nih.gov/.**

Introduction

Large-scale DNA sequencing efforts such as the Human Genome Project churn out immense amounts of sequence data, but interpretation of the blur of A, C, G, and T is required for such efforts to have any value. One of the first types of analysis includes dividing the genome into individual genes and assigning functions to the products encoded by those genes.

Putative genes can be identified by searching for specific DNA sequences associated with genes, such as promoters, start and stop codons, and intron splice sites. Similarly, protein-coding regions are characterized by open reading frames, long stretches of DNA that don't contain a translation stop codon. However, the problem of assigning a function to the gene product can be considerably more challenging.

Once a gene has been identified, how can a function be assigned to it? There are empirical approaches that might be used, but these are expensive, time-consuming, and prone to failure. For example, you could predict the amino acid sequence, synthesize the protein, and test the produced protein using a suite of activity assays. Alternatively, you could make a knockout mutant by removing the gene from the genome of the organism, and looking for a phenotype. But how do you know what phenotype to look for? And if knocking out the gene kills the organism, how does that help you understand the function of the gene? Finally, imagine how difficult this would be with an entire genome; the simple nematode *C. elegans* contains over 18,000 genes! There is clearly a strong need for computational methods that allow gene function to be predicted from the DNA sequence.

One of the most powerful methods for identifying gene function directly from the DNA sequence is to compare a new sequence with information that is already in the databases. For example, if a new sequence is almost identical to a gene described in the databases, it is reasonable to assume that the protein encoded by the new gene may have a function similar to that of the gene described.

The databases are extensive, rapidly growing collections of DNA and protein sequencing results that are submitted by researchers worldwide. Peer-reviewed journals require that sequence data be submitted to the databases prior to manuscript publication; in addition, large amounts of otherwise unpublished data is also present in the databases. GenBank is the NIH genetic sequence database, an annotated collection of all publicly available DNA and protein sequences. GenBank is part of the International Nucleotide Sequence Database Collaboration, which is also comprised of the DNA DataBank of Japan (DDBJ), and the European Molecular Biology Laboratory (EMBL). These three organizations share information freely and completely, and exchange data on a daily basis. Public access is available free of charge.

One of the most popular tools for comparing a newly derived DNA or protein sequence with the information in the databases is BLAST, (Basic Local Alignment Search Tool). BLAST is a set of similarity search programs designed to explore all of the available sequence databases. It will match several small subsequences within your search sequence to the sequences in the database. The BLAST output describes sequences that have regions of similarity to your sequence, and provides a variety of links to those sequences and other analysis programs

Objectives

- to conduct simple BLAST searches, including blastn, blastp, and blastx queries.
- to interpret the output of BLAST queries.
- to use the program ORFfinder to find open reading frames.

Project

For this project, we are going to use sequences that were uncovered during the analysis of the genome of the organism *Deinococcus radiodurans*. This unusual eubacterium is famous for its ability to withstand radiation, and it has been isolated from atomic reactors and uranium mines. *D. radiodurans* cells can survive ionizing radiation of up to 3,000,000 rads (humans can be killed by exposure to less than 500 rad). This kind of radiation is sufficient to break the DNA of this organism into several hundred pieces, and yet it has such efficient DNA repair mechanisms that it is able to survive such threats. It is easy to isolate (simply expose a sample to high radiation—almost everything else is killed) but difficult to study genetically. Why do you think that this organism might be difficult to study genetically? Of course! It's hard to generate *D. radiodurans* mutants!

Open the NCBI homepage at **http://www.ncbi.nih.gov/**. There are an incredible variety of tools and information available from this starting point. You can access the most used programs from the toolbar at the top of the page, where sites such as PubMed, Entrez, BLAST, and OMIM are listed. You might want to return to this site later and spend some time exploring on your own as well—it's easy to navigate and there's lots of background information and tutorials on the different tools. For now, we want to go to the BLAST site. Select **BLAST** on the toolbar at the top of the NCBI homepage.

BLAST searches can be conducted using DNA (nucleotide) or protein (amino acid) sequences. Or, you can have BLAST translate your DNA sequence into an amino acid sequence and then search the deduced protein sequence against protein sequences in the databases. The output of a BLAST search lets you quickly see whether or not your sequence is related to previously described sequences and also whether the similarity is over the entire length of the sequence or just a small part of the sequence. Because BLAST searches subsequences within your sequence, it can detect small regions of similarity within otherwise dissimilar sequences, which can be useful for finding out the function of a particular part of your protein (for example, an ATP-binding domain, or a DNA-binding domain). The three BLAST programs that we are going to use in this project are compared in the following table.

Program	Input sequence	Databases examined
blastn	nucleotide	DNA
blastp	amino acid	Protein
blastx	nucleotide	Protein

Blastn Search

Assume that you are sequencing the genome of *D. radiodurans*, and you have just found the gene listed in the link below. We are going to use BLAST to see if we can find similar sequences in the databases, and thereby assign a putative function to the protein encoded by the gene.

From the BLAST main page, you can select several different kinds of BLAST searches, including nucleotide, protein, and translation searches. We are going to begin by carrying out a search that will compare your nucleotide query with all of the nucleotide sequences in the various databases. Select **Standard nucleotide-nucleotide BLAST [blastn]**

This should take you to a page with a search box. The nucleotide sequence of the gene is listed below (do you see what was meant about the blur of A, C, G and T?). Copy the nucleotide sequence and paste it into the search box. Leave the default settings the way they are, and select the **BLAST!** button.

ATGTCCATCGACCACTCCCGCATCTCCACCCAGCCTTTCCCCGCGTCCGA
GAAGCGCTACCTCGGCGGCAGCCTCTTTCCTGAAGTGCGCGTGCCAGTGC
GCGCCATTACGCTTTCGCCCACCGTGGAGGCGTTTAGCGGAGTGACGCGC
ACCCTGCCCAACCCGCCGCTGCTCGTGCCCGACACCAGCGGCGCCTACAC
CGACCCCGCCGTGGCCACCGACCTGCAACGGGGCCTGCCGCACGCCCGGC
CCTGGCTGGCGCGCAGCGAGCGCACCGAGGAAACGGCGCGTTTTTCCGCG
CTGGCAGATATGGACGGGCCACTGCCGTATCCGCACGTCCTCACGCCGCG
CCGCGCCCGCAGTGAAAGCCGGGGGGGGCAGGGCATCACCCAGATGCAGG
CGGCCCGGCGCGGTGAAATTACCCCGGAAATGGAATTCGTCGCTCTGCGC
GAGAACCTGCGCCAGCAGGAGCAGGCCGAACTCGCGCACCAGCACCCCGG
CGAGAGCTTCGGCGCCAGCCTTCCGCGCCTCTACACCCCCGAATTCGTGC
GGGACGAGGTGGCGCGGGGCCGGGCGGTCATTCCGGCGAATGTCAACCAC
CCCGAGCTGGAGCCGACCATCATCGGGCGCAATTTCCGGGTCAAAATCAA
TGCGAACCTCGGCACGTCCATCGTCACGTCGAGCATCGAGGAAGAAGTCG
AGAAGATGGTCTGGGCGACGCGCTGGGGGGCCGACACGGTGATGGACCTC
TCGACCGGGCGGCACATCCACCAGACGCGCGAATGGATTGTGCGCAACAG
CCCGGTGCCGATTGGGACGGTGCCGATTTATCAGGCGCTCGAAAAGGTGG
ACGGCAAGGCCGAGGAACTGACCTGGGAGGTGTACCGCGACACGCTGATC
GAGCAGGCCGAGCAGGGCGTGGACTACTTCACCATTCACGCGGGGGTGCG
GCTGGCGCACGTGCCGTTGACCGCCCGGCGGCGCACCGGCATCGTGTCGC
GCGGGGGCAGCATCCTGGCGAAGTGGTGCCTCGCGCACCATCAGGAGAAC
TTTCTCTACACCCGTTTTGCGGAGATTTGCGAAATCATGGCCGCTTACGA
CGTCACCTTCAGCCTGGGCGACGGCCTGCGCCCAGGCAGCATCGAGGACG
CCAACGACGCCGCGCAGTTTGCCGAACTGGGCACGCTGGGCGAACTGACG

CGGGTGGCCTGGGACCACGGCGTACAGACCATGATCGAGGGGCCAGGGCA
CGTGCCCATGCAGCTTATCCGCGAGAACATGACCCGGCAGCTCGACGTGT
GCCAGGAGGCGCCGTTTTACACGCTGGGGCCGCTGACCACCGACATCGCG
CCGGGGTACGACCACATCACCTCCGCCATCGGTGCGGCGCAAATCGCATG
GTACGGCACCGCCATGCTGTGCTACGTGACGCCCAAAGAGCACCTCGGGC
TGCCCGACAAGCAGGACGTGCGTGACGGCGTGATCACCTACAAAATCGCC
GCGCACGCCGCCGACCTCGCCAAAGGGCACCCCGGCGCGCAGGCCCGCGA
CAACGCGATTTCCAAGGCGCGCTTCGAGTTCCGCTGGCAAGACCAGTTCA
ATCTGGCCCTCGACCCCGAAAAGGCCCGCGCCCTGCACGACGAGACGCTG
CCCGCCGACGCCGCCAAGACCGCGCACTTTTGCAGCATGTGCGGCCCGGC
CTTTTGCTCAATGAAGCTGAGCCACGACATCCGCGCGGGCGACGTGCTGG
CCGGTCTGGAAGCGAAGGCGCGGGAATTCCGCGAGCTGGGCGGCGAGATT
TATGTAGAAGGGTCGGATGACTGA

This will bring you to another page, which has various settings that you can use to change the format of your blastn search results. There is also a message giving you the approximate amount of time it will take before the search is completed. This time will vary from a few seconds to as much as several minutes, depending on how many people are using the site. Leave the default settings the way that they are, and select the **FORMAT!** button.

This will open a new window in your browser. If your results are not yet ready, the page is largely empty, with a message that it will automatically reload shortly. Once your results are ready, click next.

Interpreting the Output of Your Blastn Search

The BLAST output shows you the database sequences that have shared regions of sequence identity with your nucleotide query, what organisms the matching sequences came from, and how closely they match. This information is given to you in three different ways: a graphical, color-coded view; a set of one-line summaries; and a detailed match description that includes alignments and comparisons of your query with the matching sequences.

The color-coded view is at the top of the page, and consists of a series of different colored lines. The red bar at the top of the figure (in this case, it's numbered from 0 to 1750) represents your query sequence, and the numbers represent the nucleotides in your query. The different colored bars below that line represent the matches that were found in the databases. Very close matches are shown in red, relatively close matches in lavender, and so on, with black regions indicating possible low matches and white regions indicating areas of no sequence identity. You should see between 30 and 40 bars, with a bright red bar on top indicating a perfect match across the entire length of the sequence. As of this writing (remember that the databases are constantly updated), the next twelve bars also contain regions of red.

We'll use the third closest match to discuss what is shown in this output. This match is represented by a bar that has red regions representing very close matches from nucleotides 700 - 1200 and from 1300 - 1650. In the first 600 nucleotides there are short regions with some identity, represented by short blue bars, and some regions of no identity, in which the bar is white. If you **roll your mouse** over the bar, the name of the organism and the sequence that it came from are shown in the box above the figure. Moving from one bar to the next, you can see what organisms/sequences are depicted. As you move towards the bottom of the figure, you will see that regions that match are shorter, and the quality of the match becomes poorer and poorer.

Below this figure, a slightly more detailed version of the same information is presented in a series of one-line summaries. **Scroll down** the page to this region of the output. Each line has three parts: a hyperlink to the GenBank entry where the matching sequence can be found; the name of the organism that the sequence came from; and a set of values indicating how closely the sequence matches with the query. A high score value and/or a low E value indicate close matches. If you are really super-curious about how the values are calculated, the formulas are described in the 'BLAST Course' that can be accessed from the BLAST information page. Selecting the link on the left will take you to the GenBank entry associated with the matching sequence (there is a tutorial that introduces GenBank and GenBank entries at the end of Chapter 10 in your text). Selecting the score will take you to the third section of the BLAST output.

Select the score for the third-closest match (as of this writing, from *Ralstonia solanacearum*, with a score of **266**).

This takes you to the alignments section of the BLAST output, which provides detailed information on sequences that are similar to your query sequence. You also could have reached this information by scrolling down from the top of the page, where the color-coded figure is. On this page you can see an alignment of sections of the query and match sequences. The vertical lines represent nucleotides that are identical in each of the sequences. The numbering of the 'Subject,' or match sequence, refers to the numbering of the sequence in its GenBank entry. There is also a listing of percent sequence identity, which is 83% in the first region, with 314 of 374 nucleotides matching in the two sequences.

Now let's revisit our original question. We wanted to know what protein this gene encodes; that is, we wanted to know what the protein does. We were hoping that our blastn search would give us a close match with a described gene (one that someone had already determined the function of), and then we could assume that our new gene carries out a similar function. However, if you look through the descriptions of the matches listed here, they mostly consist of sequences from other genome sequencing projects. The kind of functional studies that are required for assigning a particular activity to the sequences haven't been done on these sequences. Therefore, in this instance we can’t answer our question with this search. Let's compare these results with the results we can obtain using another type of search: blastp.

Close the window that opened with your blastn search results.

Blastp Search

When you carried out a blastn search, you used a program that tried to match the subsequences of your nucleotide query with other nucleotide sequences that are in the databases. BLAST searches can also be carried out on amino acid sequences. Sometimes these searches can be more successful than blastn searches, for reasons that have to do with the degenerate nature of protein-coding sequences.

The same amino acid can be encoded by different codons. Because a codon consists of three nucleotides, and because there are four different types of nucleotides, there are 4^3 or 64 different possible codons. 61 of these encode the primary amino acids and the remaining three are stop codons. Because there are only 20 amino acids, most amino acids are encoded by more than one codon. To encode a specific amino acid, an organism will not use each of the possible codons with the same frequency. Rather, species exhibit *codon bias*, by using one particular codon for a given amino acid much more frequently than alternative codons.

There are a variety of reasons for this. For example, organisms also exhibit bias in the amount of the different tRNAs that they make for a particular amino acid, and might slow down or speed up the rate of translation of a particular mRNA by using different codons within the corresponding gene. Different organisms exhibit different codon biases. One species might use G- or C- containing codons preferentially to A- or T- containing codons, or vice versa. In eubacteria, the mean percentage of guanine + cytosine in genomic sequences, or the GC content, varies from about 25% to 70%. The %G+C of bacterial genomes is sometimes used as a distinguishing characteristic, because in many cases bacterial GC content seems to be related to phylogeny, with closely related bacteria having similar GC contents.

One implication of codon degeneracy is that similar amino acid sequences can be encoded by nucleotide sequences that are quite different. We will now do a BLAST search that compares your *amino acid sequence* (i.e., NOT your nucleotide sequence) against the protein sequences in the database. To do this you can use the amino acid sequence here, which is simply the translated form of the nucleotide sequence that you have been using. This sequence uses the single letter amino acid code, in which M stands for methionine, S represents serine, C represents cysteine, and so on.

To carry out your blastp analysis, click the large **BLAST** icon on the upper right corner of the Web page to return to the original BLAST page where you began your BLAST search. Under 'Protein Blast', select **Standard protein-protein BLAST [blastp]**. Cut and paste the following amino acid sequence into the query box of the blastp program, and select the **BLAST!** button.

MSIDHSRISTQPFPASEKRYLGGSLFPEVRVPVRAITLSPTVEAFSGVTRTLPNPPLLVP
DTSGAYTDPAVATDLQRGLPHARPWLARSERTEETARFSALADMDGPLPYPHVLTP
RRARSESRGGQGITQMQAARRGEITPEMEFVALRENLRQQEQAELAHQHPGESFGAS
LPRLYTPEFVRDEVARGRAVIPANVNHPELEPTIIGRNFRVKINANLGTSIVTSSIEEEV
EKMVWATRWGADTVMDLSTGRHIHQTREWIVRNSPVPIGTVPIYQALEKVDGKAEE

LTWEVYRDTLIEQAEQGVDYFTIHAGVRLAHVPLTARRRTGIVSRGGSILAKWCLAH
HQENFLYTRFAEICEIMAAYDVTFSLGDGLRPGSIEDANDAAQFAELGTLGELTRVA
WDHGVQTMIEGPGHVPMQLIRENMTRQLDVCQEAPFYTLGPLTTDIAPGYDHITSAI
GAAQIAWYGTAMLCYVTPKEHLGLPDKQDVRDGVITYKIAAHAADLAKGHPGAQA
RDNAISKARFEFRWQDQFNLALDPEKARALHDETLPADAAKTAHFCSMCGPAFCSM
KLSHDIRAGDVLAGLEAKAREFRELGGEIYVEGSDD

Submitting your blastp search should bring you to a format page similar to the one from your blastn search. This time, though, in addition to formatting options, there should be a message that reads 'See conserved domains from CDD' and a figure with a blue bar indicating the putative conserved domains. Although we'll return to this figure, for right now select the **Format!** button. This will open a new window in your browser. If your results are not yet ready, the page is largely empty, with a message that it will automatically reload shortly. Once your results are ready, proceed on to the next section.

Interpreting the Output of Your Blastp Search

The BLAST output shows you what protein sequences share regions of identity with your protein query, what organisms the matching sequences come from, and how closely they match. The format is similar to that for blastn, and the information is given to you in three different ways: a graphical, color-coded view; a set of one-line summaries; and a detailed match description that includes alignments and comparisons of your query and the matching sequence.

The first thing you should notice is that this search found far more close matches than the blastn search. This is due to the fact that several different nucleotide sequences can encode the same protein. The second thing you should notice is that these results allow you to assign a function to your new gene. Some of the closely related sequences have been shown experimentally to be involved in thiamine synthesis (thiamine is also known as vitamin B1). Based on the results of this search, you can tentatively conclude that your protein is the thiamine biosynthesis protein ThiC.

As with the blastn output, you can scroll down this page and you will find first a series of one-line summaries, and then a set of more detailed alignments. Go to the color-coded set of bars at the top of the page, move your cursor to the lowest red bar on the diagram, and **click** on the bar to select it. This will take you down in the page to the alignment that corresponds with that sequence.

This amino acid alignment is slightly different than that of the nucleotide alignments. Instead of a series of vertical lines indicating identical amino acids, the letter representing that amino acid is repeated in between the two sequences. In several of the positions there is a '+' sign between the two sequences. This represents a site where the amino acids are different but very similar; for example, an acidic amino acid in the query sequence lines up with a different, but still acidic, amino acid in the subject sequence. The rationale here is that substituting a similar amino acid in a protein is less likely to result in a change in the shape or function of the protein. The summary at the top of the alignment reflects this; if you are looking at the alignment with the *Arabidopsis thaliana* ThiC-1 protein, the summary shows that 317 of 509 amino acids are identical, and that 360 of 509 are similar. If you had done a blastn search, and were looking at a nucleotide alignment, a match of 62% would not have a very high score, and it wouldn't be represented by a red bar—why is this different for protein alignments?

Now, **close** the window with the blastp results in it.

Using ORFfinder to Find Open Reading Frames

Before, you compared the blastn search of a nucleotide sequence with the blastp search of the deduced protein sequence. However, this begs the question of how you go from a nucleotide sequence to a protein sequence. How do you know where a protein-coding sequence starts?

This is actually a more complex problem than it might seem. For example, imagine a nucleotide sequence in which the first 12 bases are 5' - AAAGGGTTTCCC- 3'. There are three different possible reading frames for this sequence - it might be read as "AAA-GGG-TTT..." or as "AAG-GGT-TTC..." or as "AGG-GTT-TCC...". And, because DNA exists as a complementary double-stranded molecule, a sequence of 5' - AAAGGGTTTCCC- 3' must be paired with a sequence of 5' - GGGAAACCCTTT - 3', and this sequence also has three separate possible reading frames. Therefore there are 6 different possible ways of translating any given nucleotide sequence.

Let's go now to a program at NCBI that is useful for finding open reading frames (ORFs) in nucleotide sequences. Remember that an ORF is a series of nucleotides that is grouped into codons such that there is a long stretch of nucleotides that encode amino acids, without intervening translation stop codons. Because 3 of the 64 codons are stop codons, you would expect a random (non-protein-coding) sequence to have a stop codon about every 20 codons, on average.

Click the **NCBI icon** at the top left of the page return to the NCBI home page.

On the right side of the page, under the heading 'Hot Spots', there is a series of links to different tools and features. Scroll down and select **ORF finder.**

If you enter any sequence, this program will search each of the six possible reading frames and translate them into amino acid sequences. In the output returned to you, the regions that contain coding sequences are depicted in color.

To try it out, cut and paste the sequence below into the ORF Finder query box. Then select **OrfFind**.

ACTCTTCTGGTCCCCACAGACTCAGAGAGAACCCACCATGGTGCTGTCTC
CTGCCGACAAGACCAACGTCAAGGCCGCCTGGGGTAAGGTCGGCGCGCAC
GCTGGCGAGTATGGTGCGGAGGCCCTGGAGAGGATGTTCCTGTCCTTCCC
CACCACCAAGACCTACTTCCCGCACTTCGACCTGAGCCACGGCTCTGCCC

AGGTTAAGGGCCACGGCAAGAAGGTGGCCGACGCGCTGACCAACGCCGTG
GCGCACGTGGACGACATGCCCAACGCGCTGTCCGCCCTGAGCGACCTGCA
CGCGCACAAGCTTCGGGTGGACCCGGTCAACTTCAAGCTCCTAAGCCACT
GCCTGCTGGTGACCCTGGCCGCCCACCTCCCCGCCGAGTTCACCCCTGCG
GTGCACGCCTCCCTGGACAAGTTCCTGGCTTCTGTGAGCACCGTGCTGAC
CTCCAAATACCGTTAAGCTGGAGCCTCGGTGGCCATGCTTCTTGCCCCTT
GGGCCTCCCCCCAGCCCCTCCTCCCCTTCCTGCACCCGTACCCCCGTGGT
CTTTGAATAAAGTCTGAGTGGGCGGC

You should get an output that has six bars, each representing one of the different reading frames that your nucleotide sequence might be read in. Portions of some of the bars are colored in aqua, and these regions represent open reading frames detected by the program. If you click one of the bars, the deduced protein from that ORF will be shown.

Of course, all of these open reading frames are not used to make proteins. But how do you know which ORF is the 'real' one? One way might be to select the amino acids from each of the different reading frames and then conduct a separate blastp search on each of these possible proteins. If one of the ORFs matches up closely with a studied protein, then you have probably identified the actual reading frame.

However, it turns out that there is an even easier way. There is another type of BLAST search that does all of this for you, called a blastx search. This search will translate a nucleotide sequence into its six different possible reading frames, and then search each of those potential protein coding sequences against the protein sequences in the database.

Blastx Search

Return to the NCBI BLAST homepage by clicking **BLAST** from the toolbar at the top of the ORF Finder page, and this time select **Nucleotide query - Protein db [blastx]** under Translated BLAST Searches. This search is different from the blastn search in that this time you are comparing your query sequence with the protein databases. It's also different from the blastp search in that you are using a nucleotide sequence as the query

Copy the nucleotide sequence below into the search box and press **BLAST!**

Select the **Format!** button, and when the BLAST output has loaded, scroll through the results. Did multiple ORFs match up with known proteins, or was only one of the ORFs likely to be an actual protein? What does the protein do?

ACTCTTCTGGTCCCCACAGACTCAGAGAGAACCCACCATGGTGCTGTCTC
CTGCCGACAAGACCAACGTCAAGGCCGCCTGGGGTAAGGTCGGCGCGCAC
GCTGGCGAGTATGGTGCGGAGGCCCTGGAGAGGATGTTCCTGTCCTTCCC
CACCACCAAGACCTACTTCCCGCACTTCGACCTGAGCCACGGCTCTGCCC
AGGTTAAGGGCCACGGCAAGAAGGTGGCCGACGCGCTGACCAACGCCGTG
GCGCACGTGGACGACATGCCCAACGCGCTGTCCGCCCTGAGCGACCTGCA
CGCGCACAAGCTTCGGGTGGACCCGGTCAACTTCAAGCTCCTAAGCCACT

GCCTGCTGGTGACCCTGGCCGCCCACCTCCCCGCCGAGTTCACCCCTGCG
GTGCACGCCTCCCTGGACAAGTTCCTGGCTTCTGTGAGCACCGTGCTGAC
CTCCAAATACCGTTAAGCTGGAGCCTCGGTGGCCATGCTTCTTGCCCCTT
GGGCCTCCCCCCAGCCCCTCCTCCCCTTCCTGCACCCGTACCCCCGTGGT
CTTTGAATAAAGTCTGAGTGGGCGGC

Now you are acquainted with the basics of BLAST searching, including blastn, blastp, and blastx searches. These three searches are the most commonly used types of BLAST queries. As you can imagine, there are some times when you still can't assign a function to a sequence using these programs. It is possible to do more advanced BLAST searches, in which you change some of the settings, and some of the strategies, in order to find more distant matches. If you are interested, you can now take your foundation in BLAST and explore some of these more advanced approaches on your own! Additionally, the following review questions will help to reinforce and extend what you've learned in this tutorial.

Review Questions

1. How are blastn, blastp, and blastx searches different?
2. Briefly compare and contrast blastn with blastp. When would it be advantageous to use a blastp search? When would it be advantageous to use a blastn search?
3. What do you think is the main limitation to the approach in which functions are assigned to sequences by searching for identity with previously described sequences?
4. What is an open reading frame?
5. Why do we say that a particular nucleotide sequence has six possible reading frames?

Thought and Application Questions

1. The sequences below were uncovered during the genome sequencing of *D. radiodurans*. Attempt to assign a function to these sequences using BLAST, and answer the following questions for each sequence.
    a. Which BLAST search did you use? Why?
    b. What were the three closest matches?
    c. What do you think the function of this sequence is?

    Sequence A
    ATGGCTTACACTCTTCCCCAACTGCCCTACGCTTACGACGCGCTTGAGCC
    CCATATCGACGCCCGCACGATGGAAATTCACCACACCAAGCATCACCAGA
    CCTACGTGGACAACGCCAACAAGGCGCTCGAAGGCACCGAATTCGCCGAC
    CTGCCCGTCGAACAACTCATTCAGCAGCTCGACCGCGTGCCCGCCGACAA
    GAAGGGCGCCCTGCGCAACAATGCGGGCGGCCACGCCAACCACAGCATGT
    TCTGGCAGATCATGGGCCAGGGTCAGGGCCAGAACGGCGCCAACCAGCCC
    AGCGGCGAACTGCTGGACGCCATCAACAGCGCCTTCGGCAGCTTCGACGC
    CTTCAAGCAGAAGTTCGAAGACGCCGCCAAGACCCGCTTCGGCTCGGGCT
    GGGCGTGGCTGGTCGTCAAGGACGGCAAGCTCGATGTCGTGTCCACCGCC
    AACCAGGACAACCCGCTGATGGGCGAAGCCATCGCGGGCGTCAGCGGCAC
    CCCGATTCTGGGTGTGGACGTGTGGGAACACGCCTACTACCTCAACTACC

AGAACCGCCGCCCCGACTACCTCGCCGCCTTCTGGAACGTGGTGAACTGG
GACGAAGTCTCGAAGCGCTACGCCGCCGCGAAGTAA

Sequence B
ATACCCCTGATGAAAATCAAGCTGACCGCCTACCTGACCAGCCTCGCGCT
GCTGTCGTCCGCCGCCGTCGCGCTGCCGCGTGCCCTCCCGAGCGCTCCCG
TATCGCAGGCGGCGGGCAAGCTCGAAGTCGTGCACCGCTTCTACGGCCAC
ATGCCCATCGGCGTGACCGTCAACTCGCAGGGCCGCATGTTCGTCTCCTA
CCCCAACTGGGAAGACGACGTGCCCTTCTCGATTGCCGAGATCAAGGGCG
GGCGGGAAGTGCCCTACCCCAACCGCGCCATCAACACCCGTGACCTATCG
AAGCCCGACACCACCTTCATCGGCGTGCAGGGGCTGCTGGTGGACGCCCG
GGACCGGCTGTGGGTGCTCGACACCGGCACCAGGAACCTGGGACCGATTC
TCGACCAGCGGGCGGTCAAGCTGGTGGGCATCGACACCCACACCAACGAA
GTCGTCAAGACCATCCACTTTCCTGCCGACGTGGCGCTGAAAAATACTTA
CCTCAACGACCTGCGCATCGACCTGCGTCAGGGCACGGGCGGCGTGGCAT
ACATCACCGACTCGGGGGCCAAGAGCGGCTCGGGGCTGATCGTGGTGGAC
CTCGCCAGCGGCAAATCGTGGCGCAAGCTGACGGGCGACGAGACCGTTAA
GCCCGTGCCTGGCTTCGTGCCCTACGTGGAGGGGCAGGCCCTCTTCCAGC
GGCCCAAGGGCGGCCCGGCCACCCACCTGGGCTTCGGGGCCGACTCGCTT
GCCATCAGCCCCGACGGCGCCACGCTCTACTACGCGCCCACCGCCTCGCG
CCGGCTCTACGCCGTGCCGACCGCCGCGCTGCGTGACCAGGGCCTCAGCG
ACGCCGAGGTGAAAAAGCAGGTGAAGGACCTGGGCGAGAAGGGCGCCGCC
GACGGCCTGGCCGAGGACACCGCCGGAAACATCTACATCACCAACTACGA
GCAGGGCGCCCTGGTGCGCCGCCTGCCGACCGGCGAACTGCAAACGCTGG
CGCGTGACCCCCGCCTGATCTGGCCCGATACGCTGGCAATACAGGGCAGC
TACCTGTACGTCCTCAACGACCAGCTCAACCGTCAGGGCGGCTACCACTT
CGGCAAGGACCAGCGCGTCAAGCCCTACACCCTGCTGCGGATGAAGCTCG
ATGCCAAGCCGGTGCTGCTGAAATAAGCTG

2. The sequence below was uncovered during the genome sequencing of *D. radiodurans*. What reading frame is this sequence probably read in? Why and how did you pick this?

<u>Notation you might use in your answer</u>
+ = The sequence that is shown.
- = The reverse complement of the sequence that is shown.
1 = The reading frame that starts with the first nucleotide of the sequence.
2 = The reading frame that starts with the second nucleotide of the sequence.
3 = The reading frame that starts with the third nucleotide of the sequence.
The six possible reading frames, then are +1, +2, +3, -1, -2, and -3.

ACCTTGACGCCTGGTTCGCGCTGCAAAACGTGGCTCCCGGCAAGCGGCAA
AAAGCCGTGCTGGCGGGCATGATCGAGGGCGTGGTCAAAAAGCCCACCAA
GTCGGGCGGCATGATGGCCCGTTTATCCTGGCCGACGAATCGGGGCAGA
TGGAACTGGTGGCCTTTTCGCGCGCCTATGACCGGATTGAGCCCAAGCTG
GTCAACGACACGCCCGCGCTGGTCATCGTGGAACTGGAAGCCGAGGACGG
CGGCCTGCGCGCCATCGCCGAGGAAATCGTGAGCATCGAGCAACTGTCCG

AGGTGCCCAAGGTCATGTACGTGACCATCGACCTCGAAACGGCCAGCCCT
GACGCGCTGGGCGACTTTCAGAGCGTGCTGGACGAGTACGCCGGCAGTAT
GCCCACCTACCTGCGGCTGGAAACCCCTGAGCAGTTCGTGGTGTACCAGC
TCGACCACGGCATGGGCAGCCCCGAGGCGATTCGGGCGCTGAACCAGACC
TTCGCCTGGGCCGACGCCCACCTCGCCTACGACCAGCAGACGATTCTGGG
CCGGTTTGCGCCCAAGCCGCCCGCCTGGATGAACCGGCAGCAGGGGGGCG
GGATGCGGGCGTGA

3. Assume that you are studying three closely related organisms, and you are trying to figure out which two are the most closely related. That is, you are trying to determine which two of the organisms most recently shared a common ancestor. Would it be more useful to compare nucleotide sequences between the organisms, or protein sequences? Explain your answer.

14. Mining Genomes: 16S Ribosomal RNA Studies

Starting link: **http://rdp.cme.msu.edu/html/analyses.html**.

Introduction

The diversity of life on our planet is astounding and magical, full of beauty and mystery. Throughout history, people have been fascinated with trying to understand the relationships between living things. For example, consider the breathtaking array of colors, forms, and lifestyles of fish inhabiting a coral reef—while it is easy enough to group some of the fish together, how do those groups then relate to one another? Furthermore, how are these marine fishes related to freshwater fishes? To the marine invertebrates around them?

Through the nineteenth century, people divided living things into two groups: plants and animals. While this division worked on a simplistic level, it also raised some important questions, such as whether and how plants were related to animals. Some organisms were discovered that were difficult to fit into this scheme—to which of these two groups do fungi, protozoa, or bacteria belong, and why? Darwin's and Wallace's insights into evolution by natural selection allowed people to understand that all living things are related by common descent. However, piecing together ancient relationships is difficult, and some relationships are only now being understood, with the advent of bioinformatics approaches

These same approaches are being used to identify newly discovered organisms. In this tutorial, you are cast in the role of a bioinformatics specialist working for a biotechnology company. Researchers in the company have just isolated a new microorganism that makes an array of novel, wide-spectrum antibiotics. The organism was isolated from hot, sulfur-rich sediments in Hawaii Volcanoes National Park, and the identity of the organism is unknown. You will explore the identity of this organism as well as some of the relationships between all known life forms.

Objectives

- to introduce the use of ribosomal RNA sequences in taxonomic analysis.
- to introduce molecular phylogeny.
- to acquaint you with the tools that are available for use at the Ribosomal Database Project II, which is managed by the Center for Microbial Ecology at Michigan State University.

The Use of Ribosomal RNA Sequences in Molecular Taxonomy and Phylogeny

One ancestral species can give rise to two or more progeny species. However, speciation events don't involve the vast majority of the genes in a genome, and for most of the genome, the progeny species inherit identical sequences from the ancestor. Following speciation, these sequences evolve independently in the separate lineages. Changes occur in the genetic material by processes that include base substitutions, deletions, insertions, rearrangements, and duplications. Because genomes are so large, sequence information constitutes an enormous data set that can be used to understand our evolutionary history.

Species that are closely related share very similar sequences; species with a more ancient divergence share related but less similar sequences. Therefore, it is possible to understand the identity of a newly discovered organism by comparison with sequences of known organisms, and it is possible to gain insights into the evolutionary history of all life by comparison with sequences of diverse organisms. Ribosomal RNA sequences are particularly useful and have been used for a wide range of different studies.

Ribosomes are complexes of protein and RNA that are involved in the process of translation. There are ribosomal RNA (rRNA) molecules in every ribosome. In eukaryotes, the different rRNA molecules are called 28S, 18S, 5.8S and 5S rRNA, and in prokaryotes they are called 23S, 16S, and 5S rRNA. Ribosomal RNA sequences, particularly the 16S and 18S rRNA sequences, are widely used for molecular comparisons in taxonomy studies. Some of the reasons why rRNA is commonly used include: (1) rRNA is found in all living organisms, and has the same function in all organisms; (2) 16S and 18S rRNA molecules are relatively long (~1500 and 1800 nucleotides, respectively) and information-rich; (3) the sequence and structure of rRNA is conserved in all living things, and therefore some regions can be aligned accurately while other, more variable, regions can be used for phylogenetic analysis; and (4) rRNA is abundant and relatively easy to study, for example by using PCR primers targeted towards universally conserved sequences.

The conservation of structure and sequence among rRNA molecules allows us to look at the relationships between all living things, regardless of how different they are. Because people find the idea of understanding these relationships so compelling, ribosomal RNA sequences have been determined for an incredible variety of organisms, and more than 20,000 such sequences are now available. In addition to describing how known organisms are interrelated, this data can also be used to understand the identity of newly discovered organisms.

The Ribosomal Database Project II

In 1989, Carl Woese and his collaborator Gary J. Olsen began The Ribosomal Database Project (RDP), which provides access to rRNA sequences and tools for their analysis. This project expanded rapidly, and is now managed by the Center for Microbial Ecology at Michigan State University. In this project we will explore the suite of tools that are available at the RDP II, using these tools to identify an unknown organism. Open your browser to the RDP II at **http://rdp.cme.msu.edu/html/analyses.html**.

The programs that we will use at the RDP are:

Sequence Match: Matches your sequence with the closest sequences in RDP
Sequence Aligner: Aligns your sequence with the closest sequences in RDP
Similarity Matrix: Calculates percent identity of your sequence and RDP sequences

Our Project

In this project, you will use rRNA analysis to study a recently isolated microorganism that makes several antibiotics new to medicine. You have determined part of the new organism's 16S rRNA sequence, and are going to use the RDP to try to figure out its identity, or at least what kind of organism it is. In addition to using the sequence of your new organism, you will use the 16S rRNA sequence from *Bacillus cereus*, a common soil microorganism, so that you can understand what kind of results you should expect from each of the tools.

Ribosomal Database Project II - Sequence Match

The Sequence Match program provides a rapid means of comparing your new sequence with sequences that are already in the database. The program breaks your sequence into several short (7 nucleotide) *oligomers,* and then uses a statistical analysis to determine how many oligomers are shared between your sequence and the database sequences. The best matches are displayed in the output.

From the Online Analyses page at RDP II, select the program Sequence Match by clicking on the **arrow** in the 'Run' column adjacent to the program name. This will take you to a form. **Highlight** the partial 16S sequence from *B. cereus*, and **paste** the following sequence into the box in the form labeled 'Cut & Paste a sequence from your machine'. Leave the default settings the way they are, and click on the **Submit Sequences** button to start the program. Note the format of the form for this program is very similar to that used for the other programs, although the programs themselves and their output differ significantly.

B. cereus 16S rRNA, partial sequence:
GGACGGGTGAGTAACACGTGGGTAACCTGCCCATAAGACTGGGATAACTC
CGGGAAACCGGGGCTAATACCGGATAACATTTTGAACCGCATGGTTCGAA
ATTGAAAGGCGGCTTCGGCTGTCACTTATGGATGGACCCGCGTCGCATTA
GCTAGTTGGTGAGGTAACGGCTCACCAAGGCAACGATGCGTAGCCGACCT
GAGAGGGTGATCGGCCACACTGGGACTGAGACACGGCCCAGACTCCTACG
GGAGGCAGCAGTAGGGAATCTTCCGCAATGGACGAAAGTCTGACGGAGCA
ACGCCGCGTGAGTGATGAAGGCTTTCGGGTCGTAAAACTCTGTTGTTAGG
GAAGAACAAGTGCTAGTTGAATAAGCTGGCACCTTGACGGTACCTAACCA
GAAAGCCACGGCTAACTACGTGCCAGCAGCCGCGGTAATACGTAGGTGGC
AAGCGTTATCCGGAATTATTGGGCGTAAAGCGCGCGCAGGTGGTTTCTTA
AGTCTGATGTGAAAGCCCACGGCTCAACCGTGGAGGGTCATTGGAAACTG
GGAGACTTGAGTGCAGAAGAGGAAAGTGGAATTCCATGTGTAGCGGTGAA
ATGCGTAGAGATATGGAGGAACACCAGTGGCGAAGGCGACTTTCTGGTCT
GTAACTGACACTGAGGCGCGAAAGCGTGGGGAGCAAACAGGATTAGATAC
CCTGGTAGTCCACGCCGTAAACGATGAGTGCTAAGTGTTAGAGGGTTTCC
GCCCTTTAGTGCTGAAGTTAACGCATTAAGCACTCCGCCTGGGGAGTACG
GCCGCAAGGCTGAAACTCAAAGGAATTGACGGGGGCCCGCACAAGCGGTG
GAGCATGTGGTTTAATTCGAAGCAACGCGAAGAACCTTACCAGGTCTTGA
CATCCTCTGACAACCCTAGAGATAGGGCTTCTCCTTCGGGAGCAGAGTGA
CAGGTGGTGCATGGTTGTCGTCAGCTCGTGTCGTGAGATGTTGGGTTAAG

This should result in a display that includes matches in a color-coded hierarchical taxonomic display.

As expected, the program has determined that your input sequence matches closely to *B. cereus* sequences in the database. The taxonomy of the closest matching sequences is displayed, along with the organism that the sequence came from (in blue) and a number indicating how close the match is (in red). The number does not represent the percent identity between the query and subject sequences, but rather is a statistical value. High values, close to 1, represent better matches than low values. Note also that the output is broken into two groups. The top sequences represent matches from aligned sequences, while the bottom group represents matches from sequences that have yet to be aligned. This will be explained more fully in the section below on the Sequence Aligner program.

Now use the Sequence Match program to see if there are any matches in the database for the sequence from your new organism. Go to the RDP homepage by clicking **Return to main analysis page**. Then click the **arrow** next to Sequence Match, like you did before, and **paste** your new rRNA sequence (below) from the organism isolated by the biotechnology company. Then press **Submit Sequence**.

New rRNA sequence, from organism isolated by biotechnology company:
TGTGCCAGCAGCCGCGTAATACCCGCAGCTCAAGTGTTGCCACTATTATT
GAGCCTAAAGCGTCCGTAGCCGGTCTTGTAAATCTCTGGGTAAATCCTGC
CGCTCAACGGTAGGAATGCTGGAGAGACTGCAATACTAGGGATCGGGTGA
AGTAGAAGGTACTCCTGGGGTAGGGGTAAAATCCTGTAATCCTGGGGGGA
CGACCGGTGGCGAAGGCGTTCTACTAGAACGACTCCGACGGTGAGGGACG
AAGGCTAGGGGAGCAAACCGGATCAGATACCCGGGTAGTCCTAGCTGTAA
ACGCTGCCCACTTGGTGTTGCCCTTTCTTCGGGAAAGGGCAGTGCCGGAG
CGAAGGTGTTAAGTGGGCCGCTTGGGGAGTATGGTCGCAAGGCTGAAACT
TAACGGAATTGGCGGGAGAGCACCGCAACGGGAGGAGCGTGCGGTTTAAT
TGGATTCAACGCCGGAAAACTCACTAGGGAAGACCATGGTATGAGAGCCA
GCCTGAAGAGCTTACTCGATAACATGGAGAAGTGGTGCATGGCCGTCGTC
AGCTCGTACCGTAAAGCGTTCACTTAAGTGTGATAACGAGCGAGACCCCC
ATCCACTTTTGCTAGTAATGTCGCGAGACAATACGCACTATGTGGAGACT
GCCAGTGTTAAACTGGAGGAAGGAGAGGTCAACGGCAGGTCAGTATGCCC
CGAATTCCCTGGGCTACACGCGCGCTACAAAGGACGGAACAATGGGCTGC
GACTCTGAAAAGAGGAGCCAATCTCGAAATCCGTTCGTAGTTCAGACTGA
GGGCTGTAATTCGCCCTCACGAAGCTGGATTCCGTAGTAATCGCGAGTCA
ACAGCTCGCGGTGAATATGCCCCTGCTCTTT

When you carry out this analysis, you should get very different results. For example, none of the matches have a very high value associated with them, indicating that there are no close matches in the database. Also, the matches that have been returned don't cluster into a single taxonomic grouping, as they did with the *B. cereus* sequence. It appears that this new organism is very unique.

Ribosomal RNA Structure and Sequence Alignments

The goal of phylogenetic sequence alignments is to arrange sequences so that you can compare homologous nucleotides or amino acids. Homologous sequences are sequences that have descended from a common ancestral sequence. You can't meaningfully compare sequences unless they are homologous. There are various programs available for aligning sequences, and some are more useful than others for phylogenetic analyses.

When aligning 16S sequences, you have more information than simply the sequence of nucleotides. Remember that rRNAs have a conserved shape and function as well as sequence; these conserved shape structures are useful in constructing alignments that reflect homologous relationships. Let's begin by taking a look at the structure of rRNA from a microorganism named *Sulfolobus acidocaldarius*.

Depictions of the primary and secondary structure of 16S rRNAs from several organisms can be found at the site of the rRNA server at the University of Antwerp in Belgium (**http://rrna.uia.ac.be**). Pull up the structure of the *Sulfolobus* 16S rRNA by going to **http://rrna.uia.ac.be/secmodel/Saci_SSU.html**.

This figure takes a moment to get used to; it is a two-dimensional representation of how the linear rRNA molecule folds into space to form a functional structure. Some of the regions are held together into a double-stranded conformation (similar to the structure of double-stranded DNA) by complementary hydrogen bonding between nucleotides. There are also several single-stranded regions. If you were to look at the 16S rRNA structure of a closely related organism, it would have a nearly identical arrangement of single- and double-stranded regions. In fact, some of these structures are common to rRNA molecules from all known organisms. These conserved structures provide additional information that can be used in constructing alignments that reflect homology.

This diagram also provides some insight on why there are regions of rRNA sequences that are conserved among all living things. Translation is a fundamental process of life, and accurate translation requires a specific shape of rRNA molecule. If a mutation were to occur in the DNA encoding an rRNA nucleotide that is in a double-stranded region, the change would disrupt the structure of the resulting rRNA. In order for the double-stranded structure to be maintained, another mutation would have to occur simultaneously, in the exact nucleotide that base-pairs with the first nucleotide. For example, in *Escherichia coli* 's 16S rRNA, nucleotide 1430 (an 'A') base-pairs with nucleotide 1470 (a 'U'). If nucleotide 1430 were mutated, then the structure of the rRNA would be altered, unless nucleotide 1470 were simultaneously mutated. Because mutations at a given site are infrequent, the probability of simultaneous mutations occurring is very small. Therefore, these double-stranded regions evolve at a relatively slow rate.

Ribosomal Database Project II - Sequence Aligner

The RDP II maintains an aligned sequence database, which can be very helpful for biologists aligning a new sequence. In conjunction with a secondary structure diagram of the 16S rRNA of a related organism, this database can be used to construct alignments that reflect homologous relationships between new sequences and those already in the database.

Go back to the RDP II Online Analyses page by going to http://rdp.cme.msu.edu/html/analyses.html.

Select the program Sequence Aligner by clicking on the **arrow** in the 'Run' column adjacent to the program name. This will take you to a form. **Copy** the partial 16S sequence from *B. cereus* from earlier in the tutorial and **paste** the sequence into the box in the form labeled 'Cut & Paste a sequence from your machine'. This time we will change one of the default settings. Under 'Sequence Aligner Output Options,' **change** the 'number of related sequences to include' value from 1 to 5. Leave the other default settings the way they are, and click on the **'Submit Sequences'** button in order to start the program.

The output that is returned to you is an alignment between your sequence and the 5 sequences in the RDP database that are most similar to your sequence. The database sequences used are more than 90% complete and have been previously aligned with other sequences. In the alignment, aligned bases of the user sequence and the closest match are shown in green. Aligned bases of the next most similar sequences are shown in yellow. Regions that are not highlighted indicate regions that could not be aligned unambiguously. These non-highlighted regions should be aligned by hand using a secondary structure diagram or they should be excluded from the analysis. In this output, there are no non-highlighted regions.

Go **back** to the Sequence Aligner program at **http://rdp.cme.msu.edu/cgis/seq_align.cgi?su=SSU.** Now, try the program again to see if the newly discovered sequence can be aligned with database sequences. Use the new rRNA sequence from earlier in the tutorial.

Unlike with the Sequence Match program, when you carry out this analysis you should get results that are similar to those seen with the *B. cereus* sequence. There are only small regions that are not highlighted. This indicates that there is enough sequence and structural conservation between your new sequence and the database sequences that they could be successfully aligned, which is a requirement for carrying out the taxonomic analysis.

The Ribosomal Database Project II - Similarity Matrix

The aligned sequences can also be used to assess how similar the user's sequence is to described sequences in the database, by employing the Similarity Matrix program. This program uses an alignment process similar to that of the Sequencer Aligner program, and after the alignment is constructed, a mask is created that marks the positions that could be unambiguously aligned in all of the sequences. Only the masked positions are considered in the calculation.

Go to the Online Analyses page at RDP II at **http://rdp.cme.msu.edu/html/analyses.html**. Select the Similarity Matrix program by clicking on the **arrow** in the 'Run' column adjacent to the program name. This will take you to a form. **Copy** the partial 16S sequence from *B. cereus* from earlier in the tutorial and **paste** the sequence into the box in the form labeled 'Cut & Paste a sequence from your machine.' This time we will change one of the default settings. In the 'Similarity Matrix Options,' **change** the value in the category 'For each of my sequences, include 1 most similar RDP sequences' from 1 to 10. Leave the other default settings the way they are, and click on the Calculate Matrix button to start the program.

You should see an output that looks something like a rectangle that is composed of an upper and a lower triangle. If this isn't what you see, make sure that you changed the 'For each of my sequences, include 1 most similar RDP sequences' value from 1 to 10. The numerical values in the matrix reflect the percent identity (upper triangle) and percent non-identity (lower triangle) of each of the sequences compared against each of the other sequences. For example, the fraction of masked nucleotides common to the user sequence (your sequence, sequence 1) and sequence 5 is .999, which you can determine by looking at the intersection of lines 1 and 5 in the upper triangle. To the left of the matrix is an abbreviated name of the organism whose 16S sequence is represented, and clicking on the name will take you the actual sequence entry.

The diagonal line from the top left of the triangle to the bottom right is empty, because this line is a comparison of each of the sequences with itself. There are some database sequences that match your sequence exactly, and some that match it in 99.8 % of the positions. How closely will the new query sequence match with database sequences?

Carry out a Similarity Matrix analysis of the new sequence. Go back to the Similarity Matrix site at **http://rdp.cme.msu.edu/cgis/sim_matrix.cgi?su=SSU**, and click use the sequence from earlier in the tutorial.

In looking at the results of this analysis, you can see that the Similarity Matrix program was able to match the new sequence with some sequences in the database. The closest match has identical nucleotides in about 85% of the aligned positions. Unfortunately, there is no clear-cut distinction between percent similarity and taxonomic divisions, meaning that you can't tell whether a particular sequence comes from a new species or genus on the basis of a similarity matrix.

In fact, with microorganisms, it is unclear what is even *meant* by the term species. However, as a rough ballpark estimate, you could imagine that microorganisms whose 16S rRNA differ by 3% are different species. Bacteria whose 16S rRNA differ by 15-20% are probably in different kingdoms or divisions (the taxonomic level immediately beneath domain) within the domain Bacteria.

Click on the name of the closest match. The journal article associated with this sequence is entitled *Picrophilus oshimae and Picrophilus torridus fam. nov., gen. nov., sp. nov., two species of hyperacidophilic, thermophilic, heterotrophic, aerobic archaea*. The 'fam. nov ....' part of this title means that these species were new species in a new genus in a new family—i.e., they were pretty different than previously described organisms. And your new organism is only 85% identical to this already novel family! This means that this new organism may be important to science. Because the closest match is Archaeal, it also suggests that the new isolate is in the Domain Archaea.

To get to this site, you selected the name of the organism whose rRNA sequence was the closest match to your new sequence. Several of the other entries didn't have organisms names associated with them. For example, one of the matches was an entry labeled 'AB019742'. This sequence came from a study in which bulk DNA was extracted from a deep-sea hydrothermal vent sample, and 16S rRNA genes in the DNA extract were analyzed. Many studies such as this have been performed using DNA from different environments, and they have all suggested that the actual diversity of microorganisms is greatly underestimated by current culture collections. The fact that both of the closest matches are from sequences from high temperature environments suggests that organisms living at high temperatures may be phylogenetically related

Phylogenetic Relationships Between All Known Life Forms

Because all known living things have ribosomes that contain similar rRNA molecules, rRNA sequences can be used to study the evolutionary relationships between all known life forms. In the introduction we said that more than 20,000 rRNA sequences have been determined, representing the known diversity of life. Alignments and analyses of these sequences have offered many new insights into the evolutionary history of our planet. Clearly, understanding the evolutionary history of all the major groups of organisms on planet Earth is an ambitious undertaking. It is important to realize that the resulting depiction of relationships (the phylogenetic tree) will be an estimate that is likely to contain some imprecise parts. This is true of attempts to study history at any level.

Let's take a look at an rRNA phylogenetic tree depicting the evolutionary relationships between all known living things. A depiction of this tree is at the Web site maintained by the University of California's Museum of Comparative Paleontology at **http://www.ucmp.berkeley.edu/alllife/threedomains.html**.

The results of this large-scale phylogenetic analysis are depicted in the figure at the top of the page. The figure is called a phylogenetic tree, although in this case it looks something like the root structure of a seedling. In this figure, currently existing organisms are represented at the tips of the lines. A branchpoint, where two lines diverge, represents a single ancestor species that differentiated into two distinct lineages. This analysis was pioneered in the 1970s by Carl R. Woese at the University of Illinois. As you can see, all currently known forms of life fall into one of three domains—Eukaryota, Bacteria, and Archaea.

The length of the branching represents evolutionary distance, so that two closely related organisms are connected by a smaller distance than two less closely related organisms. Compare the distances between plants and animals and fungi with the distances between thermophilic Archaea and cyanobacteria. The studies conducted to create this phylogenetic tree have convinced scientists that much of the biodiversity on our planet is microbial.

This is a fun site to explore, and it's worth spending some time here.

This completes your introduction to ribosomal RNA studies. During the course of this tutorial, you have been introduced to the use of sequence data in studies of molecular taxonomy and phylogeny. We've focused on rRNA sequences and programs available at the RDP II. The tutorial for Chapter 17, on Molecular Evolution, extends and complements this introduction using protein sequences and other analysis programs.

Review Questions

1. In your own words, what is the biological meaning of the word *homology?*
2. Why are homologous sequences from closely related organisms likely to share more sequence identity than homologous sequences from distantly related organisms?
3. What is ribosomal RNA?
4. Why are 16S and 18S ribosomal RNA sequences used for molecular phylogeny studies?

Thought and Application Questions

1. What would be the consequences of carrying out molecular phylogenetic analysis on a set of sequences that includes some sequences that were misaligned?
2. You work for a company trying to develop new and useful biological recombinant products (enzymes, vaccines, drugs, etc...). Your job is to obtain 16S rRNA gene sequences from unknowns in a variety of natural samples in order to determine their taxonomic identity. Unfortunately, you can no longer read your writing on a set of tubes and can't figure out which DNA came from which natural sample. You decided to forge ahead (in spite of this blunder), and you sequence the 16S rRNA genes from your now-unlabeled genomic DNA.
   For the sequences below, you should determine:
   a) What organism did the sequence come from?
   b) From what kind of sample was the organism likely isolated?
   c) Why might a biotechnology company be interested in this organism?
   You may need to use resources outside of the RDP to answer the last part of the question.

   16S Sequence 1:
   CGAACGCTGGCGGCGTGCCTAATGCATGCAAGTCGAGCGAAGTTATTGCT
   TTCGGGTAATAACTCAGCGGCGAACGGGTGAGTAACACGTAGGTGATCTA
   CCCTTCAGAGGGGGACAACCTTGGGAAACCGAGGCTAATCCCACATGAGT
   TCGTTGGCAGGGATGCTGATGAAGAAAAGGTTACTGAGAAGTAATCGCTG
   AAGGATGAGCCTGCGGCCCATCAGGTAGTTGGCGAGGTAAAGGCTCACCA
   AGCCGAAGACGGGTAACCGGTCTGAGAGGATGTACGGTCACACTGGGACT
   GAGACACGGCCCAGACTCTACGGGAGGCAGCAGCAGGGAATTTTGGGCAA
   TGGGCGAAAGCCTGACCCAGCGACGCCGCGTGGAGGAAGAAATCTTCGGG
   ATGTAAACTCCTGGACTGGGGGACGAGAGAGGACGGTACTCCAGTAGAAA
   GGGACGGCTAACTACGTGCCAGCAGCCGCGGTAAAACGTAGGTCCCGACG
   TTGTTCGGATTCACTGGGTGTAAAGGGCGCGTAGGCGGTTGTTTGTGTCT
   CATGTGAAAACTCAGGGCTCAACTCTGAGATTGCGTGGGAAACTAAGCAG
   CTAAGAGGACGGTAGAGGGGAGTGGAATTCCCGGTGTAGCGGTAAAATGC
   GTAGATATCGGGAAGAACACCGGTGGCGAAGGCGGCTCTCTGGGCCGATC
   CTGACGCTGAGGCGCGAAAGCTAGGGGAGCGAACCGGATTAGATACCCGG

GTAGTCCTAGCCGTAAACGATGGGTGCTAGGTGTTGGGGGGTAAAGACCC
TTCAGTGCCGGAGCGAACGCGATAAGCATCCCACCTGGGGAGTACGGCCG
CAAGGTTGAAACTCAAAGGAATTGGCGGGGGTCCGCACAAGCGGCGGAGC
ATGTGGTTTAATTCGATGCTGCCCGAAGAACCTTACCAGGGTTTGACATG
ATGGTAGTAGTAACCTGAAAGGGGAACGACCTGACCGTAAGGGAAGGAGC
CATTGCAGGTGTTGCATGGCTGTCGTCAGCTCGTGCCGTGAGGTGTCGGC
TTAAGTGCCGTAACGAGCGCAACCCTCATCCCTAATTGCTACTTACGAAG
AGTGAGAGCACAATAGGGAGACCGCTGGCGAAGAGCCAGAGGAAGGGGAG
GATGACGTCAAGTCAGCATGGCCTTTATGCCCTGGGCTACACACATGCTA
CAATGCAGGGTACAATGGGAAGCGAAGCCGCGAGGTGGAGCAAATCCCGA
AAAGCCCTGCTCAGTTCGGATCGTACGCTGCAATTCGCGTGCGTGAAGTC
GGAGTCGCTAGTAACCGTGGATCAGCAAAGCCGCGGTGAATACGTTCTCG
GGCCTTGCACACACCGCCCGTCAAGCCACCCGAGTCGGGTTCACCAGAAG
GCCGGTAACCGAAAGGGCCCGGACGACGGTGTGCCTGGTAAGGAGG

16S Sequence 2:
TTGATCCTGGCTCNGGACGAACGCTGGCGGCGTGCTTAACACATGCAAGT
CGAGCGNTGGAGTTCCTTCGGGGACGGATTAGCGGCAGACGGGTGAGTAA
CACGTGGGCAACCTACCTCGGAGTGGGGGATAACCTTCCGAAAGGGAGAT
TAATACCGCATAACATATTTTTTTACGCATGTGAGAAATATTAAAGATTT
ATTGCTTCGAGATGGGCCCGCGGCGCATTAGCTAGTTGGTGAGGTAACGG
CTCACCAAGGCAACGATGCGTAGCCGACCTGAGAGGGTGATCGGCCACAT
TGGAACTGAGACACGGTCCAGACTCCTACGGGAGGCAGCAGTGGGGAATA
TTGCGCAATGGGGGAAACCCTGACGCAGCAACGCCGCGTGAGTGATGAAG
GTTTTCGGATTGTAAAACTCTGTCTTTGGGGACGATAATGACGGTACCCA
AGGAGGAAGCCACGGCTAACTACGTGCCAGCAGCCGCGGTAATACGTAGG
TGGCAAGCGTTGTCCGGATTTACTGGGCGTAAAGAGTATGTAGGCGGATG
CTTAAGTCAGATGTGAAATCCCCGGGCTCAACCTGGGGGCTGCATTTGAA
ACTGGGCATCTGGAGTGCAGGAGAGGAAAGTGGAATTCCTAGTGTAGCGG
TGAAATGCGTAGATATTAGGAAGAACATCAGTGGCGAAGGCGACTTTCTG
GACTGTAACTGACGCTGAGATACGAAAGCGTGGGGAGCGAACAGGATTAG
ATACCCTGGTAGTCCACGCCGTAAACGATGAATACTAGGTGTCGGGGGGT
ACCACCCTCGGTGCCGCAGCTAACGCAATAAGTATTCCGCCTGGGGAGTA
CGGTCGCAAGATTAAAACTCAAAGGAATTGACGGGGACCCGCACAAGCAG
CGGAGCATGTGGTTTAATTCGAAGCAACGCGAAGAACCTTACCTAGACTT
GACATCTTCTGCATTACTCTTAATCGAGGAAATCCCTTCGGGGACAGAAT
GACAGGTGGTGCATGGTTGTCGTCAGCTCGTGTCGTGAGATGTTGGGTTA
AGTCCCGCAACGAGCGCAACCCCTATTGTTAGTTGCTACCATTAAGTTGA
GCACTCTAGCGAGACTGCCTGGGTTAACCAGGAGGAAGGTGGGGACGACG
TCAAATCATCATGCCCCTTATGTCTAGGGCTACACACGTGCTACAATGGC
CGGTACAGAGAGTTGCAATACCGCGAGGTGGAGCTAATCTTTAAAGCCGG
TCCCAGTTCGGATTGCAGGCTGAAACTCGCCTGCATGAAGTTGGAGTTGC
TAGTAATCGCGAATCAGAATGTCGCGGTGAATGCGTTCCCGGGTCTTGTA
CACACCGCCCGTCACACCATGAGAGTTGGTAACACCCGAAGCCTGTGAGG
TAACCGTAAGGAGCCAGCAGTCGAAGGTGGGATCGATGATTGGGGTGAAG
TCGTAACAAGGTAGCCGTAGGAGAACCTGCGGCTGGATCACCTC

15. Mining Genomes: Three-Dimensional Protein Structure

Starting link: **http://www.rcsb.org/pdb/.**

The function of proteins and other biological macromolecules is directly related to their shape, and understanding the three-dimensional shape of proteins is an important developing field in bioinformatics. This exercise uses tools available at the Molecular Modeling Database at the National Center for Biotechnology Information to visualize the three-dimensional shape of some interesting proteins.

Introduction

The shape of biological macromolecules determines their function. That is, biological macromolecules such as proteins and nucleic acids do what they do because of their structure. One of the most exciting areas of bioinformatics is concerned with understanding the three-dimensional shape of these molecules. The interesting topics within this field include determining the structure of known macromolecules; accurately predicting the three-dimensional structure of molecules from their amino acid or nucleic acid sequences; and using principles derived from the first two areas to engineer proteins that have new desirable features, such as pharmacological properties.

In this tutorial we will use some of the features of the Molecular Modeling Database at the National Center for Biotechnology Information. This database is the repository of experimentally (and sometimes theoretically) determined three-dimensional macromolecular structures, and also contains tools for the visualization and comparative analysis of these structures.

Objectives

- to review the principles of three-dimensional protein structure.
- to acquaint you with the tools that are available for use at NCBI to visualize three-dimensional protein structures.
- to introduce the structure-function relationships of hemoglobin and the structural basis of sickle cell anemia.

Protein Structure - Primary Structure

Polypeptides are linear chains of amino acids. These chains fold in three-dimensional space into specific structures to form functional proteins. Protein structure is described in a hierarchical fashion, using the concepts of primary, secondary, tertiary and quaternary structure.

Primary structure consists of the linear arrangement of amino acids that make up a particular protein. Different polypeptides have different numbers and arrangements of amino acids.

There are 20 common amino acids, and they can be broken into 4 groups based on charge: hydrophobic, polar, charged-acidic, and charged-basic.

Protein Structure – Secondary Structure

Each amino acid has the same *backbone* structure, consisting of a central carbon that is bound to the nitrogen atom of an amino group on one side and the carbon atom of a carboxyl group on the other side. Thus all polypeptides have a repeating backbone structure of -N-C-C-N-C-C-N-C-C-.... The secondary structure of a protein is the result of the formation of hydrogen bonds between these backbone atoms. There are two common secondary structures, the alpha helix and the beta-pleated sheet.

Protein Structure – Tertiary Structure

Different amino acids have different R-groups attached to their central carbon, and these R-groups can interact in several ways. The tertiary structure of a protein consists of interactions between side groups of amino acids. Such interactions can include hydrophobic and van der Waals interactions as well as hydrogen, ionic, and covalent bonding. For example, different regions of the protein might be held together into a specific structure by a covalent disulfide linkage between the sulfur atoms of two cysteine amino acids, or by an ionic bond between the charged-acidic amino acid lysine and the charged-basic amino acid aspartate.

Protein Structure - Quaternary Structure

Many functional proteins consist of two or more (sometimes many more) polypeptides that are associated with one another. The quaternary structure of a protein refers to the three-dimensional shape of the associated protein complex. For example, hemoglobin is a protein with quaternary structure, and is composed of four subunit polypeptides (two each of two different types). Thus, a single hemoglobin molecule has two alpha chains and two beta chains.

Sickle Cell Anemia

In this project, we are going to look at the structure of normal human hemoglobin and hemoglobin from a person with sickle cell anemia. The difference between these two forms of hemoglobin is very minor. The sixth amino acid of the beta chain of normal hemoglobin is a glutamic acid; in the most common form of sickle cell anemia, a single nucleotide alteration changes this residue to a valine.

What Determines the Three-Dimensional Shape of a Protein?

Clearly the primary structure of a polypeptide is important in determining the final shape of a protein, because it determines what the arrangement of secondary and tertiary structures can be. However, the actual determinants of three-dimensional shape include other factors as well. Evidence of this includes observations that when a protein is expressed in a foreign host it is not always made in its active conformation, and when proteins are denatured by forces such as heat they do not always renature into their active forms when the suitable environment is restored.

Biologists have solved the structures of well over 10,000 different proteins, and know the primary sequence of more than 100,000 proteins, but still the rules for determining three-dimensional protein structure remain elusive. It is not now possible to determine the three-dimensional shape of a protein molecule from the primary sequence alone. Many lab groups are working on this challenge, and predictive techniques are constantly improving. Much of this work is based on studies correlating primary sequences with 3-D structures that have been solved using X-ray crystallography and nuclear magnetic resonance (NMR) techniques.

The Protein Databank and The Molecular Modeling Database (MMDB)

Two important resources in the analysis of 3-D structures are The Protein Databank (PDB) and The Molecular Modeling Database (MMDB). The PDB's goal is to improve the understanding of the 3-D nature of biological macromolecules by providing the public, free of charge, with a repository of structures and the tools to analyze, visualize, and compare those structures. The PDB is managed by a collaboration that includes Rutgers University, The San Diego Supercomputing Center, and the National Institute for Standards and Technology.

The MMDB is a related set of programs managed by the National Center for Biotechnology Information. Both of these sites provide excellent tools for the visualization and comparison of 3-D structures; because the MMDB is integrated into the other NCBI tools, such as Entrez, we will focus on that site.

The Growth of the PDB

Open the PDB on your browser at **http://www.rcsb.org/pdb/**.

One thing that most students have difficulty with is a historical perspective on biology and bioinformatics. It is therefore useful to look at the statistics related to the growth and history of databases such as the PDB. On the left side of the PDB homepage, in the blue section under Current Holdings, click the link entitled **PDB Statistics**.

Approximately how many structures are in the database now? Approximately how many structures were in the database when your instructor was an undergraduate?

If you scroll down to the bottom of that page, you will find a table that classifies the database structures as to how many proteins, protein/nucleic acid complexes, nucleic acids, and carbohydrates are in the databases. Which of those categories has the largest number of deposited structures? Why do you suppose that this is the largest category?

The Molecular Modeling Database (MMDB)

The MMDB uses the experimentally determined (as opposed to the theoretically determined) structures in the PDB, and in addition to visualization tools based on a program called Cn3D (get it?), it provides a means to connect structure inquiries with relevant tools at NCBI. These tools include links to bibliographic information, to the sequence databases, and to the NCBI taxonomy. One of the potentially most useful tools at the MMDB is the VAST program. VAST (Vector Alignment Search Tool) carries out similarity searches similar to BLAST, but VAST searches for structures that are similar to the query structure. Using VAST, you can identify proteins that have similar structures, even if the amount of sequence identity between the two proteins is low.

Now go to the NCBI homepage at **http://www.ncbi.nih.gov**. Then click **Structure** on the top menu. From there, click **MMDB** on the left menu to get to the MMDB homepage.

From this page you can begin a study of known 3-D structures for a particular molecule by entering the name of the molecule into the search box. You can also use Entrez to search for structures. To introduce you to MMDB, we are going to look at normal human hemoglobin, and then at hemoglobin from an individual with sickle cell anemia.

Start by entering the term **hemoglobin** into the search field, and clicking **Go**.

As you can see, you have a gotten a list of a wide variety of different structures from different organisms. Many people have studied the structure of hemoglobin and hemoglobin-like molecules. Note also that each entry has a 4-character identifier. This 4-character code is a system created by the PDB for identifying the different structures with a unique name; the structure that we want is 1HAB. It will probably be simpler to enter **1HAB** in the search field and repeat the search.

This should take you to an entry with the 1HAB name, and the description 'Crosslinked Haemoglobin'. Select the **1HAB** link to take you to the MMDB structure summary site for this structure.

A Tour of the MMDB Structure Summary Site

There are a lot of different kinds of information at this site. We will start from the top and work our way down.

At the top of the page is a description of the entry along with the MMDB and PDB identifiers for this structure. Selecting the MMDB link will return you to this page while the PDB link will take you to the original PDB site for the structure. There is also a link to Medline, the publications database at NCBI. Selecting the PubMed link will take you to PubMed entries that are linked to this structure, in this case the entry for the *Nature* paper in which this structure was reported and interpreted.

Also notice the link to the taxonomy browser at NCBI, where you can figure out what kind of organism the protein came from. Although *Homo sapiens* is obvious to us all, this link can be invaluable for figuring out the identity of organisms such as *Eptatretus burgeri, Scapharca inaequvalvis,* or *Anser anser.*

Below this is a list of the protein chains in the structure. There are four chains in our structure, listed as A, B, C, and D.

We are interested in Chain B. By selecting the *Protein* Link or *CDs* link for Chain B, you would be taken to, respectively, the Entrez sequence site and the Entrez Conserved Domains site. Using the *Chain B* and *globin* links, you can carry out a VAST search to find matches for the 3-D structures of this chain (Structure Neighbors) and BLAST searches to find matches for the amino acid sequences of the chains (Sequence Neighbors).

Viewing the 3-D Structure of Hemoglobin

Now let's take a look at the 3-D structure of hemoglobin. You will need to load the program Cn3D onto your computer in order to see these structures (unless your instructor has already downloaded the program for you). To do this, select **Get Cn3D 4.0!** and follow the instructions for downloading Cn3D. If your computer is not particularly powerful, it will probably be helpful to download the version for 256-color monitors.

Then, select the gray **View 3D Structure** button, and assuming that you have configured the program appropriately, the structure should appear in a new window. Another new window, the 'OneD-Viewer', will open in the background.

The default structure emphasizes the secondary structure of the molecule. Recall that alpha helices and beta-pleated sheets are protein secondary structures that result from hydrogen bonding between backbone atoms of the chain. Here, the structure looks like a complex of wires wrapped around green dowels—the dowels represent regions of alpha helix structure. Also note the iron-containing heme groups.

You can **rotate** this structure by dragging it with your mouse in order to view it from a different perspective.

Now we'll use some of the tools that are available to manipulate the way that we view the structure. In the Cn3D window, select **Color:Cycle Chain**. This will make each of the four chains a different color, so that you can differentiate them a little bit more easily.

Note that the hemoglobin looks as if it were made up of four identical polypeptides; the alpha and beta chains are very similar in sequence and arose via a gene duplication event of a single ancestral gene.

Now go to the OneD-Viewer window. The sequence of the alpha-chain is displayed. In this window, select **File:Sequence List**. In the window that opens, select **1HAB_B(protein)** and then press **OK.** This will cause the B-chain sequence to appear in the OneD-Viewer window.

If you highlight a section of the sequence, this section will be colored yellow in the Cn3D window.

Now, let's try to figure out specifically where the sixth amino acid of the beta chain is. **Highlight** a large section of the sequence, the first ten amino acids' worth. Then find the highlighted region in the Cn3D window. Note which chain this sequence in, and the location of the region. Next, highlight just the sixth amino acid (vhltp**E**eksa). Then find the highlighted region in the Cn3D window. You will probably need to 'grab' the molecule with your cursor and spin it around in space so that you can easily see this amino acid. With some manipulation, you should be able to see that this amino acid is located close to the end of that chain, and also that it is located on the outside of the 3-D molecule

Before we proceed on, this would be a good time for you to play around a little with some of the Cn3D viewing options. For example, look at some of the different styles of structure models available under the style menu, and zoom in and zoom out on parts of the molecule using the View menu. If you downloaded the version of Cn3D for 256-color monitors, you can zoom in by holding down your shift key and using your mouse to define the zoom area.

Note that some of the options on the Cn3D viewer require a lot of memory to run and may crash your computer. For example, Style: Spacefill. If your computer is crashing frequently, it will probably help to download the version for 256-color monitors.

After you've finished exploring, close the Cn3D windows, so that the structure summary window is again visible.

The Structure of Sickle Cell Hemoglobin

Now let's examine the structure of sickle cell hemoglobin.

Return to the MMDB site by clicking the **MMDB Structure Summary** icon at the top of the page.

Search the structure database using the MMDB identifier **6228**. This should take you to an entry labeled '2HBS. The High Resolution Crystal Structure of Deoxyhemoglobin S [6228]'. Select **2HBS** to go to the structure summary site.

One of the first things that you should notice here is that there are eight chains in this structure (A-H) instead of four. Select **View 3D Structure** to examine the 3-D structure.

This structure looks very similar to the structure of normal hemoglobin, except that it is composed of twice as many chains. Under low oxygen conditions, such as at high elevation or during physical exertion, sickle hemoglobin molecules tend to clump together into long fibrous chains. These chains distort the shape of the red blood cell, causing it to form the characteristic sickle shape.

Select **Color: Cyle Domain** so that all of the chains appear a pink salmon color.

Then, go to the One-D viewer window. In this structure, the mutant beta hemoglobin (6E->V) chains are chains B, D, F and H. These chains are identical to the wild-type beta chains except for the alteration of a single amino acid—the sixth amino acid in the mutant chain is a valine instead of the wild-type glutamic acid. This substitution is the result of a single nucleotide change in the gene encoding this polypeptide.

In the One-D viewer, select **File: Sequence list: 2HBS_H (protein).** Then, **highlight** the mutated amino acid (#6, vhltp**V**eksa).

Locate the altered residue (valine, V) in the Cn3D viewer by moving the image around until you can see the highlighted yellow section.

As you discover, this amino acid is located at the junction between the two hemoglobin molecules. The change from the (charged-acidic) glutamic acid to the (nonpolar/hydrophobic) valine changes the energetics of the formation of this structure such that the polymerized form is more energetically favorable. This causes the individual hemoglobin molecules to form strands. In an person that is homozygous for the sickle allele, these strands will associate with each other, so that 14 strands are clumped together. In addition to distorting the shape of the red blood cell, these clumped strands are less efficient at transporting oxygen. If you highlight the mutated amino acid in the B, D and F chains, you will see that these amino acids are on the outside of the hemoglobin molecule. Interactions between these uncharged amino acids are responsible for the strand formation.

This project has introduced you to the MMDB using the structure of normal and sickle cell hemoglobin. If you would like to learn more about sickle cell anemia, an excellent starting point is the OMIM site on this disease, which can be found by searching OMIM for the OMIM entry # **603903.** Additionally, the following review questions will help to reinforce and extend what you've learned in this tutorial.

Review Questions

1. Which database contains more structures—the PDB or the MMDB? Why are there a different number of structures in the two databases?
2. What are the two main techniques that are used to determine protein structures?
3. Differentiate between protein secondary and tertiary structure.

Thought and Application Questions

1. Based on the structures that you viewed during the course of this tutorial, suggest an explanation for why people who are heterozygous for the sickle cell allele are less affected than people who are homozygous.
2. Select a molecule of the month from the Web site at **http://www.rcsb.org/pdb/molecules/molecule_list.html**. These articles are an interesting and fun feature of the Protein Databank. Go through the article, and then answer the following questions.
   (a) What molecule did you choose?
   (b) What is unique about the structure of this molecule that allows it to perform its biological function?
   (c) Using the name of the molecule as a search term, how many structures can you locate in the MMDB database?
   (d) For one of these structures, describe the following:
      (i) MMDB identifier
      (ii) PDB identifier
      (iii) Organism
      (iv) Short description
      (v) Briefly compare and contrast this structure with another related structure in the MMDB database.

16. Mining Genomes: Analysis of Gene Expression and the Cancer Genome Anatomy Project (CGAP)

Starting Link: **http://www.sagenet.org/home/Description.htm**.

This exercise introduces the powerful technique of microarray analysis, one of the most potent tools in bioinformatics. You will use SAGE (Serial Analysis of Gene Expression) to try to identify what genes are important in the development of specific diseases, and you will explore the Cancer Genome Anatomy Project (CGAP).

Introduction

Genomics studies have indicated that there are about 30,000 different genes in the human genome, and because of alternate RNA splicing possibilities, these can make about 100,000 different proteins. Not all of these proteins are made in any particular tissue, and the amount of a protein made in one tissue varies with different developmental and physical states. For example, consider the changes that take place during the course of pregnancy, in both the mother and the developing child. The regulation and coordinate expression of genes occurs with an exact and complex timing and progression.

In this chapter, you studied the regulation of the *lac* operon in *Escherichia coli*. Many groups of researchers worked for years to elucidate the details of how this operon is regulated. Today, new technologies are being developed that allow researchers to easily and rapidly study the coordinate expression of thousands of genes simultaneously. In this tutorial, you will be introduced to one of these technologies, named SAGE (serial analysis of gene expression).

Of all the new '-omics' (genomics, proteomics, lipomics), the most exciting and the most promising is the field of transcriptomics—the study of gene expression. Some of the very interesting questions that can be asked with these techniques include basic questions about how our bodies work, as well as specific questions concerning the causes and development of diseases. Some of these questions follow.

How many genes are expressed in different tissue types?
How many genes are expressed at a high level in different tissue types?
What do we mean by a high level of expression, anyway—10 transcripts/cell? 100 transcripts/cell? 10,000?
What fraction of the total genes are expressed in a given cell type?
How does gene expression change as a cell progresses through the cell cycle?
How does gene expression differ in a disease state? For example, how does gene expression in normal brain cells differ from gene expression in a brain tumor?

Many other types of questions can be asked as well. For example, a large fraction of the total genes in the human genome do not have an assigned function; by understanding how a gene is regulated and comparing its expression to known genes, it may be possible to gain insights into its function. And if the transcriptome of normal and diseased states is known, it may be possible to individually tailor a treatment program for a patient by monitoring how gene expression changes during the course of treatment, and responding accordingly.

Objectives

- to introduce the use of SAGE (serial analysis of gene expression).
- to introduce the tools available at NCBI for examining gene expression, including Gene Expression Omnibus, SAGEmap, and UniGene.

Project

We are going to look at SAGE (serial analysis of gene expression), a powerful means of studying gene expression. This is a method that can be used to quantitatively study the expression of thousands of genes simultaneously. During this portion of the project, you can design hypotheses and ask your own questions about patterns of gene expression.

Serial Analysis of Gene Expression

Microarrays are one approach that can be used to study gene expression (details are available in your text). SAGE is a clever approach that uses a method fundamentally different than that used in microarrays. While microarrays rely on hybridization, SAGE is a sequencing-based approach. A web site with a pictorial description of this method should be open on the left side of your browser window.

Essentially, molecular methods are used to isolate a single, specific 10-14 base-pair fragment of every mRNA that is expressed in a cell. The unique fragment of each mRNA is called the SAGEtag of that transcript. These fragments are then linked together into long strands such that each of the fragments is separated by a short 'punctuation' sequence. The nucleotide sequence of one of these strands is determined using high-throughput sequencers. If a gene is turned on, then you will be able to find the unique *SAGEtag* of its 10-14 bp sequence—and the number of times that you find that SAGEtag is a direct measure of the abundance of that transcript! This method is slower than microarray hybridization, but as sequencing methods improve, it will become more manageable as a research tool.

Go to the National Center for Biotechnology site at **http://www.ncbi.nlm.nih.gov/**, where you will be able to analyze SAGE data.

Gene Expression Omnibus (GEO)

On the right side of the page is a set of links under the heading *Hot Spots*. Select the link for **Gene expression omnibus.**

GEO is a site dedicated to providing public access to gene expression data. In addition to serving as a database and repository of different sets of gene expression data, it provides a set of tools that can be used to query the database, or to analyze gene expression data that you have generated. The design of these tools is a challenging undertaking because the data comes from many different labs in many different formats, and is generated using a variety of approaches. Like many of the new fields in bioinformatics, it is likely that the development of flexible intuitive tools will rapidly increase knowledge and understanding in ways that the tool developers frequently can't predict.

SAGEmap

There is a bar on the top of the page that has links to a variety of tools available at the GEO site. Select **SAGEmap.**

This site is a public repository of SAGE data. Quantitative, whole-genome gene expression profiles, called *libraries*, are stored here for researchers to analyze. These libraries are primarily from human tissues, and include specific tissues such as brain, colon, kidney, and pancreas. They also include similar tissues that were collected from different sources, such as males and females, diseased and normal tissue, or individuals of different ages. In some cases, gene expression profiles are from pooled tissue samples (for example, cancerous ovarian tissues from young women) in order to reduce effects from individual variations. Because the data is submitted while the projects are ongoing, some libraries are more complete than others.

Using SAGE Resources at NCBI: xProfiler

We are going to start by using a program called xProfiler. This program allows you to pick two different tissue types and compare total gene expression in those tissues. For example, you could compare gene expression in normal brain tissue and in cancerous brain tissue. Or, you might choose to compare gene expression in brain tissue from a developing fetal female and a young adult female.

Along the left side of the page, select the link labeled **Analyze...by library.** This will take you to the xProfiler site.

The samples that you compare are simply referred to as Groups A and B. There are a variety of options in choosing the samples for comparison. You could select samples from the list on this page. Alternatively, you could choose the libraries from a full list of the available libraries, by using the link to the *Library Browser* that is located on the top right of the page.

We are going to select two different libraries that should have some clear-cut differences in gene expression: leukocytes (white blood cells) and epithelial tissue. At the top of the page, enter **white blood cell** in the box for 'Group A name' and **epithelial** in the box for group B name.

In green Column A, under the heading *Homo sapiens*, **check** the box for
SAGE Duke leukocyte (48523 tags)
white blood cells normal SAGE CGAP non-normalized SAGE library method bulk

(Note that the libraries are listed in alphabetical order.)

The (48523 tags) portion of the entry refers to the total number of SAGEtags that have been sequenced, which is also the total number of specific transcripts that have been analyzed. To get 48,000 tags probably required about 1500 different sequencing reactions. Note that the number of tags may have changed, if more sequences have been obtained for this library, but the rest of the library name should be the same.

Next, select two libraries for your B sample by checking the boxes in the blue B column for

SAGE NC1 (50179 tags)
epithelium normal colon SAGE CGAP non-normalized SAGE library method bulk

SAGE NC2 (49593 tags)
epithelium normal colon SAGE CGAP non-normalized SAGE library method bulk

Now, go to the top of the page. The xProfiler program will compare the A library with the B libraries, and return to you a list of SAGEtags that are present at different frequencies in the different libraries. The default setting for the difference in frequencies is a factor of two. For our purposes, we want to **change this value to 10**.

After changing the default to 10, press **Calculate.**

The xProfiler output will load in the same window. If your query gets queued, then you may need to select the button labeled **GO**. This may take up to a few minutes, depending several factors such as how much activity the site is getting.

The SAGEtags are presented as a table with six columns. The first column is the nucleotide sequence of the tag. This is followed by three columns of numbers. The first column of numbers is the number of times that SAGEtag was found in library A, whereas the second column of numbers is the number of times that SAGEtag was found in library B. The next column is a statistical value to help evaluate whether or not the observed difference in expression is due to sampling error, and the fifth column is the UniGene cluster (more on that below) that matches the SAGEtag.

The final column in the xProfiler output describes the gene that matches the particular SAGEtag in that row. If the gene has been well characterized, then these descriptions are very specific. However, the descriptions are somewhat vague for genes that haven't been studied very thoroughly. Many of these genes will of course be new to you, but some of them will be familiar. Scroll down the list slowly, and look for proteins that you know. While you are doing this, also try to think about whether or not that pattern of gene expression makes sense, given what you know about leukocytes (white blood cells) and colons.

For example, you should see that keratin and cytokeratin are expressed more highly in the colonic tissue, and hemoglobin alpha and hemoglobin beta are expressed more highly in the blood cells. Recall that keratin is a nonenzymatic protein that has structural role in epithelial cells (and makes up the bulk of animal hair); a higher rate of expression would be expected in the colonic tissue. Similarly, hemoglobin is an oxygen-carrying protein present in blood that gives red blood cells their characteristic color. The higher rate of expression in this case is not wholly unexpected, but does raise some questions. Is hemoglobin made in white blood cells? Or, was the white blood cell sample from which this library was derived contaminated with red blood cells?

Look at the last column and find a protein that you have heard of, near the top of the list. For our purposes, find '**hemoglobin, beta**'. Select the tag sequence of that tag by clicking on the words **MORE UNIGENES** below the tag sequence.

Using SAGE Resources at NCBI: SAGE Tag to Gene Mapping

This takes you to a site called the SAGE Tag to Gene Mapping page. This site gives you details on all of the different SAGE libraries that this tag sequence has been found in, including the library name and information on the frequency with which that tag is represented in the library. The simplest way to quickly assess this information is to look at the shaded ovals in the third column of the table labeled 'SAGE library data for this tag'. The darker ovals represent high rates of expression, the light ovals represent low levels of expression, and the gray-shaded ovals represent intermediate levels of expression.

Because the different libraries are at different levels of completeness, this data is normalized to the expected number of tags that would be found in a library containing a million total tags. The shading is meant to simulate the type of results you might see if you were to carry out a quantitative hybridization analysis, comparing the expression of that gene in the different libraries. The 'Tag counts' and 'Total tags' columns describe how many times the tag has been found in that library, and how many total tags have been found in the library. As of this writing, the only SAGE library that this tag has been found in more than once is the SAGE Duke leukocyte library.

At the top of the page is the *UniGene* Identification number for this gene, in this case Hs.155376. UniGene is a system for partitioning GenBank sequences into a non-redundant set of gene and gene-like entries. Whereas a particular gene may have been sequenced and entered into GenBank a number of different times, from different organisms and by different researchers, that gene will have only one entry in UniGene. This UniGene entry will list the product and function of the gene, if known. The UniGene entry also has information such as the tissue types in which the gene has been expressed, and the map location of the gene.

The xProfiler program allows you to compare the expression of all the genes in different tissue samples. As the number of SAGE libraries increases, you can also ask questions about individual genes, for example, concerning which tissues a gene is highly expressed in and which tissues a gene is not expressed in. One of the ways that this can be done is using the SAGE Tag to Gene Mapping site, which you are at now. A more comprehensive look at individual genes can be obtained at the UniGene site for a gene.

UniGene

Select the **UniGene** link on the tool bar at the top of the page. This takes you to a page that has a description of UniGene, as well as a search function. Look over the descriptive material. Then, in the search box enter **hemoglobin.** This search will give you several hits. Select the UniGene entry for hemoglobin beta (symbol = HBB; Entry # **Hs.155376**). This takes you to the UniGene entry for this gene.

At the top of the entry are links to other NCBI sites that have information on this gene, such as LocusLink and OMIM (Online Mendelian Inheritance in Man). Below this are similar proteins that are found in other organisms, such as mice (*Mus musculus)*, and information about how similar the proteins are. For example, the beta hemoglobin of the mouse is 82% identical to the human beta hemoglobin. Below this, in the *Mapping Information* section, the location of the gene in the genome is identified and linked. Below this is a summary of gene expression information, including SAGE data.

Using SAGE Resources at NCBI: SAGE Gene to Tag Mapping

Select the SAGE **Gene to Tag mapping** link in the middle of the page. This will take you to a list of SAGEtags for beta hemoglobin - select the first one, tag **GCAAGAAAGT.**

This page depicts what is known about the expression of this gene, based on available SAGE data. The information on this page was described earlier. Here, as you can see, the gene for the beta chain of hemoglobin is expressed in many libraries, but only at a high level in the leukocyte library.

The Cancer Genome Anatomy Project

Many of the SAGE projects have been carried out as a part of the Cancer Genome Anatomy Project (CGAP), a large effort to determine the differences in gene expression between normal and cancerous tissues and cells. We will finish up this project by using a few of the CGAP tools. To get to the CGAP home page, select the **NCBI icon** at the top of this page, then select the link to **the Cancer Genome Anatomy Project** at the top right of the NCBI home page. This page describes some of the different ways that NCBI is helping to make CGAP data available to the public. From this page, select the link to the **Cancer Genome Anatomy Project** in the first sentence. The URL for the CGAP home page is **http://cgap.nci.nih.gov/**.

There are several interesting links available from this page. Select the link to **SAGE Genie**.

SAGE Genie

The SAGE Genie site lets you query the SAGE libraries in some ways that xProfiler doesn't let you. For example, you can input a specific gene, and then ask how that gene is expressed in different tissues. The output shows expression levels in both normal and cancerous tissues. Let's do this.

Select the **SAGE anatomic viewer** from the middle of the page. In the box next to 'Tag (sequence of 10 bases)', enter the Tag for beta hemoglobin, **GCAAGAAAGT**. Then click **Go**. At the *top* of page, select the **drawing of the human figure** under the words 'SAGE Anatomic Viewer.' This will return the data for tissue samples.

The output portrays the different tissue systems of the human body. The color associated with each tissue is keyed to the relative level of expression in that tissue. Again, you can see that the mRNA for beta hemoglobin was found in many different libraries, but its expression was the highest in the leukocyte library. You can also see that not only is this gene expressed in different tissues, but its expression is different in some cancerous tissues than in the corresponding normal tissues.

Selecting the link to the **Brain** displays the relative expression in different regions of the brain, as well as in different types of brain cancers. As you can see, differential expression can be seen here as well.

Now you have been introduced to the world of transcriptomics, the study of simultaneous gene expression in whole genomes. We have focused on the use of SAGE, a sequencing-based approach that quantifies small, ~10 nucleotide tags of different mRNAs. CGAP has many other tools that utilize different approaches than SAGE that you can explore from the CGAP home page at **http://cgap.nci.nih.gov/**.

Also, Johns Hopkins University has a site dedicated to SAGE that you might want to investigate at **http://www.sagenet.org/.**

Review Questions

1. What, specifically, is meant by the term 'gene expression'? Can you have 'gene expression' of a particular gene, but not make the encoded protein? Explain.
2. Why is it necessary to regulate gene expression in a cell?
3. Define the word 'transcriptomics'.

Thought and Application Questions

1. You are carrying out the analysis of a particular tissue type, and you come across the following SAGEtag as one that is highly expressed in your tissue.
   AGTGTGTGGA
   (a) What protein does this SAGEtag correspond with?
   (b) What kind of tissue are you most likely studying?
2. Use UniGene to investigate the gene expression of two proteins that you are familiar with. These should be proteins whose expression you should be able to predict something about. For example, you might pick insulin, a hormone secreted by the pancreas that is important in regulating blood glucose levels. For the proteins you choose, answer the following questions:
   (a) What is the name of the protein?

(b) Briefly, make predictions about the expression patterns that you expect to find for this protein.
(c) What is the UniGene symbol for your protein? What is the UniGene entry number?
(d) Are homologs found in other animals? Which animals, and what is the percent identity?
(e) Based on the Tag to Gene Mapping data, is the gene highly expressed in a particular library? Which one(s)? How many of the tags were found in those libraries, and how many total tags are in the library?

3. Use the xProfiler to design an experiment of your choosing, in which you compare gene expression in two different tissue types. For your experiment, describe the following:
   (a) What libraries did you compare?
   (b) What were some of the genes that were transcribed differently in the different libraries--did you recognize any of these genes? Based on what you know about the tissues, do these patterns of expression make sense?

17. Mining Genomes: Molecular Evolution

This exercise will introduce you to some of the basic principles of molecular evolution, the study of the ways in which molecules evolve and the reconstruction of the evolutionary history of molecules and organisms. You will use several of the internet tools most frequently used by contemporary molecular geneticists to analyze analogous sequences from related organisms.

Note:
Netscape Navigator or Netscape Communicator works better at The Biology Workbench site than Internet Explorer, so if you have the option, it is probably a good idea to carry out this project using Netscape.

Introduction

Molecular evolution is the study of how proteins and nucleic acids evolve. Included in this field are studies of mutations and chromosomal rearrangements, the evolutionary process, the identification of sequence patterns conferring function in proteins and nucleic acids, and the reconstruction of the phylogenetic history of organisms and the molecules that they make. All of these studies rely on comparisons of nucleotide or amino acid sequences.

In this tutorial, you will be introduced to the fundamental principles of molecular evolution and the types of bioinformatics tools that are used in evolutionary studies. We will begin by carrying out a manual sequence comparison, so that the basic concepts can be introduced, and the remainder of the project will be carried out at The Biology Workbench, a set of bioinformatics analysis programs managed by The San Diego Supercomputing Center at the University of California, San Diego.

Objectives

- to introduce the principles of molecular evolution.
- to acquaint you with the tools that are available at The Biology Workbench to compare nucleotide and amino acid sequences.
- to learn about the use of protein sequences in phylogenetic reconstructions.

Project

Branching evolution occurs when one ancestral species gives rise to two or more progeny species. However, the initial separation into two species involves only a small fraction of the total genome. Most of the genes inherited by each progeny species are identical to those in the ancestor species. Following speciation, these genes evolve independently in the separate lineages. Studies of molecular evolution therefore rely heavily on comparisons of related sequences from different organisms.

Let's jump right in. Shown below is an alignment of two homologous sequences that we will use as a starting place. Homologous sequences are sequences that have descended from a common ancestral sequence. You can't meaningfully compare sequences unless they are homologous. This alignment uses the single letter amino acid code, in which G represents glycine, Q represents glutamine, etc. The aligned proteins have been shown to be involved in the metabolism of similar, but different, toxic compounds. As you can see, these amino acid sequences are very similar and it is easy to recognize that they are related by common descent.

```
dntAc KMGVDDEVIVSRQNDGSVR
nahAc KMGIDDEVIVSRQSDGSIR
```

An expanded version of this alignment, in which both the amino acids and the corresponding DNA nucleotides are shown, follows. For ease of analysis, the codons have been broken into separate entries in a table.

Alignment of nahAc and dntAc sequences.

	K	M	G	V	D	E	V	I	V	S	R	Q	N	D
dntAc	AAA	ATG	GGC	GTC	GAT	GAA	GTC	ATC	GTC	TCC	CGC	CAG	AAC	GAT
nahAc	AAA	ATG	GGT	ATT	GAC	GAG	GTC	ATC	GTC	TCT	CGG	CAG	AGC	GAC
	K	M	G	I	D	E	V	I	V	S	R	Q	S	D

	G	S	V	R	A	F	L	N	V	C	R	H	R	G
dntAc	GGC	TCG	GTG	CGA	GCC	TTT	TTG	AAT	GTT	TGC	CGT	CAC	CGG	GGC
nahAc	GGT	TCG	ATT	CGT	GCC	TTC	CTG	AAC	GTT	TGT	CGG	CAC	CGT	GGC
	G	S	I	R	A	F	L	N	V	C	R	H	R	G

	K	T	I	V	D	A	E	A	G	N	A	K	G	F
dntAc	AAG	ACA	ATA	GTT	GAC	GCT	GAA	GCC	GGA	AAT	GCG	AAA	GGC	TTT
nahAc	AAG	ACG	CTG	GTT	AAC	GCG	GAA	GCC	GGC	AAT	GCC	AAA	GGT	TTC
	K	T	L	V	N	A	E	A	G	N	A	K	G	F

	V	C	G	Y	H	G	W	G	Y	G	S	N
dntAc	GTG	TGC	GGT	TAC	CAC	GGC	TGG	GGC	TAT	GGC	TCC	AAC
nahAc	GTT	TGC	AGC	TAT	CAC	GGC	TGG	GGC	TTC	GGC	TCC	AAC
	V	C	S	Y	H	G	W	G	F	G	S	N

This region was chosen at random to represent the changes that take place in nucleotide sequences over time. Answer the questions below by manually comparing these sequences. You may like to use the table of the single letter amino acid code and the codon usage table from Figure 15.12 in your text, which follow.

Single-Letter Amino Acid Code

G	Glycine	Gly
A	Alanine	Ala
L	Leucine	Leu
M	Methionine	Met
F	Phenylalanine	Phe
W	Tryptophan	Trp
K	Lysine	Lys
Q	Glutamine	Gln
E	Glutamic Acid	Glu
S	Serine	Ser
P	Proline	Pro
V	Valine	Val
I	Isoleucine	Ile
C	Cysteine	Cys
Y	Tyrosine	Tyr
H	Histidine	His
R	Arginine	Arg
N	Asparagine	Asn
D	Aspartic Acid	Asp
T	Threonine	Thr

The Genetic Code (Figure 15.12)

First base	Second base: U	C	A	G	Third base
U	UUU Phe	UCU Ser	UAU Tyr	UGU Cys	U
	UUC Phe	UCC Ser	UAC Tyr	UGC Cys	C
	UUA Leu	UCA Ser	UAA Stop	UGA Stop	A
	UUG Leu	UCG Ser	UAG Stop	UGG Trp	G
C	CUU Leu	CCU Pro	CAU His	CGU Arg	U
	CUC Leu	CCC Pro	CAC His	CGC Arg	C
	CUA Leu	CCA Pro	CAA Gln	CGA Arg	A
	CUG Leu	CCG Pro	CAG Gln	CGG Arg	G
A	AUU Ile	ACU Thr	AAU Asn	AGU Ser	U
	AUC Ile	ACC Thr	AAC Asn	AGC Ser	C
	AUA Ile	ACA Thr	AAA Lys	AGA Arg	A
	AUG Met	ACG Thr	AAG Lys	AGG Arg	G
G	GUU Val	GCU Ala	GAU Asp	GGU Gly	U
	GUC Val	GCC Ala	GAC Asp	GGC Gly	C
	GUA Val	GCA Ala	GAA Glu	GGA Gly	A
	GUG Val	GCG Ala	GAG Glu	GGG Gly	G

a. Assuming that the dntAc sequence represents the ancestral sequence, how many nucleotide changes (mutations) have occurred in this region to create the nahAc nucleotide sequence? Remember that in actuality neither sequence represents the ancestral sequence.

b. Of these nucleotide changes, how many of these changed the amino acid encoded by that codon (i.e, how many were nonsynonymous changes)?
c. How many nucleotide changes were in the first codon position? How many of these altered the encoded amino acid?
d. How many nucleotide changes were in the second codon position? How many of these altered the encoded amino acid?
e. How many of the nucleotide changes were in the third codon position? How many of these altered the encoded amino acid?
f. The sixth amino acid of the dntAc sequence is a glutamic acid, encoded by GAA. What would happen if the G in this codon was converted to a T?
g. Compare the % identity of these two sequences at the nucleotide vs. protein level. Percent identity is equal to # of positions in common / total # positions) × 100.
h. Mutations do not occur randomly throughout a sequence. Some areas are more prone to mutate than others, and these are called *hotspots* of mutation. One such hotspot of mutation is the dinucleotide 5'–CG–3', in which the cytosine is frequently methylated and replicated with error, changing it to 5'–TG–3'. In the aligned sequences, can you find examples where this particular mutation may have occurred? Remember that in actuality neither sequence represents the ancestral sequence.

The manual analysis that you just carried out introduced you to some of the ways that molecules evolve. The purpose of that manual analysis was to get you thinking about the mechanisms by which genes and proteins change over time, and the types of forces that control those changes.

Several computer tools have been developed that allow you to quickly retrieve, align, and compare genes and proteins from different organisms. The remaining portion of this tutorial will be carried out using The Biology Workbench, a set of bioinformatics analysis programs managed by The San Diego Supercomputing Center at the University of California, San Diego. The Biology Workbench integrates a wide variety of different programs, and this site can be used for many different kinds of analyses in addition to molecular evolution studies.

Description of Your Project

You are going to retrieve, align, and compare hemoglobin protein sequences from a variety of animals. Hemoglobin is a protein in red blood cells that is involved in oxygen transport. It belongs to a family of related globin oxygen-binding genes that has evolved through a number of speciation and gene duplication events.

The *Mining Genomes* tutorial at the end of Chapter 15 of your text examines the three-dimensional structure of hemoglobin.

Computer-Aided Sequence Comparisons

First, you will need to access The Biology Workbench at:
http://bsw.ncsa.uiuc.edu/cgi-bin/sib.py.

In order to use the site, you need to set up a free account. To do this, select **Registration** and follow the instructions.

You will be taken to a new session. You can carry out several different projects at the same time using The Biology Workbench, and it keeps your different projects in different folders, which are referred to as *sessions*.

Now, select **Protein Tools** at the top of the page to begin.

Retrieving Protein Sequences Using The Biology Workbench

You are going to retrieve, align, and compare hemoglobin proteins from a variety of different animals. You will do this using the tool on the right part of the page, named Ndjinn.

First, you need to select the database to use for the search. In the box labeled *Select one or more Databases*, scroll down the list and **highlight** Swissprot database**.** SWISSPROT is a protein sequence database that was begun in 1986 and is maintained collaboratively by the Swiss Institute for Bioinformatics and the European Bioinformatics Institute.

Second, in the search box, enter the term **beta hemoglobin**. Then select **Ndjinn.**

A list of sequences will be returned (lots and lots of hemoglobin sequences are in the databases, which is one of the reasons why we are using this protein). You want to select the box next to the entry labeled

SWISSPROT:HBB_HUMAN Hemoglobin Beta Chain [Homo sapiens (Human), Pan troglodytes (Chimpanzee), and Pan paniscus (Pygmy chimpanzee) (Bonobo)]

This should be the result at the top of the list. The names of three different species are listed simply because the sequence is identical in each of these three closely-related species. Make sure that you selected the sequence for the beta chain, and not the alpha or delta sequences.

This page has a long list of sequences. You also want to import 4 more hemoglobin beta sequences into your session—from gorilla (*Gorilla gorilla)*, mouse (*Mus musculus*), chicken (*Gallus gallus*), and bullfrog (*Rana catesbeiana*).

SWISSPROT:HBB1_MOUSE
HEMOGLOBIN BETA-1 CHAIN (B1) (MAJOR) [Mus musculus (Mouse)]
(this may be #5 on the list)

SWISSPROT:HBB_CHICK
HEMOGLOBIN BETA CHAIN [Gallus gallus (Chicken)]
(this may be #6 on the list)

SWISSPROT:HBB_RANCA
HEMOGLOBIN BETA CHAIN [Rana catesbeiana (Bull frog)]
(this may be #56 on the list)

SWISSPROT:HBB_GORGO
(P02024) HEMOGLOBIN BETA CHAIN [Gorilla gorilla]
(this may be #118 on the list)

Select these sequences by **selecting the box** next to each sequence. Then scroll down to the bottom of the page and select **Import Sequences.** (Note: if you have difficulty locating the sequences that you want, you can always go back and use Ndjinn with the appropriate search terms. Alternatively, you could use your browser's 'search in page' feature).

This will take you to a page similar to the one that you were at before, but now there should be five different hemoglobin sequences listed at the bottom of the page. These organisms are related in a way that should be pretty clear to you. The human hemoglobin sequence is identical to that from chimpanzees. Which of the remaining four amino acid sequences do you think will be most like the human sequence? Which sequence will be the next closest? Which sequence do you think will be the most different from the human sequence? Why?

Aligning the Sequences

Now we will carry out an alignment of the five hemoglobin sequences that you have retrieved. We will use a program called ClustalW to perform the alignment. ClustalW is a multiple sequence alignment tool that searches out the best global alignment—that is, the alignment with the most identity across the entire range of all of the sequences. The program is adjustable in that you can give different weights to different types of mismatches.

Select the boxes of all five of the hemoglobin sequences. Then select **CLUSTALW** from the numerous options on the right side of the page.

This will take you to a new page. **Scroll** down the page to the section called 'Sequence alignment'. Each of the amino acid sequences that you retrieved is shown in the alignment.

The alignment is color-coded, so that you can see different types of changes without looking at the specific amino acids in the sequence. Four colors are used to indicate varying levels of sequence conservation. The bright blue areas are areas that are identical across all 5 sequences. The green and dark blue areas are conserved but not identical. This means that although different amino acids are in this position, the amino acids are similar to one another—for example an amino acid with a hydrophobic R group has been replaced by another amino acid that also has a hydrophobic R group. The black regions are unconserved—different amino acids with different properties are present in this position in the different proteins.

Mutations happen randomly, however, and without regard to the type of change in the encoded protein. When changes occur that produce a nonfunctional protein, the organisms that have that mutation are likely to be eliminated via natural selection. The selection pressure against organisms with nonfunctional or less-functional hemoglobin is likely to be high. The highly conserved regions aren't conserved because mutations never occurred in these positions, but rather because natural selection eliminated them once they did occur.

One thing that you can see in this alignment is that there are some regions in the protein that are highly conserved, and others that are conserved to a lesser extent. This reflects the fact that some parts of a protein can and do evolve at different rates than other parts of a protein. The highly conserved regions are likely to be directly involved in the functioning of the protein, which in this case might be in binding the heme group or in interchain interactions. In an enzyme, amino acids located at the active site are likely to be highly conserved. The less conserved regions are unlikely to be directly involved in the functioning of the protein, for example, they may have a 'spacer' function to separate two other regions.

The Phylogenetic Tree Created from the Alignment

The ClustalW program also gives you another means of viewing the information in the alignment—it depicts it in the form of a phylogenetic tree. From the alignment page, select **Import Alignment.**

Scroll to the bottom of the page and **select the box next to your alignment.** Then go to the middle of the page and select **DRAWGRAM.** This will take you to a dendrogram that depicts the relationships between the sequences. In this figure, the sequences are represented at the tips of the lines. A branchpoint, where two lines diverge, represents a single ancestor sequence. Two sequences that have a single branchpoint between them are more closely related to each other than to the other sequences. The length of the branches represents evolutionary distance (related to amount of dissimilar sequence), so that two sequences with a high amount of similarity are connected by a smaller distance than two sequences that are less closely related.

This type of diagram can be used to infer the evolutionary relationships between the sequences because you would expect that sequences that are very similar shared a recent common ancestor. Similarly, the evolutionary relationships between the source organisms can be inferred from the relationships between the sequences.

Here, you can see the relationship that you intuitively expected—the gorilla sequence is very similar to the human sequence because these two organisms are both primates and therefore shared a common ancestor more recently than humans and mice did. Similarly, the mouse, a mammal, shared a common ancestor with the primates more recently than it shared a common ancestor with the chicken or frog.

Different models for evolutionary mechanisms might produce different evolutionary distances than those shown here.

The Relationship Between Alpha and Beta Globin Sequences.

Hemoglobin is a tetramer (4 subunits) of two different polypeptides named alpha and beta hemoglobin. Alpha and beta hemoglobin are related to each other via an ancient gene duplication event. That is, there was one gene (the ancestral globin), and it duplicated to form two genes in the same organism (alpha and beta globins), and then these genes underwent independent evolution as the progeny of that organism replicated. This gene duplication event occurred before mammals diverged, Therefore, an alpha globin gene in a mouse is more closely related to (or, more recently diverged from) an alpha globin gene from a chimpanzee than it is to a beta globin gene from the same mouse. Gene duplication is a very important evolutionary process, and it is clear that it has happened numerous times on an evolutionary timescale.

The alpha and beta sequences arose in an ancestor common to all of the animals that we are looking at. That is, this gene duplication event occurred before the divergence of the lineages that led to modern primates, mammals, birds, and amphibians.

If you constructed a dendrogram of alpha and beta globin sequences from the five organisms that we are looking at, what would the shape of that dendrogram be? Would the alpha sequences cluster together separately from the beta sequences? Would the branching pattern among the alpha sequences be similar to the branching pattern among the beta sequences?

Delta and Gamma Globins

In addition to the alpha and beta globins, humans also make delta and gamma globins. The gamma globin chains are expressed early in development, while the delta chains are expressed at a low level in adults. Gamma and delta globins also arose via gene duplication events. An analysis similar to the one we just carried out might help you to decide when this gene duplication event occurred. If you would like a succinct explanation of hemoglobin synthesis and the expression of the different globin genes during human development, go to: **http://sickle.bwh.harvard.edu/hbsynthesis.html.**

Independent Projects

Look through the projects described below, and pick one of them that you find interesting. Carry out that project. If you want to design a different project on your own, that's ok too, but get approval from your instructor. Your instructor may want you to attach a **brief** interpretation of results (1 page) and a copy of the phylogenetic tree that you generate to this exercise.

Project 1: ***Extending the Analysis by Adding Alpha Globin Sequences to the Alignment***

To import the alpha globin sequences into your session, press the gray **Return** button on the screen, then **deselect** the alignment, and then select **Protein Tools.** Then use **Ndjinn** or **Blastp** to import the following sequences. If you carry out a search for 'hemoglobin alpha' in the SWISSPROT database using Ndjinn, you will have to search through several hundred responses—it may help to use the 'find in page' feature of your internet browser.

SWISSPROT:HBA_HUMAN
HEMOGLOBIN ALPHA CHAIN [Homo sapiens (Human), Pan troglodytes (Chimpanzee), and Pan paniscus (Pygmy chimpanzee) (Bonobo)]

SWISSPROT:HBA_CHICK
HEMOGLOBIN ALPHA-A CHAIN [Gallus gallus (Chicken)]

SWISSPROT:HBA_MOUSE
HEMOGLOBIN ALPHA CHAIN [Mus musculus (Mouse)]

SWISSPROT:HBAB_RANCA
HEMOGLOBIN ALPHA-B CHAIN [Rana catesbeiana (Bull frog)]

SWISSPROT:HBA_GORGO
HEMOGLOBIN ALPHA CHAIN [Gorilla gorilla gorilla (Lowland gorilla)]

Once you have all the sequences imported, select all ten sequences (your 5 beta sequences and 5 alpha sequences) and perform another Clustal W alignment of the sequences. Make sure that the entire results page has loaded (this takes a little bit of time), and then examine the unrooted phylogenetic tree of the sequences. Were your predictions correct? Do the alpha sequences cluster, and are the branching patterns among the alphas similar to those among the betas?

Project 2: ***Estimating Relationships Between Existing Mammals***

In the analysis that we carried out, it was fairly easy to predict the results because we understand the relationships between humans, gorillas, mice, chickens, and frogs. The way in which other animals are related can be difficult to understand, however. Even among the mammals, it may be difficult to see how different organisms are related to each other. Using beta hemoglobin sequences, investigate the relationships between the organisms below.

Minke Whale
Harbor Seal
Indian Elephant
White Rhinoceros
Brazilian Manatee
European River Otter
Polar Bear
Hippopotamus

Before carrying out the analysis, predict the relationships between the different organisms.

1. Are whales and seals more closely related to one another than to any of these other species?
2. To which organism is the manatee most closely related?
3. To which organism is the elephant most closely related?
4. To which organism is the otter most closely related?

After carrying out the analysis, answer the following questions.

1. Are whales and seals more closely related to one another than to any of these other species?
2. To which organism is the manatee most closely related?
3. To which organism is the elephant most closely related?
4. To which organism is the otter most closely related?

Would you predict that you would see the same results if you used a different protein for the analysis, for example the cytochrome c protein? Why or why not?

This project has introduced you to some of the principles of molecular evolution, and some of the ways that modern researchers study molecular evolution. We've focused on the evolution of globin genes, and the alignment and comparison of globin genes using tools at the Biology Workbench. The following review questions will help to reinforce and extend what you've learned in this tutorial.

Review Questions

1. Why was it necessary to carry out an alignment of the sequences we analyzed?

Thought and Application Questions

1. Nucleotide substitutions might lead to changes in the sequence of amino acids in a protein, which can be seen by comparing homologous positions in a sequence alignment. How would deletions or insertions appear in a sequence alignment?

2. Sometimes gene duplications result in the formation of a pseudogene, a sequence of nucleotides that is very similar to a gene but isn't expressed. This can occur in a variety of ways, for example if the gene is duplicated but a regulatory region or the promoter region for that gene is not duplicated. Which would evolve (accumulate nucleotide substitutions) more rapidly, a psuedogene or a duplicated gene that was expressed? Explain.

18. Mining Genomes: Recombinant DNA Project: Designing Biotechnology Experiments

Starting link: **http://www.tigr.org/tigr-scripts/CMR2/CMRHomePage.spl**.

This exercise casts you in the role of research geneticist. Your job is to plan a project to clone a specific gene into a plasmid vector, selecting the restriction enzymes and vector that you will use. You will also design primers to be used in a polymerase chain reaction (PCR). You will utilize the web sites of some of the major suppliers of biotechnology reagents in the process.

Introduction

In Chapter 18 you studied a number of modern genetics techniques, including cloning and the polymerase chain reaction (PCR). In this tutorial you will be introduced to a variety of sites that researchers use in planning and carrying out experiments that use these techniques. You will plan two different experiments that allow you to clone a new thermostable DNA polymerase from an organism whose genome has recently been sequenced. First, you'll identify restriction enzymes that would allow you to isolate and clone the polymerase. Next, you'll learn how to design and evaluate PCR primers that would allow you to amplify the desired gene.

Objectives

- to further introduce widely used techniques such as plasmid cloning and the polymerase chain reaction.
- to introduce some of the features of Web sites managed by some of the major suppliers of biotechnology reagents.
- to introduce the principles and some of the programs that are used for PCR primer design and restriction enzyme mapping.

Tutorial

PCR reactions are frequently carried out using a DNA polymerase from a thermophilic (high-temperature-loving) bacterium named *Thermus aquaticus*. *T. aquaticus* was originally isolated from a hot spring in Yellowstone National Park, and has subsequently been shown to be widely distributed—for example, you could probably isolate this organism from your hot water heater. The DNA polymerase from this organism (*Taq* DNA polymerase) is used because it can withstand the temperature fluctuations that are required for PCR reactions to proceed. The production of this enzyme is so widespread that it now represents a billion-dollar-a year industry! However, *Taq* DNA polymerase has a disadvantage for some applications of PCR in that, at a relatively high frequency, it sometimes adds a non-complementary nucleotide to the newly synthesized strand. This is due to the fact that *Taq* DNA polymerase has 5'→3' polymerase activity but does not have a proofreading function that many DNA polymerase enzymes possess (a 3'→5' exonuclease function).

In the first part of this project, you will use several different Web sites to design a project in which you clone the thermostable DNA polymerase from another thermophilic organism whose complete genome was recently sequenced. This organism is an Archaea named *Pyrococcus abyssi*. *P. abyssi* was isolated from samples taken close to a hot spring situated 3500 meters deep in the Pacific ocean. This interesting organism is highly adapted to growth in the deep hot environment from which it was isolated, and grows optimally at 103°C (above boiling, at sea level!) and 200 atmospheres pressure. You will retrieve the relevant sequence of the organism and then identify restriction enzymes that might be used to clone the DNA polymerase. In the second part of this project, you will design PCR primers that would allow you to amplify a portion of this gene.

Outline of First Project: Cloning of DNA Polymerase from Pyrococcus abyssi

To refresh your memory as to how cloning works, go to **http://www.ornl.gov/hgmis/publicat/primer/fig11a.html** to see a figure that depicts a simplified schematic of plasmid cloning.

In this figure, the gray plasmid DNA was cut with the same restriction enzyme that the DNA to be cloned was cut with. As discussed in your text, class II restriction enzymes cut at specific DNA sequences, and some of these enzymes cut the DNA in a staggered fashion, thus leaving overhanging single-stranded regions at the ends. As a result of being cut with the same restriction enzyme, both DNA molecules have complementary single-stranded overhangs and therefore can be joined into a single recombinant DNA molecule.

Cloning of DNA polymerase from *P. abyssi* can be broken down into several steps that would be similar for the cloning of any sequence from any organism. The steps of this process include:

1. Retrieving the target sequence and the adjacent nucleotide sequences;
2. Identifying a restriction enzyme that cuts on either side of the target sequence, but not in the middle of the sequence;
3. Digesting genomic DNA from *P. abyssi* with the restriction enzyme identified in #2;
4. Isolating the restriction fragment of the correct size using agarose gel electrophoresis;
5. Digesting a plasmid cloning vector with the same restriction enzyme, or a restriction enzyme that generates identical sticky ends;
6. Ligating the isolated fragment into the plasmid; transforming *E. coli* with the ligation mix; and selecting colonies that are derived from bacteria transformed with the desired recombinant plasmid.

The details of steps 3-6 are described in Chapter 18 of your text. In this project we will be focusing on the first two steps.

Retrieving the Sequence of DNA Polymerase from P. abyssi

We will begin by retrieving the sequence that encodes DNA polymerase I from *P. abyssi*. Although there are several different ways that we could get this sequence, we will use a genomics resource maintained by The Institute for Genomic Research (TIGR) called the Comprehensive Microbial Resource (CMR). The CMR is a tool that provides access to all of the bacterial genome sequences completed to date. Let's begin by going to TIGR's CMR homepage at **http://www.tigr.org/tigr-scripts/CMR2/CMRHomePage.spl**.

Scroll down the page. As you can see, there are many different programs for analyzing genomic sequences here. We will explore some of these programs more fully in the *Mining Genomes* tutorial for Chapter 19. For now, though, we want to focus on getting the sequence of DNA polymerase I (and the adjacent sequences on either side) from *P. abyssi.* About halfway down the page, there is a pull-down menu labeled *Genome Pages*. Use this pull-down menu to go to the homepage for the genome that we are interested in, that of **Pyrococcus abyssi GE5.**

There are many different ways of looking at the information contained within a genome. One way is the circular map that is presented on the right side of the genome homepage. In this image map, each of the genes is represented, in order, by a colored line. Genes placed in the outer circle are transcribed on the + strand while genes in the next inner circle are transcribed on the - strand. Functional RNAs such as tRNAs and rRNAs are shown on the very inner circles as scattered pink and red lines. The colors of the lines refer to the function that the gene product plays in the cell, for instance, bright yellow is for genes involved in DNA replication, brick red for genes that are involved in protein synthesis, etc.

We want to figure out where the gene encoding DNA polymerase I is located. At the top of the page there are a number of links, and **mousing over** these links will display a description of the programs. To find out where our gene is, select **Overview.** This will take you to a page that has links to a large number of different programs that can be used to analyze this genome. After spending a few moments looking over the available tools, scroll down to the heading *Genome Lists by Category.* Under this heading, select **Gene List by Role Category.**

This takes you to a page where all of the genes that are present in this organism are listed in groups according to their function. Scroll down to the *DNA metabolism* group, and then select **DNA replication, recombination, and repair**. This will display a table listing all of the genes that are involved in these related processes. Scroll down the page to the bottom of the table, and you will see the gene that we are interested in, DNA polymerase I. You may recall that in *E. coli* there are at least three different DNA polymerases, named DNA pol I, DNA pol II, and DNA pol III. These were named for the order in which they were discovered. In *E. coli*, DNA pol I is primarily involved in DNA repair, DNA pol III is involved in DNA replication, and the function of DNA pol II is unknown.

To find out where the gene is located, select the link to the gene, labeled **PAB1128**.

This takes you to a page that summarizes the information known about this gene and its encoded protein, and which has links for various analysis tools. The relevant part for us is the box that tells us the coordinates of the gene. Specifically, we are interested in the fact that this 2313 nucleotide-long gene is encoded by nucleotides 1695183 to 1697495 of the genome (the entire genome is 1765118 nucleotides long).

Now we are going to retrieve that nucleotide sequence, along with a region of nucleotides on either side of the sequence. Remember, our goal is to identify a restriction enzyme that will cut on either side of the gene, but not in the center of the gene itself. Once we retrieve the sequence that we are interested in, we'll use it as the input for a program that searches for restriction enzyme sites. There are a few different ways that we could get the sequence data; we'll do it by going to the *Pyrococcus abyssi* site managed by Genoscope, the French Government group that sequenced the genome of this organism. To get to their site, go to the top of the page and select **Overview**. From the Genome Overview page, scroll to the bottom of the page and select the (misspelled) link **The Genscope P. abyssi Home Page.**

Several different kinds of information could be retrieved from this page; to get the information that we want, scroll down to the middle of the page to the area labeled
Pyrococcus abyssi genome sequence
Get a genome sequence fragment from Pyrococcus abyssi

We know that we want nucleotides 1695183 to 1697495 of the genome, along with a region on each side of the gene. 3000 bp on either side of the gene will probably be sufficient for our purposes, so therefore we want to retrieve nucleotides 1692183 - 1700495. **Enter those two values into the search boxes.** The search box should be set to retrieve *From: 1692183 To: 1700495*. Then select the **Get sequence** button.

This will return a very large block of sequence data (As, Cs, Gs and Ts) to you. Looking at the sequence for a moment will remind you why it is that computers are so useful for this type of analysis.

Now that we have the sequence we are interested in, we are going to examine it for restriction enzyme sites. Use your mouse to **highlight the nucleotide sequence data,** and then use your browser controls to **copy the selected sequence data** (Edit-Copy, or Control-C). We will paste the copied sequence data into a program designed to search for restriction enzyme sites.

Identifying the Location of Restriction Enzyme Cut Sites in the Retrieved Sequence

There are a variety of programs that we might use to analyze for the presence of restriction enzyme sites. Many of these programs could be found easily by carrying out an internet search for '*restriction enzyme mapping*' or '*program to find restriction sites*' or something similar. All of these programs do pretty much the same thing, but the types of output are slightly different from program to program. We'll use one that has a couple of nice features, which can be found at **http://arbl.cvmbs.colostate.edu/molkit/mapper/**.

Scroll down the page to get to the large white text-entry box. Now, **paste the sequence that you retrieved into the white box.** Don't worry about changing the spaces and returns; the program deals with them just fine. Select the **Create Map** button.

Beneath the white box, on the left side, is a small pull-down menu that is set to the default 'all restriction enzymes'. There are a lot of different enzymes, and some of them are pretty difficult (and expensive) to buy, so we will change the setting to **Core set of enzymes**. This way the program will only search enzymes that are common, usually inexpensive, and frequently found on the multiple cloning sites of popular plasmid cloning vectors.

The program will rapidly analyze the input sequence and return an output of the sequence (in the white box at the top) and the restriction enzymes present in that sequence (in the black box below). Remember, our goal is to identify a restriction enzyme that will cut on either side of the gene, but not in the center of the gene itself. That means that we want a restriction enzyme that cuts once in between nucleotides 1 and 3000 of our input sequence, and once between nucleotides 5314 and 8313, but doesn't cut between nucleotides 3001 and 5313.

Let's look at the first three restriction enzymes that are listed: *Bam*HI, *Bgl*II, and *Cla*I.

The positions (numbers) of the nucleotides are shown at the top of the figure in the black box. The long vertical white lines mark off 1000-base increments, whereas the short vertical yellow bars show where each restriction enzyme cuts.

Examining the figure, we see that we didn't get lucky with the first three restriction enzymes. None of these enzymes meet our requirements. The enzyme *Bam*HI cuts the sequence three times, and though none of these cut sites are in the middle of gene, they all occur on the right side of the gene sequence. We want one cut site on the left side as well, somewhere between nucleotides 1-3000. *Bgl*II and *Cla*I only cut within nucleotides 1-3000, and don't meet our requirement of also cutting between 5314 and 8313.

However, if we scroll down the list, we see that the enzyme *Eco*RI fulfills all of our needs—it cuts on either side of the gene sequence but not within the gene sequence itself. Also, the enzyme *Xba*I might fulfill our requirements, although it looks like the right cut site is very close to the end of the gene sequence, and might even be within that sequence. Furthermore, the enzyme *Hinc*II could also be used to clone our DNA polymerase gene. Although this restriction enzyme cuts the input sequence multiple times, it does cut on either side of the gene sequence without cutting within the gene sequence.

Although it might seem simply that we were lucky in identifying these restriction enzymes, it should be pointed out that the odds were in our favor. For one thing, modern cloning vectors contain many unique restriction sites within their polylinkers, or multiple cloning sites. Also, we are trying to clone a fragment that is about 2000 basepairs long. Restriction enzymes that recognize six-basepair sequences will cut random DNA sequences on average once every 4096 basepairs. And of course, if we didn't find a suitable restriction enzyme within this core group of enzymes, we could have repeated the analysis using a larger group of enzymes.

Now, let's change the way that the output information is displayed, by selecting the **Text Display** button above the black box. This changes the output so that it displays the same information in a different format. Examining this output, we see that *Eco*RI will cut our output sequence at positions 781 and 7548, thus creating a 6767-basepair fragment containing our gene. The enzyme *Xba*I cuts at positions 2612 and 5312. Position 5312 is just within our gene sequence, so we probably don't want to use that enzyme after all. The enzyme *Hinc*II will cut at positions 1895 and 5404, thus creating a 3509-basepair fragment that contains the gene for DNA polymerase I from *P. abyssi.* It appears that the enzyme *Hinc*II will be the most useful for our purposes.

Cloning - Odds and Ends

So far, we accomplished our first objective—we identified a restriction enzyme that would allow us to isolate and clone the gene for DNA polymerase I from *P. abyssi.* In the next part of this tutorial we will design PCR primers that would allow us to amplify a region of the gene using PCR. Assuming that you are the curious sort, however, you no doubt have some questions about the cloning project that we've been working on. Let's address some of the obvious questions briefly, before moving on.

Obvious question 1. *Where can we get the restriction enzyme Hinc*II?
Obvious question 2. *How do we find a cloning vector that has a single Hinc*II *site in it?*
Obvious question 3. *How can we get genomic DNA from* P. abyssi*?*

The answers to the first two questions can be found by searching the websites of some of the companies that supply biotechnology reagents to researchers. These sites are keyword-searchable and will contain lots of information on available cloning vectors and restriction enzymes. Some of these companies are listed in the table below (there are many others, several of which provide excellent products and service).

Company	**Web site**
Amersham Life Science	http://www.amersham.com
Ambion	http://www.ambion.com
Bio-Rad Laboratories	http://www.biorad.com
Clontech laboratories	http://www.clontech.com
Invitrogen	http://www.invitrogen.com
New England Biolabs	http://www.neb.com
Promega	http://www.promega.com
Quiagen	http://www.qiagen.com

To give you a feel for this, go to the Promega site's technical services page by going to **http://www.promega.com/techserv/Default.htm**.

On this page you can see links to many of the different products available from Promega, including restriction enzymes and cloning vectors. Select the link under Top Picks to **Vectors.** From the list of vectors available, select **Cloning vectors.** Because we are interested in finding a cloning vector that we can clone a 3509-basepair *Hinc*II fragment into, we want to scroll down to the middle of the list to the cloning vectors such as pGEM 1 and pGEM 2, etc. On the right side of the table we see the *MCS enzymes* listed (MCS = multiple cloning site). *Hinc*II can be used to clone into any of these pGEM plasmids. With a little searching you will see that both the cloning vector and the restriction enzyme can be ordered from this company for ~$100.

So that gets us to the third question, where can you get a sample of *P. abyssi*? It turns out that this might take some effort. Most microorganisms have been deposited in one of two international collections, The American Type Culture Collection or its German counterpart, the Deutsche Sammlung von Mikroorganismen und Zellkulturen. However, as of this writing, neither of these collections had a sample of *P. abyssi* available for purchase. This means that you would probably have to directly contact one of the researchers working with this organism, and ask them to send you a sample of the organism or its DNA. This type of free exchange of materials and organisms is standard, and a hallmark of contemporary scientific inquiry.

Project 2: Designing PCR Primers

The polymerase chain reaction (PCR) is a rapid means of making numerous copies of a specific target DNA sequence. It has a wide variety of uses, from diagnostic (i.e., is a particular pathogen present in a given biopsy/hamburger/mayonnaise sample?) to forensic (i.e., whose sperm is on the blue dress? are Sally Hemming's descendants genetically related to Thomas Jefferson's descendants?) to a variety of research applications, including DNA sequencing, cloning, and site-specific mutation. The diversity of applications derive from the ability of PCR to: a) distinguish a specific target sequence from a large excess of background, non-target sequences, and b) produce a large number of copies of a specific sequence from a very small initial amount (as low as a single molecule of the target sequence). The extreme specificity and sensitivity of PCR have allowed it to be used to amplify and clone DNA from such sources as mummified human tissues and samples of extinct plants and animals.

The specificity of a PCR reaction results from the use of specific *primers*. Primers are short (20-30 bases), single-stranded, synthetically made polynucleotides that are complementary to the sequences flanking the target sequence.

In a PCR reaction, a target DNA sample is mixed with two different primers. This mixture is heated to denature the double-stranded target DNA. When the mixture is then cooled, the primers can 'anneal' to the sequences that they are complementary to, that is, they can form a double-stranded DNA molecule consisting of the primer hydrogen-bonded to a complementary target DNA sequence. The primer thus provides a 3'-OH group for the DNA polymerase, and a copy of the region adjacent to the primer-binding-site can be synthesized.

Characteristics of Good PCR Primers

In general, useful PCR primers should have the following characteristics.

Specific - will bind only to the target sequence. Thus, you want primers that are perfectly complementary to your target but not complementary to any other sites on the template DNA. It is also important that the primers don't have regions that are self-complementary (which will form secondary structures) or where one primer is complementary to the other primer (which will cause primer-dimers).

Proper length - length will affect the probability that they will bind to other sites on the template, and it will affect the Tm (see below). Usually primers between 17 and 26 nucleotides are used.

Proper melting temperature (Tm) - The Tm of the two primers should be about the same. Also, the Tm should be 55°C or above to provide an annealing temperature of 50°C or more. Note: you may be wondering what the heck Tm means—see below.

The Tm of a DNA molecule is a value that reflects the stability of that molecule. For example, if you were to slowly increase the temperature of a solution containing a double-stranded DNA molecule, then a more stable molecule would retain its double-stranded form at a higher temperature than would a less stable molecule. Specifically, Tm is defined as the temperature at which 50% of a given oligonucleotide is in double-stranded form. The stability is a result of many features of the DNA molecule, such as length (long molecules are more stable) and percent G + C nucleotides (high G+C molecules are more stable, because G-C pairs are held together by three hydrogen bonds while A-T pairs are held together by only two).

The Tm of DNA molecules is a very important concept in many biotechnology procedures, including PCR reactions, in situ hybridizations, and Southern and Northern hybridizations. It can be calculated from the sequence of the oligonucleotide, and in fact most PCR primer-design software and probe-design software will calculate the value for you. For our purposes just remember 55°C or above, and about the same value for each of the two primers.

If you are interested, more information on the theory of primer design can be found at **http://www.alkami.com/primers/refprmr.htm**.

Designing PCR Primers to Amplify DNA Polymerase I from P. abyssi

There are a variety of programs that we might use to design PCR primers. Many of these programs could be found easily by carrying out an internet search for 'PCR primer design' or something similar. All of these programs do pretty much the same thing, but the sophistication of the analysis and the types of output are slightly different from program to program. A list of these programs can also be found at **http://www.cbi.pku.edu.cn/help/pd.html**.

We'll use one that has a couple of nice features, called Web Primer, at Stanford University at **http://genome-www2.stanford.edu/cgi-bin/SGD/web-primer.**

When we were looking for restriction enzymes to use in cloning the DNA polymerase I from *P. abyssi*, we used the gene sequence and 3000 bp of the sequence flanking either side of the gene. For our purposes, we won't target an area that big for our PCR reactions. In fact, let's simply use the gene sequence itself. For most applications, we won't need to amplify the whole gene, anyway. You could **retrieve** the gene sequence from the Genoscope site that we visited earlier, or you could simply **copy** it from here.

ATGATAATCGATGCTGATTACATAACGGAAGATGGCAAGCCGATAATAAG
GATATTCAAAAAGGAAAAGGGAGAGTTTAAGGTAGAATACGATAGGACGT
TTAGACCCTACATTTATGCTCTTTTAAAGGATGATTCGGCCATAGATGAG
GTTAAGAAGATAACCGCCGAGAGGCACGGAAAGATAGTCAGGATAACCGA
GGTTGAGAAAGTCCAGAAGAAATTCCTAGGAAGGCCAATAGAAGTCTGGA
AGCTCTATCTTGAGCATCCCCAGGATGTTCCAGCCATAAGAGAGAAGATA
AGGGAACATCCAGCTGTAGTTGATATATTTGAATACGACATACCCTTTGC
GAAGCGCTACCTCATAGACAAGGGATTGACTCCAATGGAGGGGAACGAGG
AGCTAACGTTTCTAGCCGTTGATATAGAAACATTGTACCATGAAGGAGAG
GAGTTCGGGAAAGGGCCAATAATAATGATCAGCTACGCCGACGAGGAAGG
GGCCAAGGTGATAACTTGGAAGAGCATAGACTTACCTTACGTTGAAGTGG
TTTCGAGCGAGAGGGAGATGATAAAGAGGCTCGTGAAGGTAATTAGAGAG
AAAGATCCCGACGTGATAATAACGTACAATGGTGATAATTTCGACTTTCC
GTACCTCTTAAAGAGGGCTGAAAAGCTCGGAATAAAGCTCCCCCTTGGAA
GGGACAATAGCGAGCCGAAAATGCAGAGGATGGGGGATTCATTAGCCGTA
GAGATAAAGGGCAGAATACACTTCGATTTATTCCCCGTCATAAGAAGAAC
GATCAACCTTCCAACATACACCCTCGAAGCGGTTTATGAGGCTATATTTG
GAAAGTCTAAGGAGAAAGTCTATGCCCATGAGATAGCTGAGGCCTGGGAA
ACCGGGAAAGGGCTAGAGAGGGTAGCTAAGTATTCAATGGAAGATGCGAA
GGTAACCTTTGAGCTCGGAAAGGAGTTCTTCCCGATGGAAGCCCAGCTAG
CTAGGCTCGTTGGCCAGCCAGTTTGGGACGTTTCAAGGTCGAGCACCGGA
AACCTCGTTGAGTGGTTTCTCCTTAGGAAGGCCTACGAGAGAAATGAGCT
CGCGCCCAATAAACCGGACGAGAGGGAATACGAGAGAAGGCTAAGAGAGA
GCTATGAAGGGGGTTACGTTAAGGAGCCAGAGAAGGGATTGTGGGAAGGG
ATAGTCAGCTTAGACTTTAGGTCCCTATATCCCTCTATAATTATAACTCA
CAACGTCTCACCAGACACTTTGAATAGAGAAAATTGCAAGGAATATGACG
TTGCCCCCCAAGTGGGGCACAGATTCTGCAAGGATTTCCCAGGATTCATA
CCAAGCTTACTGGGTAACCTACTGGAGGAGAGACAAAAGATAAAAAAGAG
AATGAAAGAAAGTAAAGATCCCGTCGAGAAGAAACTCCTTGATTACAGAC
AGAGAGCTATAAAAATACTTGCAAACAGCTATTATGGCTATTATGGATAT
GCAAAGGCCAGATGGTACTGTAAAGAGTGTGCAGAGAGCGTAACCGCATG
GGGAAGGCAGTACATAGACCTGGTTAGGAGGGAACTTGAGAGCAGAGGAT
TTAAAGTTCTCTACATAGACACAGATGGCCTCTACGCAACGATTCCTGGA
GCCAAGCATGAGGAAATAAAAGAGAAGGCATTGAAGTTCGTCGAGTACAT
AAACTCCAAGTTACCTGGGCTTCTTGAATTGGAATACGAAGGTTTCTACG
CGAGAGGGTTCTTCGTGACGAAGAAAAAGTACGCACTAATCGACGAGGAA
GGAAAGATAGTTACGAGGGGGCTCGAAATAGTAAGGAGAGATTGGAGTGA
AATAGCAAAGGAGACCCAGGCCAAGGTTCTCGAGGCAATACTCAAGCACG
GTAACGTTGATGAGGCCGTAAAAATAGTAAAGGAGGTTACAGAAAAACTC
AGTAAATATGAAATACCACCCGAAAAGCTTGTAATTTATGAGCAGATAAC
GAGGCCTCTGAGCGAGTATAAAGCGATAGGCCCTCACGTTGCAGTAGCTA
AAAGGCTCGCAGCGAAGGGAGTAAAAGTTAAGCCAGGGATGGTTATCGGT
TACATAGTTTTGAGGGGAGACGGGCCAATAAGCAAGAGGGCCATAGCTAT
AGAGGAGTTCGATCCCAAAAAGCATAAGTACGATGCCGAATACTACATAG
AGAACCAAGTTCTGCCAGCGGTGGAGAGGATATTGAGAGCATTTGGTTAT
CGCAAGGAGGATTTGAAGTATCAAAAAACTAAACAAGTGGGCCTTGGAGC
ATGGCTTAAGTTC

Paste this sequence into the query text box at the Web Primer site. Then, select the **Submit** button.

This will take you to a default page for the parameters of PCR primers. As you can see, most of the default values are very similar to the ones that we discussed above. If you wish, you can change some of these default values to more closely mimic the parameters that we described. Then, select the **Submit** button.

This will return you to a page that, unless the conditions you picked were too stringent, should describe the 'Best' pair of PCR primers and list a number of different possible PCR primer sets. Select **This is the BEST pair of primers**. How does it meet with the criteria that we described? Is it within the length and Tm range that we described, and are the Tm values of the two primers about the same? Are there primer pairs within the 'Valid' list that better meet those criteria? How is the output affected by changing the primer design parameters?

Ultimately, how 'good' a primer pair is has to do with how well it allows you to amplify your target sequence from the template DNA that you are using. Because there are many different factors that affect the ability of primers to produce large amounts of a specific product, you ultimately need to empirically determine whether or not a particular primer pair will work or not. That is, you need to try it and see how well it works. Also, the thermalcycler program that you use and the reaction conditions dramatically affect the results of a PCR reaction, and these frequently need to be optimized as well. Sometimes you need to try out several primer pairs, and several programs, before finding a combination that works well for a particular application. As with the reagents for cloning, PCR primers and PCR reagents can be purchased from a number of different retailers. A pair of PCR primers can usually be purchased for ~$50 or less.

That brings this tutorial to a close. In this tutorial, you learned about designing and planning commonplace biotechnology experiments, including plasmid cloning and PCR primer design. You also learned how to find and retrieve specific sequences from completed bacterial genomes. Hopefully, you feel a little more comfortable designing your own experiments now. Also, hopefully this tutorial has helped you to better visualize both how these experiments work and how researchers go about performing these techniques. The following questions will help to reinforce and further your understanding of these principles.

Review Questions

1. When discussing restriction enzymes, it was stated that a restriction enzyme with a 6-bp recognition site will cut a random DNA sequence every 4096 basepairs, on average. Why is that?
2. Primers A and B are each twenty bases long. In primer A, 6 of the bases are A or T and 14 of the bases are G or C. In Primer B, 10 of the bases are A or T, and ten of the bases are G or C. Which of the primers has a higher Tm? Why?

Thought and Application Questions

1. How many *Bgl*II sites are there within the DNA polymerase (Family X) gene of *Thermoplasma volcanium* strain GSS1? How many *Acc*I sites?
2. Let's say that you have designed a set of PCR primers to amplify a portion of a human gene. What bioinformatics programs would you use to test whether or not these primers were specific for the targeted gene? That is, how could you use *in silico* approaches to find out whether or not your primers are complementary to other sites in the human genome?

19. Mining Genomes: Genome Analysis and Comparative Genomics

Starting link: http://www.tigr.org/.

Introduction

A genome is defined as all the genetic material in the chromosomes of a particular organism. Genomics is the study of genomes, including the analysis of the specific nucleotide sequence of chromosomes, but also encompassing a variety of other topics such as the arrangement of genes and other sequences on chromosomes, the evolution of chromosomes, and intra-species sequence variation. Genomics research is usually promoted because of the healthcare benefits that are certain to result from such analyses, but additionally we are learning an immense amount of basic information about life, diversity, and evolution. Genomics analysis is also certain to open up currently unidentified new fields of study in biology, offering as yet unshaped insights into the nature of life.

The study of microbial genomes is significantly more advanced than the study of eukaryotic genomes for a number of reasons. Microbial genomes are generally smaller (*Escherichia coli* has a 5 million base pair genome size; *Homo sapiens* has 3 billion base pairs), more gene-rich (5% of the *E. coli* genome is noncoding; 95% of human genome is non-coding), and less complex (prokaryotes rarely have introns). Microbial genetics has also been studied more extensively (microbes grow quickly, are easy to manipulate genetically, and few ethical issues are generally raised by such studies), and therefore the functions of more of the genes are understood. Because they are also considerably less expensive to sequence and annotate, many microbial genome sequencing projects have been completed, which allows for *comparative* genomic studies.

In this project you will be introduced to some of the tools that contemporary scientists use to study microbial genomes. Much of our time will be spent at the Comprehensive Microbial Resource, a site managed by the not-for-profit research group The Institute for Genomic Research (TIGR). TIGR was founded in 1992 by J. Craig Venter (who later founded Celera to sequence the human genome) and in its short existence has made numerous seminal advances in basic biology and medicine.

Objectives

- to explore microbial genomes and comparative genomics.
- -to introduce some of the genomics tools available at the Comprehensive Microbial Resource and The Institute for Genomic Research.

Project

Progress in genomics is coming at a remarkable pace. In 1995, the first complete sequence of a genome was reported, that of the bacterium *Haemophilus influenzae* (1,830,137 base pairs, or 1.8 Mb [1Mb = 1 Megabase = 1 million bases]). This organism was sequenced by a private not-for-profit group called The Institute for Genomic Research (TIGR). TIGR is a large, diverse, active, and productive research institute that is involved in a wide array of research projects. Founded in 1992, TIGR followed the *H. influenzae* sequence by finishing another bacterial genome the same year (*Mycoplasma genitalium)* and the first archaeal genome (*Methanococcus jannaschii*) in 1996. They have completed the sequencing of many important pathogens, including the organisms that cause cholera, tuberculosis, meningitis, syphilis, Lyme disease, and ulcers, as well as a number of diverse microorganisms of environmental significance. Several more microbial genomes are underway.

The first eukaryotic genomes were finished shortly thereafter (*Saccharomyces cerevisiae,* 13 Mb, in 1997, and *C. elegans*, 100Mb, in December 1998). Today, more than 100 organisms have genomes available! Whole or partial genomes of over 800 organisms can be found in Entrez Genomes at the National Center for Biotechnology Information (NCBI). All three main domains of life — Bacteria, Archaea, and Eukaryota — are represented, as well as many viruses and organelles. To help you understand the breadth and depth of this information, go to **http://www.ncbi.nlm.nih.gov/PMGifs/Genomes/allorg.html** to scroll through the list of genomes available at the National Center for Biotechnology Information;

The Institute for Genomic Research

Now, we are going to begin our project at The Institute for Genomic Research (TIGR). Go to TIGR's homepage at http://www.tigr.org/.

After playing a major role in determining the draft sequence of the human genome, TIGR is currently sequencing genomes of important plant species (potato, rice, and the widely studied *Arabidopsis thaliana)* as well as of important human and agricultural pathogens. In addition to their sequencing efforts, TIGR scientists have developed software for genome assembly, analysis, and annotation. Much of the software and genomics data is distributed freely to the scientific research community, and TIGR has devoted significant resources to developing educational and training programs.

While we are going to focus on microbial genomics in this project, you should be aware that there are many interesting regions within the TIGR site that are worth exploring, as should be evident from looking over the TIGR homepage.

How Many Microbial Genomes have been Sequenced? - The Microbial Database

Let's begin by looking at the genomes that have been completed. On the left-hand side of the TIGR homepage is a list of links to different regions within their site. Select the top link, to **Genome Databases.**

There are a number of different databases maintained by TIGR, including sequence, microarray, protein structure, and historical databases, such as 'World Record Holder for the Longest Contiguous DNA Sequence' at the bottom of the page, which traces the development of large-scale sequencing accomplishments.

Underneath the heading of The Comprehensive Microbial Resource (CMR), select **world-wide completed microbial genome sequencing.**

This table presents the (more than 60) microbial genomes that have been completed, along with links to the sequence data, the papers describing the genome, the groups that carried out the sequencing, and taxonomic information on the organism.

One of the first organisms to have its complete genome sequenced was *Mycoplasma genitalium*. This unsavory-sounding organism is a Gram-positive bacterium that lives in human mucosa of the respiratory and genital tract. It is one of the smallest cell types and it may have the smallest genome of the Bacteria. This organism's genome was sequenced because it is difficult to study *Mycoplasma* in other ways—the UGA codon in several *Mycoplasma* species encodes tryptophan instead of a translational stop, and therefore many *M. genitalium* proteins are prematurely terminated when expressed in *E. coli* or other cells.

Another reason that it was sequenced was to learn what the minimum set of genetic information might be—that is, what the least amount of genetic information an entity needs to be called 'living'. Because this organism requires other living things to make a lot of the starting materials it requires (i.e, amino acids, vitamins, etc), it cannot answer this question definitively, but it is a good starting place.

Twenty-nine scientists are listed as authors on the report detailing this sequence, which required a total of 9846 sequencing reactions carried out by 5 people using eight automated sequencers (for about two months). The accuracy of 99% of the sequence was confirmed by sequencing the same region more than once, and any given nucleotide was determined an average of 6.5 times. The initial sequencing effort produced 39 large stretches of DNA sequence that were then connected by filling in the gaps.

Scroll down the page to the *M. genitalium* entry. **Select the link on the far left** side (it looks like a square with a colored circle in it) to go to the *M. genitalium* page at The Comprehensive Microbial Resource (CMR), which is the main site for microbial genomics analyses at TIGR. The CMR allows you to carry out a variety of analyses of individual genomes as well as comparative studies of multiple genomes. It is an extensive set of tools that is relatively intuitive to use, and contains TIGR genomes as well as genomes that have been finished by other groups.

The Comprehensive Microbial Resource

There are many different ways of looking at the information contained within a genome. One way is the circular map that is presented on the right side of the page. To enlarge this map, select **Circular Display** underneath the small picture, and then select **Main *Mycoplasma genitalium*** from the drop-down menu**.**

Here it is—a whole genome encapsulated into a single figure. In this image map, each of the genes is represented, in order, by a colored line. Genes placed in the outer circle are transcribed on the + strand while genes in the next inner circle are transcribed on the - strand.

Functional RNAs, such as tRNAs and rRNAs, are shown on the very inner circles as scattered pink and red lines. The colors of the lines refer to the function that the gene product plays in the cell. For instance, bright yellow is for genes involved in DNA metabolism, brick red for genes that are involved in protein synthesis, etc. The color code is outlined along the right side of the page.

This representation allows you to observe several features of genome structure. For example, is one strand used preferentially for coding? That is, are most of the genes located on one particular strand? How are the genes distributed?

It seems clear that one of the strands is primarily the coding strand for the first half of the genome, and then the other strand is primarily the coding strand for the second half of the genome. Can you think of any selective advantage that such an arrangement might have? Where do you suppose the origin of replication for this chromosome is located? Which of the strands do you think is the leading strand in DNA replication? Would you predict that this would be a general feature of microbial chromosomes?

You probably also noticed that there aren't many areas that aren't completely full of genes. Some sections of the genome that don't encode anything may still be involved in regulation, which means that even less of the genome is 'junk' DNA. If you were to look at a linear, more detailed version of the map, you would see that about 5% of the *M. genitalium* genome is non-coding. Does this suggest that selection pressure is acting to minimize non-coding regions in this organism? What might contribute to this selective pressure? How much of the human genome is non-coding?

We said earlier that the colors of the lines represent the functional class of the proteins. If you envision the circle as a clock, there is a group of red lines at four o'clock. These clustered genes (28 of them) are all involved in the process of translation. Why would it be advantageous to cluster genes involved in a particular process? Are other groups of genes clustered according to function? Prokaryotes don't always use a separate promoter for individual genes—that is, several different proteins will be translated from the same transcript. Clustering genes with the same function allows them to be coordinately regulated; they are all turned on and off simultaneously by regulating a single promoter.

Now, scroll up to the top of this page. You should see several hyperlinks to the *Mycoplasma genitalium* genome information, including *Overview, Analyses*, *List by Category, Searches,* and *Related links*.

Select **Overview**. This will take you to a very long list of possible things to explore.

Select **DNA molecule Information,** near the top of the page. How many total genes are present on this genome? How many have been assigned a function? How are they assigned functions?

Go back to the *M. genitalium* Overview page at **http://www.tigr.org/tigr-scripts/CMR2/GenomeTabs.spl?database=gmg#1**.

Select **Condensed Genome Display**.

This takes you to another map view of the genome, in which the individual genes are displayed as colored stars. The colors again refer to the function of the genes. You should again be able to see the cluster of red proteins that are involved in translation. **Mousing over** an individual star reveals information about the name, functional role, and accession number for that protein. Mousing over the red stars in the cluster shows that these proteins are part of the ribosome, the protein/RNA complex directly involved in translation.

You can also use this viewer to zoom in on a particular region within the genome. Directly above the figure, **change** the radio button selection from Gene Page to **Zoom on Region.** Then, **select the first red star** in the cluster of stars. This will open a new window.

In this view, the individual genes are depicted as arrows, with the size of the arrow corresponding to the length of the gene and the direction corresponding to the direction of transcription. If you move your cursor onto one of the genes, its description will be entered into the display above the map, including the gene locus, the nucleotide numbers, the gene name, and the functional class of the protein.

How do you know where a protein-coding sequence starts? In other words, if you are starting with a long sequence of A, C, G, and T, how do you know how to divide it into protein-coding regions? You certainly have one clue because prokaryotic proteins always begin with an ATG codon for methionine. Because three of the 64 codons are typically stop codons, you might suspect that any long open reading frames that begin with an ATG start codon could be protein-coding sequences. That is, in a random nucleotide sequence (a non-coding sequence), the probability that any codon is a stop codon is 3/64, or ~ 1/21. It is unlikely that you might have a stretch of several hundred codons, none of which are stop codons, in a random sequence; therefore it is reasonable to assume that most long open reading frames are protein-coding.

This is actually a more complex problem than it might seem. For example, imagine a nucleotide sequence in which the first 12 bases are: 5'–AAAGGGTTTCCC–3'. There are three different possible reading frames for this sequence - it might be read as “AAA-GGG-TTT...” or as “AAG-GGT-TTC...” or as “AGG-GTT-TCC...”. And, because DNA exists as a complementary double-stranded molecule, a sequence of 5'–AAAGGGTTTCCC–3' must be paired with a sequence of 5'–GGGAAACCCTTT–3', and this sequence also has three separate possible reading frames. Therefore there are six different possible ways of translating any given nucleotide sequence.

If you scroll to the bottom of the page, there is a complex-looking figure that depicts the presence of possible stop and start codons in all six different translational reading frames for the genome in this region. The long black lines represent stop sites while the short black lines represent start sites. Do you think that the presence of an open reading frame is sufficient evidence of a protein-coding region? What more should be done to verify that the potential gene really is a gene?

Of course, you could do experiments to show that mRNA transcripts from a potential gene are actually made, or use reporter gene constructs to demonstrate transcription, but those types of experiments are far beyond the scope of a genome-sequencing project. At this level, potential genes are compared with sequences from other organisms and other projects, and assigned into categories as a result of these comparisons.

Select one of the proteins by clicking on its arrow. This takes you to the TIGR gene page for that gene, which contains a more thorough description of the protein, including such features as its molecular weight and length.

Let's say that you were to get really excited about studying the protein that you selected, and you wanted to get a clone of that portion of the genome for laboratory work. Select the link on the bottom of the page labeled ***View Surrounding Clones***. This takes you to a map that displays the available clones that encompass this genomic region. These clones can be ordered (inexpensively) through the American Type Culture Collection.

Go back to the *M. genitalium* Overview page, by closing the window that we've been working in, and then going to **http://www.tigr.org/tigr-scripts/CMR2/GenomeTabs.spl?database=gmg#1**.

Select **Role Category Graph**. This will display a pie chart of the number of genes that are identified as being involved in different processes.

How many genes are involved in protein synthesis? What is the largest category? What is a 'hypothetical protein'? How is this different from a 'protein of unknown function'?

The category assignments all come from the comparison of *M. genitalium* sequences with sequences already in the databases. For example, if a new *M. genitalium* gene is essentially identical to a previously characterized gene in the database, the database gene's function is assumed. *M. genitalium* genes with protein matches that aren't identical are named after the database entries as "putative ZZZ", or "ZZZ-related", to indicate similarity, and assigned into a similar category as the database protein. *M. genitalium* genes with matches to unknown proteins are named as "similar to unknown protein".

If there is no evidence from any studies that a gene is actually transcribed, it is called a hypothetical conserved protein if it is similar to putative proteins from other organisms, and it is called simply a hypothetical protein if it doesn't match anything in the databases.

Go **back** to the *M. genitalium* Overview page.

You may recall from the Molecular Evolution project that when you did a blastn search (nucleotide vs nucleotide) you got fewer matches than when you did a blastp search (nucleotide translated into protein checked against protein databases). We said that this was because you can encode the exact same protein sequence using a variety of nucleotide sequences, as a result of codon degeneracy. That is, some amino acids are encoded by multiple (as many as six) different codons.

Let's look at codon usage by *M. genitalium.* Select **Codon Usage Chart**. This will present a table of all of the possible codons, and how many times each codon is used in the *M. genitalium* genome. For example, there are two codons that encode phenylalanine, TTT and TTC. Next to each codon are two numbers, one in green and one in blue. The greenish value is the fraction of all codons that are that specific codon, while the value in blue represents the total number of times that codon is used. How many times is the codon TTT used in *M. genitalium's* genome? How many times is TTC used?

One reason why organisms use one codon preferentially over another is that the tRNAs corresponding to each codon are not present in the same concentrations. For example, there might be many more tRNAs corresponding to the TTT codon than there are for the TTC codon. This allows an organism to regulate the rate at which a particular transcript is transcribed—if you use codons that have a low concentration of corresponding tRNA, then that transcript won't be translated as rapidly as one that contains a different codon. You might hypothesize then that genes involved in one type of process have a different codon usage than genes involved in a different type of process. To test this hypothesis, let's look at codon usage for genes that are involved in protein synthesis. At the bottom of the page, select the category **'Protein synthesis,'** the select **'Recalculate.'**

How many ORFs are involved in protein synthesis? How many TTT codons are there in these ORFs, and how many TTC?

Scroll down to the bottom of the page and **deselect** the protein synthesis category and select the **'amino acid biosynthesis'** category, and then select **Recalculate**. What happened? Why?

Genome to Genome Comparisons

When a new gene sequence is determined, the function of the encoded protein is inferred by comparing the sequence to known sequences already in the databases. Similar, although more computationally difficult, comparisons can be made between whole genomes. These genome-to-genome comparisons allow scientists to study the way in which entire genomes evolve as speciation and differentiation occur. It takes some effort to understand how to interpret and visualize these comparisons. We will go through some simple examples first.

Whole Genome Alignments

At the top of the page, select **Align Genomes.**

This is the home page for the web-based version of MUMmer, a tool for the pairwise alignment and comparison of very large DNA sequences. This program is designed for high resolution comparison of genome length sequences, and has as an output a base-to-base alignment of the input genomes. This output makes it possible to identify all single nucleotide polymorphisms (individual nucleotide differences between genomes), inserts and deletions, tandem repeat alterations, and inversions between two genomes. It is powerful enough to use for comparing eukaryotic chromosomes as well as microbial genomes.

Comparing M. genitalium with Itself

To better understand the output, first run a comparison between *M. genitalium* G-37 and itself. Using the select organisms pull-down menus, select ***M. genitalium* G-37** to be displayed on both the X- and Y-axis. Leaving the minimum alignment length at the default of 100 bp, select **Launch Alignment**.

Once the alignment has been completed, scroll down the page to see the output. You should be looking at a single red line going diagonally from the bottom left of the graph to the top right. The line is composed of points, where each point is a *MUM*, a maximally unique matching sequence, which is essentially a matching sequence from each of the two query genomes. Because the two query genomes are identical, the sequences in the beginning of genome 1 match the sequences in the beginning of genome 2, and the sequences in the end of genome 1 match the sequences in the end of genome 2, etc., which is why you have a single unbroken line with this orientation. The bar of colored stripes on either axis represent the specific genes that have been identified in the genomes, again color-coded to depict function.

The points in different comparisons will be either red or green. Alignments with the same orientation are shown in red and alignments with opposite orientations are shown in green. That is, if nucleotides 100-200 of genome 1 aligned with nucleotides 100-200 of genome 2, you would get a red point. However, if nucleotides 100-200 of genome 1 aligned with nucleotides 200-100 of genome 2, you would get a green point. It will probably be easiest to envision this by carrying out another whole genome comparison.

Comparison of Helicobacter pylori J99 with Helicobacter pylori 26695

The bacterium *Helicobacter pylori* has been shown to be important in the formation of many stomach ulcers. This identification has led to successful new treatments, including relatively simple antibiotic regimes for stomach ulcers once thought untreatable. Let's compare the genomes of two closely related isolates, *H. pylori* J99 and *H. pylori* 26695.

Scroll to the bottom of the page, and select these two genomes using the pull down menus. Leaving the minimum alignment length at 100 bp, select **Launch Alignment.**

This alignment should look similar to but different from the comparison of *M. genitalium* with itself. Most of the genomes are identical, which results in the portions of the red line that are similar to our first comparison. However, several changes are also evident. For example, there are two inverted translocations that are evident. These are the regions of opposite orientation represented by the green lines perpendicular to the red line. The fact that they are green tells us that they are inverted; the fact that they are displaced from the central axis tells us that these represent translocations. Spend some time looking at this output and try to understand what is being portrayed. Can you envision how these changes may have occurred?

Comparison of M. genitalium G-37 with M. pneumonia M129

Now carry out a similar comparison using the genomes of two related, but different *Mycoplasma* species, *M. genitalium* G-37 *and M. pneumoniae* M129. Scroll to the bottom of the page to change the settings using the pull down menus. What happens? What does this mean?

Now, carry out this same comparison, but reset the minimum alignment length to 25 bp. What happens? What does this mean?

Now, carry out this same comparison, but reset the minimum alignment length to 15 bp. What happens? What does this mean?

Lateral Transfer

One result of comparative genomics is the discovery that lateral transfer of pieces of genomes occurs at a surprising rate, and has been an important factor in the evolution of microbial genomes. 'Lateral' transfer refers to DNA that has been inherited from a non-relative, that is, not from mother-to-daughter cell. Lateral transfer can be detected when a comparison such as the MUMmer analysis indicates that a portion of the genome is closely related to a portion of a distantly related organism. That lateral transfer has in fact taken place can be supported by showing that the transferred piece is not present in close relatives of the recipient. Analyses of things such as codon bias in the putatively transferred piece can also support such designations.

In most microorganisms that have been sequenced, between 5% and 25% of their genomes appear to have been acquired through lateral transfer in the past 100 million years. This result was unexpected, and has inspired a large number of studies to understand lateral transfer better.

Genome Structure and Evolution: Lateral Transfer Gone Mad!

We are going to close this project with a brief look at cryptomonads, members of the vast group of organisms referred to as Protists. Protists are organisms that we don't mention to undergraduates very often because they defy easy understanding. But as long as you're going to be confused by genomics, you may as well be confused by cryptomonads as well! These organisms are the almost inconceivable result of multiple episodes of **mass lateral transfer**—the fusion of complete genomes from two distantly related organisms to form a completely new organism. You've learned about the endosymbiotic origin of organelles such as chloroplasts and mitochondria, and this is an interesting extension of that. In addition to chloroplasts and mitochondria, cryptomonads have an additional 'organelle' of endosymbiotic origin —the remains of an endosymbiotic eukaryote that retains vestiges of its own former nucleus (now called a nucleomorph), including 3 chromosomes that are dependent on the primary nucleus for functioning! Go to **http://www.nature.com/nature/journal/v410/n6832/fig_tab/4101091a0_F1.html** to see a diagram of this difficult to understand evolutionary history.

The complete text of the article describing this genome project can be found at **http://www.nature.com/cgi-taf/DynaPage.taf?file=/nature/journal/v410/n6832/full/4101091a0_fs.html.**

As the cryptomonad genome so strikingly demonstrates, there are a lot of surprises ahead in the field of genomics. In this project, you have examined several microbial genomes and used many of the tools available for analyzing genomes at The Institute For Genomics Research. Soon, similar analyses will be possible with large numbers of eukaryotic genomes. The following review questions will help to reinforce and extend what you've learned in this tutorial.

Review Questions

1. What is TIGR? Why was TIGR founded?
2. What is a genome? How many genomes have been sequenced to date?
3. In your own words, explain what is meant by the term lateral transfer. .

Thought and Application Questions

1. Agrobacterium tumefaciens is important as a plant pathogen and has been widely used for biotechnological applications. Compare the genome of *Agrobacterium tumefaciens* C58 Cereon with genome of *Mycoplasma genitalium.*
   How is the genome of *A. tumefaciens* structured - Is there a single circular chromosome?
   Are the genes split so that the coding strand is the leading strand?
   Are the ribosomal proteins lumped in a large cluster of genes?
   How large is the genome? How many genes?
2. Use the MUMmer program to compare the genome structure of two organisms with in the genus *Thermoplasma.* Briefly describe, compare, and contrast the output that you see at each of the following settings:
   Minimum match 100 bp
   Minimum match 80 bp
   Minimum match 25 bp
   Minimum match 20 bp
   Minimum match 15 bp